FORCE

1 lb = 4.448 221 6 N
   = 444,822.2 dynes
1 poundal = 0.138 255 N

1 N = 100,0
1 dyne = 0.

STRESS or PRESSURE

1 lb/in. = 175.126 83 N/m
1 lb/ft = 14.593 903 N/m
1 lb/in.^{2} = psi = 6,894.757 2 N/m^2 (Pa)

1 lb/ft^2 = 47.880 259 N/m^2 (Pa)
1 atmosphere = 14.695 9 lb/in.^{2} = 101,325.0 N/m^2 (Pa)
1 in. water (39.2°F) = 0.036 126 lb/in.^{2} = 249.082 N/m^2 (Pa)
1 ft. water (39.2°F) = 0.433 52 lb/in.^{2} = 62.426 lb/ft^2 = 2,988.98 N/m^2 (Pa)

1 N/m = 0.005 710 147 lb/in.
   = 0.068 521 76 lb/ft
1 N/m^2 = 1 Pa
   = 0.000 145 037 74 lb/in.^{2}
   = 0.020 885 43 lb/ft^2

MASS

1 lbm = 0.453 592 37 kg
1 slug = 32.174 05 lbm = 14.593 903 kg
1 ton (short) = 2,000 lb = 907.184 74 kg
1 ton (long) = 2,240 lb = 1,016.046 909 kg

1 kg = 2.204 622 6 lbm
1 kg = 0.068 521 76 slug

DENSITY

1 lbm/in.^{3} = 27,679.905 kg/m^3
1 lbm/ft^3 = 16.018 46 kg/m^3
1 slug/ft^3 = 515.379 kg/m^3

1 kg/m^3 = 0.000 036 127 29 lbm/in.^{3}
   = 0.062 427 95 lbm/ft^3
   = 0.001 940 321 slug/ft^3

WORK and ENERGY

1 in. lb = 0.112 984 8 Nm (J) work
1 ft lb = 1.355 82 Nm (J)
1 Btu (59°F) = 777.980 5 ft lb = 1,054.80 Nm (J)
1 calorie (59°F) = 3.087 29 ft lb = 4.185 80 Nm (J)
1 Btu = 251.995 calories

1 Nm = 1 J = 10^7 ergs = 8.850 746 in. lb
   = 0.737 562 ft lb

POWER

1 hp = 550 ft lb/sec = 0.745 7 kW (kN/sec,   1 kW = 1.3410 hp)

MISCELLANEOUS

Free fall, g = 32.174 0 ft/sec^2 = 386.09 in./sec^2 = 9.806 65 m/sec^2 acceleration
1 mi/gal = 0.425 144 km/liter,   1 km/liter = 2.352 144 mi/gal

Design
of Machine
Elements

6th Edition

Design of Machine Elements

Incorporates both U.S. customary and SI units

M.F. Spotts

Mechanical Engineering Department
The Technological Institute
Northwestern University

Prentice-Hall, Inc., Englewood Cliffs, New Jersey 07632

Library of Congress Cataloging in Publication Data

Spotts, Merhyle Franklin (date)
  Design of machine elements.

  Includes bibliographies and index.
  1. Machinery—Design.   I. Title.
TJ230.S82   1985   621.8′15   84–13377
ISBN 0–13–200593–X

Editorial/production supervision: *Nicholas C. Romanelli*
Interior and cover design: *Anne T. Bonanno*
Manufacturing buyer: *Anthony Caruso*

Printed in the United States of America

10   9   8   7   6   5   4   3   2   1

ISBN 0-13-200593-X

Prentice-Hall International, Inc., *London*
Prentice-Hall of Australia Pty. Limited, *Sydney*
Editora Prentice-Hall do Brasil, Ltda., *Rio de Janeiro*
Prentice-Hall Canada Inc., *Toronto*
Prentice-Hall of India Private Limited, *New Delhi*
Prentice-Hall of Japan, Inc., *Tokyo*
Prentice-Hall of Southeast Asia Pte. Ltd., *Singapore*
Whitehall Books Limited, *Wellington, New Zealand*

Throughout the wide fields of engineering, the most fundamental requirement is a sound knowledge of the first principles of mechanics coupled with an intimate understanding of the properties of materials. More problems arise from weaknesses in mechanical design than from any other cause, impressing upon us the necessity of being well versed in the basic principles of mechanical engineering.

Sir George H. Nelson
Institution of Electrical Engineers

Contents

2   Working Stresses   *91*

3   Shafting   *148*

8 Lubrication *381*

9 Ball and Roller Bearings *423*

12 Miscellaneous Machine Elements *548*

13 Dimensioning and Details *610*

14 Engineering Materials *660*

Index

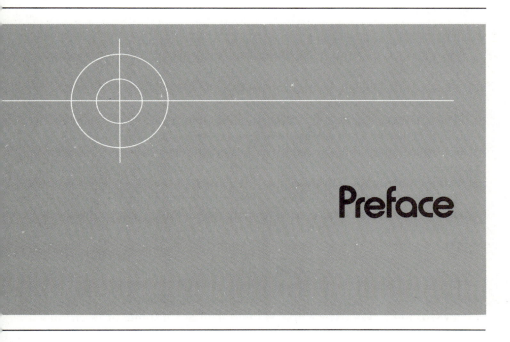

Preface

The essential features of the previous editions have been retained in this, the sixth edition. Improvements have been made throughout the book in simplifying design methods and making them more understandable and easier to use. Wherever appropriate, new material has been added to various chapters including over 200 new problems with answers.

As for previous editions the object of the book is to give training in the making of design calculations for the mechanical elements in universal use. In fact one cannot call oneself a mechanical engineer at the professional level unless familiar to some extent with basic theories of this kind. Much more, of course, could be included for each element, but space limitations have made it necessary to present what is felt to be the central and most important way to handle each of the design requirements. Special attention has been given to cases where the loads are continuously fluctuating and stress concentrations may be present. Almost all mathematical derivations have been given in full to make the book as practical and useful, as well as authoritative and adaptable, as possible.

Material on the International System of Units, commonly called SI, has been retained, but in a form that can be easily omitted if not desired. Sufficient material is included to permit design calculations to be made in the metric system whenever needed. Helpful references are appended at the end of this preface.

After leaving school, students familiar with the methods presented here will be able to make adjustments necessary to fit the particular conditions with which they are faced. The book can also serve as preliminary training for more advanced study for those students who wish to become specialists in some chosen field of mechanical design.

The notation is simple and unified, and the chapters are largely independent of each other. They can be studied in any order with virtually no need of preceeding study. All principles are illustrated by worked-out numerical examples. Chapter 1 presents a review of the essentials of fundamental mechanics used throughout the book.

The book represents experience of over 30 years in industry and the classroom. The principles presented here are fundamental to the entire field of mechanical design.

REFERENCES

1. *AN American National Standard ASTM/IEEE Standard Metric Practice, ANSI Z 210.1-1976*, New York, 345 East 47th St.: Institute of Electrical and Electronics Engineers, Inc.

2. *The International System of Units, NASA SP-7012*, Washington, D.C.: National Aeronautics and Space Administration, 1973.

3. *ASTM Standard Metric Practice Guide, E380-72*, Philadelphia, Pa., 1916 Race St.: American Society for Testing and Materials.

4. *Dual Dimensioning*—SAE J390, New York, 2 Pennsylvania Plaza: Society of Automotive Engineers, 1970.

5. *ASME Orientation and Guide for Use of Metric Units*, E58, New York, 345 East 47th St.: American Society of Mechanical Engineers, 1972.

Evanston, Illinois   M. F. Spotts

Design of Machine Elements

Introduction

1. Machine Design

Machine design is the art of planning or devising new or improved machines to accomplish specific purposes. In general, a machine will consist of a combination of several different mechanical elements properly designed and arranged to work together, as a whole. During the initial planning of a machine, fundamental decisions must be made concerning loading, type of kinematic elements to be used, and correct utilization of the properties of engineering materials. Economic considerations are usually of prime importance when the design of new machinery is undertaken. In general, the lowest over-all cost is desired. Consideration should be given not only to the cost of design, manufacture, sale, and installation, but also to the cost of servicing. The machine should of course incorporate the necessary safety features and be of pleasing external appearance. The objective is to produce a machine which is not only sufficiently rugged to function properly for a reasonable life, but is at the same time cheap enough to be economically feasible.

The engineer in charge of the design of a machine should not only have adequate technical training, but must be a man of sound judgment and wide experience, qualities which are usually acquired only after considerable time has been spent in actual professional work. A start in this direction can be made with a good teacher while the student is yet at the university.

However, the would-be designer must expect to get a substantial portion of his training after leaving school through further reading and study, and especially by being associated in his work with competent engineers.

2. *Design of Machine Elements*

This book, as the title indicates, will not deal with the broader aspects of the design of complete machines, but will attempt to explain the fundamental principles required for the correct design of the separate elements which compose the machine.

The principles of design are, of course, universal. The same theory or equations may be applied to a very small part, as in an instrument, or to a larger but similar part used in a piece of heavy equipment. In no case, however, should mathematical calculations be looked upon as absolute and final. They are all subject to the accuracy of the various assumptions which must necessarily be made in engineering work. Sometimes only a portion of the total number of parts in a machine are designed on the basis of analytic calculations. The form and size of the remaining parts are then usually determined by practical considerations. On the other hand, if the machine is very expensive, or if weight is a factor, as in airplanes, design computations may then be made for almost all the parts.

The purpose of the design calculations is, of course, to attempt to predict the stress or deformation in the part in order that it may safely carry the loads which will be imposed upon it, and that it may last for the expected life of the machine. All calculations are, of course, dependent on the physical properties of the construction materials as determined by laboratory tests. A rational method of design attempts to take the results of relatively simple and fundamental tests such as tension, compression, torsion, and fatigue and apply them to all the complicated and involved situations encountered in present-day machinery.

In addition, it has been amply proved that such details as surface condition, fillets, notches, manufacturing tolerances, and heat treatment have a marked effect on the strength and useful life of a machine part. The design and drafting departments must specify completely all such particulars, and thus exercise the necessary close control over the finished product.

Training in rapid and accurate numerical work is invaluable to the designer. The designer should keep an accurate notebook, as it is frequently necessary for him to refer to work which he has done in the past. A sketch, carefully drawn to scale, is also a necessity, and provides a convenient place for putting down a portion of the data used in connection with the problem.

It goes without saying that all data, assumptions, equations, and calculations should be written down in full in order to be intelligible when referred to at a later date. The student should start forming such habits, and it is recommended that the problems in this book be worked out and preserved as reference material.

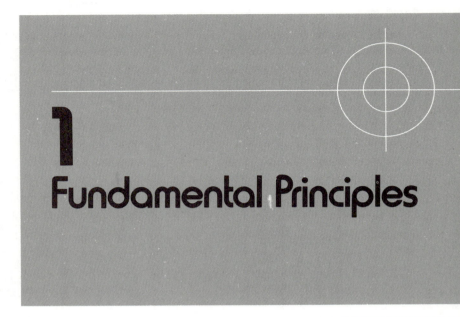

1
Fundamental Principles

Design methods for the various machine elements are founded on the theories of mechanics and strength of materials. The scope of such theories is very extensive, and the purpose of this chapter is to present, for review and ready reference, those topics that are generally used by designers, and that will be referred to throughout the book. These theories are more or less simplified approximations, and attention should be directed to the limitations imposed by the assumptions that had to be made in arriving at working formulas.

A thorough grounding in these fundamentals will prove of great value in attacking new and unfamiliar problems. In fact, only after such theories have become working tools is it possible to achieve the broad perspective and balanced judgment that must be expressed by the really competent machine designer.

A,	area	F_s,	factor of safety
b,	width of cross section parallel to neutral axis	g,	acceleration due to gravity $= 32.174 \text{ ft/sec}^2$ $= 386 \text{ in./sec}^2$ $= 9,807 \text{ mm/sec}^2$
c,	maximum distance, neutral axis to edge of cross section		
d,	distance; diameter of circle	G,	modulus of elasticity for shear stress
E,	modulus of elasticity for normal stress		
F,	force or load	h,	height of cross section perpendicular to neutral axis

i,	radius of gyration		beam
in.,	inch, 25.4 mm	v,	distance on cross section of
in. lb,	inch pound, 113 Nmm		beam perpendicular to neutral
I,	moment of inertia		axis
J	joule, or newton-meter, Nm	V,	total shear force on cross
k,	kilo (prefix), 1,000		section
kg,	kilogram, 2.2046 lb (mass);	w,	distributed load per unit length
	1 lb (mass) = 0.4536 kg	y,	deflection of beam
l,	length	γ,	(gamma) shearing
lb,	pound;		deformation, weight per unit
	1 lb (force) = 4.4482 N		volume of material
	1 lb (mass) = 0.4536 kg	δ,	(delta) axial deformation
m,	meter, 39.370 in.	ϵ,	(epsilon) strain or elongation
m,	milli (prefix), one thousandth	μ,	(mu) Poisson's ratio; for
mm,	millimeter, 0.039 37 in.,		metals, μ is usually taken
	1 in. = 25.4 mm		as 0.3
M,	mega (prefix), 1 million	σ,	(sigma) normal stress
M,	bending moment	σ_1,	maximum normal stress
MPa,	megapascal, 1 N/mm^2;	σ_2,	minimum normal stress
	1 MPa = 145 psi	σ_x,	normal stress, x-direction
N,	newton, 0.2248 lb (force);	σ_y,	normal stress, y-direction
	1 lb (force) = 4.448 N	σ_{yp},	yield point stress
P,	load	τ,	(tau) shear stress
Pa,	pascal, 1 N/m^2	τ_{max},	maximum shear stress
r,	radius of circle, radius to	τ_{xy},	shear stress, x- and y-directions
	center of curvature of deflected		

1. Statical Equilibrium

When a body is at rest, or in motion with constant velocity, the external forces acting upon it are in equilibrium. This statement applies to the body as a whole or to any portion of it. When a force analysis is to be made, it is sometimes advantageous to consider only a portion of the body, which can be obtained by assuming that cutting planes are passed through the body at the desired locations. The internal forces that were acting at the locations of the cuts must then be represented as a system of external forces properly distributed to maintain equilibrium of the separate parts and to preserve the original state of stress in the material. When a problem is analyzed in this manner the loading will consist entirely of external forces and moments. It is not necessary to consider the internal stresses.

Statical equilibrium means that both forces and moments are in balance. When a body is in equilibrium, the sum of the components of the forces in any given direction must be equal to zero. Likewise the moments about any given line as an axis must be equal to zero. If the body is undergoing acceleration, the inertia forces must be included in the equilibrium equations.

2. Engineering Materials

The mathematical equations used in designing are derived for an idealized material, which is assumed to have the following properties.

(a) *Perfect Elasticity.* Loads or forces acting on a body cause changes in its shape and dimensions. A perfectly elastic material is one that returns to its original form immediately upon removal of the loads. The equations used in designing are nearly always derived on the assumption of perfect elasticity. If the material is such that this assumption cannot be made, the mathematical complications, in many cases, become too great for practical calculations. It should never be forgotten, however, that in some cases there may be a considerable variation between the actual stresses in the body and the stresses obtained from the equations for an idealized substance. A material may exhibit a high degree of elasticity for small loads, but may retain a permanent deformation when the loads become sufficiently great.

(b) *Homogeneity.* A homogeneous body is one that has the same properties throughout its entire extent.

(c) *Isotropy.* An isotropic material is one whose elastic properties are the same in all directions.

Actually, a metal is not a homogeneous substance. It consists of an aggregate of very small crystals whose strength depends upon their orientation with respect to the applied force. When the minute crystals have a random orientation, the location in the body, or the inclination at which a test specimen is taken, has no effect on the results of the test. The assumption that the material is homogeneous and isotropic, for all practical purposes, is fulfilled. This is true for cast, hot-rolled, or annealed metals. In contrast, materials that have been cold rolled or drawn may have a preferred orientation of crystals, and may exhibit a definite grain effect with a variation in strength depending on the direction of the applied load. The assumption cannot be made that such materials are homogeneous and isotropic.

3. Tension and Compression Stress

The eyebar, in Fig. 1-1(a), which supports the load P, is said to be in *tension*, or to have an internal force of tension. Such a force causes an increase in the length of the bar. A solution for the stress can be effected by means of cutting planes as described in Section 1.

If the bar, shown in Fig. 1-1(a), is cut normal to its axis, as shown in

FIGURE 1-1   Eye-bar loaded in tension.

Fig. 1-1(b), equal and opposite tension forces, uniformly distributed, must be applied to the cut surfaces. In magnitude, each must be equivalent to the load *P*. The average stress *σ*, or force per unit of cross-sectional area *A*, is then equal to *P* divided by *A*. Hence

$$\sigma = \frac{P}{A} \qquad (1)$$

Thus the magnitude of the external forces on the cross sections in Fig. 1-1(b) constitutes the measure of the internal force in the bar shown in Fig. 1-1(a). Forces are usually expressed in pounds and areas in square inches.[1]

It is, of course, obvious that this solution is correct only if the assumption regarding the distribution of stress on the cut surfaces is correct. Had the bar been cut near one end where the shape is no longer a prism, it is apparent that the situation would be more complicated, and the stress system would no longer be simple tension uniformly distributed over the cross section.

Since the assumption relative to homogeneity of the material is never exactly fulfilled, the stresses on the cross section will not be entirely uniform but will be subject to small local variations. Equation (1) does, however, give the average value of the tensile stress over the whole cross section.

A *compressive stress* is one that causes a decrease in the length of the body in the direction of the force.

The total change of length in a uniform body caused by an axial load is called the *deformation, δ*. If the deformation is divided by the original length, *l*, of the body, the result is the deformation per unit length, and is called *elongation* or *strain, ε*. It can be represented mathematically by the

[1]Except for inches (in.) engineering abbreviations do not have periods.

equation

$$\epsilon = \frac{\delta}{l} \tag{2}$$

Although elongation ϵ is a dimensionless number, it is customary to speak of it in terms of inches per inch, or mm per mm.

For most materials used in engineering, stress and strain are directly proportional; when this condition exists, the material is said to follow *Hooke's law*. The linear relationship between stress and strain can be represented by an equation if a constant of proportionality E is introduced as follows:

$$\sigma = \epsilon E \quad \text{or} \quad \epsilon = \frac{\sigma}{E} \tag{3}$$

Constant E is called the *modulus of elasticity*, or *Young's modulus*, for the material. It has the dimension of stress and can be visualized as the tensile stress that would cause a body to double in length, $\epsilon = 1$, provided the material would remain elastic with such excessive loading. Values of E for common materials used in engineering are given in Table 2-1 of the following chapter.

Substitution of Eqs. (1) and (3) into Eq. (2) gives the important relationship

$$\delta = \frac{Pl}{AE} \tag{4}$$

Equations (3) and (4) are valid either for tension or compression. Tension stress and increase of length are considered positive, whereas compressive stress and decrease in length are considered negative.

Example 1. In Fig. 1-1, let load P be equal to 5,000 lb, and let the bar be 3 in. wide and 0.5 in. thick. The uniform portion of the bar is 60 in. long. The material is steel.

(a) Find the stress in the uniform portion of the bar.
(b) Find the deformation of the uniform portion of the bar.

Solution. Cross-sectional area: $A = 3 \times 0.5 = 1.5$ in.^{2}

(a) Stress, by Eq. (1): $\sigma = \dfrac{5,000}{1.5} = 3,333$ psi

(b) Deformation, by Eq. (4): $\delta = \dfrac{5,000 \times 60}{1.5 \times 30,000,000} = 0.00667$ in.

For statical equilibrium, the summation of the forces in any direction must equal zero, and the summation of the moments about any axis also must equal zero. The following equations must therefore be fulfilled.

$$\Sigma F = 0, \quad \Sigma M = 0$$

4. Tension and Compression in SI Units

The SI unit for length is the meter, m. For machine elements it is more convenient to use the millimeter, mm, which is one one-thousandth of a meter.[2] The inch and millimeter are related by the following equation:

$$1 \text{ in.} = 25.4 \text{ mm exactly}$$

or

$$1 \text{ mm} = 0.039\ 370 \ldots \text{ in.}$$

The SI unit for force is the newton, N. Its value in pounds is

$$1 \text{ N} = 0.2248 \ldots \text{ lb}$$

or

$$1 \text{ lb} = 4.448 \ldots \text{ N}$$

The unit for stress is the pascal, Pa, which is a stress equal to 1 N/m². This is a very small stress and the megapascal, MPa, which is 1 million newtons per square meter or 1 N/mm², is a more practical unit.

Example 1A. Let the suspended body in Fig. 1-1 have a gravity attraction of 22,500 N. The uniform portion of the bar is 75 mm wide, 13 mm thick, and 1,500 mm long. The material is steel. Find the stress and deformation in the uniform portion. See Table 2-1A for material constants.

Solution. By Eq. (1): $\sigma = \dfrac{P}{A} = \dfrac{22,500}{75 \times 13} = 23.08 \dfrac{N}{mm^2} = 23.08$ MPa

By Eq. (4): $\delta = \dfrac{Pl}{AE} = \dfrac{22,500 \times 1,500}{75 \times 13 \times 206,900} = 0.167$ mm

Conversion factors for complex terms can be easily obtained by substituting unit values. Several widely used factors are derived as follows.

Moment, torque, work: $1 \text{ in. lb} = (25.4 \text{ mm})(4.4482 \text{ N})$

$$= 113 \text{ Nmm} = 0.113 \text{ Nm}$$

Velocity: $1 \dfrac{ft}{min} = \dfrac{12 \times 25.4 \text{ mm}}{60 \text{ sec}} = 5.08 \dfrac{mm}{sec}$

Stress: $1 \text{ MPa} = 1 \dfrac{N}{mm^2} = \dfrac{0.2248 \text{ lb}}{(in./25.4)^2} = 145 \dfrac{lb}{in.^2}$

Mass: $1 \dfrac{lb \ sec^2}{in.} = \dfrac{4.448 \text{ N sec}^2}{25.4 \text{ mm}} = 0.17513 \dfrac{N \ sec^2}{mm}$

[2]On engineering drawings, the metric unit for lineal dimensions is the millimeter. See p. 8 of Reference 1 and p. 1 of Reference 4, listed under Preface to Sixth Edition.

Power:
$$1 \text{ hp} = 33,000 \frac{\text{ft lb}}{\text{min}}$$

$$= 33,000 \frac{(12 \times 0.0254 \text{ m})(4.448 \text{ N})}{60 \text{ sec}}$$

$$= 745.7 \frac{\text{Nm}}{\text{sec}} = 745.7 \text{ W} = 0.7457 \text{ kW}$$

$$1 \text{ kW} = \frac{1}{0.7457} \text{ hp} = 1.341 \text{ hp}$$

Viscosity:
$$1 \frac{\text{lb sec}}{\text{in.}^2} = \frac{(4.4482 \text{ N}) \text{ sec}}{(25.4 \text{ mm})^2} = \frac{1}{145} \frac{\text{N sec}}{\text{mm}^2}$$

5. Force and Mass

The relationship between force and mass is determined by Newton's second law of motion.

$$\text{force} = \text{mass} \times \text{acceleration}$$

In free fall, the acceleration is that due to gravity, usually designated g. If weight, or force, is called P, the mass of a body is

$$\text{mass} = \frac{P}{g} \frac{\text{lb sec}^2}{\text{ft}} \quad \text{or} \quad \frac{\text{N sec}^2}{\text{m}}$$

In U.S. customary units, g has the value 32.174 ft/sec². A unit mass, commonly called the slug, then has a weight P of 32.174 lb. In SI units the acceleration g due to gravity is 9.8066 m/sec². The unit mass, the kilogram, then has a weight of 9.8066 N.

The prototype for the kilogram is a specific cylinder of platinum–iridium alloy kept in a vault at Sèvres, near Paris, France. In U.S. customary units, the kilogram weighs 2.2046 pounds. The pound can then represent either force or mass, and it may be necessary to designate which quantity is referred to. Thus

$$1 \text{ kg} = 2.2046 \text{ lb (mass)}$$

$$1 \text{ lb (mass)} = \frac{1}{2.2046} = 0.4536 \text{ kg}$$

The gravity force acting on a 1 kg (2.2046 lb) mass is 9.8066 newtons. The force acting on 1 lb is 1/2.2046 of this amount. Then

$$1 \text{ lb (force)} = \frac{1}{2.2046} \times 9.8066 = 4.448 \cdots \text{ N}$$

$$1 \text{ N} = 0.2248 \cdots \text{ lb (force)}$$

The foregoing quantities are illustrated by Fig. 1-2.

In inch units the acceleration of gravity in free fall is $32.174 \times 12 = 386$ in./sec². A mass of 386 lb would then be accelerated at the rate of 1 in./sec² by a force of 1 pound.

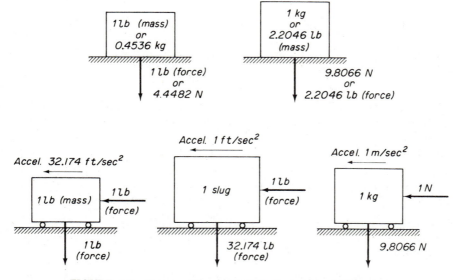

FIGURE 1-2 Forces and masses in U.S. Customary and SI units.

6. Statically Indeterminate Problems in Tension and Compression

Machine parts are sometimes arranged in a manner whereby the axial forces cannot be determined by the equations of statics alone. Such force systems are said to be statically indeterminate. They are characterized by the presence of more supports or members than the minimum required for the equilibrium of the structure. For such situations, the deformations of the parts must be taken into consideration. The following example will illustrate a typical method for solving such problems.

Example 2. Find the force in each of the vertical bars in Fig. 1-3. The weight can be assumed to be rigid and to maintain the connections for the three vertical bars in a straight line. Assume the support at the top to be rigid also.

FIGURE 1-3   Example 2.

Solution. Because both geometry and loading are symmetrical, the forces will be equal in the two outer bars. From statical equilibrium, the following equation can be written:

$$2F_1 + F_2 = 5,000 \tag{a}$$

Since two unknowns are present, it is necessary to obtain another equation to effect a solution. This can be done by considering the deformations of the bars. From the given data, all bars will have the same amount of deformation. Hence

$$\delta_1 = \delta_2 \quad \text{or} \quad \frac{F_1 l_1}{A_1 E_1} = \frac{F_2 l_2}{A_2 E_2}$$

Numerical values should be substituted into the equation above.

$$\frac{F_1 36}{0.2 \times 30,000,000} = \frac{F_2 36}{0.3 \times 15,000,000} \quad \text{or} \quad 3F_1 = 4F_2 \tag{b}$$

Equations (a) and (b) should now be solved simultaneously to give

$$F_1 = 1,818 \text{ lb} \quad \text{and} \quad F_2 = 1,364 \text{ lb}$$

The distribution of the forces in an indeterminate structure is sensitive to small variations in the dimensions of the parts. A small undetected error in machining a dimension can cause a large change in the distribution of the loads. Any calculations made by the designer will not be valid unless the fit of the parts can be rigidly maintained as originally planned.

A variation in temperature may change the values of the forces in an indeterminate system. If temperature causes a relative change between the lengths of the various parts of the assembly, the effect can be similar to that of an error in machining. The designer must therefore consider temperature variations as well as dimensional errors in his calculations for an indeterminate structure.

In the solution of indeterminate problems, it is usually necessary to assume that certain members or supports are rigid. Since complete rigidity cannot be achieved, any deformation of such assumed rigid elements will cause changes in the calculated values of the forces.

7. Center of Gravity

An equation for finding the center of gravity of an area can be derived from Fig. 1-4(a). The distance from the x-axis to the center of gravity is called $\bar{y}$, and the distance to an element dA parallel to the x-axis is called y. The moment arm for dA about a horizontal axis through the center of gravity C is equal to $(y - \bar{y})$. Mathematically, center of gravity means that the sum of the moments of the areas about the axis through C must be equal to zero as indicated by the following equation:

$$\int (y - \bar{y})\, dA = 0 \quad \text{or} \quad \bar{y} = \frac{\int y\, dA}{\int dA} = \frac{\int y\, dA}{A} \tag{5}$$

A similar equation can be written for the moments about the y-axis, which permits $\bar{x}$ to be found.

(a) Center of gravity.   (b) Center of gravity of composite section.

FIGURE 1-4

A composite figure can usually be divided into simple areas, and Eq. (5) can be applied by making the numerator equal to the sum of the $\int y\, dA$ terms of the separate parts. The denominator is the total area. If the location of the center of gravity for each of the separate areas is known, it is not necessary to perform mathematical integrations. Each $\int y\, dA$ term can be replaced by the product of the area times the distance from the axis to its center of gravity as is shown in Fig. 1-4(b). Thus

$$\bar{y} = \frac{A_1 \bar{y}_1 + A_2 \bar{y}_2 + \cdots}{A_1 + A_2 + \cdots} \tag{6}$$

Example 3. Find the location of the center of gravity of the T-shaped cross section in Fig. 1-21(b).

Solution. Divide the area into two parts of 6 in.^{2} and 4 in.^{2} by extending the vertical sides of the stem to the bottom of the cross section. Let the axis of moments be taken along the bottom of the T.

In Eq. (6): $\bar{y} = \dfrac{6 \times 3 + 4 \times 0.5}{6 + 4} = 2$ in.

The center of gravity is located 2 in. up from the bottom.

Example 3A. Let the T-shaped cross section of Fig. 1-21 have a height of 150 mm and a maximum width of 125 mm. Width of the stem is 25 mm and the thickness of flange is 25 mm. Find the location of the center of gravity.

Solution. Assume the stem extends through the flange to the bottom.

$$\text{area of stem} \quad = 150 \times 25 = 3{,}750 \text{ mm}^2$$
$$\text{area of flange} = 100 \times 25 = 2{,}500 \text{ mm}^2$$

Take the axis along the bottom of the T.

By Eq. (6): $\bar{y} = \dfrac{3{,}750 \times 75 + 2{,}500 \times 12.5}{3{,}750 + 2{,}500} = 50$ mm

8. Bending of Beams

Suppose a long, thin, straight beam is bent into a curve by moments M applied at the ends, as shown in Fig. 1-5(a). The beam and moments lie in the xy-plane with the origin at the left end and the y-axis positive downward. At distance x from the left end, the deflection of the beam is given by distance y, as shown. Figure 1-5(b) shows, enlarged, a slice AB of differential length dx cut from the beam at location x.

The planes cutting the right and left end surfaces of AB are taken perpendicular to the longitudinal axis of the originally straight beam. It is customary to assume that these cross sections will remain plane and perpendicular to the longitudinal elements of the beam after moments M are applied. Laboratory experiments have in general verified this assumption. After bending, some of the elements have been lengthened, some have been shortened, and at one location, called the *neutral surface*, no change in length has taken place.

The loading of Fig. 1-5 is called *pure bending.* No shear or tangential stress will exist on the end surfaces of AB, and the only stress will be σ, acting normally to the surfaces as shown. An equation can be derived for giving the value of this bending stress at any desired distance v from the neutral surface. Let O_1 be the center of curvature for slice AB of the deflected beam. Let $d\varphi$ be the small angle included between the cutting planes, and let r be the radius of curvature. Consider a horizontal element located a distance v below the neutral surface. Draw line BC parallel to O_1A. Angle AO_1B is equal to angle CBD and the following proportion can be written.

$$\frac{v}{r} = \frac{v\,d\varphi}{dx} = \epsilon \tag{a}$$

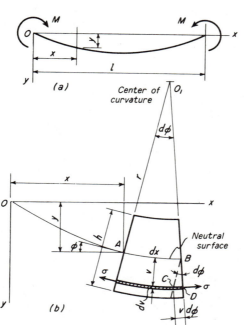

FIGURE 1-5 **Bending stress for beam loaded by moments at ends.**

Since the total deformation of the element $v\,d\varphi$ divided by the original length dx is the unit deformation or strain ϵ, Eq. (a) indicates that the elongation of the element varies directly with the distance v from the neutral surface. Let it be assumed that the material of the beam follows Hooke's law. Substitution of $\epsilon = \sigma/E$ into Eq. (a) gives

$$\frac{\sigma}{E} = \frac{v}{r} \quad \text{or} \quad \sigma = \frac{E}{r}v \tag{7}$$

Thus the stress also varies directly with the distance from the neutral surface, becoming larger as v increases. Equation (7) is of course valid only for stresses in the elastic range of the material. Above the neutral surface, for

negative values of v, the stress is compressive and increases uniformly with
the distance from the neutral surface. Equation (7), obtained from the geometry of the deformed beam, gives only a portion of the solution. It is now
necessary to consider the equilibrium of the beam.

Figure 1-6(a) shows the beam after the left-hand portion has been
removed by passing a cutting plane at A. It will be observed that the stresses
on the left-hand cross section are distributed in accordance with Eq. (7),
and that the given moment M is acting at the right-hand end. A perspective
of the stress system is shown in Fig. 1-6(b). The intersection of the neutral
surface with the cross section upon which the stresses are acting is called the
neutral axis.

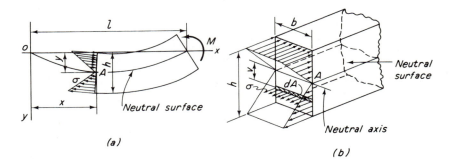

(a)

(b)

FIGURE 1-6   Stress caused by bending moment.

The portion of the beam in Fig. 1-6(a) must be in equilibrium with
respect to both forces and moments. Since the given loading M is a pure
moment, it is necessary that the resultant horizontal forces on the left end
surface be zero. The force on an element of area dA, shown in Fig. 1-6(b),
is equal to $\sigma\, dA$. When this force is added up for the entire area, $\int \sigma\, dA$,
the result must be equal to zero. The value of σ from Eq. (7) should now be
substituted as follows:

$$\int \sigma\, dA = \int \frac{E}{r} v\, dA = \frac{E}{r} \int v\, dA = 0 \qquad (b)$$

or
$$\int v\, dA = \bar{v} A = 0 \qquad (8)$$

The integral of Eq. (8) represents the total moment of area about the
neutral axis. As was shown in Section 5, it can be set equal to the product
of the area A of the cross section times the distance $\bar{v}$ of its center of gravity
from the neutral axis. The only way in which this product can be zero is for
$\bar{v}$ to have a zero value. It must therefore be concluded that the neutral axis
passes through the center of gravity of the cross section.

Since the beam in Fig. 1-6(a) is in equilibrium, the moment of the stresses of the left-hand end surface must be equal to the applied moment load M. The force on an element of area, $\sigma \, dA$, when multiplied by the distance v to the neutral axis, and integrated over the entire area, is equal to the moment M. Hence

$$M = \int \sigma v \, dA = \int \frac{E}{r} v^2 \, dA = \frac{E}{r} \int v^2 \, dA = \frac{EI}{r} \qquad (9)$$

The integral $\int v^2 \, dA$ is customarily called the *moment of inertia of the area* or sometimes the *second moment of area* and is represented by the letter I. This substitution has been made in the last form of Eq. (9). If the radius of curvature r is eliminated between Eqs. (7) and (9), the following important equation results.

$$\sigma = \frac{Mv}{I} \qquad (10)$$

This equation gives the value of the bending stress at any distance v from the neutral axis. The greatest stress is found at the location on the cross section where v is the largest. This maximum value of v is usually denoted by c, and the equation for the maximum bending stress then becomes

$$\sigma = \frac{Mc}{I} \qquad (11)$$

It should be noted that the magnitude of the stress σ given by Eq. (11) is independent of the kind of material composing the beam. The ratio I/c is called the *section modulus of the cross section*.

Although Fig. 1-6 illustrates a rectangular beam, the foregoing theory is valid for any shape of cross section. The maximum stress is located on the boundary at the point farthest removed from the neutral axis.

According to the original assumption, this theory should be applied only to long, thin beams loaded in pure bending. However, under most conditions, the equations give satisfactory results for bending stress when the bending moment is caused by transverse forces applied to the beam. Transverse forces also cause compressive stress between the elements in the neighborhood of the loads.

If the material in the beam does not follow Hooke's law, the magnitude of the stress is no longer proportional to the distance from the neutral axis. If Eq. (11) is applied to such beams, the results may be only approximate.

Bending moment in SI has the dimension of newton millimeters, Nmm. Conversion to U.S. customary units can be made as follows:

$$1 \text{ in. lb} = (25.4 \text{ mm})(4.4482 \text{ N}) = 113 \text{ Nmm}$$

9. Moment of Inertia

The integral $\int v^2 \, dA$ appeared in connection with the bending theory of beams. For convenience in writing, this integral, as was mentioned in the foregoing section, is usually replaced by the symbol I and is called the *moment of inertia*. Expressions for I for the commonly used geometric shapes of cross sections can be found in the mechanical engineering handbooks. See also Fig. 1-7.

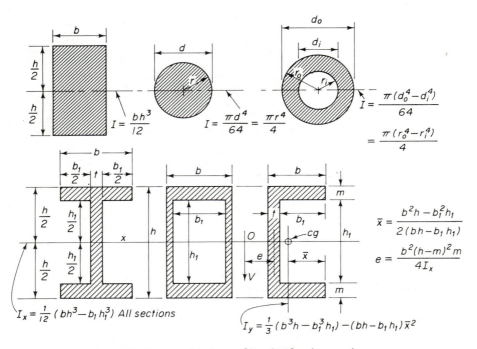

FIGURE 1-7   Moments of inertia of various sections.

If the integral is computed for a rectangle of width b and depth h about an axis parallel to side b through the center of gravity, as shown in Fig. 1-7, it is found to be equal to

$$I = \frac{bh^3}{12} \tag{12}$$

Width b is parallel to the axis about which the moment tries to rotate the cross section.

Example 4. Let the beam in Fig. 1-5 be 2 in. wide and 3 in. deep. Let the bending moment M at each end be equal to 40,000 in. lb. Find the value of the bending stress.

Solution. By Eq. (12): $I = \dfrac{bh^3}{12} = \dfrac{2 \times 3^3}{12} = 4.5$ in.⁴

By Eq. (11): $\sigma = \dfrac{Mc}{I} = \dfrac{40,000 \times 1.5}{4.5} = 13,330$ psi

Example 4A. Let the beam in Fig. 1-5 be 50 mm wide and 75 mm deep. Let the bending moment M at each end be equal to 5,000,000 Nmm. Find the value of the bending stress.

Solution. By Eq. (12): $I = \dfrac{bh^3}{12} = \dfrac{50 \times 75^3}{12} = 1,757,800$ mm⁴

By Eq. (11): $\sigma = \dfrac{Mc}{I} = \dfrac{5,000,000 \times 37.5}{1,757,800} = 106.7$ MPa

Example 5. A steel bandsaw blade is 0.028 in. thick. Find the value of the bending stress when the blade is passing around a pulley of 18-in. diam. $E = 30,000,000$ psi.

Solution. By Eq. (7): $\sigma = \dfrac{Ev}{r} = \dfrac{30,000,000 \times 0.014}{9} = 46,670$ psi

Example 5A. A steel saw blade is 0.7 mm thick. Find the value of the bending stress when the blade is passing around a pulley of 450 mm diameter. $E = 206,900$ MPa.

Solution. By Eq. (7): $\sigma = \dfrac{Ev}{r} = \dfrac{206,900 \times 0.35}{225} = 321.8$ MPa

Another expression for the bending stress of a beam of rectangular cross section can be secured by substituting Eq. (12) into Eq. (11).

$$\sigma = \frac{6M}{bh^2} \tag{13}$$

For a circle of diameter d or radius r about a diameter through the center, the value of the integral $\int v^2 \, dA$ becomes

$$I = \frac{\pi d^4}{64} = \frac{\pi r^4}{4} \tag{14}$$

When Eq. (14) is substituted into Eq. (11), the following expression for the bending stress of a beam of circular cross section is obtained.

$$\sigma = \frac{32M}{\pi d^3} \tag{15}$$

If a beam with a channel section, the last sketch of Fig. 1-7, is loaded with a vertical force through the center of gravity, it is found that the beam not only deflects but the cross sections will twist and the ordinary equation for bending, Mc/I, cannot be applied. To have simple bending without twisting the vertical load must be applied at a point such as O a distance e from the center of the web. Twist occurs for any other location of the load. Point O is called the *center of twist*. The equation for e is shown in the figure[3].

10. Transfer of Axis for Moment of Inertia

Area A

FIGURE 1-8   Moment of inertia about parallel axis.

Suppose that the moment of inertia for the shaded rectangle in Fig. 1-8 is desired about axis 1. Axis 0 should be drawn through the center of gravity parallel to axis 1 about which the moment of inertia is desired. The integral of distance squared times differential area then becomes

$$I_1 = \int v + \bar{y})^2 \, dA \tag{a}$$

$$= \int (v^2 + 2\bar{y}v + \bar{y}^2) \, dA$$

The integration can be made term by term and is to extend over the shaded area. The first integral, $\int v^2 \, dA$, is equal to the moment of inertia about axis 0 and may be written I_0. The second term, $2\bar{y} \int v \, dA$, is equal to zero since it represents merely the moment of the area about an axis through the center of gravity. The third term is seen to be equal to $A\bar{y}^2$, where A is the area. The moment of inertia for area A about axis 1 is then equal to

$$I_1 = I_0 + A\bar{y}^2 \tag{16}$$

This equation is known as the *parallel axis theorem*. The moment of inertia for a composite cross section can be found by dividing it into elementary parts, finding the moment of inertia about the desired axis for each by Eq. (16), and then adding the results together. Axis 0, however, must be taken through the center of gravity of the area under consideration and must be parallel to the axis about which the total value of I is desired.

The equation for moment of inertia is sometimes written

$$I = i^2 A \tag{b}$$

where i is called the *radius of gyration* for the area. It is hypothetical distance at which the entire area A would have to be concentrated in order to give the same value for I as determined by the integral.

---

[3]For derivation, see p. 52 of Reference 5, end of chapter.

Example 6. Find the value of the moment of inertia of the T-shaped cross section of Fig. 1-21(b).

Solution. By Example 3, the center of gravity is located 2 in. up from the bottom.

By Eq. (16), for the stem: $I = \dfrac{1 \times 6^3}{12} + 6 \times 1^2 = 24$ in.^{4}

By Eq. (16), for the T: $I = \dfrac{4 \times 1^3}{12} + 4 \times 1.5^2 = 9.33$ in.^{4}

Total: $I = 24 + 9.33 = 33.33$ in.^{4}

Example 6A. Find the moment of inertia for the T-shaped cross section of Fig. 1-21 about a horizontal axis through the center of gravity. Use the dimensions of Example 3A.

Solution. By Example 3A the center of gravity is 50 mm up from the bottom. The distance from the center of the stem to the center of gravity of the section is $75 - 50$, or 25 mm. The distance from the center of the flange to the center of gravity of the section is $50 - 12.5$, or 37.5 mm. A sketch should be made showing these dimensions.

Stem, by Eq. (16): $I = \dfrac{25 \times 150^3}{12} + 3{,}750 \times 25^2$

$$= 7{,}031{,}000 + 2{,}344{,}000 = 9{,}375{,}000 \text{ mm}^4$$

Flange, by Eq. (16): $I = \dfrac{100 \times 25^3}{12} + 2{,}500 \times 37.5^2$

$$= 130{,}000 + 3{,}516{,}000 = 3{,}646{,}000 \text{ mm}^4$$

Total: $I = 9{,}375{,}000 + 3{,}646{,}000 = 13{,}021{,}000 \text{ mm}^4$

11. Principle of Superposition

Stresses and deformations are produced in a body by the forces that are exerted upon it. It is natural to assume that the resultant effect at any chosen point is the sum of the effects of the various loads. In general, experiments have shown that this is so. The idea that the resultant effect is the sum of the separate effects is known as the *principle of superposition.* In general, it is valid for cases of loading only where the magnitude of the stress and deflection is directly proportional to the load.

Example 7. Calculate and plot the distribution of stress over a cross section of the offset link shown in Fig. 1-9(a). The main body of the link is straight and is $\frac{3}{4}$ in. thick.

It is obvious that the loading of Fig. 1-9(a) produces both direct tension and bending stress on the cross section. The principle of superposition applies, and each stress can be computed separately. The two stresses can then be added algebraically to obtain the resultant.

Offset link   Equivalent loading   Resultant stress

FIGURE 1-9   Offset link with tension load.

At this point, good use may be made of a principle from statics whereby a given force may be resolved into a parallel force and a couple. By doing so, the equivalent loading on the cross section shown by Fig. 1-9(b) is secured. The moment arm is equal to the distance from the line of action of the force to the center of gravity of the cross section.

Solution. The computations are as follows.

$$A = \frac{3}{4} \times 2 = 1.5 \text{ in.}^2$$

$$I = \frac{bh^3}{12} = \frac{3}{4} \times 8 \times \frac{1}{12} = 0.5 \text{ in.}^4$$

In Eq. (1): $\sigma = \frac{P}{A} = \frac{3,600}{1.5} = 2,400$ psi direct stress

The moment arm is the distance from the line of action of the force to the center of gravity of the cross-sectional area.

In Eq. (11): $\sigma = \frac{Mc}{I} = \frac{3,600 \times 1.125 \times 1}{0.5} = 8,100$ psi bending stress

Superposition applies and the resultant stress is given by

$$\sigma = \frac{P}{A} + \frac{Mc}{I} \tag{17}$$

On the inside edge: $\sigma = 2,400 + 8,100 = 10,500$ psi tension

On the outside edge: $\sigma = 2,400 - 8,100 = -5,700$ psi compression

Views (c) and (d) show the effect of moment and direct stress separately; the resultant stress on a cross section through the main body of the link is given by (e).

Note that views (b) and (e) are in static equilibrium. Also note that the line of zero stress no longer passes through the center of gravity of the cross section.

Superposition cannot be applied if the loads produce deflections that are so great that the basic configuration of the system is thereby changed. Take, for example, the leaf spring with the load on the end shown in Fig. 1-10(a). Suppose, upon doubling the load, that the deflection, shown by the dashed outline, becomes so large that the moment arm of the loads is reduced. The stress and deflection will not be twice as great as for a single load.

(a) Beam with large deflection

(b) Deflection causes change in span length

FIGURE 1-10 Examples where deflection changes original geometry; superposition does not apply.

Another example where the fundamental configuration of the system is changed by the application of a load is given in Fig. 1-10(b). A change in loading causes a change in deflection, which in turn causes a change in the length of the span. The load and deflection therefore are not proportional to each other.

In general, superposition is not valid for slender members loaded in compression. After the load reaches a value known as the *critical* or *buckling load*, a large lateral deflection of the member results from a small additional increase in the axial load.

12. Additional Beam Equations

For most beams used in engineering, the slope φ, shown in Fig. 1-5, has a very small value. Therefore, the tangent dy/dx can be considered as having a value very close to the angle φ. Then

$$\varphi = \frac{dy}{dx} \quad \text{and} \quad \frac{d\varphi}{dx} = \frac{d^2 y}{dx^2}$$

The slope φ is reduced in Fig. 1-5 in passing from A to B, and the increment $d\varphi$ of the angle is thus a negative quantity. From the figure, $dx = -r\,d\varphi$. Then

$$\frac{d\varphi}{dx} = -\frac{1}{r} = \frac{d^2y}{dx^2} \tag{18}$$

The negative sign is inserted because both dx and r are positive quantities. Substitution of Eq. (9) gives

$$\frac{d^2y}{dx^2} = -\frac{M}{EI} \tag{19}$$

Equation (19) is the fundamental equation of beam theory. For the general case of loading, moment M is a function of x and not a constant as in Fig. 1-5. If the beam has a large deflection, as, for example, a leaf spring in some types of service, the approximation $\varphi = dy/dx$ cannot be used, and the mathematics of the solution becomes very complicated.

When the bending moment of the beam is not constant, but varies with x, the loading for slice AB is more complicated than in the case of pure bending previously considered. Shearing forces, as well as moments, exist on the end surfaces of the slice. In general the top surface of the beam is acted upon by transverse loads.

A slice from such a beam is shown in Fig. 1-11. The rate at which the bending moment is changing in value is equal to dM/dx, and the distance over which this change takes place in passing from A to B is dx. The change in value for the moment then is $(dM/dx)\,dx$. The moment on the right end is equal to the moment M of the left end plus the increment or change. Similarly, the shear force, as a function of x, changes from a value V on the left to the amount on the right, as shown. Should either the moment or shear be decreasing with increasing x, the corresponding derivative would be negative, and would thus effect a reduction for the right-hand value. All shears and moments in Fig. 1-11 have positive directions. Note, however, that the arrows have opposite directions on ends A and B.

FIGURE 1-11   **Element loaded by forces and moments.**

The moments on the A–B element must be in equilibrium. The equation for moments about point A, for example, is

$$M + \frac{dM}{dx}\,dx - \left(V + \frac{dV}{dx}\,dx\right)dx - w\,dx \times \frac{dx}{2} - M = 0$$

When differentials of higher order are neglected, this equation reduces to

$$\frac{dM}{dx} = V \tag{20}$$

The location of the point of maximum bending moment is found by differentiating M with respect to x and setting the result equal to zero. However, dM/dx is equal to V by Eq. (20). Therefore, a maximum (or a minimum) value of the bending moment occurs at those points for which the shear V is equal to zero. The load w on the top surface of the beam may be uniform, may vary with x, or may be equal to zero except at the points where concentrated forces or moments are acting.

13. Deflection of Beams

The foregoing equations can be used for deriving equations for the deflections of beams. If the expression for the bending moment, as a function of x, is substituted in Eq. (19), and if two integrations are performed and the constants of integration are evaluated, the equation for the deflection y at location x is obtained. The process is illustrated by the following example.

Example 8. Derive the equation for the deflection y for values of x located between the supports for the beam shown in Fig. 1-12.

FIGURE 1-12   Example 8.

Solution. By taking moments, the values of the reactions are found to be $R_1 = 3wl/8$ and $R_2 = 9wl/8$.

At location x:

$$M = \frac{3}{8}wlx - \frac{wx^2}{2}$$

In Eq. (19):

$$EI\frac{d^2y}{dx^2} = -M = -\frac{3}{8}wlx + \frac{wx^2}{2}$$

Integrate:

$$EI\frac{dy}{dx} = -\frac{3wlx^2}{16} + \frac{wx^3}{6} + C_1$$

Since the slope dy/dx is not known for any point on the beam, constant C_1 must be retained and evaluated later.

Integrate:
$$Ely = -\frac{3wlx^3}{48} + \frac{wx^4}{24} + C_1x + C_2$$

At the left end, $x = 0$ and $y = 0$. When these are substituted, $C_2 = 0$. At the right end, $x = l$ and $y = 0$. When these are substituted,

$$C_1 = \frac{wl^3}{48}$$

The equation for the deflection thus is

$$y = \frac{w}{48EI}(2x^4 - 3lx^3 + l^3x)$$

A transverse load on a beam causes a deflection or change in elevation at the point of application. However, a reaction generally remains fixed in elevation even though it carries a force.

Figures 1-13 to 1-16 give the shears, moments, and deflections for a number of beams of various types of loads and conditions of support. In these figures, a simple support is assumed to offer no resistance to lateral motion or to rotation in the plane of the moments.

Example 9. Suppose it is specified that the deflection from its own weight at the center of a simply supported steel shaft should not exceed 0.010 in. per foot of span.

(a) Find the maximum permissible length for a 3-in.-diam shaft.

(b) Find the stress caused by the weight of the shaft. For steel, $\gamma = 0.283$ lb/in.^{3}, $E = 30,000,000$ psi.

Solution. (a) By the conditions of the problem the deflection at the center is $0.010l/12$, where l is the length in inches. Hence by No. 8 of Fig. 1-14,

$$\frac{5wl^4}{384EI} = \frac{0.010l}{12} \quad \text{or} \quad l^3 = \frac{0.010 \times 384EI}{12 \times 5w}$$

By Eq. (14):
$$I = \frac{\pi d^4}{64} = \frac{\pi 3^4}{64} = 3.976 \text{ in.}^4$$

$$w = \gamma A = 0.283\frac{\pi 3^2}{4} = 2.00 \text{ lb/in.}$$

Substitution of these values gives

$$l^3 = 3,817,000 \quad \text{or} \quad l = 156.3 \text{ in.} = 13.02 \text{ ft}$$

(b)
$$M = \frac{wl^2}{8} = \frac{2 \times 156.3^2}{8} = 6,106 \text{ in. lb}$$

$$\sigma = \frac{Mc}{I} = \frac{6,106 \times 1.5}{3.976} = 2,300 \text{ psi}$$

Example 10. Show that the spring constants, load for a unit deflection, for the two beams of Fig. 1-17 are the same.

1.

$$\theta_1 = \theta_2 = \frac{M_1 l}{2EI}; \quad y = \frac{M_1}{2EI}(lx - x^2); \quad y_{max} = \frac{M_1 l^2}{8EI}$$

2.

$$\theta_2 = \frac{M_1 l}{EI}; \quad y = \frac{M_1 x^2}{2EI}; \quad y_{max} = \frac{M_1 l^2}{2EI}$$

3.

$$\theta_1 = \frac{M_1 l}{3EI}; \quad \theta_2 = \frac{M_1 l}{6EI}; \quad y_{max} = \frac{M_1 l^2}{9\sqrt{3}EI}$$

$$y = \frac{M_1}{6lEI}(2l^2 x - 3lx^2 + x^3)$$

4.

$$\theta_1 = \frac{M_1 l}{6EI}; \quad \theta_2 = \frac{M_1 l}{3EI}; \quad y_{max} = \frac{M_1 l^2}{9\sqrt{3}EI}$$

$$y = \frac{M_1}{6lEI}(l^2 x - x^3)$$

5.

$$\theta_2 = \frac{Pl^2}{2EI}; \quad y = \frac{P}{6EI}(3lx^2 - x^3); \quad y_{max} = \frac{Pl^3}{3EI}$$

6.

$$\theta_1 = \theta_2 = \frac{Pl^2}{16EI}; \quad y_{max} = \frac{Pl^3}{48EI}$$

For $0 < x \leqslant \frac{l}{2}$; $\quad y = \frac{P}{48EI}(3l^2 x - 4x^3)$

FIGURE 1-13 Shear, moment, and deflection in beams of uniform cross section.

FIGURE 1-14   Shear, moment, and deflection in beams of uniform cross section.

13.

$$\theta_1 = \theta_2 = \frac{Pa(l-a)}{2EI}$$

For $o < x \leqslant a$, $\; y_1 = \frac{Px}{6EI}\left[3a(l-a)-x^2\right]$

For $a \leqslant x \leqslant (l-a)$, $\; y_2 = \frac{Pa}{6EI}\left[3x(l-x)-a^2\right]$

$$y_{max} = \frac{Pa}{24EI}(3l^2-4a^2)$$

14.

$$\theta_1 = \frac{Pab}{6EI}, \quad \theta_2 = \frac{Pab}{3EI}, \quad \theta_3 = \frac{Pb}{6EI}(2l+b)$$

For $o < x \leqslant a$; $\; y_1 = \frac{Pbx}{6aEI}(x^2-a^2)$

For $o < z \leqslant b$; $\; y_2 = \frac{P}{6EI}\left[z^3-b(2l+b)z+2b^2l\right]$

$$y_{max} = \frac{Pb^2l}{3EI}$$

15.

$$P_{cr} = \frac{\pi^2EI}{4l^2} \qquad P_e = \frac{\pi^2EI}{l^2} \qquad P_{cr} = \frac{4\pi^2EI}{l^2}$$

Critical or buckling loads for centrally loaded columns.

16.

$$y_{max} = e\!\left(\sec \frac{l}{2}\sqrt{\frac{P}{EI}} - 1\right)$$

Maximum moment:

$$M_{max} = P(e + y_{max})$$

$$= Pe \, \sec \frac{l}{2}\sqrt{\frac{P}{EI}}$$

Maximum stress:

$$s_{max} = \frac{P}{A}\left(1 + \frac{ec}{i^2}\sec\frac{l}{2i}\sqrt{\frac{P}{AE}}\right)$$

Where A = area of cross section
$i = \sqrt{I/A}$, radius of gyration
c = distance from neutral axis to edge of section

Eccentricity

17.

For $o < x \leqslant (l-c)$ $\qquad p^2 = \frac{S}{EI}$

$$y = \frac{P\sin pc}{Sp \sin pl}\sin px - \frac{Pc}{Sl}x$$

Equations for slope and moment can be found by differentiation.

18.

For $o < x \leqslant (l-c)$ $\qquad p^2 = \frac{S}{EI}$

$$y = -\frac{P\sinh pc}{Sp \sinh pl}\sinh px + \frac{Pc}{Sl}x$$

Equations for slope and moment can be found by differentiation.

FIGURE 1-15 Shear, moment, and deflection in beams and columns.

FIGURE 1-16   Shear, moment, and deflection in beams of uniform cross section.

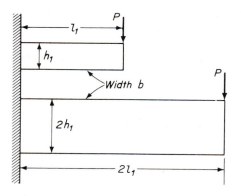

FIGURE 1-17   Example 10.

Solution. The spring constant can be derived from the deflection equation for the beam by setting the load P equal to k when the deflection y becomes equal to unity.

By No. 5 of Fig. 1-13:

$$1 = \frac{kl^3}{3EI} \qquad \text{or} \qquad k = \frac{3EI}{l^3} = \frac{3E}{l^3}\frac{bh^3}{12} = \frac{Ebh^3}{4l^3}$$

For sketch (a):

$$k = \frac{Ebh_1^3}{4l_1^3}$$

For sketch (b):

$$k = \frac{Eb(2h_1)^3}{4(2l_1)^3} = \frac{Ebh_1^3}{4l_1^3}$$

If the beam has a cross-sectional width 8 or 10 or more times as great as the thickness, the beam is stiffer, and the deflection is less than that indicated by the equation for a narrow beam. The large cross-section width prevents the lateral expansion and contraction of the material, and the deflection is thereby reduced. A better value for the deflection of a "wide" beam is obtained by multiplying the result given in the equation for a narrow beam by $(1 - \mu^2)$ where μ is Poisson's ratio. Poisson's ratio is the ratio of the unit lateral contraction to the unit axial elongation experienced by a bar in tension.

Data for the stress and deflection of flat plates are given in Chapter 12.

14. Effect of Ribs on Castings

Ribs are sometimes added to the webs of castings to give greater strength and rigidity. It is possible, however, that the addition of a shallow rib to a body loaded in bending may cause an increase in stress rather than a decrease. The low rib gives a small increase in the moment of inertia, but the distance from the neutral axis to the edge of the cross section becomes relatively

greater, and the stress is accordingly increased. The situation is illustrated
by the following example.

Example 11. Figure 1-18 represents the cross section through a simply supported
beam 60 in. long that carries a 200-lb load at the center. $E = 15,000,000$ psi.

FIGURE 1-18   Cross section for ribbed beam, Example 11.

(a) Find the value of the bending stress and the deflection at the center if the
ribs were omitted.

(b) Find the value of the bending stress and the deflection at the center when the
ribs are present.

Solution. (a) $I = \dfrac{bh^3}{12} = \dfrac{4.2 \times 1^3}{12} = 0.35$ in.⁴

$$M = \frac{Pl}{4} = \frac{200 \times 60}{4} = 3,000 \text{ in. lb}$$

$$\sigma = \frac{Mc}{I} = \frac{3,000 \times 0.5}{0.35} = 4,290 \text{ psi}$$

$$y = \frac{Pl^3}{48EI} = \frac{200 \times 60^3}{48 \times 15,000,000 \times 0.35} = 0.1714 \text{ in.}$$

(b) Area: $A = 4.2 \times 1 + 2 \times 0.2 \times 0.6 = 4.44$ in.²

To find $\bar{y}$, take the axis along top of the cross section.

$$4.44\bar{y} = 4.2 \times 0.5 + 0.24 \times 1.1 = 2.364$$

$$\bar{y} = 0.5324 \text{ in.}$$

For the main area: $I = \dfrac{bh^3}{12} + A\bar{y}^2 = 0.35 + 4.2 \times 0.0324^2 = 0.3544$ in.⁴

For the ribs: $I = \dfrac{1.2 \times 0.2^3}{12} + 0.24 \times 0.5676^2 = 0.0781$ in.⁴

Total: $I = 0.3544 + 0.0781 = 0.4325$ in.⁴

$$\sigma = \frac{Mc}{I} = \frac{3,000 \times 0.6676}{0.4325} = 4,630 \text{ psi}$$

$$y = \frac{Pl^3}{48EI} = \frac{200 \times 60^3}{48 \times 15,000,000 \times 0.4325} = 0.1387 \text{ in.}$$

As shown above, shallow ribs cause an increase in the bending stress. The deflection, however, has been decreased. If the ribs were made somewhat larger, the stress would be decreased, and the beam would be stronger.[4]

Ribbed structures are frequently made of a brittle material, such as cast iron, which is weak in tension. If possible, the ribs should be in compression. When a cast-iron body with parallel ribs is bent and the ribs are in tension, care must be exercised to make certain that all ribs are of the same height, or they may fail progressively beginning with the highest, and the full strength of the body could not be realized.

15. Shearing Stress

Suppose an element is loaded by shearing stresses acting tangentially to its sides as shown in perspective in Fig. 1-19(a) or in plan in Fig. 1-19(b) and (c). Such loading causes no change in the length of the sides of the element, but merely produces a distortion or change in the value of the 90° angles in the corners.

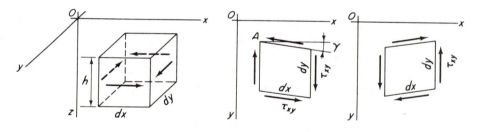

(a) Element loaded in shear   (b) Positive shear   (c) Negative shear

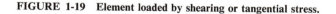

FIGURE 1-19 **Element loaded by shearing or tangential stress.**

Shearing stresses are usually denoted by double subscripts. The first subscript indicates the direction of the normal to the plane under consideration, and the second subscript indicates the direction of the stress. Hence stress τ_{xy} lies in a plane whose normal is in the x-direction, while the stress acts in the y-direction. For similar reasons τ_{yx} indicates that the stress is in a plane perpendicular to the y-axis, and is parallel to the x-axis. Since the element is in equilibrium, the moments of the forces about a point, say A,

[4]See Marin, J., "Stiffness of Ribbed Plates," *Machine Design*, **19**, May 1947, p. 145; and Radich, E. A., "Strength and Stiffness of Ribbed Plates," *Machine Design*, **21**, Sept. 1949, p. 149.

must add up to zero. The stress should be multiplied by the area and then by the moment arm to give

$$\tau_{xy}\,dx\,dy\,h - \tau_{yx}\,dx\,dy\,h = 0$$

or
$$\tau_{xy} = \tau_{yx} \tag{21}$$

Equation (21) shows that the shearing stresses in two perpendicular directions at a point are equal. Usually no distinction in notation is made, and both would be represented by the same symbol.

It should be noted, however, that four arrows are necessary to specify a state of shear for an element, and for equilibrium these arrows must be arranged either as in Fig. 1-19(b) for positive shear, or as in Fig. 1-19(c) for negative shear. Thus if the direction of one arrow is reversed, all four must be reversed. In other words, if shearing stress exists on one side of an element, then, in general, shearing stress must exist on all four sides, as shown in Fig. 1-19.

The shearing strain or angular deformation γ is proportional to the shearing stress for values within the elastic range, and Hooke's law for shear becomes

$$\tau_{xy} = \gamma G \tag{22}$$

The constant of proportionality, G, is called the *modulus of elasticity in shear*. It has dimensions of psi. In magnitude it would be equal to the stress that would cause the angular deformation to become equal to one radian, provided Hooke's law is valid for such imaginary loading. The mathematical relationship between the three elastic constants E, G, and μ is given by

$$G = \frac{E}{2(1 + \mu)} \tag{23}$$

where μ is Poisson's ratio.

16. Transverse Shearing Stress in Beams

In addition to the bending stresses, the loads on a beam may also cause shearing stresses between the elements. The designer is interested in the magnitude of these stresses, since machine parts made of ductile materials are usually designed on the basis of shearing stress.

If a vertical load is supported by a stack of laminated strips, the shearing effect is as shown in Fig. 1-20(a). In a solid beam the elements do not slide on each other, but the shearing stress tending to make them do so is present.

Figure 1-20(b) shows a portion *ABCD* cut from a beam of uniform cross section by two adjacent vertical planes and a horizontal plane, located at a

(a) Sliding of laminated strips
under transverse shear loads

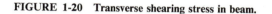

(b) Side view of beam   (c) Section A-A

FIGURE 1-20   Transverse shearing stress in beam.

distance v_1 below the neutral surface. Moment M is acting on the cross section of the left end of the slice, while moment $M + dM$ is acting on the right.

To maintain equilibrium in the horizontal direction, shearing stress τ must act toward the left on surface AB. This shear is necessary because the normal stresses from the bending moment are assumed to be larger on surface BC on the right than on surface AD on the left.

An element of area dA on surface AD has the normal force $\sigma\, dA$ or $Mv\, dA/I$ acting as shown in Fig. 1-20(b). The total force on the left end extending from A to D is then

$$\int_{v_1}^{c_1} \frac{Mv}{I}\, dA \qquad\qquad (a)$$

Similarly, the total force on the right end BC is

$$\int_{v_1}^{c_1} \frac{(M + dM)v}{I}\, dA \qquad\qquad (b)$$

The shearing force on the horizontal surface AB is $\tau b\, dx$, where b is the width of the beam at the location where the shear stress τ is acting.

The equilibrium equation for horizontal forces for $ABCD$ is then

$$\tau b \, dx + \int_{v_1}^{c_1} \frac{Mv \, dA}{I} = \int_{v_1}^{c_1} \frac{(M + dM)v \, dA}{I} \qquad (c)$$

or

$$\tau = \frac{1}{b} \int_{v_1}^{c_1} \frac{dM}{dx} \frac{v \, dA}{I} = \frac{V}{Ib} \int_{v_1}^{c_1} v \, dA \qquad (24)$$

In the last form of Eq. (24), shear V has been substituted for dM/dx. In Eq. (24), $v \, dA$ represents the moment of the area of the element about the neutral axis. This is integrated over the entire surface from v_1, the location where the shearing stress τ is desired, to the outer edge. This integral can also be written $\bar{v} A_a$, where A_a is the shaded area of view A-A, and $\bar{v}$ is the distance from its center of gravity to the neutral axis. Equation (24) can then be written

$$\tau = \frac{V}{Ib} \bar{v} A_a \qquad (25)$$

As we proved in the preceding section, the shearing stress on the vertical end surfaces at distance v_1 from the neutral axis is also equal to the horizontal shear stress τ as determined by Eq. (24) or (25).

For composite cross sections it is convenient to divide area A_a into several parts, find $\bar{v} A_a$ for each of them, and then add together for the final result. For such beams, Eq. (25) is written

$$\tau = \frac{V}{Ib} \sum \bar{v} A_a \qquad (26)$$

The total shear force on the cross section is represented by V. The distance from the neutral axis to the point where the shearing stress is desired is given by v_1.

Example 12. Find the transverse shear in the material 3 in. from the top surface for the beam of Fig. 1-21(a).

Solution. As is shown in Example 3, the center of gravity of the cross section in Fig. 1-21(b) is found to be 2 in. up from the bottom. As shown in Example 6, the moment of inertia about the horizontal axis through the center of gravity is found to be 33.33 in.[4]

Referring to Fig. 1-21(c), it is seen for location 3 in. from the top that $\bar{v} = 2.5$ in. and $A_a = 3$ in.[2]. Substitution in Eq. (25) gives

$$\tau = \frac{10,000}{33.33 \times 1} \times 2.5 \times 3 = 2,250 \text{ psi}$$

It is of course immaterial whether A_a is taken above or below the location at which the stress is desired. Equation (26) gives

$$\tau = \frac{10,000}{33.33 \times 1} (1.5 \times 4 + 0.5 \times 3) = 2,250 \text{ psi}$$

FIGURE 1-21 Examples 3, 6, and 12.

Example 12A. Find the transverse shear stress in the material 75 mm from the top surface for the beam of Fig. 1-21(a); but use the dimension given for Example 3A. The force at the end of the beam is 45,000 N.

Solution. A large-size sketch should be made and dimensioned for the cross section in mm. As found in previous examples, $\bar{y} = 50$ mm and $I = 13,021,000$ mm⁴. When the area above 75 mm is considered:

$$\bar{v} = 62.5 \text{ mm}, \qquad A_a = 1,875 \text{ mm}^2$$

By Eq. (25): $\tau = \dfrac{V}{I_b}\bar{v}A_a = \dfrac{45,000}{13,021,000 \times 25} \times 62.5 \times 1,875 = 16.2$ MPa

When the areas below 75 mm are considered:

Let the vertical stem extend to the bottom, $\bar{v} = 12.5$ mm and $A_a = 1,875$ mm². For the remainder of the flange, $\bar{v} = 37.5$ mm and $A_a = 2,500$ mm².

By Eq. (26): $\tau = \dfrac{45,000}{13,021,000 \times 25}(12.5 \times 1,875 + 37.5 \times 2,500)$

$$= 16.2 \text{ MPa}$$

When Eq. (24) is applied to rectangular cross section, $dA = b\,dv$ and $c_1 = h/2$. After making these substitutions and integrating, the following result is obtained.

$$\tau = \frac{V}{2I}\left(\frac{h^2}{4} - v_1^2\right) = \frac{3V}{2A}\left(1 - \frac{4v_1^2}{h^2}\right) \tag{27}$$

The transverse shearing stress in a rectangular beam thus varies as a second-degree parabola in v_1. Its value is proportional to the length of the arrows in Fig. 1-22(a). The maximum value occurs at the neutral axis, where $v_1 = 0$, and is equal to

$$\tau_{max} = \frac{3V}{2A} \tag{28}$$

The shear decreases both above and below the axis in accordance with Eq. (27) until at the upper and lower edges it becomes zero.[5] The end surface of the beam in Fig. 1-22(a) is also acted upon by the system of normal stresses caused by the bending moment, as shown by Fig. 1-6(b). In Fig. 1-22(a) these have been omitted for greater clarity.

FIGURE 1-22   Distribution of shearing stress over cross section of beam.

Within reasonable limits, the presence of the shearing stress has no effect on the value of the bending stress, and vice versa.

The shearing deformation also varies over the surface of a cross section; it is maximum at the neutral axis and zero at the top and bottom. The shearing stress causes a warping of the cross sections, which were originally plane and perpendicular to the longitudinal elements of the beam.[6] For a cantilever with a load on the end, the situation is as shown in Fig. 1-22(c).

For a solid circular cross section, Eq. (24) gives the following value for the maximum transverse shear, which occurs at the neutral axis.[7]

$$\tau_{max} = \frac{4}{3} \cdot \frac{V}{A} \tag{29}$$

[5] See p. 116 of Reference 4, end of chapter.

[6] For deflection of beams caused by shear, see p. 170 of Reference 4, end of chapter.

[7] A more exact analysis gives values of 1.38 P/A and 1.23 P/A for the shearing stress at the center and ends, respectively, of the neutral axis.

For a circular tube with very thin walls, the maximum transverse shear stress at the neutral axis is given by

$$\tau_{max} = 2\frac{V}{A} \tag{30}$$

The distribution of transverse shear stress for an I-beam is shown in Fig. 1-22(d). The stress is practically uniform except in the regions near the top and bottom. A good approximate value for the stress is obtained by dividing the total shearing force V by the area of the web td with the web considered as extending the entire depth of the beam.

17. Shear and Bending Moment Diagrams

The effects of the forces and moments that act upon the different parts of a machine are of primary importance to the machine designer. Forces arise from a variety of causes. A force may be due to weight or to inertia if a body is being accelerated. A force may be transmitted to the body by another member of the machine at the point where the two parts are in contact. It is common practice to represent forces by means of arrows on sketches in solving force problems. It is very important, except for gravity and inertia, to keep in mind that the body must have contact with the rest of the structure at the point where a force, represented by an arrow, is considered as acting. As an example, consider the shaft of Fig. 1-23(a). The entire loading occurs at point A, where the bracket is attached to the shaft. The equivalent loading diagram is shown in Fig. 1-23(b). The diagram of Fig. 1-23(c) is incorrect because forces P are not in contact with the shaft.

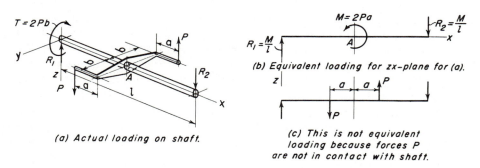

(a) Actual loading on shaft.

(b) Equivalent loading for zx-plane for (a).

(c) This is not equivalent loading because forces P are not in contact with shaft.

FIGURE 1-23   Actual and equivalent load diagram for shaft.

Moment loads, such as those described in Fig. 1-23(b), frequently occur in machine parts. Thus in the tank shown in Fig. 1-24(a) the floor beam not only resists the vertical pressure of the fluid, but has moments applied to it at the ends by the uprights, which are acted upon by the horizontal pressure of the fluid. The equivalent loading is shown in Fig. 1-24(b). In Fig. 1-24(c) the vertical bar is loaded at the base by the moment Wa as well as by the force P. The equivalent loading is shown in Fig. 1-24(d).

Equivalent loading
for floor beam

FIGURE 1-24   Examples of beams with moment loads.

In solving most stress problems it is first necessary to find the reactions that the given loading produces. It is customary to make use of the following equations from statics.

$$\sum F_x = 0, \qquad \sum F_y = 0, \qquad \sum F_z = 0$$
$$\sum M_x = 0, \qquad \sum M_y = 0, \qquad \sum M_z = 0$$

According to these equations, the components of the forces, as well as the moments, in each of the three coordinate directions must add to zero. For static equilibrium, each of these equations must be satisfied. It should be noted that each is satisfied independently of the others. So far as the reactions are concerned, a force can be considered to be acting anywhere along its line of action, and a moment load can be considered to be acting anywhere in its plane. The state of stress in the material, however, is determined by the location, on the line of action, at which the force is considered to be applied.

The shearing force and bending moment, acting internally on the cross section of a beam at any point, are equal to the force and moment required for the equilibrium of each portion of the beam after it has been cut in two at the given point. For example, consider the simply supported beam, shown in Fig. 1-25(a), with load P and reactions R_1 and R_2. In (b) and (c), after cutting at distance x from the left end, shear V_1 and moment M_1 are required to maintain equilibrium of each portion. Reactions R_1 and R_2 in (b) and (c) are the same as in (a). Shear V_1 is found by summation of vertical forces. Moment M_1 is found by making a summation of moments, with the location of the cut taken as the moment center. The shear and moment diagrams shown in (d) and (e) are constructed by taking different values of x and finding V and M for each until a sufficient number of values has been found to plot the diagrams. It is generally assumed that a reaction does not change

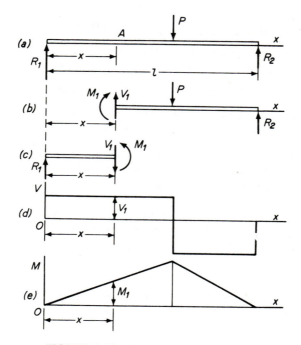

FIGURE 1-25 Shear and moment in beam.

in elevation. Other transverse loads deflect the beam, and the elevations of the loads are thereby changed.

Moments and shears can, of course, be either positive or negative, depending on the direction in which they act. The shear and moment at point A for the beam shown in Fig. 1-25 are both positive. Positive directions for the arrows representing these quantities are accordingly shown in the figures. Even though both shear and moment are positive at A, note that a reversal in direction for both the V_1 and M_1 arrows occurs in (b) and (c). The direction of the arrow thus depends on whether the portion of the beam to the right of the cut, as in (b), is considered, or whether the portion to the left, as in (c), is considered. Note that both (b) and (c) are in agreement with Fig. 1-19(b) with respect to direction of positive shear. Although the y-axis is taken positive downward, both the V- and M-axes are taken positive upward.

Depending on the loading and method of supporting the beam, the shear at a cross section may have one sign, and the moment may have the opposite sign.

Another method for determining the sign of the moment is shown in Fig. 1-26. Here a moment tending to bow the beam concavely upward is

Positive
moment

Negative
moment

FIGURE 1-26   Sign convention for moments.

considered positive, whereas a moment tending to make the beam concave downward is negative.

Example 13. A simply supported beam is loaded by a concentrated moment of 25,000 in. lb as shown in Fig. 1-27. Find values of reactions R_1 and R_2, and draw and dimension the bending moment diagram.

FIGURE 1-27   Example 13.

Solution. Reaction R_1 can be found by writing a moment equation for the entire beam with center of moments taken at R_2. The resulting equation is

$$64R_1 = 25,000; \qquad R_2 = 390.6 \text{ lb}$$

Reaction R_2 can be found by a moment equation with R_1 taken as the center.

$$64R_2 = 25,000; \qquad R_2 = 390.6 \text{ lb}$$

The moment diagram consists of segments of straight lines. Let it be assumed that the beam is cut just to the left of the point of application of the moment load. The bending moment is positive and is equal to

$$M = 36R_1 = 36 \times 390.6 = 14,062.5 \text{ in. lb}$$

Again let it be assumed that the beam is cut just to the right of the point of application of the load. The moment is negative.

$$M = 28R_2 = 28 \times 390.6 = 10,937.5 \text{ in. lb}$$

18. Slender Compression Members or Columns

When a short block is loaded in compression as in Fig. 1-28(a), the average compressive stress in the material is found by dividing the load by the cross-sectional area. However, when the member is long and slender, as that shown

(a) Short block   (b) Long slender member

FIGURE 1-28 Bodies under compressive loading.

in Fig. 1-28(b), the situation is complicated by the possibility of lateral buckling. Buckling does not occur when the bar is straight and a load, smaller than the critical value, is centrally applied. Such a column is stable; the bar, if given a lateral deflection, returns to its originally straight condition upon removal of the lateral force.

Force P in Fig. 1-28(b) can be increased until the straight form of the bar becomes unstable, and the axial loads will then maintain the bar in a curved form. The smallest load capable of maintaining the bar in a slightly bent form is called the *critical* or *buckling load*. A load only slightly greater can cause a relatively large lateral deflection. After buckling the stress increases rapidly; in addition to the direct compression, a bending stress from moment Py is present.

Hence the phenomenon of stability or buckling is quite different from the phenomenon of bending. A beam with lateral load starts to deflect as soon as any load is present, and the deflection is directly proportional to the magnitude of the load. A slender member in compression, in contrast, exhibits no lateral deflection until the critical or buckling load has been reached. Any increase of load then causes a large increase of deflection with accompanying danger of failure. It is obvious that the principle of superposition does not apply to columns.

For the straight column with hinged ends loaded by centrally applied forces, it can be shown that the critical buckling load is given by

$$P_e = \frac{\pi^2 EI}{l^2} \tag{31}$$

This is commonly known as the Euler equation. See No. 15 of Fig. 1-15. Moment of inertia I should be about the axis of rotation of the cross section.

Sometimes a column may not be perfectly straight, but has some initial crookedness a, as shown in Fig. 1-29. The bending moment then will be

FIGURE 1-29 Column with initial crookedness a.

larger, and the carrying capacity will be reduced. It can be shown[8] that the working load P for such a column with factor of safety F_s is given by

$$P^2 - \left[\sigma_{yp}A + \left(1 + \frac{ac}{i^2}\right)P_e\right]\frac{P}{F_s} + \frac{\sigma_{yp}AP_e}{(F_s)^2} = 0 \tag{32}$$

where σ_{yp} = yield point stress for the material

$\quad\quad\quad A$ = area of cross section

$\quad\quad\quad c$ = distance from neutral axis to edge of cross section

$$i = \sqrt{I/A} \tag{33}$$

Here i is called the *radius of gyration* of the cross section.[9]

As for other machine elements, a factor of safety must be used in column design. Because the stress is not proportional to the load, the factor of safety F_s is applied to the load rather than to the stress. Let P_{yp} be the load on the column that causes the maximum stress to be equal to the yield point value σ_{yp}. Let the working load P be equal to P_{yp} divided by the factor of safety.

$$P = \frac{P_{yp}}{F_s} \tag{34}$$

When this equation is substituted into Eq. (32), the result is

$$P_{yp}^2 - \left[\sigma_{yp}A + \left(1 + \frac{ac}{i^2}\right)P_e\right]P_{yp} + \sigma_{yp}AP_e = 0 \tag{35}$$

It can also be shown that the maximum stress in the material is given by

$$\sigma = \frac{P}{A}\left(1 + \frac{ac}{i^2} \times \frac{P_e}{P_e - P}\right) \tag{36}$$

If P in this equation is increased to the yield point value P_{yp}, then stress σ becomes equal to the yield point stress σ_{yp}.

Example 14. A solid circular steel column with hinged ends is 36 in. long and $2\frac{5}{8}$ in. in diameter. Yield point stress for the material is 50,000 psi. Initial crookedness a is $\frac{1}{16}$ in. Find the load capacity for an F_s of 4.

Solution. Area: $A = \dfrac{\pi d^2}{4} = \dfrac{\pi 2.625^2}{4} = 5.412$ in.^{2}

Moment of inertia: $I = \dfrac{\pi d^4}{64} = \dfrac{\pi 2.625^4}{64} = 2.331$ in.^{4}

Radius of gyration: $i^2 = \dfrac{I}{A} = \dfrac{d^2}{16} = \dfrac{2.625^2}{16} = 0.431$ in.^{2}

Euler load: $P_e = \dfrac{\pi^2 EI}{l^2} = \dfrac{\pi^2 E 2.331}{36^2} = 532,500$ lb

[8] See p. 31 of Reference 6 and p. 13 of Reference 3, end of chapter.

[9] Some books represent the radius of gyration by r. When this is done, care must be taken not to confuse the radius of gyration with the radius r of a round cross section.

In Eq. (32):

$$P^2 - \left[50,000 \times 5.412 + \left(1 + \frac{0.0625 \times 1.3125}{0.431} \right) 532,500 \right] \frac{P}{4}$$
$$+ \frac{50,000 \times 5.412 \times 532,500}{4^2} = 0$$

$$P^2 - 226,100P + 9,005,400,000 = 0$$

Upon solution, this quadratic gives

$$P = 51,600 \text{ lb}, \qquad \text{working load}$$

For hinged or pin-connected ends, the assumption is made that there is no restraint against rotation at the ends of the column. For square or fixed ends, all rotation is prevented as shown by No. 15(c) of Fig. 1-15. The deflection curve consists of two quarter-waves and one half-wave. Bending moments are present at the ends, but no moments act at the inflection points. Calculations are usually made by taking one-half the actual length for l in the equations for columns with hinged ends. For many practical cases, the end conditions are intermediate between being completely hinged and completely restrained.

The well-known secant equation for the maximum stress in a column with an eccentrically applied load is given in No. 16 of Fig. 1-15. For small values of e and a, the secant equation gives practically the same result as the equation for the column with initial crookedness. The secant equation, however, is very inconvenient for design calculations.

An actual column under load may behave differently from an ideal column, and the design of this element therefore presents many difficulties. Uncertainties such as amount of restraint at the ends, eccentricity of the load, initial crookedness, nonhomogeneity of the material, and deflection caused by the load can produce large variations in the behavior of a column. The choice of a suitable value for the factor of safety also presents great difficulties. Empirical equations are frequently used for column designing. The possibility of buckling about both of the principal axes of the cross section should be investigated.

Sometimes only a portion of a stressed body is in compression, as, for example, the compression flange of a beam. The danger of buckling may be present here if sufficient lateral support is lacking. Such action has been the cause of failures and should be carefully guarded against by the designer.[10]

If the load P approaches the critical load P_e, Eq. (36) indicates that the stress becomes infinite. Although there can be no such stress, it is characteristic of column equations to indicate the buckling phenomenon in this manner.

The hydraulic cylinder with extended piston rod forms a column.[11]

---

[10] A complete treatment of this subject may be found in the reference works at the end of the chapter.

[11] For design equations see p. 136 of Reference 3, end of chapter.

19. Stresses in Any Given Direction

The stresses in a body, as found by the equations of this chapter, have definite directions. It is sometimes necessary to have the stresses at directions other than those given by the equations.

Figure 1-30(a) shows an element of a plate with the vertical surfaces subjected to the general two-dimensional state of stress. The element has been cut from a larger plate so that stresses σ_x, σ_y, and τ_{xy} represent the effect of the surrounding material on the element. A plan view of the element is shown in Fig. 1-30(b). Suppose stresses σ_x, σ_y, and τ_{xy} are known, and that it is necessary to find the values of the stresses on an inclined surface whose normal makes an angle φ with the x-axis as shown in Fig. 1-30(c). Angle φ is an arbitrarily chosen angle and determines the directions of the u- and v-axes.

Assume that stress σ_r must be applied to the cut surface in order to maintain equilibrium of the remaining portion of the plate. Resultant stress σ_r can be resolved into the components of normal stress σ_u and shear stress τ_{uv} as shown.

(a) Two-dimensional stress

(b) Plan view of element shown in (a)

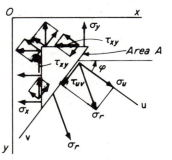

(c) Components of stress in directions u and v

FIGURE 1-30 Shear and normal stress on element at any angle φ.

If the area of the inclined surface is A, then the area of the horizontal side of the body will be $A \sin \varphi$, and the area of the vertical side, $A \cos \varphi$. Since the plate of Fig. 1-30(c) is in equilibrium, the projections of the forces on the perpendicular to the inclined surface must be in equilibrium. Multiplication of stress by area and then by the appropriate trigonometric function gives the following equation for σ_u.

$$\sigma_u = 2\tau_{xy} \sin \varphi \cos \varphi + \sigma_x \cos^2 \varphi + \sigma_y \sin^2 \varphi$$

The trigonometric terms should be changed by the substitution of the equations involving the double angles. Then,

$$\sigma_u = \frac{\sigma_x + \sigma_y}{2} + \frac{\sigma_x - \sigma_y}{2} \cos 2\varphi + \tau_{xy} \sin 2\varphi \qquad (37)$$

If the element in Fig. 1-30 is cut at $90°$ to the direction in sketch (c), summation of the forces will give the equation for the normal stress in the v direction.

$$\sigma_v = \frac{\sigma_x + \sigma_y}{2} - \frac{\sigma_x - \sigma_y}{2} \cos 2\varphi - \tau_{xy} \sin 2\varphi \qquad (38)$$

Thus the normal stresses in the material at any desired angle φ can be found by use of the above equations. Should the equation give a negative result, the corresponding stress is compressive.

In a similar manner, τ_{uv} can be found by making the sum of the projections of all forces parallel to the cut surface equal to zero. Hence,

$$\tau_{uv} = \tau_{xy}(\cos^2 \varphi - \sin^2 \varphi) - (\sigma_x - \sigma_y) \sin \varphi \cos \varphi$$

or
$$\tau_{uv} = \tau_{xy} \cos 2\varphi - \frac{\sigma_x - \sigma_y}{2} \sin 2\varphi \qquad (39)$$

The shear stress τ_{uv} at any desired angle φ can thus be found by Eq. (39). A positive result for τ_{uv} means that the stress is directed as in Fig. 1-30(c), and a negative result means that the stress is directed oppositely.

Angle φ is positive when taken clockwise from the x-axis.

20. The Mohr Circle

A graphical solution to the combined stress problem, known as the Mohr circle, will now be given. Use of this method rather than the previously derived equations usually effects a considerable saving in time. However, certain conventions regarding signs and directions must be understood and carefully followed.

Figure 1-31 shows the perpendicular axes σ and τ. Normal stresses, regardless of the inclination of the surface on which they act, are plotted horizontally—positive, or tension, to the right of the origin, and negative,

$$\tau_{uv} = a \sin (2\theta - 2\phi)$$
$$= a \left[\sin 2\theta \cos 2\phi - \cos 2\theta \sin 2\phi \right]$$
$$= a \left[\frac{\tau_{xy}}{a} \cos 2\phi - \frac{\sigma_x - \sigma_y}{2a} \sin 2\phi \right]$$
$$= \tau_{xy} \cos 2\phi - \frac{\sigma_x - \sigma_y}{2} \sin 2\phi$$

(b)

$$a \cos (2\theta - 2\phi) = a (\cos 2\theta \cos 2\phi + \sin 2\theta \sin 2\phi)$$
$$= a \left[\frac{\sigma_x - \sigma_y}{2a} \cos 2\phi + \frac{\tau_{xy}}{a} \sin 2\phi \right]$$
$$= \frac{\sigma_x - \sigma_y}{2} \cos 2\phi + \tau_{xy} \sin 2\phi$$

(a) Given state of stress

(c) Stresses on element oriented at angle ϕ

(d) Principal stresses

(e) Maximum shear stress

FIGURE 1-31 Mohr circle for two-dimensional stress.

or compression, to the left. Shear stresses are plotted vertically upward or downward on the diagram. The normal and shear stresses at a point in the body thus become the coordinates of a point on the circle.

Stresses σ_x and τ_{xy} acting on the right and left edges of the plate in Fig. 1-30(b) locate point A in Fig. 1-31. Tension σ_x is plotted to the right in accordance with the above-mentioned rule. Since shear stress τ_{xy} tends to rotate the element in a clockwise direction, it is plotted upward. Stresses σ_y and τ_{xy} of the upper and lower edges of the plate shown in Fig. 1-30(b) locate point B in Fig. 1-31. Tension σ_y is plotted to the right. Since shear stress τ_{xy} on these surfaces tends to produce counterclockwise rotation, it is plotted downward. The Mohr circle is drawn with line AB as a diameter. Greater facility in the determination of angles will be obtained if radii AC and BC are marked x-axis and y-axis, respectively.

To find the stresses on an element oriented at angle φ, as shown in Fig. 1-31(c), the angle 2φ is laid off from CA in the same direction as angle φ is turned in the body. Diameter DE is thus located.

The horizontal projection of CD has the value shown in the figure. When this is added to OC, the result is the value of σ_u as given by Eq. (37). The vertical projection of CD has the value shown on the figure. This is equal to τ_{uv} as given by Eq. (39). It is plain that the coordinates of point D of the circle are equal to the normal and shear stresses as found by the combined stress equations.

Stresses σ_v and τ_{uv} for the surface, in Fig. 1-31(c), whose normal lies at angle $(90° + \varphi)$ from the x-axis, are given by the coordinates of point E.

A clockwise angle φ on the body corresponds to a clockwise angle of 2φ on the circle, and vice versa.

Values of stresses σ_u, σ_v, and τ_{uv} change as angle φ is changed. The maximum and minimum values of the normal stresses are called the principal stresses, and are designated σ_1 and σ_2, respectively. Their values can be found from the abscissas for points F and G in Fig. 1-31(b). The element for the principal stresses is oriented at angle θ to the x-axis as shown in Fig. 1-31(d). As shown by the circle, the value of θ can be found by the following equation.

$$\tan 2\theta = \frac{2\tau_{xy}}{\sigma_x - \sigma_y}, \qquad \text{for principal stresses} \qquad (40)$$

The radius of the circle has the value shown. The equations for σ_1 and σ_2 are as follows.

$$\sigma_1 = \frac{\sigma_x + \sigma_y}{2} + \sqrt{\left(\frac{\sigma_x - \sigma_y}{2}\right)^2 + \tau_{xy}^2} \qquad (41)$$

$$\sigma_2 = \frac{\sigma_x + \sigma_y}{2} - \sqrt{\left(\frac{\sigma_x - \sigma_y}{2}\right)^2 + \tau_{xy}^2} \qquad (42)$$

It should be noted that the sides of the element for principal stresses are free from shearing stress. If shear stress τ_{xy} should be equal to zero, stresses σ_x and σ_y would become the principal stresses, σ_1 and σ_2.[12]

The maximum shearing stress to which the material is subjected has a value equal to the radius of the circle. On the circle, point H is located 90° from points F and G for principal stresses. In the body, the surfaces for maximum shear stress are thus inclined 45° to the surfaces for the principal stresses. The element of maximum shearing stress, as shown in Fig. 1-31(e),

---

[12]When the stress consists only of simple tension, it can be designated σ rather than σ_1.

is inclined at θ_1 to the x-axis. As shown by the circle, the value of θ_1 can be found by the following equation.

$$\tan 2\theta_1 = -\frac{\sigma_x - \sigma_y}{2\tau_{xy}}, \qquad \text{for maximum shear stress} \qquad (43)$$

The value of the maximum shearing stress is

$$\tau_{max} = \sqrt{\left(\frac{\sigma_x - \sigma_y}{2}\right)^2 + \tau_{xy}^2} \qquad (44)$$

The circle of Fig. 1-31 indicates that at points of maximum shear, such as at H, normal stresses $\sigma_{1'}$ and $\sigma_{2'}$ are present whose value is given by the equation

$$\sigma_{1'} = \sigma_{2'} = \frac{\sigma_x + \sigma_y}{2} \qquad (45)$$

When shear stress τ is equal to zero, the radius of the circle, or the maximum shearing stress is equal to

$$\tau_{max} = \tfrac{1}{2}(\sigma_1 - \sigma_2) \qquad (46)$$

Example 15. Let the state of stress at some point in a body be defined as follows.

$$\sigma_x = 20{,}000 \text{ psi}, \qquad \sigma_y = -4{,}000 \text{ psi}, \qquad \tau_{xy} = 5{,}000 \text{ psi}$$

(a) Draw the view of the element for the given state of stress and mark values thereon.

(b) Draw the Mohr circle for the given state of stress and mark completely.

(c) Draw the element oriented 30° clockwise from the x-axis and show values of all stresses.

(d) Draw the element correctly oriented for principal stresses and show values.

(e) Draw the element for maximum shear stress and mark values of all stresses.

Solution. (a), (b). The given state of stress and the Mohr circle are shown in Fig. 1-32(a) and (b), respectively.

(c) Diameter ECD should be drawn at 60° clockwise to the x-axis of the circle, and stresses σ_u and σ_v scaled and placed on the element of Fig. 1-32(c). Since point D lies below the σ-axis, shear stress τ_{uv} crosses the u-axis of Fig. 1-32(c) in the direction that causes a counterclockwise moment on the element. Likewise, since E lies above the σ-axis, stress τ_{uv} crosses the v-axis in sketch (c) in the direction that causes a clockwise moment on the element.

(d) Principal stresses σ_1 and σ_2, together with their angle of inclination, are scaled directly from the circle, and are shown acting on an element properly oriented in Fig. 1-32(d).

(e) The maximum shear stress τ_{max} and the corresponding normal stress $\sigma_{1'}$ are shown on the element of Fig. 1-32(e). The arrows are directed in accordance with the previously explained rules.

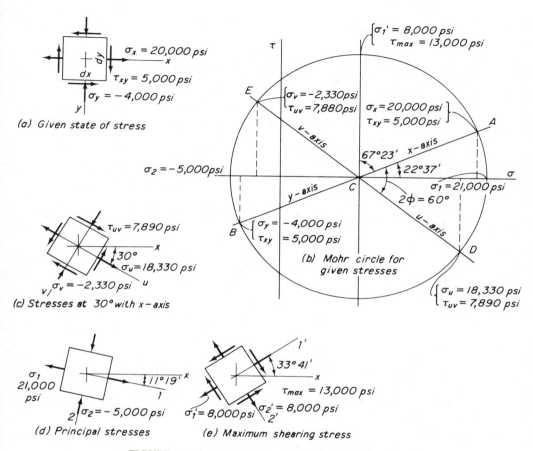

FIGURE 1-32 **Solution of Example 15 by Mohr circle.**

The advantages of the graphical method for solving combined stress problems should now be apparent. Not only is the method more rapid, but the state of stress for any direction can be scaled directly from the circle. When the equations are used for solving a problem, a separate computation must be made for each desired direction. The Mohr circle also aids in forming a mental picture of the state of stress in the body. In working problems, care must be exercised that all necessary information is placed on the drawing for the circle as well as on the views of the various elements.

The reader should check all values shown in Fig. 1-32(c), (d), and (e) by using the appropriate equations. Note that for $\varphi = 30°$, Eq. (39) gives a negative result for τ_{uv}. This result checks with the circle and indicates that the shear stress for this direction is acting oppositely to that shown in Fig. 1-30(c).

21. Stresses and Deformations in Two Directions

The elongation ϵ_x in the x-direction caused by the tensile stress σ_x is accompanied by a decrease in the width of the body at right angles to the stress, as shown in Fig. 1-33. This decrease of width is a definite proportion of the increase of length and is expressed by the equation

$$\epsilon_y = -\mu \frac{\sigma_x}{E} \text{ in./in.} \tag{47}$$

where the factor μ is known as *Poisson's ratio.* In a similar manner, a tensile stress in the y-direction causes a decrease of the length in the x-direction equal to

$$\epsilon_x = -\mu \frac{\sigma_y}{E} \tag{48}$$

FIGURE 1-33   Deflection due to stress σ_x.

Conversely, a compressive stress causes an increase in the width at right angles to the stress. This result is confirmed by Eqs. (47) and (48), since compressive or negative stress values are substituted in the equations. For most metals in engineering service, μ has a value between 0.25 and 0.30.

Superposition of the stresses shown in Figs. 1-33 and 1-34 gives those shown in Fig. 1-35. Stress σ_x causes an increase in length in the x-direction,

FIGURE 1-34   Deflection due to stress σ_y.

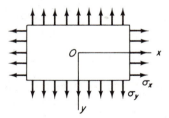

FIGURE 1-35   Tension in two directions.

whereas in this same direction a shortening occurs because of σ_y. The net effect is given by

$$\epsilon_x = \frac{1}{E}(\sigma_x - \mu\sigma_y) \text{ in./in.} \tag{49}$$

In the y-direction a similar equation applies.

$$\epsilon_y = \frac{1}{E}(\sigma_y - \mu\sigma_x) \tag{50}$$

Equations (49) and (50) represent Hooke's law for two-dimensional stress. These equations can be solved simultaneously for the stresses to give

$$\sigma_x = \frac{E}{1 - \mu^2}(\epsilon_x + \mu\epsilon_y) \tag{51}$$

$$\sigma_y = \frac{E}{1 - \mu^2}(\epsilon_y + \mu\epsilon_x) \tag{52}$$

22. Deflection of Beam from Shearing Stress

In addition to the bending, warping of the cross sections, shown in Fig. 1-22(c), from the shearing stress causes additional deflection of a beam. The total deflection at the center for a simply supported beam carrying a uniform load w per unit length is given by

$$y = \frac{5wl^4}{384EI}\left(1 + \frac{25\alpha I}{l^2 A}\right) \tag{53}$$

The shear increment is represented by the second term of the parentheses. Here α represents the ratio of the maximum shearing stress to the average shearing stress, V/A, on the cross section. By Eq. (28) α is equal to $\frac{3}{2}$ for rectangular cross sections, and by Eq. (29) α is equal to $\frac{4}{3}$ for circular cross sections. For other shapes α must be computed. In Eq. (53) it is assumed that there is no restraint to the free warping of the cross sections.

For a simply supported beam carrying a concentrated load P at the center, the equation is

$$y = \frac{Pl^3}{48EI}\left(1 + \frac{31.2\alpha I}{l^2 A}\right) \tag{54}$$

Example 16. Find the percent of increase in deflection because of shear for a simply supported 8-in. I-beam with load P at the center. Area A is 7.09 in.^{2}. The thickness of the web is 0.24 in. The moment of inertia I is 83.4 in.^{4}. Length of the span is equal to 10 times the depth of the beam.

Solution. In accord with Section 13, the maximum shearing stress will be taken as the shear force V divided by the area of the web.

$$\tau_{max} = \frac{V}{0.24 \times 8} = \frac{V}{1.92}$$

The average shearing stress: $\qquad \tau_{av} = \frac{V}{A} = \frac{V}{7.09}$

Then: $\qquad \alpha = \frac{\tau_{max}}{\tau_{av}} = \frac{7.09}{1.92} = 3.69$

By Eq. (54):

$$y = \frac{Pl^3}{48EI}\left(1 + \frac{31.2 \times 3.69 \times 83.4}{80^2 \times 7.09}\right)$$

$$= \frac{Pl^3}{48EI}(1 + 0.212)$$

The deflection due to shear is thus 21.2% as great as that due to bending.[13]

23. Principle of St. Venant

It was pointed out in Section 3 that equation $\sigma = P/A$ was not applicable in the region close to the eye of the eyebar. At some distance away, however, the stress distribution becomes simple tension uniformly distributed, and the use of the equation for finding the value of the stress is justified.

A similar situation prevails for the use of equation $\sigma = Mc/I$ for finding bending stress. The moments M in Fig. 1-5 could be applied to the beam in a number of ways as shown in Fig. 1-36. It is obvious that the state of stress near the end of the beam in Fig. 1-36(a) is not the same as that in Fig. 1-6(b), which represents the distribution for $\sigma = Mc/I$. The same is true for Fig. 1-36(b). Equation (11) might be applied all the way to the end of the beam for sketch (c) because the basic assumption that the cross sections are planes is substantially fulfilled. At locations sufficiently far from the ends in Fig. 1-36(a) and (b), the use of Eq. (11) for finding the bending stress is satisfactory.

FIGURE 1-36   Methods for applying moment to end of beam.

The remarks made above can be applied to the stress situation in the neighborhood of a concentrated force (either load or reaction). In Fig. 1-37 the stress situation for cross section OO' at the load is complicated by the vertical compressive stresses between the elements from the load. For cross section NN' located a distance equal to or greater than the depth of the beam from OO', the stress situation is approximately the same as that represented by Fig. 1-6(b) for bending stress and Fig. 1-22(b) for shear stress.

[13]For derivation of Eqs. (53) and (54) as well as for a more precise method for computing α, see Chapter 5 of Reference 4, end of chapter.

FIGURE 1-37 Beam with concentrated load.

The phenomenon that the stress system tends to become regular at distances removed from the disturbance is known as the principle of St. Venant. This principle states that if the forces acting on a small region of the body are replaced by a different, but statically equivalent system, no change in the stress or deformation will be experienced at points sufficiently far removed from the loads. It is valid for both normal and shear stress.

A sudden change in shape or form, arising from notches, holes, or fillets, causes the stresses to be increased beyond the values indicated by the equations of this chapter. This increase in stress is usually local in nature, and occurs in the immediate neighborhood of the discontinuity. It is taken care of in the computations by multiplying the stress, as given in the usual equation, by a stress concentration factor, which will be explained in the following chapter.

When a metal is stressed beyond the elastic limit in tension, it will subsequently be found that the elastic limit in compression is reduced. Conversely, the elastic limit in tension is reduced for a material strained beyond the elastic limit in compression. This phenomenon is known as the *Bauschinger effect*.

It must always be kept in mind that the equations in this book are derived on the assumption that the material is perfectly elastic, and that at all times the stress is proportional to the strain.

REFERENCES

1. FAUPEL, J. H., and F. E. FISHER, *Engineering Design*, 2d ed., New York: John Wiley & Sons, Inc., 1980.
2. ROARK, R. J., and W. C. YOUNG, *Formulas of Stress and Strain*, 5th ed., New York: McGraw-Hill Book Company, 1976.
3. SPOTTS, M. F., *Mechanical Design Analysis*, Englewood Cliffs, N.J.: Prentice-Hall, Inc., 1964.
4. TIMOSHENKO, S., *Strength of Materials*, 3d ed., Part I, New York: Van Nostrand Reinhold Company, 1955.

5. TIMOSHENKO, S., *Strength of Materials*, 3d ed., Part II, New York: Van Nostrand Reinhold Company, 1956.

6. TIMOSHENKO, S., and J. M. GERE, *Theory of Elastic Stability*, 2d ed., New York: McGraw-Hill Book Company, 1961.

PROBLEMS

See Tables 2-1 and 2-1A for mechanical properties of engineering materials.

1. The lower ends of the two hangers in Fig. 1-38 were at the same elevation before the loads were applied. The horizontal member is of uniform cross section. Find the value of length l if the horizontal member (assumed to be rigid) is horizontal after all loads are acting.   *Ans.* $l = 38.3$ in.

FIGURE 1-38   Problem 1.

2. The bottom member in Fig. 1-39 is of uniform cross section and can be assumed to be rigid. Find the value of the distance x if the lower member is to be horizontal.   *Ans.* $x = 564$ mm.

FIGURE 1-39   Problem 2.

FIGURE 1-40   Problem 3.

3. The bottom member in Fig. 1-40 is of uniform cross section and can be assumed to be rigid. Its hinge is frictionless. Find the number of degrees of rotation of the lower member. *Ans.* $\varphi = 0.137°$.

4. The bottom member in Fig. 1-41 is of uniform cross section. Its hinge is frictionless. The rods are of steel. Find the distance point A drops upon attachment of the weight. *Ans.* $\delta = 0.12$ mm.

FIGURE 1-41   Problem 4.   **FIGURE 1-42   Problem 5.**

5. In Fig. 1-42, find the drop of the 500-lb weight. *Ans.* $\delta = 0.0126$ in.

6. In Fig. 1-43 the lower member is of uniform cross section and can be assumed to be rigid. Find the angular rotation of the lower member in degrees.

Ans. $\varphi = 0.029°$.

FIGURE 1-43   Problem 6.   **FIGURE 1-44   Problem 7.**

7. In Fig. 1-44 the lower member is of uniform cross section and can be assumed to be rigid. Find the change in elevation of the left end because of the stretch of the rods. *Ans.* Drop = 0.0352 in.

8. The members in Fig. 1-45 have a neat fit at the time of assembly. Find the force caused by an increase in temperature of 50°C. Supports are immovable.

Ans. $P = 96,720$ N.

FIGURE 1-45   Problem 8.

9. After being drawn up snug, the nut in Fig. 1-46 is given one-quarter additional turn. Threads on both bolt and nut are hardened. Find the force in the pipe and bolt. *Ans.* $F = 13,750$ lb.

FIGURE 1-46   Problem 9. FIGURE 1-47   Problem 10.

10. The bars in Fig. 1-47 are fitted top and bottom to immovable supports. The bars are of the same material and have the same cross section. Find the force in each bar. *Ans.* Top, 57,100 N; bottom, 42,900 N.

11. In Fig. 1-48 the outer bars are symmetrically placed with respect to the center bar. The top member is rigid and located symmetrically on the supports. Find the load carried by each of the supports. Modulus for the bars is 2,000,000 psi.

Ans. Center, 5,161.4 lb; outer, 2,419.3 lb.

FIGURE 1-48   Problem 11. FIGURE 1-49   Problem 12.

12. In Fig. 1-49, find the deflection at the load.   *Ans.* $y = 23.36$ mm.

13. Because of an error in fabrication, the center strut in Fig. 1-50 was made 0.005 in. shorter than the other two. The members on top and bottom can be considered rigid. Bars are made of the same material and have equal cross sections. $E = 2,000,000$ psi. Find the load carried by each bar.

    Ans.   Outer, 14,667 lb; inner, 10,666 lb.

FIGURE 1-50   **Problem 13.**

14. In Fig. 1-51 the three bars are of the same material and have equal cross sections and lengths. Find the force in each bar.

    Ans. $P_1 = 200,000$ N; $P_2 = 500,000$ N; $P_3 = 700,000$ N.

FIGURE 1-51   **Problem 14.**

15. Find the distance x in Fig. 1-52 that causes the 1,000-lb rigid weight to remain level if the lower ends of the hanger are at the same elevation before the weight is applied.   *Ans.* $x = 21.2$ in.

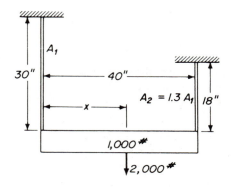

FIGURE 1-52   Problem 15.

16. Find the force in each bar in Fig. 1-53. $E_1 = 2E_2$. The 2,300-kg mass can be considered rigid.   *Ans.* $F_1 = 7,200$ N; $F_2 = 6,480$ N.

FIGURE 1-53   Problem 16.

17. The rigid beam in Fig. 1-54 was level before the load was applied. Find the force in each hanger.   *Ans.*   Steel, 10,210 lb; aluminum, 7,660 lb.

FIGURE 1-54   Problem 17.

18. The bars in Fig. 1-55 are of the same material and have equal cross-sectional areas. There is no stress in the bars before the load is applied. Find the load carried by each bar. *Ans.* Outer, 19,350 N; inner, 32,980 N.

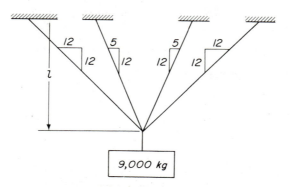

FIGURE 1-55 Problem 18.

19. The bars in Fig. 1-56 have the same cross-sectional area. There is no stress in the bars before the load is applied. Each bar is 0.5 in. square.
(a) Find the force in each bar.
(b) Find the force in each bar if the temperature drops 100°F.
Ans. (a) Steel, 8,340 lb, brass, 5,560 lb;
(b) Steel, 8,100 lb, brass, 5,970 lb.

FIGURE 1-56 Problem 19.

Draw the shear and bending moment diagrams for the beams shown in the following figures and find the values of the maximum bending stress.

20. Figure 1-57   *Ans.* $\sigma = 7.2$ MPa.

21. Figure 1-58.   *Ans.* $\sigma = 19,500$ psi.

FIGURE 1-57   Problem 20.   FIGURE 1-58   Problem 21.

22. Figure 1-59.   *Ans.* $\sigma = 81.5$ MPa.

23. Figure 1-60.   *Ans.* $\sigma = 9,450$ psi.

FIGURE 1-59   Problem 22.   FIGURE 1-60   Problem 23.

24. A steel saw blade 1.25 mm thick is bent into an arc of a circle of 600-mm radius. Find the bending stress.   *Ans.* $\sigma = 215$ MPa.

25. Find the reactions and also the value of the bending stress at a point 5 ft from the left end for the beam of Fig. 1-61.   *Ans.* $\sigma = 18,090$ psi.

FIGURE 1-61   Problem 25.

26. Find the reactions and also the value of the bending stress at a point 1,950 mm from the left end for the beam in Fig. 1-62.   *Ans.* $\sigma = 69.5$ MPa.

FIGURE 1-62   Problem 26.

27. Figure 1-63 illustrates the front side rod of a steam locomotive. Find the bending stress from inertia and dead weight for a train speed of 90 mph. Weight of steel equals 0.283 lb/in.³. Assume the rod to be a simply supported beam of uniform cross section *A-A*.   *Ans.* $\sigma = 12{,}430$ psi.

FIGURE 1-63   **Problem 27.**

28. The deflection of the beam shown in Fig. 1-64 is given by the equation

$$y = \frac{w_1}{360EIl^2}(x^6 - 5l^3x^3 + 4l^5x)$$

Find the location of the maximum bending moment and its value.

        Ans.   Max $M = 0.039w_1l^2$.

FIGURE 1-64   **Problem 28.**

29. Compute the change of elevation of the end of the light pointer attached to the beam of Fig. 1-65 upon the application of the load. The beam is 4 in. wide. $E = 1{,}500{,}000$ psi.

FIGURE 1-65   **Problem 29.**

30. Moments M_1 and M_2 are applied to the ends of the simply supported beam shown in Fig. 1-66. Light pointers of lengths $2l/3$ and $l/3$ are attached to the ends as shown and are directed along the axis of the beam before applying the moments. If the distance a is measured, and if E, I, and l are known, show that moment M_1 is equal to $6EIa/l^2$.

The idea of this problem can be used for finding the value of the unknown end moments acting on a beam. The pointers are clamped in position, and distance a is measured. Length l must be free from transverse loads.

FIGURE 1-66   Problem 30.

31. Find the deflection at the end C of the beam in Fig. 1-67. $E = 1,600,000$ psi.
Ans. $y_c = 0.17$ in., up.

FIGURE 1-67   Problem 31.

32. Find the deflection of the end A of the beam in Fig. 1-68. $E = 10,000$ MPa.
Ans. $y_a = 3.28$ mm down.

FIGURE 1-68   Problem 32.

33. Find the deflection of the end A of the beam of Fig. 1-69. $E = 1,500,000$ psi.
Ans. $y_a = 0.158$ in., down.

FIGURE 1-69   Problem 33.

34. Find the deflection at *A* in Fig. 1-70. Steel.   *Ans.* $y = 13.5$ mm.

FIGURE 1-70   **Problem 34.**   **FIGURE 1-71**   **Problem 35.**

35. The supports top and bottom in Fig. 1-71 can be considered immovable. The upper strut was found to be $\frac{1}{32}$ in. too long to be assembled without stress. If this member is driven into place, what force will be induced therein? All parts are of steel.   *Ans.* $F = 27,040$ lb.

36. The steel bar in Fig. 1-72 was welded into place with a neat fit at time of assembly. Find the force in the bar if the temperature drops 70 deg C. Supports can be considered rigid.   *Ans.* $F = 46,180$ N.

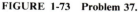

FIGURE 1-72   **Problem 36.**   **FIGURE 1-73**   **Problem 37.**

37. Find the deflection of the weight in Fig. 1-73.   *Ans.* $\delta = 0.185$ in.

38. The steel beams in Fig. 1-74 are 36 mm wide and 50 mm deep. Find the deflection of each beam.   *Ans.* Top, 4.6 mm; bottom, 21.1 mm.

FIGURE 1-74   Problem 38.

39. Find the value of the stress in the steel band when it is passing around a pulley in Fig. 1-75. Pulley bearings are frictionless.   *Ans.* $\sigma = 140,000$ psi.

FIGURE 1-75   Problem 39.   **FIGURE 1-76   Problem 40.**

40. Compute the values of the transverse shear stress at points 25, 50, 75, and 100 mm below the top surface of the beam in Fig. 1-76 for cross sections to left of the load. The beam is 150 mm wide and 200 mm deep.

Ans. $\tau = 2.1, 3.6, 4.5,$ and 4.8 MPa.

Find the value of the maximum transverse shear stress for the beams of the following figures:

41. Figure 1-77.   *Ans.* $\tau = 1,170$ psi.

FIGURE 1-77   Problem 41.

42. Figure 1-78.   *Ans.* $\tau = 17.28$ MPa.

43. Figure 1-79.   *Ans.* $\tau = 99$ psi.

FIGURE 1-78   Problem 42.   **FIGURE 1-79   Problem 43.**

44. Find the thickness of the web required to make the value of the maximum transverse shearing stress equal to 4 MPa for the beam in Fig. 1-80.

Ans. $t = 8.9$ mm.

FIGURE 1-80   Problem 44.

45. Utilizing Eq. (24), prove Eq. (29).

46. Find the maximum value of the transverse shear stress for a hollow circular shaft having outer and inner radii of r_1 and r_2, respectively.

Ans. $\tau_{max} = V(r_1^3 - r_2^3)/3I(r_1 - r_2)$.

47. Prove Eq. (30).

48. Find the values and plot the distribution of stress over the cross section of the upright of Fig. 1-81. Locate the point of zero stress.   *Ans.* $\sigma_c = 10.5$ MPa.

FIGURE 1-81   Problem 48.   FIGURE 1-82   Problem 49.

49. Find the values and plot the distribution of stress over the cross section of the upright of Fig. 1-82. Locate the point of zero stress.   *Ans.* $\sigma_t = 6,860$ psi.

50. Determine the width d of the offset link of Fig. 1-83 if the permissible value of the working stress is 55 MPa. Carry out the work in a manner similar to Example 7.   *Ans.* $d = 69.3$ mm.

FIGURE 1-83   Problem 50.

FIGURE 1-84   Problem 51.

51. The loading and general arrangement of a link are the same as for Problem

50, but the cross section is shown in Fig. 1-84. Find width of flange *b*.

Ans. *b* = 1.55 in.

52. The section shown in Fig. 1-85 is loaded in bending about a horizontal axis through the center of gravity. By the equations of Fig. 1-7 calculate the moment of inertia about this axis, also the value of *e* for the shear center.

Ans. $\bar{x}$ = 24.5 mm; *I* = 320,200 mm⁴; *e* = 17.76 mm.

FIGURE 1-85   Problem 52.

53. Make a sketch showing the equivalent loading of axial force and moments about the *x*- and *y*-axes for the bar shown in Fig. 1-86. Compute and show on iso-metric sketches the distribution of stress on cross section *ABCD* caused by each of the separate loads. Also make a sketch showing resultant distribution of stress and mark the value at each corner. Dimension the location of the neutral axis for the cross section.

Ans. σ_{max} = 600 psi.

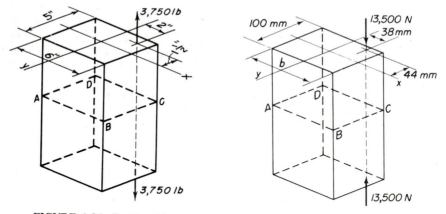

FIGURE 1-86   Problem 53.   **FIGURE 1-87   Problem 54.**

54. Make a sketch for the equivalent loading for the bar of Fig. 1-87. What must be the dimension *b* in order that the maximum stress on a typical cross section, such as *ABCD*, will be equal to 7 MPa? Make sketches and mark values for the stresses due to the separate loads. Also make the sketch for the resultant stresses and mark values at corners. Dimension the location of the neutral axis.

Ans. *b* = 110 mm.

55. The inclined load shown in view *M-M*, Fig. 1-88, is applied to the pin at the end of the beam. Resolve load into horizontal and vertical components and find the bending stress due to each. Add algebraically to get stress at points *A*, *B*, *C*, and *D*.   *Ans.*   At *A*, $\sigma = 1{,}562.5$ psi.

FIGURE 1-88   Problem 55.

56. The moment of inertia about axis 1-1 in Fig. 1-89 is equal to 50,000,000 mm⁴. The area is equal to 13,000 mm². Find the moment of inertia about axis 2-2.
                    Ans. $I_2 = 90{,}625{,}000$ mm⁴.

FIGURE 1-89   Problem 56.

57. The moment of inertia about axis 1-1 in Fig. 1-90 is equal to 540 in.⁴. Area is equal to 60 in.². Find the moment of inertia about axis 2-2.
                    Ans. $I_2 = 1{,}440$ in.⁴.

FIGURE 1-90   Problem 57.

FIGURE 1-91   Problem 58.

58. Find the value of the moment of inertia about the left edge in Fig. 1-91.

Ans. $I = 28,708,000$ mm⁴.

59. Find the reactions, draw shear and moment diagram, and dimension significant points for the beam shown in Fig. 1-92. *Ans.* $R_1 = 1,350$ lb.

FIGURE 1-92   Problem 59.

60. Repeat Problem 59 using the beam of Fig. 1-93. Does the curve for shear have a horizontal tangent at a point 6 ft from the left end? Why? Also find the point where the bending moment has the maximum value.

Ans. Max 761 mm from left end.

FIGURE 1-93   Problem 60.

61. Draw and dimension the bending moment diagram for the member shown in Fig. 1-94. *Ans.* Max $M = 28,800$ in. lb.

FIGURE 1-94   Problem 61.

FIGURE 1-95   Problem 62.

62. Repeat Problem 61 for Fig. 1-95.

63. Cut the *AB* portion of the beam of Fig. 1-96 free from the balance of the beam and place the shears and moments on the end surfaces which were acting before cutting. Is this portion of the beam now in equilibrium?

FIGURE 1-96   Problem 63.

64. Figure 1-97 shows the loads and general dimensions of a freight car. The horizontal load is due to centrifugal effects in passing around a curve. The loading shown is for one axle. Make a force diagram showing the equilibrium of a unit consisting of one axle and two wheels. The centrifugal force is assumed to be applied to the axle at the inner bearing only, and to be resisted by the flange of the outer wheel only. Also draw and dimension the bending moment diagram for the axle. Wheels are press-fitted to the axle.   *Ans.*   Max $M = 93,750,000$ Nmm.

FIGURE 1-97   Problem 64.

Draw and dimension the load and reaction diagram, the shear diagram, and the bending moment diagram for the beams shown in the following figures:

65. Figure 1-98.

66. Figure 1-99.

FIGURE 1-98   Problem 65.   FIGURE 1-99   Problem 66.

67. Figure 1-100.

FIGURE 1-100   Problem 67.

68. Figure 1-101.
69. Figure 1-102.

FIGURE 1-101   Problem 68.   **FIGURE 1-102   Problem 69.**

70. Draw and dimension the shear and load diagram for the simply supported beam whose moment diagram is given in Fig. 1-103.

FIGURE 1-103   Problem 70.

FIGURE 1-104 Problem 71.

71. If the values in Fig. 1-104 represent pounds of shear for a simply supported beam draw and dimension the corresponding load and moment diagrams for this beam.

72. Let Fig. 1-104 represent a moment diagram for a simply supported beam. Horizontal lengths are 2,500 mm and 3,750 mm, respectively. Moments are 1,080,000 Nmm and 135,000 Nmm, respectively. Draw and dimension the corresponding load and shear diagrams.

73. If the values in Fig. 1-105 represent pounds of shear for a simply supported beam, draw and dimension the corresponding load and moment diagrams for this beam.

FIGURE 1-105 Problem 73.

74. If the values in Fig. 1-105 represent foot pounds of moment, draw and dimension the corresponding load and shear diagrams.

75. Draw and dimension the load and moment diagrams for the simply supported beam whose shear diagram is represented by Fig. 1-106.

FIGURE 1-106 Problem 75.

FIGURE 1-107 Problem 76.

76. Repeat Problem 75 for the beam whose shear diagram is given in Fig. 1-107.

77. Draw and dimension the load and shear diagrams for the simply supported beam whose moment diagram is represented by Fig. 1-108.

FIGURE 1-108 Problem 77.

78. Repeat Problem 77 for the moment diagram shown in Fig. 1-109.

FIGURE 1-109 Problem 78.

79. Make an exploded isometric view of the three portions of the beam of Fig. 1-110 and show all forces and moments necessary for equilibrium.

FIGURE 1-110 Problem 79.   **FIGURE 1-111 Problem 80.**

80. Repeat Problem 79 using Fig. 1-111.

FIGURE 1-112 Problem 81.   **FIGURE 1-113 Problem 82.**

81. Draw and dimension the bending moment diagram for the beam shown in Fig. 1-112. Find the location where the moment is maximum and find its value.

Ans. Max $M = 4{,}900$ in. lb.

82. Find the deflection of point D in Fig. 1-113. All parts are steel.

Ans. 12.7 mm at D.

83. A simply supported beam is 60 in. long and carries a 5,000-lb load at the center.

(a) If the cross section is a hollow square 6 in. high on the outside and 5 in. high on the inside, find the value of the maximum bending stress.

(b) Suppose the cross section is a hollow circle of the same area as for part (a) with wall thickness equal to 0.5 in. Find the value of the maximum bending stress.

Ans. (a) $\sigma = 4{,}020$ psi;

(b) $\sigma = 4{,}090$ psi.

84. A simply supported cast-iron beam is 900 mm long and carries a 1,350 N load at the center. Note that the three cross sections in Fig. 1-114 have areas equal to one another. Find the stress and deflection at the center for each beam. $E = 103{,}400$ MPa.

Ans. (b) $\sigma = 32.11$ MPa; $y = 1.23$ mm.

FIGURE 1-114 Problem 84.

85. Three beams have the equal cross-sectional areas shown in Fig. 1-115. Find the stress in each beam for a bending moment of 30,000 in. lb.

Ans. (b) $\sigma = 8{,}820$ psi.

FIGURE 1-115 Problem 85.

86. A 50 × 50-mm steel column with hinged ends is 1,500 mm long. Yield strength of the material is 275 MPa. $F_s = 2.5$. Initial crookedness $a = 1.25$ mm. Find the load capacity. $E = 206,700$ MPa. *Ans.* $P = 153,250$ N.

87. A column is made from a piece of steel pipe 144 in. long. OD = 4.50 in. ID = 4.30 in. Yield strength of the material is 36,000 psi. $F_s = 10$. Initial crookedness $a = 0.2$ in. Find the load capacity if the column can be considered as having hinged ends. *Ans.* $P = 6,990$ lb.

88. The material in a body is subjected to the following stresses: $\sigma_x = -70$ MPa, $\sigma_y = -28$ MPa, and $\tau_{xy} = 28$ MPa. Draw the Mohr circle and mark all significant points. Draw a view of the element, properly oriented, for maximum normal stress and show values of all stresses. Do the same for the element with maximum shearing stress.
Ans. $\sigma_1 = -15.5$ MPa; $\sigma_2 = -82.5$ MPa; $\sigma_{max} = 33.5$ MPa.

89. A uniformly distributed normal stress, either tension or compression, is applied to the edges of the plate of Fig. 1-116 in the *x*-direction, and a uniform tension or compression is also applied in the *y*-direction. The dimensions of the deformed plate are 15.010 and 9.996 in. Find the stresses σ_x and σ_y. $E = 30,000,000$ psi. Poisson's ratio $\mu = 0.3$.

FIGURE 1-116 Problem 89.

90. Draw the Mohr circle for a prismatic bar loaded in tension and prove the rule: "The maximum shearing stress is equal to one-half the axial stress and is located at 45° to the direction of the axial stress."

91. Draw the Mohr circle for an element loaded on pure shear, and determine the value of the maximum normal stress and its direction.

92. An element is acted upon by the following stresses: $\sigma_x = 49$ MPa, $\sigma_y = -14$ MPa, and $\tau_{xy} = 14$ MPa.

(a) By means of the equations compute the stresses on the sides of an element oriented 30° clockwise with the x-axis.

(b) Find the value of φ for maximum and minimum normal stress and compute the values of these stresses.

(c) Repeat part (b) for maximum shear.

(d) Make a view showing the given state of stress, and draw the corresponding Mohr circle. Scale the stresses for the element oriented 30° from the x-axis, and compare them with the results secured by use of the equations. Draw a view of this element with the stresses placed thereon.

(e) Scale the values of the maximum and minimum normal stress and represent them by arrows on an element oriented at the proper angle.

(f) Repeat for the element that is subjected to the maximum shearing stress.

$\qquad$ *Ans.* Max $\sigma = 52$ MPa; max $\tau = 34.5$ MPa.

93. Repeat Problem 92 using given stresses of $\sigma_x = 1,500$ psi, $\sigma_y = 22,500$ psi, and $\tau_{xy} = -10,000$ psi. $\qquad$ *Ans.* Max $\sigma = 26,500$ psi; max $\tau = 14,500$ psi.

94. The total normal force uniformly distributed over edge *AB* of the plate of Fig. 1-117 is 67,500 N, and the total shear force is 54,000 N. For edge *BC* the normal and shear forces are 607,500 N and 135,000 N, respectively. Draw a view showing an element with the given state of stress and also draw the corresponding Mohr circle. Draw an element with sides parallel to the x- and y-axes and show the stresses acting on it. $\qquad$ *Ans.* $\sigma_x = -18.3$ MPa.

FIGURE 1-117 Problem 94. $\qquad$ FIGURE 1-118 Problem 95.

95. The uniformly distributed normal and shear forces for edge *AB* of the plate of Fig. 1-118 are 15,000 and 24,000 lb, respectively. For edge *BC* the normal and shear forces are 100,000 and 32,000 lb, respectively. Draw an element for the given state of stress; also draw the Mohr circle. Draw the element with sides parallel to the x- and y-axes and show the stresses that are acting on it. *Ans.* $\sigma_x = 5,050$ psi.

96. For the loading shown in Fig. 1-119, draw the element located at 45° from the *x*-axis and place the stresses thereon that are acting on it. *Ans.* $\tau = 30$ MPa.

FIGURE 1-119 Problem 96.

97. In Fig. 1-120 the elements show the normal stresses that are acting at the same point in a body. Determine the values of the shearing stresses for both elements and mark them with arrows properly directed.

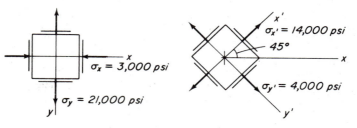

FIGURE 1-120 Problem 97.

98. Make a drawing for the element at *A* of the beam in Fig. 1-121 with horizontal and vertical sides, and show the stresses acting on it. Construct the corresponding Mohr circle. Draw the element for the principal stresses correctly oriented and show the stresses acting on it. Do the same for the element of maximum shear stress. *Ans.* Min $\sigma = -21.2$ MPa; max $\tau = 11.3$ MPa.

FIGURE 1-121 Problem 98.

FIGURE 1-122   Problem 99.

99. Repeat Problem 98 for the element at *A* of Fig. 1-122.

Ans.   Min $\sigma = -3,060$ psi; max $\tau = 2,390$ psi.

100. Draw a view of the element at *A* in Fig. 1-123 with horizontal and vertical sides and show all stresses acting on it. *Ans.* $\sigma = 0.44$ MPa.

FIGURE 1-123   Problem 100.   **FIGURE 124   Problem 101.**

101. Draw a view of the element at *A* in Fig. 1-124 with horizontal and vertical sides and show all stresses acting on it. *Ans.* $\sigma = 750$ psi; $\tau = 140.6$ psi.

102. Draw a view of the element at *A* in Fig. 1-125 with horizontal and vertical sides and show all stresses acting on it. Draw the Mohr circle for this element and determine the value of the maximum normal stress and the maximum shear stress.

Ans. $\sigma_1 = 5.09$ MPa; $\tau_{max} = 2.75$ MPa.

FIGURE 1-125   Problem 102.

103. Strain gage measurements at a certain point on a body give the elongations shown by Fig. 1-126. Draw a view of the element with horizontal and vertical sides and show all normal and shear stresses acting on it. Do the same for the element whose sides are inclined 45° with the *x*-axis. Check all results with the Mohr circle. $E = 30,000,000$ psi; $\mu = 0.3$.

   Ans. $\sigma_x = 20,770$ psi; $\sigma_y = 39,230$ psi; $\tau_{xy} = 18,460$ psi.

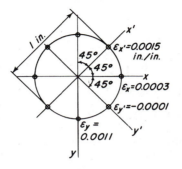

FIGURE 1-126   Problem 103.

104. Repeat Problem 103 for the strain gage readings shown by Fig. 1-127.

   Ans. $\sigma_{x'} = 72.52$ MPa; $\sigma_{y'} = 41.76$ MPa; $\tau_{x'y'} = 76.92$ MPa.

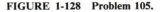

FIGURE 1-127   Problem 104.   **FIGURE 1-128   Problem 105.**

105. Prove that the strain measurements when taken as in Fig. 1-128 must fulfill the equation

$$\epsilon_x + \epsilon_y = \epsilon_{x'} + \epsilon_{y'}$$

106. When free of loads, a steel plate is 375 mm long in the *x*-direction and 300 mm long in the *y*-direction. After the loads are applied the 12-in. length becomes 200.15 mm. Find the value of σ_y if $\sigma_x = 133.33$ MPa.   *Ans.* $\sigma_y = 140$ MPa.

FIGURE 1-129   Problem 107.

FIGURE 1-130   Problem 108.

107. After loads in the *x*- and *y*- directions are applied to the steel plate of Fig. 1-129 the length in the *y*-direction is increased by 0.01 in. Find the change of length in the *x*-direction.   *Ans.* $\delta_x = -0.0184$ in.

108. A uniformly distributed stress, either tension or compression, is applied to the right and left edges of the plate shown in Fig. 1-130. A uniform tension or compression is also applied to the top and bottom edges. After the loads are acting, the final dimensions of the plate are 299.7 mm and 500.2 mm.

Find the values of σ_x and σ_y if the material is steel in the as-rolled condition.
                          Ans. $\sigma_x = 2.20$ MPa; $\sigma_y = -19.34$ MPa.

109. The moment of inertia about the 1-1 axis in Fig. 1-131 is equal to 4,320 in.⁴ The area is equal to 120 in.² Find the moment of inertia about axis 2-2.

FIGURE 1-131   Problem 109.

FIGURE 1-132   Problem 110.

110. The moment of inertia about axis *A* in Fig. 1-132 is equal to 343,750,000 mm⁴. Find the moment of inertia about axis *B*.   *Ans.* $I_B = 695,312,000$ mm⁴.

111. The load shown on the beam of Fig. 1-133 is removed and replaced by a new load *w* uniformly distributed over the entire beam. The maximum bending moments for both loadings are the same. Find the value of *w*.

112. In Fig. 1-134 let the bearings be considered simple supports. The shaft is steel. Find the deflection at point *A*.   *Ans.* $y = 0.213$ mm, up.

FIGURE 1-133   Problem 111.   FIGURE 1-134   Problem 112.

113. The pins in Fig. 1-135 are all at the same elevation. The bar is of spring steel. If the bar was originally straight, find the value of the maximum bending stress.

FIGURE 1-135   Problem 113.

114. Let the cross section through an industrial car be similar to that of the car in Fig. 1-97. The loading for a single axle is 20,000 N vertical and 5,000 N horizontal. Rails are 900 mm center to center, and journals are 1,100 mm center to center. Wheels are 400 mm in diameter. The center of gravity of load is 475 mm above the center of the axle. Draw and dimension the bending moment diagram for the axle.

115. Find the deflection at the midpoint between the bearings in Fig. 1-136. The shaft is steel.

FIGURE 1-136   Problem 115.

FIGURE 1-137   Problem 116.

116. In Fig. 1-137 assume that the forces in the oil films are symmetrically disposed about the center line of the bearing. Determine the value of the eccentricity *e* so that the slope assumed by the bearing is the same as the slope of the shaft at *A*. The shaft is steel. Modulus of elasticity for the bearing is 100,000 MPa.

Ans.   e = 1.06 mm.

117. A 12-in. standard I-beam of 31.8 lb per ft has $I_x = 215.8$ in.^{4} and $I_y = 9.5$ in.^{4} The width of the flange is 5 in. Find the percentage increase in bending stress caused by changing a vertical load to one inclined at 1° to the vertical.

118. The casting shown in section in Fig. 1-138 has its center of gravity located as shown. Assume that all loads can be considered as acting at the center of the 75-mm bearing. Find the value of the bending stress for this point.

FIGURE 1-138   Problem 118.

119. An element in two-dimensional stress has $\sigma_x = 12,000$ psi, $\sigma_y = -4,000$ psi, and $\tau_{xy} = 3,900$ psi. Draw and dimension that Mohr circle. Draw and dimension the elements for the principal stress and for the maximum shearing stress.

120. Strain gage readings for a point in a stressed body show $\epsilon_x = -0.0006$ mm/mm, $\epsilon_y = 0.0026$ mm/mm, $\epsilon'_x = 0.0022$ mm/mm, and $\epsilon'_y = -0.0002$ mm/mm. Axes *x* and *x'* are at 45° to each other. $E = 200,000$ MPa, $\mu = \frac{1}{3}$.

(a) Draw the view of an element with horizontal and vertical sides and show values of the stresses acting on it.

(b) Do the same for an element that is inclined 45°.

Check all results by making a Mohr circle.

121. The valve push rod for an overhead valve engine is $\frac{1}{4}$ in. in diameter and 14 in. long. Find the critical load when the rod is considered as a column with round ends. Assume the rod is perfectly straight.

122. A solid circular steel column with hinged ends is 50 mm in diameter and 1,400 mm long. Yield strength of the material is 250 MPa. Initial crookedness $a = 1.25$ mm. Take E for steel as 200,000 MPa. Find the yield point loading for this column.   *Ans.* $P_{yp} = 246,600$ N.

123. A 10-in. standard I-beam of 25.4 lb per ft has a cross-sectional area of 7.38 in.^{2} and a moment of inertia of 122.1 in.^{4} Suppose a steel plate 10 in. deep and $\frac{3}{4}$ in. thick is used in bending when standing on edge. Compare the relative carrying capacities for the two beams.   *Ans.*   I-beam is 1.95 times as strong.

124. A 24-in. standard I-beam has a width of flange of 7 in. and a moment of inertia of 2,087.2 in.^{4} It is reinforced on the top and bottom by a 1×10 in. plate. If the shearing force at the cross section is 100,000 lb find the shearing force per longitudinal inch at the interface between the flange and plate. When this is known, the designer can then arrange suitable welding or rivet spacing between flange and plate to carry the shearing force.   *Ans.*   2,398 lb per lineal inch.

125. Same as Problem 124 except the beam is a 20-in. I at 65.4 lb per ft, width of flange 6.25 in., and moment of inertia 1,169.5 in.^{4}

                    Ans.   3,110 lb per lineal inch.

126. Use superposition to derive the equation for reaction R_2 in No. 21 of Fig. 1-16.

127. In Fig. 1-139 let $h_1 = 2$ in., $h_2 = 8$ in., $l = 60$ in. Find the distance x from the right end at which the bending stress is a maximum. Assume elementary bending theory to be valid.   *Ans.* $x = 20$ in.

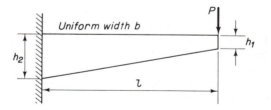

FIGURE 1-139   **Problems 127, 128, and 129.**

128. In Fig. 1-139 let $h_1 = 60$ mm, $h_2 = 180$ mm, and $l = 2,000$ mm. Find the distance x from the right end at which the bending stress is a maximum. Assume elementary bending theory to be valid.   *Ans.* $x = 1,000$ mm.

129. In Fig. 1-139 let $h_1 = 100$ mm, $h_2 = 250$ mm, and $l = 1,500$ mm. Find the distance x from the right end at which the bending stress is a maximum. Assume elementary bending theory to be valid.   *Ans.* $x = 1,000$ mm.

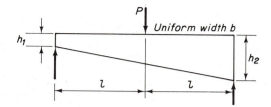

FIGURE 1-140   Problems 130 and 131.

130. In Fig. 1-140 let $h_1 = 3$ in., $h_2 = 10.5$ in., and $l = 5$ ft. Find the distance x from the left end where the bending stress is a maximum. Assume elementary bending theory to be valid. *Ans.* $x = 48$ in.

131. In Fig. 1-140 let $h_1 = 75$ mm, $h_2 = 195$ mm, and $l = 1,500$ mm. Find the distance x from the left end where the bending stress is a maximum. Assume elementary bending theory to be valid. *Ans.* $x = 1,875$ mm.

132. In Fig. 1-141 let $b_1 = 3$ in., $b_2 = 9$ in., and $l = 150$ in. Find the distance x from the left end where the bending stress is a maximum. Assume elementary bending theory to be valid. *Ans.* $x = 81.125$ in.

FIGURE 1-141   Problems 132 and 133.

133. In Fig. 1-141 let $b_1 = 75$ mm, $b_2 = 225$ mm, and $l = 3,750$ mm. Find the distance x from the left end where the bending stress is a maximum. Assume elementary bending theory to be valid. *Ans.* $x = 2,028$ mm.

134. In Fig. 1-142 the vertical beams are spaced on 5-ft centers. Assume simple supports. $h_1 = 72$ in., $h_2 = 120$ in. Find the height x from the bottom where the bending moment is a maximum. *Ans.* $x = 39.8$ in.

FIGURE 1-142   Problems 134 and 135.

135. In Fig. 1-142 the vertical beams are spaced on 4-ft centers. Assume simple supports. $h_1 = 60$ in., $h_2 = 100$ in. Find the height x from the bottom where the bending moment is a maximum.   *Ans.* $x = 33.17$ in.

136. The dimensions in Fig. 1-143 refer to the unloaded steel plate. After uniform normal stresses σ_x and σ_y are applied to the edges, the dimensions become 2.997 in. and 5.002 in. For $\mu = 0.3$, find the values of σ_x and σ_y.
Ans. $\sigma_x = 3,300$ psi; $\sigma_y = -29,010$ psi.

FIGURE 1-143   **Problem 136.**

137. Find the values for the three reactions in Fig. 1-144.   *Ans.* $R_1 = 7P/12$.

FIGURE 1-144   **Problem 137.**

138. Find the values for the three reactions in Fig. 1-145.
Ans. $R_1 = 71P/238$.

FIGURE 1-145   **Problem 138.**

139. The rectangular coverless box of Fig. 1-146 has a surface area for the four sides and bottom of 12 m². Find dimensions x and y if the contents of the box, level full, are to be a maximum.   *Ans.* $x = 2$ m; $y = 1$ m.

FIGURE 1-146   Problem 139.   **FIGURE 1-147   Problem 140.**

140. The rectangular coverless box of Fig. 1-147 has a width x, length ky (where k is a known constant), and height y. Find the relationship between x and y if the area of the plate material is to be a minimum for a given volume for the box level full.   *Ans.* $x = (1 + k)y/k$.

141. A fiber-board box of usual construction is shown in Fig. 1-148. The flaps on the bottom are arranged similarly to those on the top. Find expressions for x, y, and z in terms of volume V that will make the material for the box a minimum for any given value of V.   *Ans.* $x = \sqrt[3]{2V}$; $y = \frac{1}{2}\sqrt[3]{2V}$; $z = \sqrt[3]{2V}$.

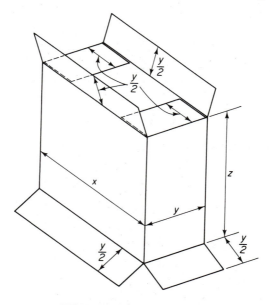

FIGURE 1-148   Problem 141.

142. A steel box has exact inside dimensions of $6 \times 8 \times 10$ in. at room temperature. Find the volume after a temperature rise of $80°F$. Note that the coefficient of increase in volume per cubic inch per degree is three times the linear coefficient.

Ans. Final volume $= 480.749$ in.^{3}

143. In Fig. 1-149 let $a = 1,875$ mm, $b = 2,500$ mm, and $P = 5,000$ N. Find the force in each member in accordance with the following assumptions: all members are of the same material, no connections where the diagonal members cross, all joints are frictionless, all members have the same cross section, and no buckling in the compression members.

Ans. $F_{ab} = 789.5$ N.

FIGURE 1-149 **Problem 143 and 144.**

144. In Fig. 1-149 let $a = b = 100$ in. and $P = 1,000$ lb. Find the force in each member in accord with the assumptions of Problem 143.

Ans. $F_{ab} = 369.8$ lb.

145. The bar in Fig. 1-150 is continuous and lies in the horizontal plane. It is

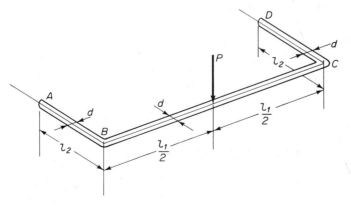

FIGURE 1-150 **Problem 145.**

welded to a vertical wall at *A* and *E*. Find the torque in the bar at points *A* and *E*.

$$Ans. \quad T = \frac{Pl_1^2}{8E\left(\dfrac{l_1}{E} + \dfrac{l_2}{G}\right)}.$$

146. Derive the equations for $\bar{x}$ and *I* in Fig. 1-7(f).

147. Same as Problem 17, Fig. 1-54, except the steel and aluminum bars are interchanged. *Ans.* Steel, 11,814 lb; aluminum 2,735 lb.

148. Same as Problem 15, Fig. 1-52, except that left bar has a cross section 1.3 times that of the right bar. *Ans.* $x = 23.708$ in.

149. Same as Problem 12, Fig. 1-49, except bar is 20 mm thick and 80 mm wide. Force at end is 2,500 N and spring rate is 30 N/mm. *Ans.* $y = 50.86$ mm.

150. Same as Problem 16, Fig. 1-53, except that the cross-sectional areas of the two bars are reversed. *Ans.* $F_1 = 5,884\text{N}; F_2 = 7,355\text{N}.$

151. Find the moment of a 1-in. circle about an axis through its center. An annulus 2 in. in diameter has the same area as the 1-in. solid round. Find the moment of inertia about the axis through the center and compare with that of the solid bar. *Ans.* $I_a/I_s = 7.$

152. Same as Problem 151 except the diameter of the solid bar is 1.5 in. and the outside diameter of the annulus is 3 in.

153. For the last sketch in Fig. 1-7 let $h = 4$ in., $b = 1.6$ in., $h_1 = 2.4$ in., and $b_1 = 1$ in. Find values for $\bar{x}$, *e*, and I_y.

$$Ans. \quad \bar{x} = 0.980 \text{ in.}; \ e = 0.710 \text{ in.}; \ I_y = 8197 \text{ in.}^4$$

154. Suppose a $\frac{1}{4} \times 10$ in. plate is formed into a channel section 6 in. high with flanges top and bottom estimated at $2\frac{1}{4}$ in. wide. If the resulting member is to be used in bending about the *y*-axis, find the values for $\bar{x}$ and I_y if the equations of Fig. 1-7 can be assumed applicable. *Ans.* $\bar{x} = 1.675$ in.; $I_y = 1.10$ in.^{4}

155. Suppose a 6×250 mm plate is formed into a channel section 150 mm high with flanges assumed to be 58 mm wide top and bottom. If the resulting member is to be used in bending about the *y*-axis, find the values for $\bar{x}$ and I_y if the equations of Fig. 1-7 can be assumed applicable. *Ans.* $\bar{x} = 43.13$ mm; $I_y = 452,700$ mm^4.

156. Find the values of $\bar{x}$ and I_y for a channel section with the following dimensions: $h = 35$, $b = 12$, $h_1 = 32$, and $b_1 = 10.5$. *Ans.* $\bar{x} = 9$; $I_y = 1,008.$

2
Working Stresses

The problem of mechanical strength is one of the most important features of the design of machine parts. Stress equations in general are applicable to idealized homogeneous materials subjected to steady loads. This chapter will extend the theories of designing to include cases of fluctuating loads, where the fatigue strength of the material has an important influence on the success of the design. Machine parts should be shaped so that stress concentrations at points of high loading are avoided as much as possible. Suitable adjustments must also be made when the material carries loads in two directions. A distinction must be made between ductile and brittle materials. Lack of knowledge or appreciation of the behavior of engineering materials under actual service conditions has been the cause of many expensive failures.

F_s,	factor of safety	mm,	millimeter, 0.039 37 in.
in.,	inch, 25.4 mm	M,	mega (prefix), 1 million
in. lb,	inch pound, 113 Nmm	M,	bending moment
J,	joule, or newton-meter, Nm	MPa,	megapascal or 1 N/mm² or
k,	kilo (prefix), 1,000		145 psi
kg,	kilogram, 2.2046 lb (mass)	N,	newton, 0.2248 lb (force)
K,	stress concentration factor	Nmm,	newton millimeter
lb,	pound;	P,	force or load
	1 lb (force) = 4.4482 N	Pa,	pascal, 1 N/m²
	1 lb (mass) = 0.4536 kg	μ,	Poisson's ratio; for metals,
m,	meter, 39.3701 in.		μ usually taken as 0.3
m,	milli (prefix), one thousandth	σ_{av},	average stress

σ_e,	endurance limit stress for reversed bending	σ_{ut},	ultimate stress, tension
σ_1, σ_2,	principal stresses	σ_{yp},	yield point stress, tension
σ_r,	range stress	τ,	shear stress
σ_{uc},	ultimate stress, compression	τ_{yp},	yield point stress, shear

1. Stress-Strain Diagrams

Much useful information concerning the behavior of materials and their suitability for engineering purposes can be obtained by making tensile tests and plotting a graph for the relationship between stress and strain. The characteristic shape of the stress-strain diagram for low-carbon steel is shown in Fig. 2-1(a). It should be noted that the material followed Hooke's law until the loading became a little more than one-half of the *ultimate strength*. This material has a well-defined *yield point* or stress at which a marked increase in elongation occurs without increase in load. The *proportional limit* marks the maximum value of the stress for which Hooke's law holds. The modulus of elasticity of the material can be found from the slope, σ/ϵ, of the straight-line portion of the curve, or from Eq. (3) of Chapter 1, $E = \sigma/\epsilon$.

After the ultimate stress is reached, soft steel specimens undergo a marked reduction in diameter, called *necking*, at some point in the stressed material. It is customary, however, to construct the diagram on the basis of stresses computed by using the original cross-sectional area. The ratio of the loss of cross-sectional area at failure to the original area of the specimen is called the *reduction of area*. This quantity, together with the elongation at failure, gives useful information concerning the *ductility* of the material. The speed at which the load is applied affects the shape of the diagram. The yield point and ultimate stresses become higher as the speed of loading increases.

Many steels, especially those that have been heat treated, do not have a well-defined elastic limit, but yield gradually after passing the proportional limit, as shown in Fig. 2-1(b). If the loading were stopped at point A at a higher stress than the elastic limit, and if the specimen were then unloaded and readings taken, the curve would follow the dashed line, and a permanent set, or plastic deformation, would exist. For such materials, the stress corresponding to some given permanent set (usually 0.001 or 0.002 in./in.) is called the *yield strength*, and is taken as the limit of the engineering usefulness of the material.

Most materials do not exhibit a permanent set if loaded slightly beyond the proportional limit. The maximum value of such stress is known as the

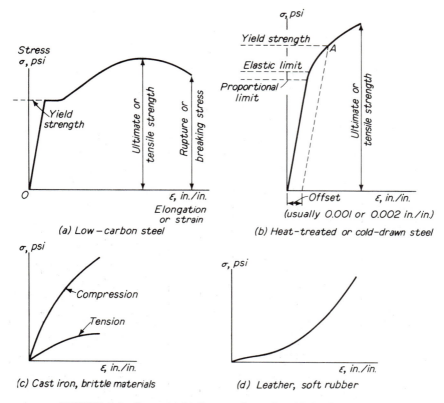

FIGURE 2-1   Stress-strain diagrams for various kinds of materials.

elastic limit, which is usually difficult to determine experimentally. *Proof loads* or *proof stresses* refer to loading that the material or part must sustain while fulfilling specified conditions relative to failure or deformation.

For ductile materials, the value of the yield strength in shear is equal to about 0.5–0.6 of the yield strength in tension.

Nonductile or brittle materials, such as cast iron and ceramics, do not follow Hooke's law to any noticeable degree. The characteristic stress–strain diagram for either tension or compression is shown in Fig. 2-1(c). Leather and rubber have diagrams similar to that in Fig. 2-1(d).

Mechanical properties for a number of widely used engineering materials are given in Tables 2-1 and 2-1A.

Elementary elastic theories as discussed in Chapter 1 apply, in general, to bodies of uniform cross section but are unable to take account of the effect of a change in shape on the resulting stresses.

TABLE 2-1   Average Values for Mechanical Properties
of Engineering Materials*

| Material | Modulus of Elasticity | | Weight, γ, lb/in.³ | Coefficient of Linear Expansion, α, in./(in. deg F) |
	Tension, E, psi	Shear, G, psi		
Cast iron	See Table 14-13		0.256	0.000 0056
Steel	30,000,000	11,500,000	0.283	0.000 0065
Stainless steel, 18-8	28,000,000	10,000,000	0.295	0.000 0096
Brass, bronze	15,000,000	5,300,000	0.30–0.32	0.000 0102
Aluminum	10,000,000	3,850,000	0.100	0.000 0128
Magnesium	6,500,000	2,400,000	0.065	0.000 0145

*Poisson's ratio, $\mu = 0.3$. Acceleration for free fall, $g = 32.174$ ft/sec² $= 386.1$ in./sec². Variations in values shown are possible, depending on composition and method of manufacture.

TABLE 2-1A   Average Values for Mechanical Properties
of Engineering Materials in SI Units*

| Material | Modulus of Elasticity | | Weight per Unit Volume, N/mm³ | Coefficient of Linear Expansion, α, mm/(mm deg C) |
	Tension-Comp, E, N/mm² or MPa	Shear, G, N/mm² or MPa		
Cast iron			0.000 0695	0.000 0100
Steel	206,900	79,300	0.000 0768	0.000 0117
Stainless steel, 18–8	193,100	69,000	0.000 0800	0.000 0173
Brass, bronze	103,400	36,600	0.000 0814–869	0.000 0184
Aluminum	69,000	26,600	0.000 0271	0.000 0230
Magnesium	44,800	16,600	0.000 0176	0.000 0261

*Stress values are those of Table 2-1 divided by 145, the conversion factor from psi to MPa. 1 lb/in.³ $= 4.4482$ N/(25.4 mm)³ $= 0.00027145$ N/mm³. Acceleration for free fall, $g = 9.8066$ m/sec².

2. Types of Failure. Ductile Materials and Brittle Materials

Two types of mechanical failures occur in materials: yielding and fracture. Yielding or permanent deformation is a pronounced sliding along certain angular planes in the material. It takes place without rupture. The engineering usefulness for most machine parts is ended after a sufficient amount of yielding has taken place. Therefore, yielding can properly be termed failure.

Fracture is a separation failure that occurs on a cross section normal to the tension stress.

A ductile material can be defined as one whose resistance to sliding is smaller than its resistance to separation. Failure takes place by yielding. Many ductile materials have the same yield point value for compression as for tension.

A brittle material is one whose resistance to separation is less than its resistance to sliding. Failure takes place by fracture. A limit of about 5% elongation is usually taken as the dividing line between ductile materials and brittle materials. Most brittle materials have a considerably higher value for the ultimate strength in compression than for tension.

Under certain conditions, a material ordinarily said to be ductile will undergo a fracture or separation failure similar to that of a brittle material. Some of these conditions are (a) cyclic loading at normal temperatures (fatigue); (b) longtime static loading at elevated temperatures (creep); (c) impact or every rapidly applied loading, especially at low temperatures; (d) work hardening by a sufficient amount of yielding; (e) severe quenching in heat treatment if not followed by tempering; and (f) a three-dimensional state of stress in which sliding is prevented, as at the bottom of the narrow groove in the bar shown in Fig. 2-2. Internal cavities or voids in castings or forgings may have a similar effect.

FIGURE 2-2 Bar in tension with deep groove.

3. Ductile Materials with Steady Stress

Under steady or static loading, a machine part made of a ductile material can undergo a considerable amount of yielding or distortion. The usefulness of a part is usually lost after a sufficient amount of yielding has taken place. The working stress is therefore based on the yield point stress.

It is possible for the yield point to be exceeded by the stress concentration as a result of a sudden change of form even though the elementary equation indicates that the average stress at the cross section has a safe value. In general, no damage occurs provided that the load is steady and the material is ductile. The material merely yields locally in the overstressed regions, and the stress is thereby relieved. Hence, it is customary for designers to neglect the effects of stress concentration when the loads are steady and the material is ductile.[1]

---

[1]See p. 10 of Reference 9, p. 447 of Reference 3, and p. 298 of Reference 4, end of chapter.

(a) *Simple Tension or Compression.* When the material is subjected to simple tension or compression stress σ, the factor of safety F_s is given by the equation

$$F_s = \frac{\sigma_{yp}}{\sigma} \tag{1}$$

where σ_{yp} is the yield point stress in simple tension.

(b) *Pure Shear.* For pure shear loading τ_{max} the equation for factor of safety is

$$F_s = \frac{\tau_{yp}}{\tau_{max}} \tag{2}$$

where τ_{yp} is the yield point stress in shear.

The expressions maximum stress or working stress are frequently used interchangeably.

4. Maximum Shear Theory of Failure

Because ductile materials fail by shearing, the *maximum shear theory of failure* is in wide use by designers. The theory is applied by first finding the maximum shearing stress for the given loading and then dividing it into the yield point stress in shear to find the factor of safety. It is thus an application of Eq. (2).

The Mohr circle in Fig. 2-3 indicates that a body with simple tension stress σ has shear stresses equal to one-half this value at directions 45° to the direction of σ.

$$\tau_{max} = \tfrac{1}{2}\sigma \tag{3}$$

If stress σ in this figure is increased to the yield point value, the maximum shear theory postulates that the material will then be at the yield point value

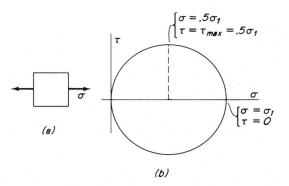

FIGURE 2-3   Element in simple tension and corresponding Mohr circle.

in shear. Thus

$$\tau_{yp} = \tfrac{1}{2}\sigma_{yp} \qquad (4)$$

This can be substituted into Eq. (2) to give

$$F_s = \frac{0.5\,\sigma_{yp}}{\tau_{max}} \qquad (5)$$

Failure in shear is assumed to occur along the 45° directions of Fig. 2-4(a). Equations (4) and (5) are valid only when it is understood that the maximum shear theory of failure is being employed.

(a)

(b)

FIGURE 2-4   Planes of maximum shear stress for bar with tensile load.

In addition to the shear stresses, an element oriented at 45° to the direction of σ in Fig. 2-3(a) has normal stresses on all sides of 0.5σ. The complete loading for this element is then given by Fig. 2-4(b).

Consider the element of Fig. 2-5(a) loaded in pure shear. The corresponding Mohr circle is shown in sketch (b). For such loading, an element oriented at 45° to the pure shearing stresses is loaded only by the normal stresses of Fig. 2-5(c). The stresses of sketches (a) and (c) can thus be considered as being interchangeable.

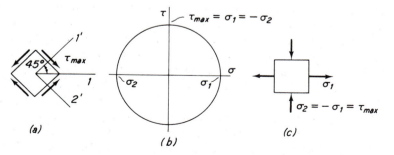

FIGURE 2-5   Elements and Mohr circle for pure shear stress.

5. Normal Stresses in Two Directions

For two- and three-dimensional states of stress, the failure of an engineering material is a complicated phenomenon. In addition, test data for combined loading is nearly always lacking, and the design must therefore be based solely on the yield point or ultimate strength values as found by the simple tension test. It is under such conditions that a theory of failure is most useful.

Let the theories for combined stress of Chapter 1 be applied to the case of general loading for stresses σ_x, σ_y, and τ_{xy}, and thus obtain the principal stresses σ_1 and σ_2 of Fig. 2-6. For convenience, these are shown in the horizontal and vertical directions. The algebraically larger of the two stresses is designated σ_1. If shear stress τ_{xy} is equal to zero, σ_x and σ_y are principal stresses.

To arrive at suitable values for the working stresses, it is necessary to know how the element in Fig. 2-6 will fail. The presence of two stresses makes the situation more complicated than the case of simple tension of Fig. 2-3. Figure 2-7 shows a perspective of the element in Fig. 2-6. Since all bodies are three-dimensional, three planes of failure must be investigated.

FIGURE 2-6 Principal stresses in two dimensions.

FIGURE 2-7 Plane of failure for two-dimensional stress. All stresses tension.

(1) Both stresses are tension as shown in Fig. 2-7. The weakest plane is *BADE* because σ_2 has no effect on this plane. Failure is determined solely by stress σ_1. Equations (1)–(5) apply.

(2) Both stresses are compression as in Fig. 2-8. Here σ_2 is numerically larger than σ_1. The weakest plane is *BCDF* because stress σ_1 has no effect on this plane. Failure is determined solely by stress σ_2. Equations (1)–(5) apply.

(3) Stress σ_1 is tension, and stress σ_2 is compression as in Fig. 2-9. The weakest plane is *ACEF*. Both stresses contribute to the shear stress on this

FIGURE 2-8   Plane of failure for two-dimensional stress. All stresses compression.

FIGURE 2-9   Plane of failure for two-dimensional stress, σ_1 tension, σ_2 compression.

plane. The value of the maximum shear stress is

$$\tau_{max} = \tfrac{1}{2}(\sigma_1 - \sigma_2) = \sqrt{\left(\frac{\sigma_x - \sigma_y}{2}\right)^2 + \tau_{xy}^2} \tag{6}$$

Equation (5) applies.

The maximum shear theory of failure is thus very easy to apply to two-dimensional stress problems.

6. Mises-Hencky or Distortion Energy Theory

Another criterion, and one that fits experimental results closely, is the *Mises-Hencky* or *distortion energy theory*. For two-dimensional stress the equation is

$$S^2 = \sigma_1^2 + \sigma_2^2 - \sigma_1\sigma_2 \tag{7}$$

Substitution of Eqs. (41) and (42) of Chapter 1 for σ_1 and σ_2 gives

$$S^2 = \sigma_x^2 - \sigma_x\sigma_y + \sigma_y^2 + 3\tau_{xy}^2 \tag{8}$$

Equivalent stress S can be considered a simple normal stress, and can be used as stress σ in Eq. (1).

Example 1. The stresses at a point in a body are $\sigma_x = 13{,}000$ psi, $\sigma_y = 3{,}000$ psi, and $\tau_{xy} = 12{,}000$ psi. The material tests $\sigma_{yp} = 40{,}000$ psi.
(a) Find the factor of safety by the maximum shear theory of failure.
(b) Find the factor of safety by the Mises–Hencky theory.

Solution. (a) By the Mohr circle or the combined stress equations the following values for the principal stresses σ_1 and σ_2 are obtained.

$$\sigma_1 = 21{,}000 \text{ psi}; \qquad \sigma_2 = -5{,}000 \text{ psi}$$

The weakest shearing plane is the one to which both σ_1 and σ_2 contribute to the shearing stress as in Fig. 2-9.

By Eq. (6): $\tau_{max} = \frac{1}{2}[21,000 - (-5,000)] = 13,000$ psi

By Eq. (4): $\tau_{yp} = 0.5 \times 40,000 = 20,000$ psi

By Eq. (2): $F_s = \dfrac{20,000}{13,000} = 1.54$

(b) By Eq. (7): $S = \sqrt{21,000^2 + (-5,000)^2 - 21,000(-5,000)}$
$$= 23,900 \text{ psi}$$

Hence $F_s = \dfrac{40,000}{23,900} = 1.67$

Example 1A. The stresses at a point in a body are $\sigma_x = 91$ MPa, $\sigma_y = 21$ MPa, and $\tau_{xy} = 84$ MPa. Material tests $\sigma_{yp} = 280$ MPa.

(a) Find the factor of safety by the maximum shear theory of failure.
(b) Find the factor of safety by the Mises–Hencky theory.

Solution. (a) By the Mohr circle, or the combined stress equations, the following values for the principal stresses σ_1 and σ_2 are obtained.

$$\sigma_1 = 147 \text{ MPa}; \qquad \sigma_2 = -35 \text{ MPa}$$

By Eq. (6): $\tau_{max} = \frac{1}{2}[147 - (-35)] = 91$ MPa

By Eq. (4): $\tau_{yp} = 0.5 \times 280 = 140$ MPa

By Eq. (2): $F_s = \dfrac{140}{91} = 1.54$

(b) By Eq. (7): $S = \sqrt{147^2 + (-35)^2 - 147(-35)} = 167.3$ MPa

Hence $F_s = \dfrac{280}{167.3} = 1.67$

Example 2. The same body as that used in Example 1 has stresses $\sigma_x = 20,000$ psi, $\sigma_y = 4,000$ psi, and $\tau_{xy} = 6,000$ psi.

(a) Find the factor of safety by the maximum shear theory of failure.
(b) Find the factor of safety by the Mises–Hencky theory.

Solution. (a) The principal stresses are found to be

$$\sigma_1 = 22,000 \text{ psi}; \qquad \sigma_2 = 2,000 \text{ psi}$$

The weakest shearing plane is the one affected by σ_1 alone as in Fig. 2-7. For this plane

By Eq. (3): $\tau_{max} = 0.5 \times 22,000 = 11,000$ psi

By Eq. (2): $F_s = \dfrac{20,000}{11,000} = 1.82$

(b) By Eq. (7): $S = \sqrt{22,000^2 + 2,000^2 - 22,000 \times 2,000}$
$$= 21,070 \text{ psi}$$

Hence $F_s = \dfrac{40,000}{21,070} = 1.90$

Example 3. A 2-in.-diam shaft is loaded statically in pure torque at a shearing stress of 10,000 psi. Find the F_s if the material is 4140 hot-rolled steel. Use the Mises–Hencky theory.

Solution. By Fig. 2-5(c): $\quad \tau_{max} = \sigma_1 = -\sigma_2 = 10,000$ psi

By Eq. (7): $\qquad\qquad\qquad S^2 = 10,000^2 + 10,000^2 - 10,000(-10,000)$

$$S = 10,000 \sqrt{3} = 17,320 \text{ psi}$$

By Table 14-5: $\qquad\qquad\qquad \sigma_{yp} = 62,000$ psi

By Eq. (1): $\qquad\qquad\qquad F_s = \dfrac{\sigma_{yp}}{S} = \dfrac{62,000}{17,320} = 3.58$

7. Endurance Limit of Materials

Working stresses that have been determined from the ultimate or yield point values of the material by a factor of safety give safe and reliable results only for static loading. Many machine parts, however, are subjected to a loading cycle in which the stress is not steady but continuously varying. Failures in machine parts are generally caused by such repeated loadings and at stresses that are considerably below the yield point. For many materials, long experience has proved that when the stress is below a certain value called the *fatigue* or *endurance limit*, the part will last indefinitely so far as ill effects from the stress are concerned. However, for slightly greater values of the stress, failure can be expected after a certain number of repetitions of the stress cycle. Fracture occurs without perceptible stretching and resembles the failure of a brittle material. Although such breaks are called fatigue failures, no change has taken place in the material except in the immediate neighborhood of the fracture itself. Failure has been brought about by a tiny crack that started at a stress concentration or at a flaw in the material at a highly stressed point. The crack itself serves as a stress concentration, and grows continually larger until the failure of the part occurs.

Methods of testing have been developed that aid in evaluating the ability of a material to resist failure by fatigue. The rotating-beam test is in widest use. It applies a pure bending moment to the specimen, shown schematically in Fig. 2-10(a). As the specimen rotates, the bending stress varies continuously from a maximum tension to a maximum compression, which can be represented on the time-stress axes by the curve of Fig. 2-10(b). The *endurance-limit stress* σ_e is frequently defined as the maximum value of the completely reversed bending stress, which a plain specimen can sustain for 10 million or more load cycles without failure. If a specimen of ferrous material lasts for this number of cycles, in general it can be assumed that it will last indefinitely.

Fatigue testing is also done for types of loading other than the completely

Specimen

To motor and counter

Load

(a) Machine for applying uniform bending moment to specimen

(b) Endurance limit stress is highest value of completely reversed bending stress for continuous operation

FIGURE 2-10   Rotating beam type of fatigue testing machine (schematic drawing).

reversed bending stress described above. Many kinds of stresses can be obtained by superposing a variable or fluctuating component on a static or steady stress. Testing machines are built that will give fluctuating axial stresses of various magnitudes to the specimen. It is customary to test spring materials with a pulsating shearing stress that varies continuously from zero to some maximum value.

Fatigue failures due to bending are the most common. Torsion failures are next, and failures due to axial loading are rare. Once started, a fatigue crack follows a general direction normal to the tension stress. A fatigue failure usually takes place across the crystals. Failure is by fracture or separation without visible yielding or distortion.

8. Factors Affecting Fatigue Strength

The value of the endurance limit is dependent on the condition of the surface of the specimen. The *endurance stress* σ_e for ground and polished specimens, when no stress concentrations are present, is frequently found to be approximately equal to one-half the ultimate strength for wrought steels. For a somewhat rougher surface, as produced by machining, the endurance limit may be only 35 or 40% of the ultimate strength. The endurance limit is further reduced if the surface is covered with scale from hot rolling or forging. Corrosion from water or acids may reduce the endurance limit to a very low value. The small pits that form on the surface act as stress raisers. The effect of the condition of the surface[2] on the endurance limit for steels of various tensile strengths is shown in Fig. 2-11. The endurance limit is reduced for temperatures above room temperature. Carbon and alloy wrought steels

[2]See Noll, G. C., and C. Lipson, "Allowable Working Stresses," *Proc. Soc. Exp. Stress Anal.*, 3(2), 1946, p. 89.

FIGURE 2-11 Relation between endurance limit and tensile strength for unnotched specimens in reversed bending.

give the most consistent results with respect to fatigue strength. For steel castings and cast iron, the endurance limit is about 40 per cent of the ultimate strength. Apparently no relationship exists between the endurance limit and the yield point, impact strength, or ductility. Experiments have shown that the endurance limit for reversed torsion is about 0.56 of that for reversed bending.[3]

Fatigue cracks can start not only at easily recognized changes of form, but also at frequently overlooked stress raisers, such as file and tool marks, accidental and grinding scratches, quenching cracks, or part number and inspection stamps, which produce a high value for the stress and serve as the starting point for the progressive failure. The attention of the designer must therefore be focused on such "sore spots" whenever they are located in a region of high tension stress.

Since fatigue cracks are due to tensile stress, a residual stress of tension on the surface of the part constitutes an additional fatigue hazard. Such a tensile stress, for example, may arise from a cold-working operation on the part without stress relieving. Parts that are finished by grinding frequently have an extremely thin surface layer, which is highly stressed in tension. Such residual stresses, combined with the tensile stress from the loading, may give a resultant stress sufficiently great to cause a fatigue crack to start.

[3]See Fig. 7, p. 7, of Reference 9, end of chapter.

Any residual surface tension should be removed, or better still, converted into a layer of compression. A prestressed surface layer of compression can be secured by such shop operations as shot blasting, peening, tumbling, or cold working by rolling between hardened steel rolls. When the surface layer is in compression, the resultant working stress in tension may have a low value. Sometimes the part can be subjected to an excessive load and the yield point stress exceeded in such a way that a residual stress of compression can be obtained at a point where the maximum working stress is tension. Any cold working of the part must, of course, be done under controlled conditions. Sand blasting must be avoided since the scratches serve as stress raisers. Carburized and nitrided parts have a compressive surface layer, which may account for the effectiveness of such parts in resisting fatigue. A finish grinding operation, however, may leave the surface in tension, as previously mentioned. Additional discussion of residual stress is given in Chapter 14.

A weak decarburized layer on the surface of a heat-treated part has a low endurance limit. This condition is especially harmful in springs. Reductions in strength are also brought about by residual stresses, such as those produced by press and shrink fits. A press-fitted, antifriction bearing race also causes a reduction in the fatigue strength of the part to which it is fitted. The rough surface of a weld or an internal void also serves as a stress concentration. A layer of hot-dipped galvanizing causes a considerable reduction in the fatigue strength. The same is true for chromium platings. Electroplated zinc coatings have been found to be harmless. Many nonferrous materials do not have a definitely defined endurance limit.

The par value for the material is the endurance limit for a plain polished specimen. Fatigue tests with various kinds of notches show how much of the potential strength is being sacrificed by a particular type of notch. It is usually difficult if not impossible to devise a notched specimen that has the same fatigue strength as a machine part of a particular shape. Apparently such fatigue values can be obtained only from tests on the actual part. The results of tests on notched specimens may, however, serve as a better guide for the selection of a suitable steel than endurance tests made with plain specimens.

If proper attention has been given to the effect of stress concentration, failure in service is due mainly to accidental overload or abuse of the part, which could not be anticipated by the designer. Thus the ability to withstand overload is a very desirable quality in engineering materials. Sometimes a material of low endurance limit exhibits better resistance to overloads in fatigue than do high-strength, heat-treated materials. The ability to resist crack propagation after a crack has started is another desirable quality. Experiments have shown that killed steels are superior to rimmed steels with respect to crack propagation. Killed steels are deoxidized in the furnace or ladle before being poured into the ingot mold. Rimmed steels are not so degassified.

The life of a part can be reduced by factors other than fatigue, such as wear, corrosion, and high temperatures. If, for example, the useful life of a machine will be limited by wear of some of the parts, it would be uneconomical to design the other parts for infinite life in fatigue. The use of higher working stresses in fatigue can sometimes be justified in order to make the life of all the parts approximately equal. Precise information on the *S-N* curve for the material is required for such exact designing, and extensive testing must be conducted on the finished product. Ball bearings, and sometimes automotive parts, are designed on the basis of finite life.

Fluctuating stresses high enough to cause failure in a few thousand cycles or less is called low-cycle fatigue. Plastic yielding at localized areas may be involved. The theory is beyond the scope of this book.

9. Stress Concentration Caused by Sudden Change in Form

Only rarely does the failure of a machine part occur because of the sudden application of a single heavy load. Breakage, in the great majority of cases, is caused by repeated or fatigue loading, and takes place at a point of stress concentration where an abrupt change in the form of the part occurs. Such failures can occur without warning or plastic deformation. The average stress for the cross section may be below the elastic limit for the material.

Consider, for example, the state of stress in the tension member of two widths illustrated in Fig. 2-12. Near each end of the bar the internal force is uniformly distributed over the cross sections. The nominal stress in the right portion can be found by dividing the total load by the smaller cross-sectional area; the stress in the left portion can be found by dividing by the larger area. However, in the region where the width is changing, a redistribution of the force within the bar must take place. In this portion the load is no longer uniform at all points on a cross section, but the material in the neighborhood of points *B* in Fig. 2-12 is stressed considerably higher than the average value. The stress situation is thus more complicated, and the elementary equation P/A is no longer valid. The maximum stress occurs at

FIGURE 2-12 Stress concentration caused by sudden change in cross section.

some point on the fillet, as at *B*, and is directed parallel to the boundary at that point.

Another example is a bar in tension with a circular hole as shown in Fig. 2-13(a). If the bar is cut on the cross section of the hole, the tension stresses will be as shown in Fig. 2-13(b). The stress distribution along the cut surface is practically uniform until the neighborhood of the hole is reached, where it suddenly increases. High stresses such as these cause a fatigue crack to start under fluctuating loading.

FIGURE 2-13   Stress concentration for bar with hole loaded in tension.

This irregularity in the stress distribution caused by abrupt changes of form is called *stress concentration*. It occurs for all kinds of stress, axial, bending, or shear, in the presence of fillets, holes, notches, keyways, splines, tool marks, or accidental scratches. Inclusions and flaws in the material or on the surface also serve as stress raisers. The maximum value of the stress at such points is found by multiplying the nominal stress as given by the elementary equation by a stress concentration factor *K*, which is defined as follows.

$$K = \frac{\text{highest value of actual stress on fillet, notch, hole, etc.}}{\text{nominal stress as given by elementary equation for minimum cross section}}$$

(9)

Values of stress concentration factors can be found experimentally by photoelastic analysis or direct strain gage measurement. For a number of cases, solutions have been obtained by mathematical analysis.

10. Stress Concentration Factors

Stress concentration factors have been determined for a wide variety of geometric shapes and types of loading. The factors[4] for rectangular bars of two widths in tension or compression are given by the curves of Fig. 2-14.

[4]Such curves, by R. E. Peterson, appeared in *Machine Design*, **23**, 1951, in the February, March, May, June, and July issues. These have been collected into an enlarged book, Reference 9, end of chapter.

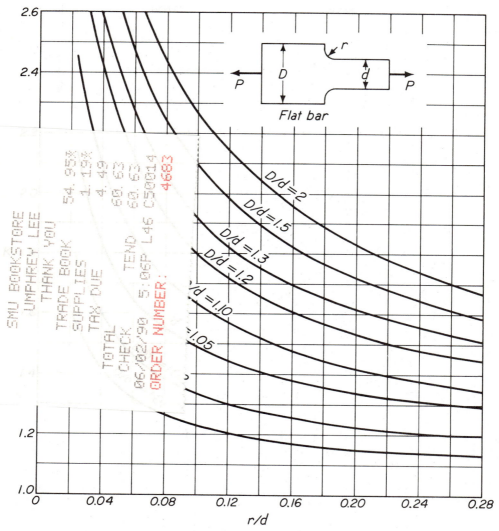

FIGURE 2-14 Factors of stress concentration K for various sizes of fillets for flat bar in tension or compression to be applied to the stress in the section of width d.

When such a bar is loaded in pure bending, the stress concentration factors can be obtained from Fig. 2-16. Corresponding curves for shafts are given by Figs. 2-15 and 2-17. The use of such curves is illustrated by the following example.

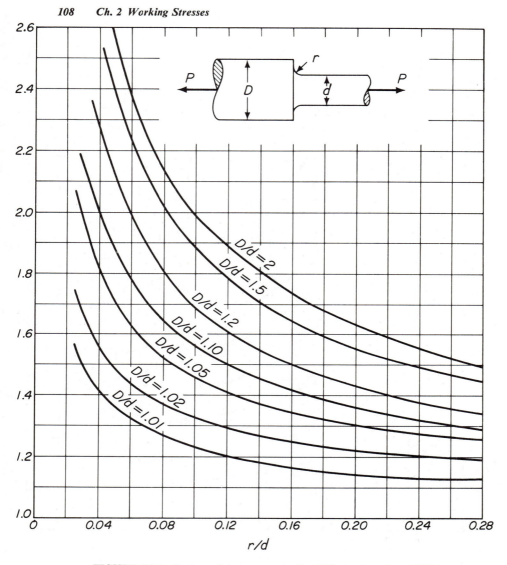

FIGURE 2-15 **Factors of stress concentration *K* for various sizes of fillets for round bar in tension or compression to be applied to the stress in the section diameter *d*.**

Example 4. Let the minor width of the flat bar in Fig. 2-12 be 1.25 in., the major width be 2.25 in., and the radius of the fillet be 0.25 in.

(a) Find the value of the stress concentration factor when the bar is loaded in tension.

(b) Find the value of the stress concentration factor if the loading is a pure moment instead of tension.

FIGURE 2-16 Factors of stress concentration *K* for various sizes of fillets for flat bar in bending to be applied to the stress in the section of width *d*.

Solution. (a) $D = 2.25$ in., $d = 1.25$ in., $r = 0.25$ in., $D/d = 1.80$, $r/d = 0.2$.
From Fig. 2-14: $\qquad\qquad\qquad K = 1.78.$

The maximum tension stress in the bar is found by taking 1.78 times the P/A value for the narrower width.

(b) From Fig. 2-16 $\qquad\qquad K = 1.48.$

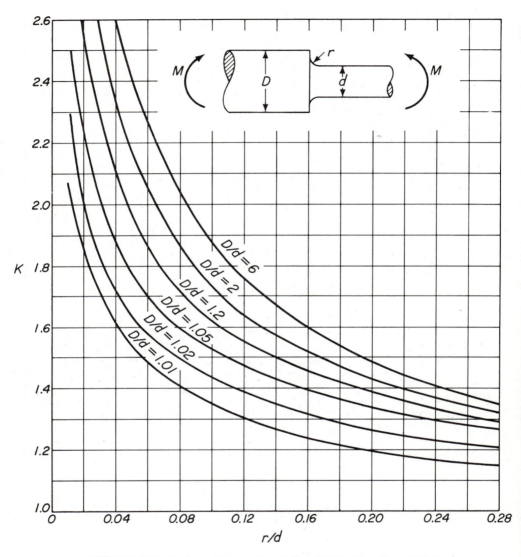

FIGURE 2-17   **Factors of stress concentration *K* for various sizes of fillets for round bar in bending to be applied to the stress in the section of diameter *d*.**

The maximum bending stress in the bar is found by taking 1.48 times the Mc/I value for the narrower width.

The stress concentration factor for this bar would be increased if the width of the left portion were increased. A decrease in fillet radius also causes an increase in the factor.

Example 4A. Work Example 4 if the minor width of the bar in Fig. 2-12 is 30 mm, the major width 54 mm, and the radius of the fillet is 6 mm.

Solution. (a) From the above, $D = 54$ mm, $d = 30$ mm, $r = 6$ mm, $D/d = 1.8$, and $r/d = 0.2$.

From Fig. 2-14: $\qquad\qquad\qquad\qquad K = 1.78$

    (b) From Fig. 2-16: $\qquad\qquad K = 1.48$

Example 5. Suppose the part of Example 4 is 0.5 in. thick and is loaded by a steady tensile force of 18,750 lb. The material is soft steel with a yield point value 45,000 psi. Find the value of the factor of safety based on the yield point.

Solution. By maximum shear theory: $\quad \tau_{yp} = 0.5 \times 45,000 = 22,500$ psi

$$A = 1.25 \times 0.5 = 0.625 \text{ in.}^2$$

In right portion: $\qquad\qquad\qquad \sigma = \dfrac{P}{A} = \dfrac{18,750}{0.625} = 30,000$ psi

By Eq. (3): $\qquad\qquad\qquad \tau_{max} = 0.5 \times 30,000 = 15,000$ psi

By Eq. (2): $\qquad\qquad\qquad F_s = \dfrac{\tau_{yp}}{\tau_{max}} = \dfrac{22,500}{15,000} = 1.5$

As found in Example 4, the stress concentration factor for the fillets is equal to 1.78. The stress thus reaches the yield point value on the fillets, and local yielding occurs in these regions. The stress for the cross section as a whole, however, remains at a safe value. It is thus justified to neglect the effect of stress concentration for ductile materials and steady loads.

In general, a stress concentration factor is applied to the stress computed for the net or smallest cross section.

The effect of notches on the stresses for tension can be found in Figs. 2-18 and 2-19, respectively. The stress concentration factors for a bar containing a circular hole[5] are shown in Fig. 2-20. If a circular hole in a plate contains a pin through which the load is applied, the stress concentration factors will be as shown in Fig. 2-21.

Inspection of the curves in Figs. 2-14 to 2-20 indicates that stress concentration factors are reduced by the use of larger fillets or by more gradual transitions from cross sections of one size to those of another. Sometimes the designer can specify the removal of material in such a way as to secure a more gradual transition in size. Thus in Fig. 2-22(b) it is easy to visualize

[5]See Frocht, M. M., "Factors of Stress Concentration Photoelastically Determined," *Trans. ASME,* **57**, 1935, p. A-67.

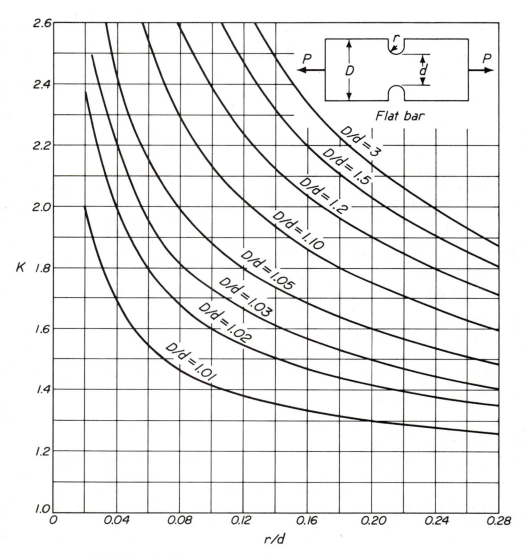

FIGURE 2-18 **Factors of stress concentration *K* for grooves of various depths in tension or compression to be applied to the stress in the section of the flat bar of width *d*.**

that the stress concentration would be less when the notches *B* are present than when the main notch stands alone. For the same reason, a bolt with a continuous thread shows fewer concentration effects than a bar with a single circumferential groove. The narrow projection of Fig. 2-22(d), into which the

FIGURE 2-19 Factors of stress concentration K for grooves of various depths for round bar in tension or compression to be applied to the stress in the section of diameter d.

force cannot spread, has less increase in stress than the wide projection of Fig. 2-22(c). It may be beneficial to use a stress-relieving groove, as shown in Fig. 2-22(f), on a shaft with a sudden change in diameter if it is not possible to use a fillet of suitable size at the junction. A reduction of stress concentra-

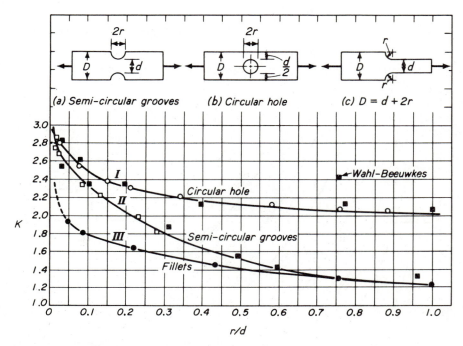

FIGURE 2-20   Invariant cases in tension and compression.

FIGURE 2-21   Stress concentration factors around a central circular hole
in a plate loaded through a pin in the hole.

FIGURE 2-22 Reduction of stress concentration by removal of material.

tion can be had by using fillets of elliptical shape as shown in Fig. 2-23. Fillets are needed only in regions of high stress. At points of low stress, undercuts may simplify machining and grinding operations. The designer can

FIGURE 2-23 Fillet of elliptical form.

frequently reduce the harmful effects of a stress concentration by carefully studying the details and by making minor changes in the outline of the parts.

11. Ductile Materials with Completely Reversing Stress

As mentioned at the beginning of Section 3, local yielding under steady load takes place if the yield point is exceeded at certain points of stress concentration. However, when the load is fluctuating, such local relief cannot be obtained, and a suitable stress concentration factor K must be applied. Fatigue failure from such loading will be by fracture.

When the load is alternating or completely reversing, the endurance-limit stress σ_e, as determined by testing, is the criterion used for determining the factor of safety. For such loading, the working stress can be called the range stress σ_r. Hence

$$F_s = \frac{\sigma_e}{K\sigma_r} \tag{10}$$

A similar equation could be written for completely reversed shearing stress.

Example 6. A Class E freight car axle, Fig. 2-24, carries a load of 52,500 lb. The material is medium-carbon forged steel. The Brinell hardness number is 180, and the ultimate strength is 90,000 psi. The material surface is in the as-forged condition.

Find the value of the F_s at the center of the axle as the car is proceeding along a smooth straight level track.

Solution.

$$I = \frac{\pi d^4}{64} = \frac{\pi 6.4375^4}{64} = 84.302 \text{ in.}^4$$

$$\sigma = \frac{Mc}{I} = \frac{26,250 \times 9.5 \times 3.219}{84.302} = 9,520 \text{ psi}$$

FIGURE 2-24   Example 6.

Since the bending stress is completely reversing, Eq. (10) will apply.

By Fig. 2-11: $\sigma_e = 19{,}500$ psi

By Eq. (10): $F_s = \dfrac{\sigma_e}{\sigma_r} = \dfrac{19{,}500}{9{,}520} = 2.05$

12. Ductile Materials with Combined Steady and Alternating Stress. Modified Goodman Diagram

In many strength problems, the major components of stress are static, with less accurately known alternating stresses superposed. Most failures originate with stresses of this type. The problem presents great difficulties because of the fundamentally different mechanisms of failure in the two sources of stress.

Suppose the tensile load P on the bar of Fig. 2-25(a) is continuously varying in magnitude. This load can be considered as being made up of two parts, the steady or average load P_{av}, and the variable or range load P_r. The maximum load is equal to the average load plus the range load; the minimum load is equal to the average minus the range load. Normal stresses σ_{av} and σ_r are found by dividing loads P_{av} and P_r by the cross-sectional area A. When the average stress is high, the material will safely carry only a small additional range component. However, if the average stress is small, a larger range component can be permitted.

To take care of the unlimited number of combinations of range and average stress, the line of failure of the material must be used. Specimens are tested with fluctuating loads that are low enough to permit continuous operation but high enough so that any increase in either the average or range load will eventually cause failure.

The results are plotted, σ_{av} horizontally and σ_r vertically, for each specimen tested. A sufficient number of test points gives the curve of failure for the material. It is a broad sweeping curve running from A to B in Fig. 2-25. For a static stress, failure occurs at the ultimate or point B in the figure. For a completely reversing stress (zero average), failure occurs at the endurance

FIGURE 2-25 Working stress diagram for fluctuating load.

limit or point A in Fig. 2-25. For combinations of average and range stress, failure occurs at intermediate points between A and B. Such experimentally determined curves in general are not available to the designer and an approximation must be used.

For this the curve ACD as drawn in the figure can be used. All points as located by average and range stress lying on ACD can be considered as equally safe if fatigue failure is not to occur. As mentioned previously, a stress concentration factor K, if present, will be applied only to the range component. Since a factor of safety is employed both components σ_{av} and $K\sigma_r$ must be multiplied by F_s to cause the resulting stress to fall on ACD, the assumed line of failure.

Thus the average and range components locating a point such as E are considered as being equivalent to the static stress OB or σ_{ult}. An equation for the equivalent static stress σ is easily derived. By similar triangles in Fig. 2-25:

$$\frac{FB}{EF} = \frac{OB}{OA} \quad \text{or} \quad FB = \frac{OB}{OA} \times FF = \frac{\sigma_{ult}}{\sigma_e} F_s K \sigma_r$$

Equivalent static stress: $\quad OB = OF + FB$

$$\sigma_{ult} = F_s \sigma_{av} + F_s K \frac{\sigma_{ult}}{\sigma_e} \sigma_r$$

$$\sigma = \frac{\sigma_{ult}}{F_s} = \sigma_{av} + K \frac{\sigma_{ult}}{\sigma_e} \sigma_r \tag{11}$$

or

$$\frac{\sigma_{av}}{\sigma_{ult}} + \frac{K\sigma_r}{\sigma_e} = \frac{1}{F_s} \tag{11a}$$

Example 7. A part with a machined surface has continuously varying tension loads $P_{max} = 45,000$ lb and $P_{min} = 15,000$ lb. Material tests $\sigma_{ult} = 90,000$ psi, and $\sigma_{yp} = 70,000$ psi. A stress concentration factor of 1.42 is present. Area of the part is 2.5 in.^{2} Find the factor of safety.

Solution. $\quad P_{av} = \dfrac{45,000 + 15,000}{2} = 30,000$ lb, $\qquad \sigma_{av} = \dfrac{30,000}{2.5} = 12,000$ psi

$$P_r = \frac{45,000 - 15,000}{2} = 15,000 \text{ lb}, \qquad \sigma_r = \frac{15,000}{2.5} = 6,000 \text{ psi}$$

By Fig. 2-11: $\quad \sigma_e = 34,000$ psi

A rough sketch indicates that the working stress point E will lie on the upper limb of the curve in Fig. 2-25.

By Eq. (11a): $\qquad\qquad \dfrac{12,000}{90,000} + \dfrac{1.42 \times 6,000}{34,000} = \dfrac{1}{F_s}$

$$0.1333 + 0.2506 = \frac{1}{F_s}$$

$$F_s = 2.60$$

Example 7A. A part with a machined surface has continuously varying tensile loads $P_{max} = 200,000$ N and $P_{min} = 68,000$ N. Material tests $\sigma_{ult} = 600$ MPa. A stress concentration factor of 1.42 is present. Area of the part is 1,600 mm^2. Find the factor of safety for the part

Solution.

$$P_{av} = \frac{200,000 + 68,000}{2} = 134,000 \text{ N} \qquad \sigma_{av} = \frac{134,000}{1,600} = 83.75 \text{ MPa}$$

$$P_r = \frac{200,000 - 68,000}{2} = 66,000 \text{ N} \qquad \sigma_r = \frac{66,000}{1,600} = 41.25 \text{ MPa}$$

$\sigma_{ult} = 145 \times 600 = 87,000$ psi

By Fig. 2-11: $\quad \sigma_e = 34,000$ psi $= 234$ MPa

By Eq. (11a): $\qquad\qquad \dfrac{83.75}{600} + \dfrac{1.42 \times 41.25}{234} = \dfrac{1}{F_s}$

$$0.1396 + 0.2503 = \frac{1}{F_s}$$

$$F_s = 2.56$$

When the average stress is greater the loading point may lie on the lower limb of the curve as at H in Fig. 2-25. Here, the equivalent static tension σ is equal to OD or σ_{yp}. Then

$$F_s(\sigma_{av} + K\sigma_r) = \sigma_{yp} \tag{12}$$

Example 8. Same data as for Example 7 except $P_{max} = 55,400$ lb and $P_{min} = 44,600$ lb. Find the F_s.

Solution. A sketch will indicate that the loading point will lie on the lower limb of the curve in Fig. 2-25.

$$P_{av} = \frac{55,400 + 44,600}{2} = 50,000 \text{ lb} \qquad \sigma_{av} = \frac{50,000}{2.5} = 20,000 \text{ psi}$$

$$P_r = \frac{55,400 - 44,600}{2} = 5,400 \text{ lb} \qquad \sigma_r = \frac{5,400}{2.5} = 2,160 \text{ psi}$$

By Eq. (12): $F_s(20,000 + 1.42 \times 2,160) = 70,000$

$$F_s = \frac{70,000}{23,067} = 3.03$$

13. Sensitivity to Stress Concentration

The actual reduction in fatigue strength, as indicated by the foregoing theoretical stress concentration factors, is approached only by large parts made of fine-grained heat-treated steel. The effect of stress concentration in coarse-grained annealed steels may be considerably less. Small specimens are affected less by stress concentration than larger parts made of the same material. The size effect in steel is attributed mainly to the grain size of the material. When the crystal size is taken into account, it is seen that there is not complete geometric similarity between large and small specimens of the same material. Although a heat-treated part of expensive alloy steel may have a higher endurance limit than one made of a softer nonheat-treated material, the advantage may be largely lost in the presence of a stress concentration.

A wide variation exists in the *notch sensitivity* of different materials. For some quenched and tempered steels the effect of a sharp notch may be so great that a high-strength material may be no better in fatigue than one of lower strength. Materials that work harden rapidly, such as the 18-8 stainless steels, may show great resistance to loss of fatigue strength due to notches. Notches have but little effect on the fatigue strength of cast iron, but may have a large effect as far as impact loads are concerned. However, the impact strength of some steels is but little affected by notches.

Methods are available for making a quantitative estimate of the sensitivity of a steel to stress concentration,[6] but the methods are beyond the scope of this book. However, when the full theoretical stress concentration factor is applied to the fluctuating component, the result will usually be on the safe side.

---

[6]See Peterson, R. E., "Relation between Life Testing and Conventional Tests of Materials," *ASTM Bull.*, **133**, Mar. 9, 1945. See also p. 8 of Reference 9, Chapter 4 of Reference 8, and Chapter 13 of Reference 11, end of chapter.

14. Interpretation of Service Fractures

The appearance of the fractured section gives information about the magnitude of the stress that caused the failure. For example, the fracture of a

shaft in bending is usually composed of smooth and coarse areas as shown in Fig. 2-26. The smooth area was caused by the opening and closing of the crack during its development. The coarse area was caused by the final rupture. A fatigue crack starts under cyclic loading if the stress exceeds the strength of the weakest grain on a cross section. Continued operation causes the crack to grow as the strength of adjacent grains is exceeded by the high stress at the end of the crack.

FIGURE 2-26   Fatigue failure of shaft.

If the stress was originally at a high value, the strength at a number of grains around the edge of the cross section was probably exceeded. Cracks started at these, and spread until they united with one another. The circumferential crack then progressed toward the center until final rupture occurred. It can therefore be concluded, when the ruptured area was near the center of the cross section, that the stress was considerably greater than the endurance limit for the material. Failure may have been caused by only a few hundred thousands of stress cycles.

When the final ruptured area was far to one side, the stress was probably only slightly above a safe value, so that the endurance limit was exceeded at but a single point on the boundary. Several million cycles may have been required to produce failure, and a small improvement in the design may make the part safe for indefinite operation.

It has been estimated that perhaps 90% of all service failures from mechanical causes are due to fatigue. Fatigue failures are particularly dangerous since they occur without warning and with no gross deformation at failure. A brittle-type fracture results.

The ability of a material to turn its mechanical energy into heat is called *internal friction* or *damping*. A high damping capacity is desirable in limiting the vibrational amplitude at resonance and thus preventing fatigue failure.

15. Brittle Materials with Steady Stress

Failure in brittle materials takes place by fracture. The ultimate strength is used as the basis for determining the working stress. It is necessary to have separate equations for the factor of safety in tension and for the factor of safety in compression. Brittle materials are unable to yield locally at points of high stress caused by changes of form, and stress concentration factors are customarily applied even when the loading is steady.

(a) *Simple Tension or Compression.* The following equations can be written.

For tension, σ_t:
$$F_s = \frac{\sigma_{ut}}{\sigma_t} \tag{13}$$

where σ_{ut} represents the ultimate stress in tension.

For compression, σ_c:
$$F_s = \frac{\sigma_{uc}}{\sigma_c} \tag{14}$$

where σ_{uc} represents the ultimate stress in compression.

(b) *Normal Stress in Two Directions.* Two cases must be considered.

(1) Both principal stresses of same sign. In accord with Figs. 2-7 and 2-8, failure is assumed to be due only to the principal stress of larger magnitude without regard to the stress at right angles thereto. Equations (13) and (14) apply.

(2) Principal stresses have opposite signs as shown in Fig. 2-27. The problem is as yet poorly understood. The best-known rational method is due to Mohr and is based on maximum shear stress theory, which yields the following equation.[7]

$$\frac{\sigma_1}{\sigma_{ut}} + \frac{\sigma_2}{\sigma_{uc}} = \frac{1}{F_s} \tag{15}$$

In using this equation, σ_{uc} must be substituted as a negative number. Stress concentration factors can be applied as shown in the following example.

FIGURE 2-27 Brittle material with tension and compression.

Example 9. Assume that the computed stresses at a point in a cast-iron body are as follows: $\sigma_x = 2,000$ psi, $\sigma_y = -6,000$ psi, and $\tau_{xy} = 3,000$ psi. The stress concentration factor is equal to 2 for all stresses. Material tests $\sigma_{ut} = 20,000$ psi and $\sigma_{uc} = -80,000$ psi. Find the value of the factor of safety.

Solution. In Eq. (41), Chapter 1: $K\sigma_1 = 2(-2,000 + \sqrt{4,000^2 + 3,000^2})$.

$$= 6,000 \text{ psi}$$

In Eq. (42), Chapter 1: $K\sigma_2 = 2(-2,000 - \sqrt{4,000^2 + 3,000^2})$

$$= -14,000 \text{ psi}$$

---

[7]For derivation, see earlier editions of this book. See also p. 438 of Reference 3, end of chapter.

In Eq. (15):

$$\frac{6,000}{20,000} + \frac{-14,000}{-80,000} = \frac{1}{F_s}$$

$$F_s = \frac{80}{38} = 2.1$$

Example 9A. Assume that the computed stresses at a point in a cast-iron body are as follows: $\sigma_x = 14$ MPa, $\sigma_y = -42$ MPa, and $\tau_{xy} = 21$ MPa. The stress concentration factor is equal to 2 for all stresses. Brittle material tests $\sigma_{ut} = 140$ MPa and $\sigma_{uc} = -560$ MPa. Find the value of the factor of safety.

Solution. In Eq. (41), Chapter 1: $\quad K\sigma_1 = 2\left[\frac{14-42}{2} + \sqrt{\left(\frac{14+42}{2}\right)^2 + 21^2}\right]$

$$= 2(-14 + 35) = 42 \text{ MPa}$$

In Eq. (42), Chapter 1: $\quad K\sigma_1 = 2(-14 - 35) = -98 \text{ MPa}$

In Eq. (15): $\quad\quad \frac{42}{140} + \frac{-98}{-560} = \frac{1}{F_s}$

$$F_s = \frac{560}{266} = 2.1$$

The mechanism of failure for brittle materials is very complex. Unfortunately, Eq. (15) does not fit test results very well, and it should therefore be considered only as a rough approximation.

(c) *Pure Shear.* Equation (15) is applicable to pure shear loading by taking $\sigma_1 = -\sigma_2 = \tau$ as indicated by Fig. 2-5.

16. Brittle Materials with Fluctuating Loads

Although successful applications can be cited, brittle materials are usually considered unsatisfactory where the load is fluctuating. Large values for the factor of safety must be used.

17. Loading in the Finite Life Range

As previously demonstrated, when stresses σ_{av} and $K\sigma_r$ are sufficient to cause the point to fall on the *ACD* line in Fig. 2-25, the material was deemed safe for continuous operation so far as fatigue failure was concerned.

However, suppose the loading is heavier so as to cause the stress point to fall outside the *ACD* line as occurs for point *M* in Fig. 2-28. As before, σ_{av} has been plotted horizontally and $K\sigma_r$ vertically. Fatigue failure can now be expected after some finite number of stress reversals.

As shown in the figure, for static loading failure will occur at stress σ_{ult}, and for a completely reversing load the safe stress is the endurance limit stress σ_e. For loading in the finite range, the completely reversing

FIGURE 2-28 Method for determining equivalent completely reversed stress σ_R for loading in the finite life range.

stress, equivalent to the combined loading at M, is considered to be ON or σ_R. This stress is found by extending the straight line from σ_{ult} through M onward until it intersects with the vertical axis. An equation for finding σ_R can be easily derived by forming a proportion of corresponding sides of similar triangles in Fig. 2-28. It is seen that

$$\frac{\sigma_R}{\sigma_{ult}} = \frac{K\sigma_r}{\sigma_{ult} - \sigma_{av}} \qquad \text{or} \qquad \frac{K\sigma_r\sigma_{ult}}{\sigma_{ult} - \sigma_{av}} \qquad (16)$$

Example 10. In Fig. 2-29 let $\sigma_{av} = 25{,}000$ psi, $\sigma_r = 21{,}667$ psi, and stress con-

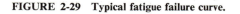

FIGURE 2-29 Typical fatigue failure curve.

centration factor $K = 1.5$. The material has $\sigma_{ult} = 90,000$ psi and $\sigma_e = 40,000$ psi. Find the value of the equivalent completely reversing stress σ_R.

Solution. By Eq. (16): $\sigma_R = \dfrac{1.5 \times 21,667 \times 90,000}{90,000 - 25,000} = 45,000$ psi

For problems in the finite life range, stresses σ_{av} and $K\sigma_r$ will be transformed into σ_R. Stress σ_R will then be used in the solution of problems.

18. *The Fatigue Failure Curve*

Tests are made to determine the number of reversals L to failure at the completely reversed stress σ_r and the results plotted. A curve similar to Fig. 2-29 results. The curve approaches asymptotically to the horizontal whose ordinate is the so-called endurance limit stress σ_e. A considerable amount of scatter is usually unavoidable in test results. As a conservative estimate the *RS* portion of the curve is assumed to join the horizontal at 10^6 cycles.

The *RS* curve plots into a straight line on log-log paper as is done in Fig. 2-30. For the *RS* curve, or the finite life range less than 10^6 cycles, it is observed that the relationship between completely reversed stress σ_R and life L, the number of reversals to failure, can be represented by the exponential equation

$$A = \sigma_R L^B \tag{17}$$

This is known as Basquin's equation. Constants A and B can be determined from experimental fatigue curves. Such curves are observed approximately to pass through the point at $0.9\sigma_{ult}$ and $L = 1,000$ reversals and σ_e and $L = 10^6$ reversals. As before σ_{ult} is the ultimate stress in tension and σ_e is the endurance limit in completely reversed bending. Determination of constants A and B is illustrated by the following example.

Example 11. The endurance limit, for a ground specimen from a steel of 90,000 psi tensile strength, is 40,000 psi. Find the value of constants A and B.

Solution. The failure curve, on log-log paper as mentioned above, extends from $L = 1,000$ cycles and $\sigma_R = 0.9 \times 90,000 = 81,000$ psi to $L = 10^6$ cycles and the endurance limit stress $\sigma_e = 40,000$ psi as shown in Fig. 2-30. Equation (16) can be written

$$\log A = \log \sigma_R + B \log L$$

The given values for σ_R and L are substituted.

$$\log A = \log 81,000 + B \log 1,000$$
$$\log A = \log 40,000 + B \log 1,000,000$$

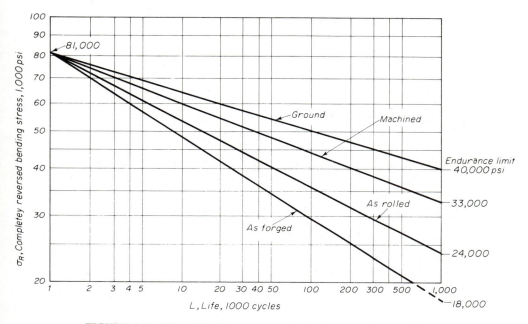

FIGURE 2-30 Typical fatigue curves: steel specimens; reversed bending; σ_{ult} 90,000 psi. Relationship between equivalent completely reversed stress σ_R and fatigue life L.

The values of the common logarithms of σ_R and L are substituted.

$$\log A = 4.90849 + 3B$$

$$\log A = 4.60206 + 6B$$

Solution of these two equations yields

$$\log A = 5.21492$$

$$B = 0.10214$$

Equation (16) then becomes

$$5.21492 = \log \sigma_R + 0.10214 \log L \qquad (18)$$

The straight line running from $\sigma_R = 81,000$ psi, $L = 1,000$, and $\sigma_e = 40,000$ psi, $L = 10^6$ represents the life L for any given value of σ_R as determined by the solution to the above example.

Example 12. Suppose as in Example 10 that the stresses σ_{av} and $K\sigma_r$ for the part are such that they give a σ_R of 45,000 psi. Calculate the expected number of reversals to failure and check by Fig. 2-30.

Solution. For $\sigma_R = 45{,}000$ psi

By Eq. (18):
$$5.21492 = \log \sigma_R + 0.10214 \log L$$
$$5.21492 = 4.65321 + 0.10214 \log L$$
$$\log L = 5.49941$$
$$L = 315{,}800 \text{ reversals}$$

The result agrees with Fig. 2-30. The curve thus gives a rough check but logarithms must usually be used to obtain sufficient accuracy.

Surface irregularities on the specimen cause a reduction in the fatigue strength. The top curve in Fig. 2-30 is for a ground surface. An identical specimen but with a machined surface would have a shortened fatigue life. Still lower values would apply to as-rolled and as-forged surfaces. The endurance limit stress for such surfaces can be had from the ultimate stress σ_{ult} and the factors in Table 2-2. Corresponding lines for these surfaces have been drawn in Fig. 2-30. Additional diagrams like Fig. 2-30 can be easily made when the ultimate strength of the steel is known using the constants of Table 2-2.

TABLE 2-2 Endurance Limit, σ_e, for Various Types of Surfaces

Surface	$\dfrac{\sigma_e}{\sigma_{ult}}$
Ground	0.44
Machined	0.37
As-rolled	0.27
As-forged	0.20

19. Miner's Equation

A machine part may have one stress for a portion of its life, another stress for another portion, and so on, as illustrated in Fig. 2-31. Each combination of σ_{av} and σ_r can be reduced to its equivalent completely reversed stress, σ_{R1}, σ_{R2}, σ_{R3}, and so on.

Let fatigue life for stress σ_{R1}, if applied alone, be L_1 reversals and let N_1 reversals at this stress be applied. Stress σ_{R1} has then consumed N_1/L_1 of the life of the part.

Similarly, let L_2 be the fatigue life at stress σ_{R2} and let N_2 reversals at this stress be applied. Stress σ_{R2} has thus consumed N_2/L_2 of the life of the part.

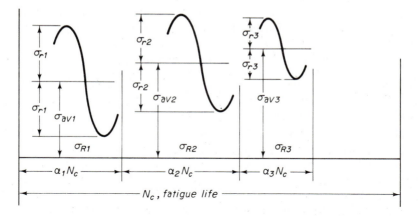

FIGURE 2-31 Average and range stresses in different parts of a load cycle. σ_R is equivalent completely reversed stress.

A similar expression can be written for each portion of the stress cycle. The sum of the proportions constitutes the entire life or unity.

$$\frac{N_1}{L_1} + \frac{N_2}{L_2} + \frac{N_3}{L_3} + \cdots = 1 \qquad (19)$$

Let the total number of reversals to failure for the combined loadings be called L_c. Then N_1 can be represented by $\alpha_1 N_c$ where α_1 is an appropriate constant. Similarly, $N_2 = \alpha_2 N_c$, $N_3 = \alpha_3 N_c$, and so on. Let these be substituted into Eq. (19) and N_c be shifted to the right.

$$\frac{\alpha_1}{L_1} + \frac{\alpha_2}{L_2} + \frac{\alpha_3}{L_3} + \cdots = \frac{1}{N_c} \qquad (20)$$

Since the sum of the αN_c terms is equal to N_c, it is obvious that the sum of the α's is equal to unity. This is known as Miner's equation.

Letter N in the SI refers to force, newtons, but the letter N has other long-established meanings in engineering. As used in the Miner equation derived above it denotes number of reversals of stress. The reader must carefully keep this in mind wherever letter N appears and be sure he is giving it the correct meaning. In this chapter, when the force newton must be designated it will be spelled out in full.

Example 13. The material is the same as for previous examples, namely σ_{ult} = 90,000 psi and σ_e = 40,000 psi for a ground surface. Suppose the loading is as follows where $K = 1.5$:

$$\sigma_{av} = 45,000 \text{ psi}, \ \sigma_r = 13,641 \text{ psi for 0.8 of the time}$$

$$\sigma_{av} = 35,000 \text{ psi}, \ \sigma_r = 20,787 \text{ psi for 0.2 of the time}$$

Find the expected number of reversals L_c to failure.

Solution.

By Eq. (16):
$$\sigma_{R1} = \frac{K\sigma_r\sigma_{ult}}{\sigma_{ult} - \sigma_{av}} = \frac{1.5 \times 13,641 \times 90,000}{90,000 - 45,000} = 40,923 \text{ psi}$$

$$\sigma_{R2} = \frac{1.5 \times 20,787 \times 90,000}{90,000 - 35,000} = 51,023 \text{ psi}$$

$$B \log L = \log A - \log \sigma_R$$

By Example 9: $B = 0.10214$, $\log A = 5.21492$

By Eq. (18): $0.10214 \log L = 5.21492 - \log \sigma_R$

For $\sigma_{R1} = 40,923$ psi: $0.10214 \log L_1 = 5.21492 - 4.61197 = 0.60295$

$$\log L_1 = 5.90317$$

$$L_1 = 800,150 \text{ reversals}$$

For $\sigma_{R2} = 51,023$ psi: $0.10214 \log L_2 = 5.21492 - 4.70777 = 0.50715$

$$\log L_2 = 4.96524$$

$$L_2 = 92,310 \text{ reversals}$$

By Eq. (20):
$$\frac{\alpha_1}{L_1} + \frac{\alpha_2}{L_2} = \frac{1}{L_c}$$

$$\frac{0.8}{800,150} + \frac{0.2}{92,310} = \frac{1}{L_c}$$

$$0.000\,000\,9998 + 0.000\,002\,1666 = \frac{1}{L_c}$$

$$0.000\,003\,1664 = \frac{1}{L_c}$$

$$L_c = 315,800 \text{ reversals}$$

20. Fatigue Life Determined by Short-Time Testing

It is widely felt that analytical results should be verified whenever possible by life testing. Some machines are designed for a very long life and such testing may be both inconvenient and expensive. The testing time can be reduced if a portion of the test is done at the service stress and the remainder is done at some exclusively high stress which produces failure in a shorter time. Miner's equation is employed using two terms.

$$\frac{N_1}{L_1} + \frac{N_2}{L_2} = 1 \tag{21}$$

Here N_1 is number of reversals at the service stress first placed on the part which has a service life L_1 reversals, which is to be found. The part is then run at the high stress which would have the life L_2 if used alone. The additional number of reversals at failure at the higher stress is N_2.

Example **14.** Same material as previously used, namely $\sigma_{ult} = 90,000$ psi, $\sigma_e = 40,000$ psi. Ground surface with stress concentration factor 1.5 at point of expected failure.

For a short time test for this part, the service stress of $\sigma_{R1} = 45,000$ psi was first applied for $N_1 = 15,000$ reversals. Then a stress σ_{R2} of 60,000 psi was applied. After $N_2 = 17,991$ additional reversals the part failed. Find the expected life L_1 of the part.

Solution. Use Eq. (21): $\dfrac{N_1}{L_1} + \dfrac{N_2}{L_2} = 1$

where $N_1 = 15,000$ reversals of $\sigma_{R1} = 45,000$ psi was first applied

   $L_1 =$ service life of part; to be found

   $N_2 =$ additional reversals before failure at the higher stress $\sigma_{R2} = 60,000$ psi

   $L_2 =$ life at stress σ_{R2} if used alone

As before, $\log A = 5.21492$, $B = 0.10214$.

By Eq. (18): $0.10214 \log L_2 = 5.21492 - \log \sigma_{R2}$

   $0.10214 \log L_2 = 5.21492 - 4.77815$

   $\log L_2 = 4.27619$

   $L_2 = 18,888$ reversals, life at stress σ_{R2}

By Eq. (21): $\dfrac{15,000}{L_1} + \dfrac{17,991}{18,888} = 1$

   $\dfrac{15,000}{L_1} = 1 - 0.95251 = 0.04749$

   $L_1 = 315,800$ reversals, expected service life

Sections 17 to 20 represent an attempt to make designs in steel for the finite life range taking account of ultimate strength, endurance limit, surface roughness, fluctuations in loading, and stress concentration. Any combination of the variables can be used. In fatigue loading there is always more scatter in the results than when the loading is static, but the method presented here should form a better basis of design than conventional methods using merely a factor of safety and making no attempt to predict the life of the part.

21. Factor of Safety

It is sometimes difficult to evaluate accurately the various factors that are involved in an engineering design problem. It is particularly difficult, in some cases, to determine the magnitude of the various forces to which a machine part is subjected. Sometimes the shape of the part is such that no design equations are available for accurate computation of the stresses. Variations and nonuniformity in the strength of the material must be kept in mind, as well as the consequences that might result from failure of the parts.

Engineers employ a so-called "factor of safety" to insure against uncertain or unknown conditions as mentioned above. The working stress is determined from the yield or ultimate strength of the material by means of a factor of safety, as demonstrated by the equations of this chapter. For columns and other elements where load and stress are not proportional, the factor of safety should be applied to the load on the member rather than to the stress.

Static loads can sometimes be determined accurately, but the values of fluctuating loads are more uncertain. The effects of impact loads and residual stress are particularly difficult to evaluate. Consideration must also be given to the long-term effects of corrosion and high temperatures. Sometimes the shape of the parts, or the method of support, can be modified in such a way that the design equations may fit the conditions more accurately. Perhaps a more uniform and reliable construction material can be used.

The strength of materials is usually obtained by laboratory tests. It must be kept in mind, however, that conventional laboratory tests rarely reproduce the conditions that the material must meet in service. Surface conditions in particular may be different for the test specimen than for an actual part. A much greater amount of knowledge of engineering materials than is now available must be gained before the designer will be able to apply test data to conditions that differ appreciably from the conditions of the test. The best method, whenever it is possible, is to make final adjustment in the proportioning of the components by tests on the completed product.

In general it is not economical to use safety factors large enough to eliminate all possibility of failure resulting from the worst possible combination of circumstances. The designer attempts to reduce the probability of failure to a suitable level, which necessarily depends on each particular application. A failure that involves only a little inconvenience or loss of time might be allowed more frequently than one involving large financial loss or human life. Provision should be made for easy and rapid replacement of failed parts. If the product operates under conditions of frequent service inspection, smaller values for the factor of safety can sometimes be used. A more thorough and detailed analysis of the problem may show that smaller factors of safety can be used, and may justify the additional engineering expense involved.

When building one general product, the engineer usually does not think in terms of the factor of safety. He has learned from experience that certain materials under certain conditions, with working stresses of given values, will lead to satisfactory results. Although the determination of suitable values for the factor of safety is a matter of great importance, the subject has been much neglected and is in an unsatisfactory state. Experience, which can only be accumulated as a result of a long period of trial and error, is the ultimate basis for the prediction of failure in engineering designs. Many failures are due to circumstances that the designer failed to consider.

22. Another Form for the Modified Goodman Diagram

The modified Goodman diagram of Fig. 2-25 can be arranged in a different form as shown in Fig. 2-32. Here average stresses are equal to the abscissa and ordinate for a point on a line drawn at 45° from the origin. Point D then represents the yield point stress and point B the ultimate. The horizontal DC corresponds to line DC of Fig. 2-25. Stresses $K\sigma_r$ are plotted vertically above and below point F which has the given value for σ_{av}. Stresses σ_{max} and σ_{min} can thus be read directly from the diagram. Figure 2-32 has been made with F_s equal to unity.

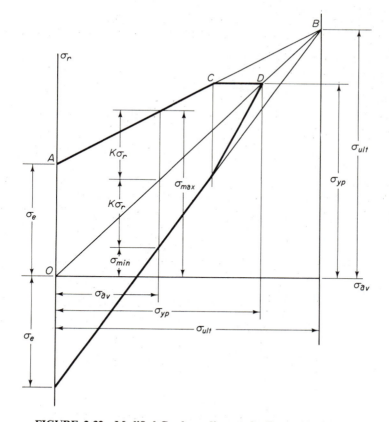

FIGURE 2-32   Modified Goodman diagram for fluctuating stress.

REFERENCES

1. BATTELLE MEMORIAL INSTITUTE, *Prevention of Fatigue of Metals*, New York: John Wiley & Sons, Inc., 1941.

2. FINDLEY, W. N., "Fatigue of Metals under Combinations of Stresses," *Trans. ASME*, **79**, 1957, p. 1337.

3. HETÉNYI, M., Ed., *Handbook of Experimental Stress Analysis*, New York: John Wiley & Sons, Inc., 1950.

4. HORGER, O. J., Ed., *Metals Engineering Design, ASME Handbook*, New York: McGraw-Hill Book Company, 1953.

5. JUVINALL, R. C., *Engineering Considerations of Stress, Strain and Strength*, New York: McGraw-Hill Book Company, 1967.

6. MARIN, J., *Mechanical Properties of Engineering Materials*, Englewood Cliffs, N.J.: Prentice-Hall, Inc., 1962.

7. MINER, M. A., "Cumulative Damage in Fatigue," *Trans. ASME*, **67**, 1945, *J. Appl. Mech.*, p. A-159.

8. MURRAY, W. M., Ed., *Fatigue and Fracture of Metals*, New York: John Wiley & Sons, Inc., 1952.

9. PETERSON, R. E., *Stress Concentration Design Factors*, New York: John Wiley & Sons, Inc., 1974.

10. *Proc. Int. Conf. Fatigue Metals*, London: Institution of Mechanical Engineers, 1956.

11. SINES, G., and J. L. WAISMAN, Eds., *Metal Fatigue*, New York: McGraw-Hill Book Company, 1959.

12. SPOTTS, M. F., *Mechanical Design Analysis*, Englewood Cliffs, N.J.: Prentice-Hall, Inc., 1964.

13. STARKEY, W. L., and S. M. MARCO, "Effects of Stress-Time Cycles on the Fatigue Properties of Metals," *Trans. ASME*, **79**, 1957, p. 1329.

PROBLEMS

1. Find the value of the stress at each hole in Fig. 2-33.

    Ans. A, $K\sigma = 29{,}710$ psi; B, $K\sigma = 31{,}330$ psi; C, $K\sigma = 35{,}680$ psi.

FIGURE 2-33   **Problem 1.**

2. The loading on an element of 1045 steel in the hot-rolled condition is shown in Fig. 2-34. Loads are steady and no stress concentrations are present. Assume maximum shear theory to be valid. Take σ_{yp} as 400 MPa.

(a) Find the value of the F_s for the *CDEF* plane.

(b) Find the value of the F_s for the *ACGE* plane.

(c) Find the value of the F_s for the *AFGD* plane.

Ans. (a) $F_s = 3.64$; (b) $F_s = 5.33$; (c) $F_s = 11.4$.

FIGURE 2-34 Problem 2.

3. A plate of 1045 steel in the hot-rolled condition is subjected to the following stresses: $\sigma_x = 3{,}300$ psi; $\sigma_y = -29{,}000$ psi; $\tau_{xy} = 0$.

(a) Find the value of the F_s by the maximum shear theory.

(b) Find the value of the F_s by the Mises–Hencky theory.

(c) Find F_s if the plate is made of Class 25 cast iron.

Ans. (a) $F_s = 1.83$; (b) $F_s = 1.92$; (c) $F_s = 2.37$.

4. A machine part has an F_s of 3 by the Mises-Hencky theory. The material tests $\sigma_{yp} = 435$ MPa. If σ_x is equal to 165 MPa, find the value of σ_y. $\tau_{xy} = 0$.

Ans. $\sigma_y = 107.1$ or 57.9 MPa.

5. A shaft is loaded by a torque of 40,000 in. lb. The material has a yield point of 50,000 psi. F_s is equal to 2.

(a) Find the required diameter by the maximum shear theory.

(b) Find the required diameter by the Mises–Hencky theory.

Ans. (a) $d = 2.535$ in.; (b) $d = 2.417$ in.

6. A 380 × 380-mm steel plate of material with $\sigma_{yp} = 400$ MPa has normal stresses only acting on all edges. Stress σ_x is tension and σ_y is compression. The length in the *y*-direction is reduced by 0.2 mm. F_s is equal to 2 by the maximum shear theory. Find the values of the stresses.

Ans. $\sigma_x = 135.3$ MPa; $\sigma_y = -64.7$ MPa.

7. A 12 × 12-in. steel plate has normal stresses only acting on all edges. Stress σ_x equals 12,000 psi tension. σ_y is compression. Increase in length in the *x*-direction is 0.006 in. If F_s is equal to 2.5 by the maximum shear theory, what is the yield-point value for the material?

Ans. $\sigma_{yp} = 55{,}000$ psi.

8. A 380 × 380-mm plate of 1045 steel in the as-rolled condition has normal stresses only acting on all edges. Stress σ_x equals 40 MPa tension. σ_y is compression. For an F_s equal to 3 by the maximum shear theory, find the change in length in the *x*-direction. Take σ_{yp} as 400 MPa.

Ans. $\delta_x = 0.129$ mm.

9. The link shown in Fig. 2-35 is subjected to a completely reversing load of 20,000 lb. Find the maximum value of stress at each hole.

Ans.   Top, $K\sigma_r = 17,880$ psi; bottom, $K\sigma_r = 17,540$ psi.

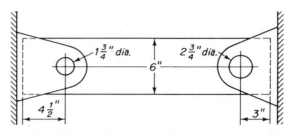

FIGURE 2-35   Problem 9.

10. The plate shown in Fig. 2-36 is 13 mm thick. Load P varies from 90,000 N to 45,000 N. Material tests $\sigma_{yp} = 290$ MPa; $\sigma_e = 165$ MPa. Factor of safety based on yield point is 2. Find the maximum permissible value of width D.

Ans. $D = 90$ mm.

FIGURE 2-36   Problem 10.   **FIGURE 2-37   Problem 11.**

11. The plate of Fig. 2-37 is 1 in. thick. The load varies from 50,000 to 30,000 lb. Factor of safety $= 2$; $\sigma_{yp} = 40,300$ psi; $\sigma_e = 28,000$ psi.

(a) Find the value of d.

(b) Find d if the minimum load is 20,000 lb, other data remaining unchanged.

Ans. $d = 2.67$ and 3.24 in.

12. Find the diameter of the hole and the total width of the plate of Fig. 2-38 if the part is to be safe for continuous operation. The load varies from 160,000 N to 89,000 N. Use stress values of $\sigma_{ult}/F_s = 280$ MPa and $\sigma_e/F_s = 110$ MPa. The plate is 25 mm thick.

Ans. $D = 56$ mm.

FIGURE 2-38   Problem 12.

FIGURE 2-39 Problem 13.

13. A part is designed as shown in Fig. 2-39. Check the design by plotting the line of failure and points representing the stress values for the material at the hole and fillet. Is the part safe for continuous operation? The load varies from 12,000 to 2,000 lb. $\sigma_{yp} = 41,000$ psi; $\sigma_e = 28,500$ psi. Plate is $\frac{7}{16}$ in. thick.

14. Figure 2-40 shows a nonrotating shaft with load P varying from 4,500 N to 13,500 N. The material tests $\sigma_{ult} = 500$ MPa, $\sigma_{yp} = 290$ MPa, and $\sigma_e = 130$ MPa. Factor of safety is equal to 2. Find the permissible value for D if stress conditions at the fillets are to be satisfactory for continuous operation. *Ans.* $D = 142.5$ mm.

FIGURE 2-40 Problem 14. **FIGURE 2-41 Problem 15.**

15. Material for the eyebolt of Fig. 2-41 tests $\sigma_{ult} = 52,000$ psi, $\sigma_{yp} = 39,000$ psi and $\sigma_e = 18,000$ psi. The factor of safety equals 2. Threads are American National. Let stress concentration factor for the threads be equal to 2.5. Upon assembly, the spring is given an initial stretch. During operation, the lower end of the spring moves 1 in. each way from its initial position. Find the permissible amount of initial stretch that may be given to the spring if the eyebolt is to be safe for continuous operation.   *Ans.*   5.05 in.

16. The shaft of Fig. 2-42 rotates but carries no torque. For the material, σ_e = 240 MPa. Determine the value of the bending stress at the fillet.

(a) Is the shaft safe for continuous operation? What is the value of the factor of safety?

(b) Suppose in the turning operation the radius of the fillet is made 3.2 mm and that the inspectors do not discover the mistake. Is the part now safe for continuous operation?   *Ans.*   (a) $F_s = 1.45$.

FIGURE 2-42   Problem 16.

17. The shaft in Fig. 2-43 rotates. Find the length l if the bending stress at the fillet is equal to the stress at the center.   *Ans.* $l = 91.4$ in.

FIGURE 2-43   Problem 17.

18. The part in Fig. 2-44 is made of 1045 steel, quenched and drawn, $\sigma_{ult} = 820$ MPa, $\sigma_{yp} = 620$ MPa. $\sigma_e = 300$ MPa. Bending moment varies from 1,000,000 to 5,000,000 Nmm; machined surfaces. Find the factor of safety.   *Ans.* $F_s = 1.80$.

FIGURE 2-44   Problem 18.   **FIGURE 2-45   Problem 19.**

19. The tensile load on the turned bolt of Fig. 2-45 varies from 12,000 to 20,000 lb. The stress concentration factor for the threads is equal to 3 and for the fillet, 1.2. Material is 8742 steel, oil quenched and drawn at 1,200°F. If threads are safe for continuous operation, find the value of area A_1 necessary to make the fillet safe for continuous operation.   *Ans.* $A_1 = 0.360$ in.^{2}

20. The beam shown in Fig. 2-46 is made of a brittle material. Find the value of the maximum bending stress at cross section A.   *Ans.* $\sigma = 45$ MPa.

FIGURE 2-46   **Problem 20.**

21. The plates in Fig. 2-47 are made of a brittle material and are identical. The loading for plate (a) consists of a uniform tension applied at each end. In plate (b) the hole contains a pin that carries the load and a uniform tension is applied to the other end of the plate. Find the maximum tensile stress for each type of loading. Note that loading through a pin as for (b) gives a higher stress. $H > D$ in Fig. 2-11.   *Ans.*   (a) $K\sigma = 4,890$ psi; (b) $K\sigma = 6,110$ psi.

FIGURE 2-47   **Problem 21.**

22. The plates shown in Fig. 2-48 are made of a brittle material. Assume each rivet transfers one-half the load. Find the value of the maximum tensile stress in the thicker plate.   *Ans.* $\sigma = 68.25$ MPa.

FIGURE 2-48   Problem 22.

23. Find the value of the maximum stress on the fillet in Fig. 2-49 if the stress concentration factor is equal to 1.6. What is the F_s if the part is made of Class 25 cast iron?   *Ans.* $F_s = 3.12$.

FIGURE 2-49   Problem 23.

24. Find the value of the maximum stress on the fillet in Fig. 2-50 if the stress concentration factor is equal to 1.75. $D/d = 1.5$. What is the F_s if the part is made of cast iron; $\sigma_{ut} = 200$ MPa?   *Ans.* $F_s = 3.58$.

FIGURE 2-50   Problem 24.

25. (a) Find the value of the F_s if the part in Fig. 2-51 is made of 1045 steel in the as-rolled condition.

(b) Find the value of the F_s if the same part in Fig. 2-51 is made of Class 20 cast iron.

FIGURE 2-51   Problem 25.

26. The element in Fig. 2-52 is located at the inner edge of a hub that has been press fitted on a shaft. Stress σ_r or p is equal to -40 MPa, and stress σ_t is equal to 70 MPa.

(a) Find the F_s by maximum shear theory if the hub is made of steel with a yield point value of 400 MPa.

(b) Find the F_s if the hub is made of cast iron, with $\sigma_{ut} = 170$ MPa and $\sigma_{uc} = 680$ MPa.   *Ans.*   (a) $F_s = 3.64$; (b) $F_s = 2.12$.

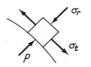

FIGURE 2-52   Problem 26.

27. The state of stress for a material is shown in Fig. 2-53.

(a) Find the F_s by the maximum shear theory if the material is 1045 steel in the hot-rolled condition.

(b) Find the F_s by the Mises–Hencky theory.

(c) Find the F_s if the material is Class 30 cast iron.

Ans. (a) $F_s = 4.54$; (b) $F_s = 5.00$; (c) $F_s = 2.77$.

$\sigma_x = 6,000\ psi$

$\tau_{xy} = 6,000\ psi$

$\sigma_y = 1,000\ psi$

FIGURE 2-53   Problem 27.

28. Same as Problem 27 except for the element of Fig. 2-54. For part (a) take $\sigma_{yp} = 400$ MPa and for part (c) take $\sigma_{ut} = 200$ MPa and $\sigma_{uc} = 760$ MPa.

Ans. (a) $F_s = 6.0$; (b) $F_s = 6.7$; (c) $F_s = 3.8$.

$\sigma_x = 40\ MPa$

$\tau_{xy} = 20\ MPa$

$\sigma_y = -13.5\ MPa$

FIGURE 2-54   Problem 28.   **FIGURE 2-55   Problem 29.**

29. The shaft in Fig. 2-55 is rigidly attached to the wall and is made of Class 25 cast iron. Stress concentration factor at the wall for bending and torsion is equal to 2. Find the value of the F_s.   *Ans.* $F_s = 3.08$.

30. The hollow cylinder in Fig. 2-56 is fixed at the wall and has an OD of 150 mm. and an ID of 113 mm. Stress concentration factor at wall is equal to 3.

(a) Find value of F_s if material is cast iron, $\sigma_{ut} = 172$ MPa, $\sigma_{uc} = 690$ MPa.

(b) Find value of F_s if material is 0.33 percent carbon cast steel normalized. $\sigma_{yp} = 345$ MPa.   *Ans.* (a) $F_s = 3.81$; (b) $F_s = 19.8$.

FIGURE 2-56 Problem 30.

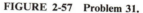

FIGURE 2-57 Problem 31.

31. The shaft in Fig. 2-57 is made of Class 20 cast iron. The ends are simply supported, but are keyed against rotation. The stress concentration factor at bracket is equal to 2. Find the F_s for the shaft on either side of the bracket.

Ans. Left, $F_s = 6.07$; right $F_s = 2.82$.

32. Find the diameter of the hole in Fig. 2-58 if the stress concentration factor there is to be the same as at the fillet. *Ans.* Dia. $= 46.7$ mm.

FIGURE 2-58 Problem 32.

FIGURE 2-59 Problem 33.

33. Find the value of the fillet radius r_1 on the left in Fig. 2-59 if the stress is to be the same as on the fillet on the right.

34. In Fig. 2-55, let the 15-in. dimension become 300 mm, and the 20-in. dimension become 250 mm. Diameter is 50 mm. Shaft material tests $\sigma_{yp} = 400$ MPa. Find the permissible load at end of arm if F_s by the Mises–Hencky theory is to be 2.5. *Ans.* $P = 5,030$ N.

35. A steel plate 20 × 10 in. has normal stresses only on the four edges. After the loads are applied, the dimensions become 20.0066 and 9.9981 in. Material is 1060 hot-rolled steel. Find the F_s by the maximum shear theory.

36. A 50-mm-diam nonrotating shaft of hot-rolled steel $\sigma_{yp} = 400$ MPa has a steady bending moment of 1,000,000 Nmm. Find the permissible steady torque that can be superposed on the bending moment if the F_s is to be 3 by the Mises–Hencky theory. *Ans.* $T = 1,495,000$ Nmm.

37. An element of 1045 hot-rolled steel has a steady axial stress of 10,000 psi. Find the permissible shearing stress that can be superposed if the F_s is to be 2.5 by the Mises–Hencky theory.

38. A 300×300-mm steel plate has normal stresses only acting on all edges. Stress σ_x equals 80 MPa tension. σ_y is compression. Increase in length in the x-direction after stresses are applied is 0.15 mm. If F_s is 2.5 by the Mises–Hencky theory what is the yield point value for the material? *Ans.* $\sigma_{yp} = 318$ MPa.

39. Work Problem 7, but with 0.006 in. as 0.0048 in. and σ_x as 9,600 psi.

40. A $375 \times 500 \times 12$-mm plate of hot-rolled steel $\sigma_{yp} = 400$ MPa carries a uniformly distributed tension of 400,000 N on the 375 mm edge. Find the compressive stress on the 500-mm edge if the F_s is 2 by the maximum shear theory. Find the dimensions of the plate after the loads are acting.
Ans. 374.74×500.31 mm.

41. In Fig. 2-57, the 16-in. dimension becomes 8 in. with the load at the end of the bracket equal to 2,500 lb. The shaft material tests 54,000 psi yield and 90,000 psi ultimate. Find the F_s by the maximum shear theory for an element on the top surface 4 in. to the left of the bracket. Do the same by the Mises–Hencky theory.

42. The part shown in Fig. 2-60 is made from quenched and tempered steel. $\sigma_{ult} = 900$, $\sigma_{yp} = 760$, and $\sigma_e = 320$ MPa. The load varies continuously from 270,000 N to 450,000 N; machined surfaces. Find the F_s. *Ans.* $F_s = 1.31$.

FIGURE 2-60 Problem 42.

43. A material carries a steady shearing stress of 8,000 psi and a fluctuating range stress of 2,000 psi. Stress concentration factor is 1.5 and factor of safety 2.5. If the endurance limit in shear is 0.3 of the ultimate stress in shear, find the value of τ_{ult}. *Ans.* $\tau_{ult} = 45,000$ psi.

44. A body loaded by normal stresses σ_1 and σ_2 has an F_s by the Mises–Hencky theory that is 10 per cent greater than the F_s when computed by the maximum shear theory. If σ_1 is 70 MPa, find the value of σ_2.
Ans. $\sigma_2 = -20.2$ MPa.

45. In Fig. 2-61, the leaf spring is straight and unstressed when the cam and shaft are removed. The material is cold-drawn 3140 steel. The stress concentration is zero. Spring surfaces are ground. If the cam rotates continuously, find F_s for the spring.

FIGURE 2-61   Problem 45.

46. An element is loaded by stresses $\sigma_x = 119$ MPa, $\sigma_y = 7$ MPa, and $\tau_{xy} = 42$ MPa. Material is cast iron with $\sigma_{ut} = 200$ MPa and $\sigma_{uc} = 730$ MPa. Find the value of the factor of safety.   *Ans. $F_s = 1.49$.*

47. For an F_s of 4, find the total force P in Fig. 2-62 if the material is Class 30 cast iron.

FIGURE 2-62   Problem 47.

48. Work Problem 30(a), but find the F_s for an element at the wall at midheight on the surface of the shaft on the side toward the observer.   *Ans. $F_s = 5.0$.*

49. A 10×20-in. plate of Class 50 cast iron has normal stresses only acting on the edges.

(a) Find the F_s if the dimensions become 9.9981×20.0066 in. after the loads are applied.

(b) Do the same, but with dimensions of 10.0011×20.0048 in.

50. Find the relationship between τ_{yp} and σ_{yp} by the Mises–Hencky theory.

Ans. $\tau_{yp} = 0.577 \, \sigma_{yp}$.

51. When the motor is turning in Fig. 2-63, it is observed that the end of the beam has a total up and down vibration of 0.25 in. The beam is of 1035 cold-rolled steel, but assume that the surface is equivalent to a hot-rolled surface. Find the factor of safety at the wall where the stress concentration factor is 1.5.

Ans. $F_s = 4.16.$

FIGURE 2-63   Problem 51.   **FIGURE 2-64   Problem 52.**

52. The hollow cylinder in Fig. 2-64 is of cast iron $\sigma_{ut} = 350$ MPa, $\sigma_{uc} = 1,100$ MPa. If the stress concentration factor is 1.5, find the value of the factor of safety.

Ans. $F_s = 4.8.$

53. A part with a machined surface is subjected to a reversed bending stress of 70,000 psi for 5 per cent of the time, a reversed bending stress of 55,000 psi for 15 per cent of the time, and a reversed bending stress of 40,000 psi for 80 per cent of the time. BHN for the material is 190.

Find the expected life of the part in cycles.   *Ans.*   36,700 cycles.

54. A part with an as-forged surface was found to have a fatigue life of 200,000 cycles. For 30 percent of the time the reversed bending stress had a value of 207 MPa, and for 60 percent of the time the stress had a value of 138 MPa.

Find the probable value of the reversed bending stress for the remainder of the time.   *Ans.* $\sigma = \pm 234$ MPa.

55. Consider the freight car axle of Example 6, Fig. 2-24. The center of gravity of the given load is located 72 in. above the top of the rails. The wheels have a diameter of 33 in. Suppose that the car is going around a sharp curve to the right and that the horizontal centrifugal force is sufficient to reduce the reaction at the right rail to zero. Find the value of the bending stress at the center of the axle.

Ans. $\sigma = 16,350$ psi.

56. A machine part is operated for time $N_1 = 4.8$ hr at the service load. The load is removed and the part operated at the heavy test load for a time $N_2 = 7.2$ hr when failure occurred. The life L_2 at the heavy test load alone is 7.5 hr. Find the value of the life L_1 at the service load.   *Ans.* $L_1 = 120$ hr.

57. A machine part is operated for time $N_1 = 49.5$ hr at the service load. The load is removed and the part operated at the heavy test load for a time $N_2 = 49.5$ hr when failure occurred. The life L_2 at the heavy test load alone is 50 hr. Find the value of the life L_1 at the service load. *Ans.* $L_1 = 4,950$ hr.

58. In Fig. 2-25 find the values for σ_{av} and $K\sigma_r$ for point C.

Ans. $\sigma_{av} = \sigma_{ult}(\sigma_{yp} - \sigma_e)/F_s(\sigma_{ult} - \sigma_e)$; $K\sigma_r = \sigma_e(\sigma_{ult} - \sigma_{yp})/F_s(\sigma_{ult} - \sigma_e)$.

59. Find the cross-sectional area required for safe continuous operation of a uniform bar in tension if $P_{max} = 50,000$ lb and $P_{min} = 20,000$ lb. Material has $\sigma_{ult} = 90,000$ psi with factor of safety equal to 1.5. Bar has a machined surface.

Ans. $A = 1.245$ in.^{2}

60. Same as Problem 59 except there is a stress concentration of 1.42 at the point of expected failure. *Ans.* $A = 1.523$ in.^{2}

61. A material in the as-rolled condition tests $\sigma_{ult} = 110,000$ psi and $\sigma_{yp} = 85,000$ psi. Stress concentration factor 1.2; factor of safety 1.75.

(a) Find the required cross-sectional area if the part carries a steady tensile load of 20,000 lb.

(b) Find the area required if the load is varying continuously from 19,000 lb to 21,000 lb.

(c) Find the area if the load is varying continuously from 18,000 lb to 22,000 lb.

Ans. (a) $A = 0.412$ in.^{2}; (b) $A = 0.502$ in.^{2}; (c) $A = 0.593$ in.^{2}

62. A steel has a tensile strength of 90,000 psi and an endurance limit of 40,000 psi when the surface is ground.

(a) Find the value of constants A and B.

(b) For a loading of $\sigma_{av} = 27,050$ psi and $\sigma_r = 20,020$ psi and stress concentration factor 1.5, find the value of σ_R the equivalent completely reversed stress.

(c) For the loading of part (b) calculate the expected number of reversals to failure. *Ans.* $\sigma_R = 42,934$ psi; $L = 500,300$ reversals.

63. Material is the same as for Problem 58, namely $\sigma_{ult} = 90,000$ psi, $\sigma_e = 40,000$ psi for a ground surface. At the point of expected failure $K = 1.5$. The loading is as follows:

$\sigma_{av} = 21,000$ psi, $\sigma_r = 21,300$ psi, for $\alpha_1 = 0.36$ of the time

$\sigma_{av} = 39,000$ psi, $\sigma_r = 15,480$ psi, for $\alpha_2 = 0.49$ of the time

$\sigma_{av} = 31,000$ psi, $\sigma_r = 20,850$ psi, for $\alpha_3 = 0.15$ of the time

(a) Find the equivalent completely reversed stresses for each stress.

(b) Find the life if each stress only would be applied and find the expected overall life L_c. *Ans.* $L_c = 500,500$ reversals.

64. A short-time test is to be performed on the part of Problems 58 and 59. Let the service reversing stress of $\sigma_{R1} = 42,934$ be first applied for $N_1 = 20,000$ reversals. Then let the stress be increased to $\sigma_{R2} = 56,000$ psi. After the application of $N_2 = 35,628$ reversals at this stress the part failed. Find the expected life L_1 of the part. *Ans.* $L_1 = 500,300$ reversals.

65. A steel has a tensile strength of 90,000 psi and an endurance limit of 33,000 psi for a machined surface.

(a) For loading in the finite life range, find the value of constants A and B.

(b) For $\sigma_{av} = 30,000$ psi, $\sigma_r = 16,565$ psi, and $K = 1.4$ find the value of σ_R, the equivalent completely reversing stress.

(c) Calculate the expected number of reversals to failure for the stresses of part (b).

Ans. $\log A = 5.29847$; $B = 0.12999$; $\sigma_R = 34,788$ psi; $L = 666,900$ reversals.

66. Material is the same as for Problem 61. Let the loading be as follows:

$$\sigma_{av} = 29,000 \text{ psi}, \quad \sigma_r = 16,216 \text{ psi}, \quad \text{for 0.45 of the time}$$

$$\sigma_{av} = 35,000 \text{ psi}, \quad \sigma_r = 14,830 \text{ psi}, \quad \text{for 0.30 of the time}$$

$$\sigma_{av} = 42,000 \text{ psi}, \quad \sigma_r = 14,147 \text{ psi}, \quad \text{for 0.25 of the time}$$

Find the expected number of reversals N_c to failure.

Ans. $N_c = 666,900$ reversals.

67. A short-time test is to be performed on the part of Problems 60 and 61. Suppose the part is first run for $N_1 = 30,000$ reversals with the service stress σ_{R1} of 34,786 psi. Then it is loaded with a σ_{R2} of 50,000 psi until failure at 39,074 additional reversals. Find the expected service life at $\sigma_{R1} = 34,786$ psi.

Ans. $L_1 = 666,700$ reversals.

68. (a) A 1×3 in. bar of 1045 hot-rolled steel is loaded by a 50,000-lb force in simple tension. Find the factor of safety based on the yield strength.

(b) Find the factor of safety if the bar is continuously loaded with $P_{max} = 50,000$ lb and $P_{min} = 40,000$ lb, $K = 1.5$.

(c) Same but the loading is $P_{max} = 50,000$ lb and $P_{min} = 30,000$ lb.

Ans.   (a) $F_s = 3.54$; (b) $F_s = 2.48$; (c) $F_s = 1.91$.

69. (a) A $\frac{3}{4} \times 3$ in. bar of 1045 hot-rolled steel is loaded by a 40,000-lb force in simple tension. Find the factor of safety based on the yield strength.

(b) Find the factor of safety if the bar is continuously loaded with $P_{max} = 40,000$ lb and $P_{min} = 30,000$ lb; $K = 1.5$.

(c) Same but the loading is $P_{max} = 40,000$ lb and $P_{min} = 20,000$ lb.

Ans.   (a) $F_s = 3.32$; (b) $F_s = 2.17$; (c) $F_s = 1.61$.

70. A beam of 8740 hot-rolled steel is subjected to a steady bending stress of 30,000 psi.

(a) Find the value of the factor of safety.

(b) Let the material be subjected to a fluctuating stress ranging from 25,000 to 30,000 psi. Stress concentration 1.2. Find the value of the factor of safety.

(c) Let the stress fluctuate between 20,000 and 30,000 psi. $K = 1.2$. Find the factor of safety.   *Ans.*   (a) $F_s = 2.13$; (b) $F_s = 1.68$; (c) $F_s = 1.38$.

71. A beam of 8740 hot-rolled steel is subjected to a steady bending stress of 25,000 psi.

(a) Find the value of the factor of safety.

(b) Let the material be subjected to a fluctuating stress ranging from 20,000 to 25,000 psi. Stress concentration 1.35. Find the value of the factor of safety.

(c) Let the stress fluctuate between 15,000 and 25,000 psi. $K = 1.35$. Find the factor of safety.   *Ans.*   (a) $F_s = 2.56$; (b) $F_s = 1.86$; (c) $F_s = 1.46$.

72. A machine member is made of an alloy steel heat treated to $\sigma_{ult} = 190,000$ psi and $\sigma_{yp} = 160,000$ psi. Machined surface. Stress concentration factor 1.25. Stresses fluctuate between 30,000 and 50,000 psi.

(a) Find the value of the factor of safety.

(b) If the surface could be ground at the point of expected failure, find the factor of safety. *Ans.* (a) $F_s = 2.05$; (b) $F_s = 2.33$.

73. A machine member is made of an alloy steel heat treated to $\sigma_{ult} = 170,000$ psi and $\sigma_{yp} = 130,000$ psi; as-forged surface; stress concentration factor 1.30. Stress fluctuates between 30,000 and 40,000 psi.

(a) Find the value of the factor of safety.

(b) If the surface could be ground at the point of expected failure, find the factor of safety. *Ans.* (a) $F_s = 1.64$; (b) $F_s = 2.62$.

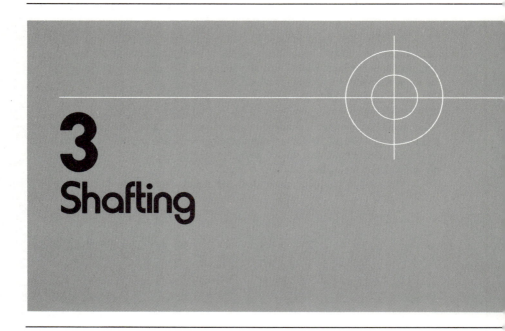

3
Shafting

Shafts are used in all kinds of machinery and mechanical equipment. Although the elementary theory for a circular shaft with static torsional loads is useful, most shafts are subjected to fluctuating loads of combined bending and torsion with various degrees of stress concentration. For such shafts the problem is fundamentally one of fatigue loading. In addition to the shaft itself, the design usually must include the calculation of the necessary keys and couplings. The normal operating speed of a shaft should not be close to a critical speed, or large vibrations are likely to develop. Equations are given for finding the deflections of shafts of nonuniform diameters. Machine parts with noncircular cross sections are sometimes loaded in torsion; the designer must therefore be able to determine the stresses and deformations sustained by such bodies.

d,	diameter	J,	joule, or newton-meter, Nm
E,	modulus of elasticity, normal stress	J,	polar moment of inertia
		k,	kilo (prefix), 1,000
fpm,	feet per minute	kg,	kilogram, 2.2046 lb (mass)
F_s,	factor of safety	kW,	kilowatt, 1.341 hp
G,	modulus of elasticity in shear	K,	stress concentration factor, normal stress
hp,	horsepower, 0.7457 kW		
in.,	inch, 25.4 mm	K_t,	stress concentration factor, shear stress
in. lb,	inch pound, 113 Nmm		
I,	moment of inertia	l,	length

lb,	pound;	rpm,	revolutions per minute
	1 lb (force) = 4.4482 N	SAE,	Society of Automotive
	1 lb (mass) = 0.4536 kg		Engineers
m,	meter, 39.370 in.	T,	torque
m,	milli (prefix), one thousandth	V,	velocity, feet per minute
mm,	millimeter, 0.039 37 in.	V',	velocity, meters per second
M,	bending moment	V_s,	velocity, fps
M,	mega (prefix), 1 million	W,	watt, J/sec, or Nm/sec
MPa,	megapascal, 1 N/mm² or	y,	deflection
	145 psi	σ_e,	(sigma) endurance limit stress,
n,	rpm, revolutions per minute		reversed bending
N,	newton, 0.2248 lb (force)	σ_{yp},	yield point stress, tension
Nmm,	newton millimeter	τ,	(tau) shearing stress
P,	load	φ,	(phi) angular deformation
Pa,	pascal, 1 N/m²	ω,	(omega) angular velocity,
r,	radius		radians per second

1. Torsion of Circular Shaft

Figure 3-1 shows a circular shaft of uniform cross section loaded at the ends by the torques T, which twist it about the longitudinal axis. The shaft is assumed to be much longer with respect to the diameter than is indicated by the figure. It can be shown experimentally that cross sections perpendicular to the axis before loading remain plane and perpendicular to the axis after the loads T have been applied. The diameter of the bar is unchanged and radial lines remain straight and radial after twisting.

The only deformation in the bar is the rotation of the cross sections with respect to each other. As shown in Fig. 3-1, the bottom cross section has been rotated with respect to the top through the angle φ.

The sides of an element on the cylindrical surface of radius r_1 are unchanged in length, but the angles in the corners are changed by angle γ from their original 90° values. The element is thus stressed in pure shear. As shown by Fig. 3-1, $r_1\varphi = l\gamma$. Substitution of

FIGURE 3-1 Circular shaft twisted by torques at ends.

Hooke's law, $\gamma = \tau/G$, where G is the modulus of elasticity in shear, gives

$$\tau = \frac{\varphi G r_1}{l} \tag{1}$$

Since φ, G, and l are constants in Fig. 3-1, the value of the shearing stress τ varies directly with the radius r_1.

If the portion of the bar above the element dA in Fig. 3-1 were removed, the torque of the shearing stress τ, if summed or integrated over the whole cross section, would be equal to the applied torque T. Hence,

$$T = \int_0^r \tau r_1 \, dA \tag{a}$$

The right side is multiplied and divided by r_1; by Eq. (1), ratio τ/r_1 is a constant and can be removed from the integral. Thus

$$T = \int_0^r \frac{\tau}{r_1} r_1^2 \, dA = \frac{\tau}{r_1} \int_0^r r_1^2 \, dA = \frac{\tau}{r_1} J \tag{b}$$

In the last form of Eq. (b), the symbol J, called *polar moment of inertia*, has been substituted for the integral $\int r_1^2 \, dA$.

The maximum value of the shearing stress occurs at the outer surface, where $r_1 = r$. Hence, from Eq. (b),

$$\tau = \frac{Tr}{J} = \frac{16T}{\pi d^3} \tag{2}$$

The similarity of Eq. (2) to Eq. (11) of Chapter 1 for bending stress, $\sigma = Mc/I$, should be noted. The ratio J/r is called the *section modulus* of the shaft.

For a solid circular cross section,

$$J = \frac{\pi d^4}{32} = \frac{\pi r^4}{2} \tag{3}$$

It should be noted that the value of J for a circle is twice as great as the corresponding value of I. For a hollow shaft with outside diameter d_o and inside diameter d_i, the net value for the polar moment of inertia is equal to the value of J for the outer circle minus the J for the inner circle. Hence, for a hollow shaft,

$$J = \frac{\pi}{32}(d_o^4 - d_i^4) = \frac{\pi}{2}(r_o^4 - r_i^4) \tag{4}$$

Elimination of τ between Eqs. (b) and (1) gives

$$\varphi = \frac{Tl}{JG} \tag{5}$$

This equation can be easily committed to memory when its analogy to Eq. (4), Chapter 1, $\delta = Pl/AE$, for axial deformation is noticed.

Angle φ is in radian measure. It should be recalled that 1° is equal to $\pi/180$ rad, or 1 rad = 57.296°.

In order that the foregoing equations may be valid in the neighborhood of the ends, moments T must be applied by means of stresses that vary in intensity with the distance from the axis. Since this condition rarely occurs in practice, Eq. (2) gives correct results only at cross sections somewhat removed from the points where loads T are applied.

Example 1. The shaft in Fig. 3-2 does not rotate; the loads are steady. Assume simple supports.

(a) Make a sketch for the element at A on the bottom surface of the shaft and show the values of the stresses.

(b) Do the same for the element at B at the elevation of the shaft axis.

FIGURE 3-2   Example 1.

Solution. (a) *Element at A.*

By Eq. (15), Chapter 1: $\sigma = \dfrac{32M}{\pi d^3} = \dfrac{32 \times 3{,}500}{\pi 8} = 4{,}460$ psi, tension

By Eq. (2): $\tau = \dfrac{16T}{\pi d^3} = \dfrac{16 \times 9{,}000}{\pi 8} = 5{,}730$ psi

If it is assumed that the shaft is cut at the right edge of the element, the given torque causes the shear stress τ to have the direction shown in Fig. 3-2(b). The transverse shear stress for element A is equal to zero.

(b) *Element at B.* The torsional shear stress has the same value as in part (a). The transverse shear stress is found by Eq. (29) of Chapter 1.

$$\tau = \frac{4V}{3A} = \frac{4 \times 1{,}000}{3\pi} = 420 \text{ psi}$$

If it is assumed that the shaft is cut at the right edge of the element, the transverse shear stress acting upon the element is directed upward. The total shear stress has the value shown in Fig. 3-2(c). The bending stress is zero for the element at B.

Example 1A. The shaft in Fig. 3-2 does not rotate. Loads are steady. Assume simple supports. Diameter is 50 mm. Length is 180 mm. Elements are located similarly to those of the figure. Load at center is 9,000 N. Torques at ends are equal to 1,000,000 Nmm each. Answer the same questions as for Example 1.

Solution. (a) *Element at A.*

By Eq. (15), Chapter 1: $\sigma = \dfrac{32M}{\pi d^3} = \dfrac{32 \times 4{,}500 \times 90}{\pi 50^3} = 33$ MPa

By Eq. (2): $\tau = \dfrac{16T}{\pi d^3} = \dfrac{16 \times 1{,}000{,}000}{\pi 50^3} = 40.7$ MPa

(b) *Element at B.* Transverse shear stress.

By Eq. (29), Chapter 1: $\tau = \dfrac{4V}{3A} = \dfrac{4 \times 4{,}500 \times 4}{3\pi 50^2} = 3.06$ MPa

Example 2. A hollow shaft must carry a torque of 30,000 in. lb at a shearing stress of 8,000 psi. The inside diameter is to be 0.65 of the outside diameter. Find the value of the outside diameter.

Solution. $d_i = 0.65 d_o$

In Eq. (4): $J = \dfrac{\pi}{32}(d_o^4 - 0.65^4 d_o^4) = 0.08065 d_o^4$

In Eq. (2): $J = \dfrac{30{,}000 \times 0.5 d_o}{8{,}000} = 0.08065 d_o^4$

$d_o^3 = 23.2486$

$d_o = 2.854$ in.

Example 2A. A hollow shaft must carry a torque of 3,400,000 Nmm at a shearing stress of 55 MPa. The inside diameter is to be 0.65 of the outside diameter. Find the value of the outside diameter.

Solution. $d_o = 0.65 d_i$

In Eq. (4): $J = \dfrac{\pi}{32}(d_o^4 - 0.65^4 d_o^4) = 0.08065 d_o^4$

In Eq. (2): $J = \dfrac{Tr}{\tau} = \dfrac{3{,}400{,}000 \times 0.5 d_o}{55} = 0.08065 d_o^4$

$d_o^3 = 383{,}250$

$d_o = 72.63$ mm

Example 3. Suppose it is specified that the angular deformation in a shaft should not exceed 1° in a length of 6 ft. The permissible shearing stress is 12,000 psi. Find the diameter of the shaft. The material is steel.

Solution. $\varphi = 1° = \dfrac{180}{\pi}$ rad

By Eqs. (2) and (5): $T = \dfrac{\tau J}{r} = \dfrac{\varphi J G}{l}$

or $r = \dfrac{\tau l}{\varphi G} = \dfrac{12{,}000 \times 72 \times 180}{11{,}500{,}000 \pi} = 4.305$ in.

$d = 8.609$ in.

Example 3A. Suppose it is specified that the angular deformation in a shaft should not exceed 1° in a length of 1,800 mm. The permissible shearing stress is 83 MPa. Find the diameter of the shaft. The material is steel.

Solution. $\qquad \phi = 1° = \dfrac{\pi}{180} = 0.01745$ radian

By Eqs. (2) and (5): $\quad T = \dfrac{\tau J}{r} = \dfrac{\phi J G}{l}$

$$r = \frac{\tau l}{\phi G} = \frac{83 \times 1,800}{0.01745 \times 77,000} = 111.2 \text{ mm}$$

Diameter: $\qquad d = 222.4$ mm

Example 4. The shaft in Fig. 3-3 carries the torque of 10,000 in. lb at the location shown. If the ends of the shaft are fixed against rotation, find the values of the torque reactions T_1 and T_2.

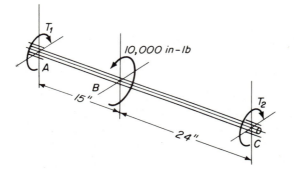

FIGURE 3-3 Example 4.

Solution. By Eq. (5), Ang. def. for *AB*: $\quad \varphi_1 = \dfrac{15T_1}{JG}$

By Eq. (5), Ang. def. for *BC*: $\qquad\qquad \varphi_2 = \dfrac{24T_2}{JG}$

The angular deformation in *AB* is equal to the angular deformation in *BC*. The values for φ_1 and φ_2 should be equated.

$$15T_1 - 24T_2 = 0 \qquad\qquad\qquad (c)$$

By statics: $\qquad\qquad\qquad T_1 + T_2 = 10,000$

The equations above should now be solved simultaneously to give

$$T_1 = 6,154 \text{ in. lb}; \qquad T_2 = 3,846 \text{ in. lb}$$

2. Power Transmitted

Power is the rate of doing work and can be represented by the equation

$$\text{Power} = \text{Force} \times \text{Velocity}$$

A horsepower is defined as 33,000 ft lb per minute. Then

$$1 \text{ hp} = 33,000 \frac{\text{ft lb}}{\text{min}}, \qquad \text{hp} = \frac{FV}{33,000} \qquad (6)$$

where force is in pounds and V is in feet per minute.

Example 5. If a draft horse walks at the rate of 3 miles per hour, what uniform force must he exert if the output power is exactly 1 horsepower?

Solution.
$$V = \frac{3 \times 5280}{60} = 264 \text{ fpm}$$

By Eq. (6):
$$F = \frac{33,000 \text{ hp}}{V} = \frac{33,000 \times 1}{264} = 125 \text{ lb}$$

For rotating systems, force F in this equation can be replaced by torque T, in. lb, divided by a radius r in inches.

$$F = \frac{T}{r} \text{ lb} \qquad (a)$$

Linear velocity V, at radius r, feet per minute, is

$$V = \frac{2\pi rn}{12} \frac{\text{ft}}{\text{min}} \qquad (b)$$

where radius r is in inches and n is the rotative speed, rev/min. Radius r in those equations, which cancels out upon substitution, can have any convenient value, such as the radius of the shaft in Fig. 3-4. Substitution gives

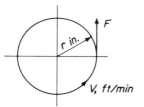

$$\text{hp} = \frac{\dfrac{T}{r} \times \dfrac{2\pi rn}{12}}{33,000} = \frac{Tn}{63,025} \qquad (7)$$

Let ω be the radius turned per second, then

$$\omega = \frac{2\pi n}{60} \quad \text{or} \quad n = \frac{60\omega}{2\pi}$$

FIGURE 3-4 Relationship for deriving horsepower equation.

Substitution into Eq. (7) gives

$$\text{hp} = \frac{T\omega}{6,600} \qquad (8)$$

Example 6. A shaft carries a torque of 10,000 in. lb and turns 900 rpm. Find the horsepower transmitted.

Solution. By Eq. (7): $\text{hp} = \dfrac{10,000 \times 900}{63,025} = 142.8$

The unit of work in SI units is the joule, J, or a newton-meter. The unit of power is the watt, W, or a joule per second. In the metric system the equation for power in watts, W, is

$$W = NV \text{ watts}$$

where force N is in newtons and velocity V is in meters per second. A more practical unit is the kilowatt, kW, or 1,000 watts. Then

$$kW = \frac{NV}{1,000} \tag{9}$$

Example 5A. If a draft horse walks at the rate of 5 kilometers an hour, what uniform force in newtons must he exert if the output power is exactly 1 kilowatt?

Solution. $\qquad V = \dfrac{5,000}{3,600} = 1.3889 \dfrac{m}{sec}$

By Eq. (9): $\qquad N = \dfrac{1,000 \text{ kW}}{V} = \dfrac{1,000}{1.3889} = 720 \text{ newtons}$

Force N in newtons is given by

$$N = \frac{T}{r} \tag{c}$$

where torque T is in newton millimeters and some radius r is in millimeters. For a rotating shaft, velocity V in m/sec is given by

$$V = \frac{2\pi r n}{60 \times 1000} \tag{d}$$

where radius r is in millimeters and n represents the speed revolutions per minute. Substitution into Eq. (9) gives

$$kW = \frac{1}{1,000} \frac{T}{r} \times \frac{2\pi r n}{60 \times 1,000} = \frac{Tn}{9,550,000} (T \text{ in Nmm}) \tag{10}$$

The relationship between horsepower and kilowatts is shown as follows:

$$1 \frac{ft}{min} = \frac{12 \text{ in.}}{min} = \frac{12 \times 25.4 \text{ mm}}{60 \text{ sec}} = 5.08 \frac{mm}{sec} = 0.00508 \frac{m}{sec}$$

$$1 \frac{ft \, lb}{min} = 0.00508 \frac{m}{sec} \times 4.4482 \text{ newtons} = 0.022 \, 597 \frac{newton \, m}{sec} \text{ or watts}$$

$$1 \text{ hp} = 33,000 \frac{ft \, lb}{min} = 33,000 \times 0.022597 = 745.7 \text{ watts or } 0.7457 \text{ kW}$$

$$1 \text{ kW} = \frac{1}{0.7457} = 1.341 \text{ hp}$$

Example 6A. The torque equivalent to that of Example 6 is 1,130,000 Nmm. For the same speed of 900 rpm, find the power transmitted.

Solution. By Eq. (10): $\quad kW = \dfrac{Tn}{9,550,000} = \dfrac{1,130,000 \times 900}{9,550,000} = 106.5$

Check: $\qquad\qquad hp = \dfrac{kW}{0.7457} = \dfrac{106.5}{0.7457} = 142.8$

3. Maximum Static Shearing Stress

Many shafts carry combined loads of bending and torque. The bending moment M causes a normal stress in the axial direction of the shaft as shown by σ in Fig. 3-5(a), and the torque T produces the shearing stress τ. The normal stress in the y-direction, or at right angles to the shaft axis, is in general equal to zero. From the Mohr circle for this element, shown in Fig. 3-5(b), the value of the maximum shearing stress for static loading is given by the equation

$$\tau_{max} = \frac{0.5\sigma_{yp}}{F_s} = \sqrt{\left(\frac{\sigma}{2}\right)^2 + \tau_2} \tag{11}$$

(a)

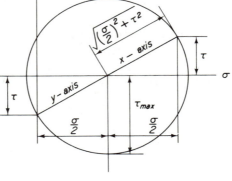

(b)

FIGURE 3-5   Stresses on element of shaft surface.

As demonstrated by Eq. (4), Chapter 2, the yield point stress in shear is equal to one-half the yield point value in simple tension. The equations for the stresses for a solid circular shaft,

$$\sigma = \frac{32M}{\pi d^3} \quad \text{and} \quad \tau = \frac{16T}{\pi d^3} \tag{12}$$

should now be substituted into Eq. (11) to give the following equation for the maximum shearing stress for *static* loads.

$$\tau_{max} = \frac{0.5\sigma_{yp}}{F_s} = \frac{16}{\pi d^3}\sqrt{M^2 + T^2} \tag{13}$$

Substitution of $\tau_{max} = 0.5\sigma_{yp}/F_s$ implies that the maximum shear theory of failure has been assumed to be applicable. This is in accord with Eq. (5) of Chapter 2.

4. ASME Code for Design of Transmission Shafting

Since the loads on most shafts in machinery are not constant, it is necessary to make proper allowance for the harmful effects of the fluctuation. The ASME Code for the Design of Transmission Shafting, B17c—1927, does this by inserting constants C_m and C_t into the equation for stress as follows:

$$\tau_{max} = \frac{0.5\sigma_{yp}}{F_s} = \sqrt{\left(\frac{C_m\sigma}{2}\right)^2 + (C_t\tau)^2} \tag{14}$$

or
$$\tau_{max} = \frac{0.5\sigma_{yp}}{F_s} = \frac{16}{\pi d^3}\sqrt{(C_mM)^2 + (C_tT)^2} \tag{15}$$

where C_m = numerical combined shock and fatigue factor to be applied in every case to the computed bending moment
  C_t = the corresponding factor to be applied to the computed torque

The recommended values for the shock and fatigue factors are given in Table 3-1. For rotating shafts, the bending stress σ is not constant; it varies continuously from maximum tension to maximum compression as the shaft rotates. For steady loads, the table indicates that suitable compensation can be made for the alternating nature of the bending stress by using a value of 1.5 for C_m. If the shaft is hollow, factor $16/\pi d^3$ can be replaced by r/J, where r is the outside radius, and J is computed for the net area.

TABLE 3-1   Constants for ASME Code

Nature of Loading	Values for	
	C_m	C_t
Stationary shafts		
Gradually applied load	1.0	1.0
Suddenly applied load	1.5–2.0	1.5–2.0
Rotating shafts		
Gradually applied or steady load	1.5	1.0
Suddenly applied loads, minor shocks only	1.5–2.0	1.0–1.5
Suddenly applied loads, heavy shocks	2.0–3.0	1.5–3.0

Example 7. Find the diameter by the ASME Code for a rotating shaft subjected to a maximum steady torque of 16,200 in. lb, and a steady bending moment of 27,000 in. lb. The shaft has a keyway.

Solution. $\tau_{max} = 0.75 \times 8,000 = 6,000$ psi. See Section 21, this chapter.

From Table 3-1: $C_m = 1.5$ and $C_t = 1.0$

In Eq. (15): $d^3 = \dfrac{16}{6,000\pi}\sqrt{(1.5 \times 27,000)^2 + 16,200^2} = 37.026$

$d = 3.333$ in.

Example 7A. Find the diameter by the ASME Code for a rotating shaft subjected to a maximum steady torque of 1,800,000 Nmm and a steady bending moment of 3,000,000 Nmm. The shaft has a keyway.

Solution. $\tau_{max} = 0.75 \times \dfrac{8,000}{145} = 41.38$ MPa. See Section 21, this chapter.

From Table 3-1: $C_m = 1.5$ and $C_t = 1.0$

By Eq. (15): $d^3 = \dfrac{16}{\pi 41.38}\sqrt{(1.5 + 3,000,000)^2 + 1,800,000^2}$

$= \dfrac{16 \times 1,000,000}{\pi 41.38}\sqrt{4.5^2 + 1.8^2}$

$= \dfrac{16 \times 1,000,0000}{\pi 41.38} \times 4.847 = 596,500$

$d = 84.18$ mm

Example 8. A 2-in.-diam rotating shaft carries a torque of 12,000 in. lb, which may be applied suddenly, and a bending moment of 8,000 in. lb which also may be applied suddenly. Material tests $\sigma_{yp} = 70,000$ psi. Find the value of the F_s by the ASME Code.

Solution. Assume that $C_m = 2$ and $C_t = 1.5$.

By Eq. (15): $\dfrac{0.5 \times 70,000}{F_s} = \dfrac{16}{\pi 2^3}\sqrt{(2 \times 8,000)^2 + (1.5 \times 12,000)^2}$

$\dfrac{35,000}{F_s} = \dfrac{2}{\pi} \times 24,083$

$F_s = 2.28$

5. Maximum Shear Theory When Loads Are Fluctuating

It will now be shown that the maximum shear theory of failure can be applied when the normal and shear stresses in a shaft are fluctuating. The loading for an element on the shaft surface is shown in Fig. 3-6. Stresses normal to the shaft axis are zero. It is assumed that the normal and shear stresses reach their maximum and minimum values simultaneously.

Equation (11) of Chapter 2 for the fluctuating normal stress of Fig. 3-6, gives the equivalent static normal stress

$\sigma_{av} \pm \sigma_r$

$\tau_{av} \pm \tau_r$

FIGURE 3-6 Element loaded by fluctuating stresses.

σ as

$$\sigma = \sigma_{av} + \frac{K\sigma_{ult}}{\sigma_e}\sigma_r \qquad\qquad (a)$$

The equivalent static shear stress τ for Fig. 3-6 has the value

$$\tau = \tau_{av} + \frac{K_t\sigma_{ult}}{\sigma_e}\tau_r \qquad\qquad (b)$$

Equations (a) and (b) are now substituted into Eq. (11) to give the resultant static shear stress τ_{max}.

$$\tau_{max} = \sqrt{\frac{1}{4}\left(\sigma_{av} + \frac{K\sigma_{ult}}{\sigma_e}\sigma_r\right)^2 + \left(\tau_{av} + \frac{K_t\sigma_{ult}}{\sigma_e}\tau_r\right)^2} \qquad\qquad (16)$$

or

$$\tau_{max} = \frac{16}{\pi d^3}\sqrt{\left(M_{av} + \frac{K\sigma_{ult}}{\sigma_e}M_r\right)^2 + \left(T_{av} + \frac{K_t\sigma_{ult}}{\sigma_e}T_r\right)^2} \qquad\qquad (17)$$

Stress concentration factors for the appropriate change of form and type of loading can be determined from the curves of Chapter 2 and from Section 8 of this chapter.

The equations above apply to solid circular shafts. When a keyway is present at the section for which the calculations are made, the strength is reduced not only because of stress concentration but because of loss of cross section as well. A theoretical determination of this latter quantity would be very involved, but its magnitude may be estimated by the principles explained in Section 16. Perhaps the best way of taking care of the situation in a design is to use a lower value for the working stress in shear. Fortunately, the form of the equation is such that this correction could be considerably in error without causing very much difference in the resulting diameter of the shaft.

Example 9. A revolving shaft carries a bending moment of 27,000 in. lb and a steady torque of 80,000 in. lb. Assume the torque fluctuates 20 percent each way from the mean value. Stress concentration factor for bending and torsion is equal to 1.35. Material has an ultimate strength of 120,000 psi, a yield strength of 90,000 psi, and an endurance limit strength of 44,000 psi. Shaft diameter is $3\frac{3}{8}$ in. Find the value of the factor of safety.

Solution. $\dfrac{K\sigma_{ult}}{\sigma_e} = \dfrac{1.35 \times 120,000}{44,000} = 3.682$ $\qquad M_{av} = 0$ (rotating shaft)

By Eq. (17):

$$\tau_{max} = \frac{16}{\pi 3.375^3}\sqrt{(0 + 3.682 \times 27,000)^2 + (80,000 + 3.682 \times 16,000)^2}$$

$$= \frac{16 \times 1,000}{\pi 38.443}\sqrt{99.414^2 + (80 + 58.912)^2}$$

$$= 132.48\sqrt{9,883 + 19,297}$$

$$= 132.48\sqrt{29,180} = 132.48 \times 170.8 = 22,630 \text{ psi}$$

$$F_s = \frac{0.5\sigma_{yp}}{\tau_{max}} = \frac{45,000}{22,630} = 1.99 \text{ based on yield point stress}$$

Example 9A. A revolving shaft with machined surface carries a bending moment of 3,000,000 Nmm and a steady torque of 9,000,000 Nmm. Assume the torque fluctuates 20% each way from the mean value. Stress concentration factor for bending and torsion is equal to 1.35. Material has an ultimate strength of 830 MPa, a yield strength of 620 MPa, and an endurance limit strength of 300 MPa. Shaft diameter is 86 mm. Find the value of the factor of safety.

Solution. $\dfrac{K\sigma_{ult}}{\sigma_e} = \dfrac{1.35 \times 830}{300} = 3.735 \qquad M_{av} = 0$ (rotating shaft)

By Eq. (17):

$$\tau_{max} = \frac{16}{\pi 86^3}\sqrt{(0 + 3.735 \times 3,000,000)^2 + (9,000,000 + 3.735 \times 1,800,000)^2}$$

$$= \frac{16 \times 1,000,000}{\pi 636,056}\sqrt{11.205^2 + (9 + 6.723)^2}$$

$$= 8.007\sqrt{372.765} = 8.007 \times 19.31 = 154.62 \text{ MPa}$$

$$F_s = \frac{0.5\sigma_{yp}}{\tau_{max}} = \frac{310}{154.62} = 2.00$$

6. Mises-Hencky Theory for Shafting

As before, the element is loaded as in Fig. 3-6. The Mises–Hencky equation, Eq. (8) of Chapter 2, for the equivalent static normal stress S for this element is

$$S^2 = \sigma^2 + 3\tau^2$$

Substitution of Eqs. (a) and (b) of the preceding section gives

$$S = \sqrt{\left(\sigma_{av} + \frac{K\sigma_{ult}}{\sigma_e}\sigma_r\right)^2 + 3\left(\tau_{av} + \frac{K_t\sigma_{ult}}{\sigma_e}\tau_r\right)^2} \tag{18}$$

$$S = \frac{32}{\pi d^3}\sqrt{\left(M_{av} + \frac{K\sigma_{ult}}{\sigma_e}M_r\right)^2 + 0.75\left(T_{av} + \frac{K_t\sigma_{ult}}{\sigma_e}T_r\right)} \tag{19}$$

Example 10. Use the data of Example 9 and calculate the required diameter by the Mises–Hencky theory.

Solution. Use numerical values from Example 9.

By Eq. (19): $S = 2 \times 132.48\sqrt{9,883 + 0.75 \times 19,297} = 264.96\sqrt{9,883 + 14,473}$

$$= 264.96\sqrt{24,356} = 264.96 \times 156.1 = 41,360 \text{ psi}$$

$$F_s = \frac{\sigma_{yp}}{S} = \frac{90,000}{41,360} = 2.18$$

A slightly larger factor of safety is obtained when the Mises–Hencky criterion is used. However, unless the torsional loading on the shaft predominates, the two methods give practically the same results.

A shaft with fluctuating bending moment is vulnerable to fatigue failure at the edge of a press-fitted hub or ring. The press fit causes compressive stress in both directions perpendicular to the axis while in the axial direction fluctuating tension stress will occur. The resulting shear stress may exceed the endurance limit of the material.[1]

7. Keys

Shafts and hubs are usually fastened together by means of keys. Several different kinds of keys are shown in Fig. 3-7. The square and flat type of keys are in wide use for general machine construction. Dimensions for square

FIGURE 3-7 Types of keys.

keys are given in Table 3-2. Kennedy keys are usually made tapered and are driven tightly into place upon assembly. They are adapted for rough, heavy service. The Woodruff key is widely used in the automotive and machine tool industries. The gib-head key facilitates removal, although the projecting head for some applications constitutes a hazard for workmen.

In addition to a key, setscrews are usually employed to keep the hub from shifting axially on the shaft. Generally, two screws are placed in the

[1] See p. 45 of Reference 7, end of chapter.

hub: one screw bears on the key and the other bears on the shaft. For light service, rotation between shaft and hub may be prevented by setscrews alone.

Movement between shaft and hub can be prevented by a taper pin driven tightly into place. The so-called "roll pin" is not solid. It has the cross section shown in Fig. 3-7. It has sufficient flexibility to accommodate itself to small amounts of misalignment and variation in hole diameters, and will not come loose under vibrating loads.

TABLE 3-2   Dimensions of Square Keys*

Diameter Shaft	Size Key	Diameter Shaft	Size Key	Diameter Shaft	Size Key
1/2–9/16	1/8	1-7/16–1-3/4	3/8	3-3/8–3-3/4	7/8
5/8–7/8	3/16	1-13/16–2-1/4	1/2	3-7/8–4-1/2	1
15/16–1-1/4	1/4	2-5/16–2-3/4	5/8	4-3/4–5-1/2	1-1/4
1-5/16–1-3/8	5/16	2-7/8–3-1/4	3/4	5-3/4–6	1-1/2

*From ASA B17.1–1943.

For high-grade construction, and for cases where axial movement between shaft and hub is required, relative rotation is prevented by means of splines machined on the shaft and into the bore. One type of spline uses the involute curve as the outline. The spline on the shaft can be cut by a hobbing process similar to that used for cutting gears.

Tables of dimensions for the foregoing machine elements may be found in engineering handbooks and catalogs of various supply houses.

The distribution of the force on the surfaces of a key is very complicated. It is dependent upon the fit of the key in the grooves of shaft and hub, as illustrated by Fig. 3-8(a) and (b), in which the distributed loads are represented by single arrows. In addition, the stresses are not uniform along the key in the axial direction; they are highest near the ends.

(a) Forces on key which fits tightly top and bottom.

(b) Forces acting on loosely fitted key.

FIGURE 3-8   Forces on key.

Because of many uncertainties, an exact analysis of the stresses usually cannot be made. Engineers commonly assume that the entire torque is carried by a tangential force F located at the shaft surface. That is,

$$T = Fr \qquad (20)$$

Shearing and compressive stresses are computed for the key from force F, and a sufficiently large factor of safety is employed.

Example 11. A $3\frac{7}{16}$-in.-diam shaft is made from material with a yield point value of 58,000 psi. A $\frac{7}{8} \times \frac{7}{8}$-in. key is to be used of material with a yield point value of 48,000 psi. Let $\tau_{yp} = 0.5\sigma_{yp}$. The factor of safety is equal to 2.

Find the required length of key based on the torque value of the gross shaft.

Solution.

Shaft: $\sigma_{yp} = 58{,}000$ psi, working stress, $\sigma = \dfrac{58{,}000}{2} = 29{,}000$ psi

$\tau_{yp} = 29{,}000$ psi, working stress, $\tau = \dfrac{29{,}000}{2} = 14{,}500$ psi

Key: $\sigma_{yp} = 48{,}000$ psi, working stress; $\sigma = \dfrac{48{,}000}{2} = 24{,}000$ psi

$\tau_{yp} = 24{,}000$ psi, working stress; $\tau = \dfrac{24{,}000}{2} = 12{,}000$ psi

$$J = \frac{\pi d^4}{32} = 13.708 \text{ in.}^4$$

In Eq. (2), torque in shaft: $T = \dfrac{\tau J}{r} = \dfrac{14{,}500 \times 13.708}{1.719} = 115{,}650$ in. lb

Force at shaft surface: $F = \dfrac{T}{r} = \dfrac{115{,}650}{1.719} = 67{,}280$ lb

For length of key:

Based on bearing on shaft: $l = \dfrac{67{,}280}{29{,}000 \times 0.438} = 5.30$ in.

Based on bearing on key: $l = \dfrac{67{,}280}{24{,}000 \times 0.438} = 6.41$ in.

Based on shear in key: $l = \dfrac{67{,}280}{12{,}000 \times 0.875} = 6.41$ in.

8. Stress Concentration

Stress concentration factors[2] for a shaft with two diameters joined by fillets and loaded in torsion are given in Fig. 3-9. When a shaft has a transverse hole with a bending load, the stress concentration factors[3] are as shown in Fig.

[2] See Peterson, R. E., *Machine Design*, **23**, June 1951, p. 177. See also p. 76 of Reference 5, end of chapter.

[3] See Peterson, R. E., *Machine Design*, **23**, July 1951, p. 156. See also p. 104 of Reference 5, end of chapter.

FIGURE 3-9 Factors of torsional stress concentration K for circular shafts of two diameters to be applied to the shear stress in section of diameter d.

FIGURE 3-10 Stress concentration factors for shaft with transverse hole loaded in bending. Based on section modulus of the net area.

3-10. The stress concentration factors[4] for a transverse hole and torsional loading are given in Fig. 3-11.

FIGURE 3-11 Stress concentration factors for shaft with transverse hole. Torsional loading. Based on full cross sectional area. No reduction for hole.

The results of some tests to determine the fatigue strength reduction factors[5] for alternating bending stresses for shafts with keyways are given in Table 3-3. Two kinds of steel were used: a medium-carbon steel and a heat-treated, chrome–nickel steel. Specimens tested were 1 in. in diameter with two types of keyways: sled-runner and profile, which are shown in Fig. 3-12. To simulate conditions at an oil hole, tests were also made on carbon steel specimens with a $\frac{1}{4}$-in. transverse hole; the results are shown in Table 3-3.

For fluctuating loads, the fatigue stress concentration factor K is defined

TABLE 3-3 Fatigue Stress Concentration Factors in Bending for Shafts with Keyways Based on Section Modulus of Full Area

Steel	Tensile Strength, psi	Yield Strength (Plastic Def. 0.2 per cent)	For Reversed Bending Stress	Chrome-nickel, Heat-treated		Medium-carbon, Normalized	
				Endur-ance Limit, psi	Stress Concen-tration Factor, K	Endur-ance Limit, psi	Stress Concen-tration Factor, K
Chrome-nickel (About SAE 3140)	103,500	70,000	No keyway, or-dinary tapered specimen	58,000		37,000	
			Sled-runner keyway	36,000	1.61	28,000	1.32
Medium-carbon (About SAE 1045)	80,000	45,000	Profile keyway	28,000	2.07	23,000	1.61
			$\frac{1}{4}$-in. transverse hole			12,100	3.06

[4]See Seely, F. B., and T. J. Dolan, *Univ. Illinois Eng. Exp. Sta. Bull. 276*, 1935.
[5]See Peterson, R. E., "Fatigue of Shafts Having Keyways," *Proc. ASTM*, **32**, Part II, 1932, p. 413.

Sled-runner keyway   *Profile keyway*

FIGURE 3-12 **Types of keyways tested for stress concentration effects.**

as follows:

$$K = \frac{\text{endurance limit for plain specimen}}{\text{endurance limit with keyway or hole}}$$

In these tests a fatigue crack started on the outer surface of the shaft near the end of the keyway. The results of the tests indicate that the sled-runner keyway is preferable to the profile. It should also be noted that the heat-treated specimens had larger stress concentration factors than the plain carbon. As was mentioned in Section 13, Chapter 2, a heat-treated steel exhibits greater sensitivity to notch effects than does plain carbon steel. This fact unfortunately causes a reduction in the advantages that would otherwise be obtained by the use of a high-strength alloy steel. The tests also show that a designer should avoid locating an oil hole at a highly stressed point on a shaft.

Because of the lack of available data, Table 3-3 may also be used for the stress concentration factors for torsion in the equations of Sections 5 and 6.

All stress concentration factors so far presented in this book are the geometric or full theoretical values. Section 13 of Chapter 2 discussed the fact that under fatigue loading the actual effect of stress concentration was usually less severe than indicated by the theoretical values.

9. Couplings

A wide variety of devices are available for connecting the ends of two shafts together. The solid coupling shown in Fig. 3-13 is a typical example. It is inexpensive and will withstand rough usage. Good alignment between the ends of the shafts is necessary, however, to avoid inducing bending stresses in the shafts or loads in the bearings.

FIGURE 3-13   Solid coupling.

Example 12. For the coupling shown in Fig. 3-13, the key is $\frac{1}{2} \times \frac{1}{2}$ in. The shaft carries a steady load of 50 hp at 150 rpm. For all parts, $\sigma_{yp} = 60,000$ psi, and $\tau_{yp} = 30,000$ psi. Find the following stresses and the F_s based on the yield point:
  (a) Shear and bearing in key.
  (b) Shear in bolts.
  (c) Bearing on bolts in flange.
  (d) Shear in flange at hub.

Solution. (a) From Eq. (7): $T = \dfrac{63,000 \text{ hp}}{n} = \dfrac{63,000 \times 50}{150} = 21,000$ in. lb

Tangential force at shaft surface: $F = \dfrac{T}{r} = \dfrac{21,000}{1.094} = 19,200$ lb

Area in bearing for key $= 0.25 \times 3.25 = 0.8125$ in.^{2}

Compressive stress: $\sigma_c = \dfrac{19,200}{0.8125} = 23,630$ psi

$$F_s = \frac{60,000}{23,630} = 2.54 \text{ in bearing}$$

Area in shear for key $= 0.5 \times 3.25 = 1.625$ in.^{2}

Shearing stress in key: $\tau = \dfrac{19,200}{1.625} = 11,820$ psi

$$F_s = \frac{30,000}{11,820} = 2.54 \text{ in shear}$$

(b)   Area in shear for bolts $= 6 \times \dfrac{\pi}{4} \times 0.625^2 = 1.841$ in.^{2}

Force at bolt circle: $F = \dfrac{21,000}{3} = 7,000$ lb

Shear stress in bolts: $\tau = \dfrac{7,000}{1,841} = 3,800$ psi

$$F_s = \frac{30,000}{3,800} = 7.89$$

(c)   Area in bearing for bolts $= 6 \times 0.625 \times 0.625 = 2.344$ in.^{2}

Compressive stress on bolts: $\sigma_c = \dfrac{7,000}{2.344} = 2,990$ psi

$$F_s = \frac{60,000}{2,990} = 20.1$$

(d)   Area in shear at edge of hub $= 4.25\pi \times 0.625 = 8.345$ in.^{2}

Force at edge of hub $= \dfrac{21,000}{2.125} = 9,880$ lb

Shear stress in web: $\tau = \dfrac{9,880}{8.345} = 1,180$ psi

$$F_s = \frac{30,000}{1,180} = 25.4$$

Many types of flexible couplings are available that provide for some misalignment. Such couplings are often provided with springs or rubber inserts to cushion the shock of suddenly applied loads. Details, dimensions, and load ratings based on long experience may be found in the catalogs of the various manufacturers.[6] Information is also given in the mechanical engineering handbooks.

10. Bending Loads in Two Planes

Shafts are sometimes subjected to loads applied at different angles. To find the resulting bending moment at any cross section, it is necessary to have the components of the loads in two perpendicular axial planes. The following example illustrates a typical method for solving such problems.

Example 13. Figure 3-14 shows a shaft with the belts making angles of 45° and 60° with the horizontal. Find the value of the maximum bending moment for the shaft.

Solution. The components of the belt forces in the horizontal plane and the corresponding bearing reactions are shown in Fig. 3-14(b). These loads give the bending moment diagram of sketch (c).

The loads and reactions for the vertical plane are given in Fig. 3-14(d), and the bending moment diagram is shown in sketch (e).

The maximum bending moment occurs at the left pulley and has the following value.

$$M_{max} = \sqrt{25,130^2 + 36,430^2} = 44,290 \text{ in. lb}$$

[6]See, for example, Reference 1, end of chapter.

FIGURE 3-14 Shaft with bending loads in two planes. Example 13.

11. Shaft on Three Supports

Shafts are sometimes supported on three bearings as illustrated in Fig. 3-15. Such a problem is statically indeterminate with three unknown reactions R_1, R_2, and R_3. It is possible to write only two independent equations from statics, one for the summation of the vertical forces and one for the summation of the moments. The additional equation required for a solution to the problem can be obtained by taking into account the deformation of the body.[7] This can be done in a variety of ways. The following example illustrates a typical method of solution.

FIGURE 3-15   Indeterminate shaft on three supports. Examples 12, 13, 14.

Example 14. In Fig. 3-15(a) let the shaft be supported by three bearings all at the same elevation. Find the value of reaction R_2.

Solution. Suppose in sketch (b) the center bearing is temporarily removed. The downward deflection at the center by No. 13 of Fig. 1-15 is

[7]See also p. 58 of Reference 7, end of chapter.

$$y = \frac{Pa}{24EI}(3l^2 - 4a^2) = \frac{1,000 \times 36(3 \times 120^2 - 4 \times 36^2)}{24 \times 120,000,000}$$

$$= 0.4752 \text{ in.}$$

Now let the unknown reaction R_2 be applied upwardly at the center. By No. 6 of Fig. 1-13:

$$y = \frac{R_2 l^3}{48EI} = \frac{R_2 \times 120^3}{48 \times 120,000,000} = 0.0003 R_2$$

To preserve the original configuration, the two figures of sketch (b) are combined. By superposition, the resulting deflection at the center is zero.

$$0.0003 R_2 = 0.4752$$

$$R_2 = 1,584 \text{ lb}$$

Example 15. Suppose the center bearing in Fig. 3-15 is made lower than the other two by 0.050 in. Find the value of R_2.

Solution. The upward deflection in Fig. 3-15(b) fails by 0.050 in. to be equal to the downward deflection from the loads. Then

$$0.0003 R_2 + 0.0500 = 0.4752$$

$$R_2 = 1,417.3 \text{ lb}$$

Thus even a small change in the elevation of the bearing has made a rather large change in the reaction. This is characteristic of indeterminate problems. When they are used, care must be exercised that all the specified dimensions are retained in practise. The idea has been used to good advantage in this example to reduce the bearing load at the center.

Example 16. Suppose the center bearing in Fig. 3-15(c) is resting on an elastic support which will deflect under load. Suppose the flexibility can be represented by a spring of rate $k = 20,000$ lb/in. Find the value of reaction R_2'.

Solution. After removing the spring the loads deflect the shaft at the center by 0.4752 in. as found in Example 14. The spring force, as represented by R_2', will deflect the shaft upwardly by $0.0003 R_2'$ as determined by the equation in No. 6 of Fig. 1-13. The difference, $0.4752 - 0.0003 R_2'$, when multiplied by k will be equal to the reaction R_2'. Thus

$$(0.4752 - 0.0003 R_2')20,000 = R_2'$$

$$R_2'(1 + 6) = 9,504$$

$$R_2' = 1,357.7 \text{ lb}.$$

This result can be easily checked for accuracy.

Upward deflection from R_2': $\quad y = 0.0003 R_2' = 0.0003 \times 1357.7$

$$= 0.4073 \text{ in.}$$

Difference in deflection times k:

$$(0.4752 - 0.4073)20,000 = 1,357.7 \text{ lb}$$

Actually, the spring in Fig. 3-15(a) may consist of an unusually elastic portion of the machine or structure where the elastic deformations of the elements supporting the bearings must be taken into account.

12. Crankshafts

To determine the stresses in a crankshaft, the loading on the separate parts of the mechanism must be secured. A typical example is shown in Fig. 3-16(a), which illustrates a single-cylinder, belt-driven air compressor. Suppose the dimensions of the machine are known, and it is desired to find the stresses in the cheek CD of the crankshaft. From the cylinder bore and the air pressure, the force on the piston can be found. By making a force triangle, the force in the connecting rod can be determined. This force also acts on the crankpin

(a) Belt driven air compressor

(b) Forces on wrist and crank pins

(c) Components of belt forces in coordinate directions

(d) Equilibrium of crankshaft

(e) Forces and moments on cross section of cheek

FIGURE 3-16   Forces and moments for crankshaft. Example 17.

at *A*. As shown in Fig. 3-16(b), it is divided into components tangential and normal to the plane of the crank. The normal component gives the torque that the belt must exert at the given crank position. The forces in the tight and slack sides of the belt are now determined from the torque, and the sum is divided into components in the coordinate directions, as shown in Fig. 3-16(c).

The free-body diagram for the crank should now be made as shown in Fig. 3-16(d), utilizing the rod and belt forces, and with the bearing reactions determined by statics. It is customary to assume that the entire bearing load acts at the center of the bearing. The crank can now be cut apart, and the forces and moments for each portion can be determined in the usual way. Thus Fig. 3-16(e) shows the cheek after cutting at the midpoint between *C* and *D* with the various forces and moments that act on the cut surface.

Example 17. Let the bore of the compressor of Fig. 3-16 be 4.75 in., the stroke 6 in., and the length of rod 12 in. For a crank angle of 30°, the air pressure is 24.7 psi gage. The belt is horizontal with tension T_1 in the tight side 3 times as great as tension T_2 in the slack side. The coordinate system is as shown. Find the forces and moments for the cross section of the cheek midway between *C* and *D*.

Solution. Force on piston $= \frac{\pi}{4} \times 4.75^2 \times 24.7 = 437.7$ lb

By taking components, the force in the rod is found to be 441.2 lb, as shown in Fig. 3-16(b). At the crankpin, the rod force divides into components 351.5 lb and 266.7 lb, tangential and normal, respectively, to the plane of the crank.

Torque on crank: $T = 266.7 \times 3 = 800$ in. lb

$$T = 4(T_1 - T_2) = 4(3T_2 - T_2) = 800 \text{ in. lb}$$

Hence $T_2 = 100$ lb

$T_1 = 300$ lb

The total belt load of 400 lb is now divided into components, as shown in Fig. 3-16(c).

The free-body diagram for the crank is shown in Fig. 3-16(d), in which the previously determined crankpin and belt loads are acting. The bearing reactions are determined in the usual way by taking moments.

The forces and moments for the desired cross section of the cheek are shown in Fig. 3-16(e). The reader should check the values shown thereon. Moment arms are the distances from the line of action of the force to the center of gravity of the cross section affected.

The arrows in Fig. 3-16(e) represent the resultant values for the forces and moments on the entire cross section. Such loads actually are distributed as stresses in some fashion over the surface of the cheek where it was cut from the balance of the shaft. Because of the irregular shape of the crank, the stress equations previously derived may or may not be applicable. If the crank of Fig. 3-16 is of lightweight construction with relatively small-diameter pins, the stress equations might be valid for the cross section at the midpoint of the cheek. However, the stresses for large, massive crankshafts, such as those used in internal combustion engines, are usually determined experimentally by direct strain gage measurements.[8]

13. Critical Speed of Rotating Shaft

Rotating shafts become dynamically unstable at certain speeds, and large vibrations are likely to develop. The speed at which this phenomenon occurs is called a *critical speed*. It is shown in books on vibration theory that the frequency for free lateral vibration when the shaft is not rotating will be the same as its critical speed.

Vibration difficulties frequently arise at the lowest or fundamental critical speed. The equation for finding this speed for a shaft on two supports is as follows.

$$f = \frac{1}{2\pi}\sqrt{\frac{g(W_1 y_1 + W_2 y_2 + W_3 y_3 + \cdots)}{W_1 y_1^2 + W_2 y_2^2 + W_3 y_3^2 + \cdots}} \text{ cycles/sec} \qquad (21)$$

where W_1, W_2, etc., represent the weights of the rotating bodies, and y_1, y_2, etc., represent the respective static deflections of the weights. The gravitational constant of 386 in./sec² is represented by g.

Example 18. Find the value of the fundamental critical speed for the shaft shown in Fig. 3-17. $E = 30,000,000$ psi.

FIGURE 3-17 Example 18.

Solution. The static deflections at the weights can be found by the equations for No. 7 of Fig. 1-14.

$$I = \frac{\pi d^4}{64} = \frac{\pi}{4} = 0.7854 \text{ in.}^4$$

At A, due to 80 lb: $y = \dfrac{80 \times 60 \times 30}{6 \times 90 \times 30,000,000 \times 0.7854}(90^2 - 60^2 - 30^2)$

$$= 0.04074 \text{ in.}$$

[8]See Timoshenko, S., "Torsion of Crankshafts," *Trans. ASME*, **44**, 1922, p. 653; **45**, 1923, p. 449.

At A, due to 120 lb:
$$y = \frac{120 \times 20 \times 30}{6 \times 90 \times 30{,}000{,}000 \times 0.7854}(90^2 - 20^2 - 30^2)$$
$$= 0.03848 \text{ in.}$$

Total deflection at A: $y = 0.04074 + 0.03848 = 0.07922$ in.

At B, due to 80 lb:
$$y = \frac{80 \times 30 \times 20}{6 \times 90 \times 30{,}000{,}000 \times 0.7854}(90^2 - 30^2 - 20^2)$$
$$= 0.02565 \text{ in.}$$

At B, due to 120 lb:
$$y = \frac{120 \times 20 \times 70}{6 \times 90 \times 30{,}000{,}000 \times 0.7854}(90^2 - 70^2 - 20^2)$$
$$= 0.03697 \text{ in.}$$

Total deflection at B: $y = 0.02565 + 0.03697 = 0.06262$ in.

In Eq. (21):
$$f = \frac{1}{2\pi}\sqrt{\frac{386(80 \times 0.07922 + 120 \times 0.06262)}{80 \times 0.07922^2 + 120 \times 0.06262^2}}$$
$$= 11.80 \text{ cycles/sec}$$

$$n_{cr} = 11.80 \times 60 = 708 \text{ rpm}, \qquad \text{critical speed}$$

The normal operating speed for a shaft should be considerably above or below the value of a critical speed.

It should be noted that Eq. (21) ignores the effect of the weight of the shaft and also assumes that all weights are concentrated. The equation does not take into account any effect of the flexibility of the bearings or supports. This additional flexibility may in some cases lower the value of the critical speed below that indicated by the equation.

A shaft will have as many critical speeds as there are rotating masses. The determination of the higher critical speeds is beyond the scope of this book.

14. Deflection of Shaft of Nonuniform Diameter

Another way to determine the deflection y of a shaft is by elastic energy. The equation is as follows.

$$EFy = \int \frac{M_p M_f \, dx}{I} \tag{22}$$

As shown in Fig. 3-18, force F represents an auxiliary load placed on the shaft at point A where the deflection y is desired; M_p is the bending moment in terms of distance x for any general point B as caused by the actual loads P; M_f is the bending moment at B caused by the auxiliary load F. The modulus of elasticity of the material is represented by E; I is the moment of inertia of the cross section. The integration must extend throughout the entire length of the shaft.

FIGURE 3-18   Deflection of shaft by elastic energy.

Equation (22) can be given a geometric interpretation[9] that eliminates the necessity of performing mathematical integrations. The integrand can be considered a solid with the three dimensions M_p/I, M_f, and dx, as shown in Fig. 3-19(a). If the moment diagrams for M_p and M_f consist of segments of straight lines, it is not necessary to consider a solid of thickness dx, but solids of finite lengths along the shaft can be taken and the volumes found by solid geometry. When this is done for the full length of the shaft, the integration of Eq. (22) will have been performed.

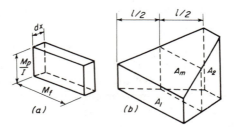

FIGURE 3-19   Deflection of shaft by elastic energy.

Odd-shaped solids will, in general, be present, and for these the prismoidal equation is useful.

$$\text{Vol.} = \frac{1}{6}(A_1 + 4A_m + A_2) \tag{23}$$

As shown in Fig. 3-19(b), A_1 and A_2 refer to the end surfaces of the solid that are perpendicular to the centerline of the shaft. Length l refers to the distance between A_1 and A_2. Area A_m is the area of the cross section midway between A_1 and A_2. Note that in general it is not $(A_1 + A_2)/2$.

Equations (22) and (23) are especially useful when the shaft is nonuniform in diameter as illustrated by the following example.

Example 19. Find the deflection for point A where the diameter of the shaft of Fig. 3-20 changes. $E = 30,000,000$ psi.

[9]See the author's article "Critical Speed of Shafts," *Prod. Eng.*, **12**, 1941, p. 20.

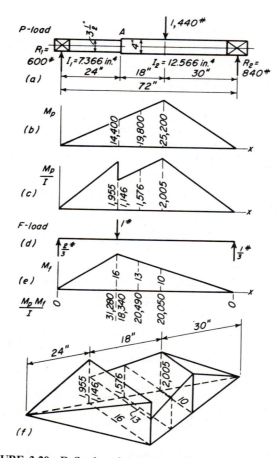

FIGURE 3-20   **Deflection of shaft of two diameters. Example 19.**

Solution. The bending moment diagram for the *P* load is shown in Fig. 3-20(b). The values of the moments are divided by the corresponding values for *I* to give the M_p/I diagram of sketch (c). Let load *F* be taken as 1 lb acting at point *A*, as shown in sketch (d). This load has the bending moment diagram of sketch (e). The various solids formed by the M_p/I and M_f diagrams are given by sketch (f). The solid on each end is a pyramid with a volume equal to one-third of the product of the base by the altitude. The prismoidal equation, however, must be applied to the solid in the middle. The calculations are as follows.

$$\tfrac{1}{3} \times 24 \times 31{,}280 = 250{,}250$$
$$\tfrac{18}{6}(18{,}340 + 4 \times 20{,}490 + 20{,}050) = 361{,}050$$
$$\tfrac{1}{3} \times 30 \times 20{,}050 = \underline{200{,}500}$$
$$Ey = 811{,}800$$

$$y = \frac{811{,}800}{30{,}000{,}000} = 0.02706 \text{ in.}$$

In solving problems, it is not necessary to make a sketch such as (f). The M_pM_f/I values can be entered on sketch (e) and used directly. It should be noted that each solid extends between concentrated loads or points where the diameter changes.

15. *Slope of Shaft by Elastic Energy*

The slope of the shaft can also be found by elastic energy, but the auxiliary load must be a moment applied at the point where the slope is desired. Thus a moment M' is applied to the shaft at point A in Fig. 3-21. The M_f diagram is given by Fig. 3-21(c). Let θ be the rotation or change of slope at A caused by load P. The external work from load M' due to rotation θ is then $M'\theta$. The expression for internal work is the same as that derived for Eq. (22). The energy equation thus becomes

$$EM'\theta = \int \frac{M_pM_f\,dx}{I} \qquad (24)$$

FIGURE 3-21   Slope of shaft by elastic energy.

Moment M' is usually taken as 1 in. lb. Positive and negative signs must be observed as indicated by Fig. 3-21(c).

16. *Torsion of Noncircular Shafts*

It is sometimes necessary to make a design for a shaft of noncircular cross section. For example, the designer might have to know the torsional stress in the rectangular cheek of a crankshaft. Also, miscellaneous machine parts, such as brackets and supports, although not shafts, are sometimes loaded in torsion.

The theory of torsion for noncircular cross section is complicated because the assumptions that are valid for circular shafts do not apply. Cross sections are no longer plane and perpendicular to the shaft axis after twisting; rather, they are warped, as shown in Fig. 3-22, and the equations for the stresses are therefore more involved.

Complicated problems may be solved experimentally by a method known as the *membrane analogy*.[10] A thin homogeneous membrane, such as a soap film, is stretched over a hole in a closed box. The hole is geometrically similar to the cross section of the shaft being studied. The film is slightly bowed by air pressure and the elevations of the resulting surface are measured. The measurements obtained will permit the contours, or lines of uniform elevations, to be plotted. For example, see Fig. 3-24(c). By mathematical analysis it can be shown that

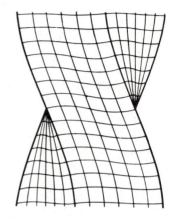

FIGURE 3-22 Rectangular bar in torsion.

(1) The maximum shearing stress in the shaft has the same direction as the contour line at the corresponding point in the film.

(2) The maximum shearing stress at any point is proportional to the slope of the membrane at right angles to the contour at the corresponding point in the film.

(3) The torque carried by the shaft is proportional to twice the volume enclosed between the membrane and the plane of its base.

Although complicated problems can be solved by taking measurements on the membrane, experimental apparatus, which is rarely available to the designer, is required. Nevertheless, the idea of the membrane analogy is very useful since it gives a mental picture of the state of stress. The designer can visualize the points of greatest slope, hence greatest stress, on the bowed film, and often, by making small changes in shape, can cause a reduction in stress.

For example, the membrane may have a steep slope, indicating a high stress at the internal corners of a keyway as shown in Fig. 3-23(a). The membrane analogy indicates that this stress concentration can be reduced by rounding off the bottom corners. However, as was mentioned in Section 8, unless the keyway runs the full length of the shaft, a still greater stress occurs on the shaft surface at the end of the keyway.

The membrane for a rectangular shaft has its greatest slope at point A_1

[10]See p. 266 of Reference 8 and p. 268 of Reference 9, end of chapter.

FIGURE 3-23   Typical cross sections of bars loaded in torsion.

in Fig. 3-23(b); hence the stress is at a maximum at this point, and not at the corners, as is sometimes supposed. In the corner, the membrane has a zero slope along both edges, indicating a zero stress.

As shown by Fig. 3-23(c), the volume enclosed by the membrane for a thin, narrow, rectangular cross section is practically the same whether the rectangle is straight or formed into a curve. The torques carried by either section are the same. However, a slight concentration of stress will exist on the inner side of the curved section. If the flat bar is bent into a full circle to make a slit tube, the deformations at the ends, when torques are applied, will be as indicated in sketch (d).

The volume enclosed by the membrane for a composite section, as in Fig. 3-23(e), is approximately equal to the sum of the volumes for the separate parts. Hence the membrane analogy permits the torque that such sections will carry to be found easily. However, a stress concentration will exist on the fillet in the reentrant corner.

17. Torsion of Wide Rectangular Bar

As was previously mentioned, a mathematical solution for the stresses in a rectangular shaft is difficult to obtain. If the bar is very wide, however, the stress situation is much simplified. This is illustrated by Fig. 3-24(a), where it is assumed that side b of the cross section is much greater than width c. The membrane for this shaft has the contour lines shown in Fig. 3-24(c). These lines are practically straight and parallel to side b for almost the entire surface. Hence, except for small regions in the ends, the shearing stress is directed parallel to side b.

In accordance with the foregoing, the only stresses that act on an element at an interior point A of the bar of Fig. 3-24(a) are those shown in Fig. 3-24(b). The maximum value of the stress occurs on the long side b of the cross section. It can be shown that the value of this stress is given by the following equation.

$$\tau_{max} = \frac{T}{0.333bc^2} \tag{25}$$

(b) Enlarged view
of element at A
showing stresses

(c) Contour lines
for shaft of (a)

(d) Cross section through
membrane of (c)

(a) Long flat bar loaded
by torques at ends

FIGURE 3-24 Torsion of thin, wide, rectangular shaft.

The angular rotation between two cross sections 1 unit distance apart is given by the following equation.

$$\theta_1 = \frac{T}{0.333 Gbc^3} \tag{26}$$

One of the equations above should now be divided by the other, to give

$$\tau_{max} = Gc\theta_1 \tag{27}$$

Reference to Eq. (1) shows that the stress in the bar shown in Fig. 3-24(a) is twice as great as for a round bar of diameter c with equal angular deformation θ_1.

18. Torsion of Rectangular Bars—General Case

When side b of the cross section is not relatively great as compared with width c, the foregoing equations cannot be used. For the bar in Fig. 3-22, the general equations for stress and deformation may be written in the following forms:

$$\tau = \frac{T}{\alpha_1 bc^2} \qquad \text{for point } A_1, \text{ Fig. 3-23(b)} \tag{28}$$

$$\tau = \frac{T}{\alpha_2 bc^2} \qquad \text{for point } A_2, \text{ Fig. 3-23(b)} \tag{29}$$

$$\theta_1 = \frac{T}{\beta Gbc^3} \qquad \text{ang. def., radians per inch of length} \tag{30}$$

Values of the constants α_1, α_2, and β have been computed for various ratios of b/c and are given in Table 3-4.

TABLE 3-4   Constants for Torsion of Rectangular Bars

b/c	1.00	1.20	1.50	1.75	2.00	2.50	3.00	4.00	5.00	6.00	8.00	10.00	∞
α_1	0.208	0.219	0.231	0.239	0.246	0.258	0.267	0.282	0.291	0.299	0.307	0.312	0.333
α_2	0.208	0.235	0.269	0.291	0.309	0.336	0.355	0.378	0.392	0.402	0.414	0.421	$\cdots$
β	0.1406	0.166	0.196	0.214	0.229	0.249	0.263	0.281	0.291	0.299	0.307	0.312	0.333

With the values from the table, computations can be made for the shear stress at the midpoint of both the long and the short sides, as well as the angular deformation, for rectangular cross sections starting from the square $b/c = 1$ to $b/c = \infty$. The maximum shear stress on the cross section occurs at the center A_1 of the long side, and is found by using α_1.

Example 20. Find the stresses at points R and S at the center of the sides of the cheek for the cross section midway between points C and D in Fig. 3-16.

Solution. Element at R:

Direct stress: $\sigma = \dfrac{P}{A} = \dfrac{351.5}{2} = 176$ psi,   compression

Bending: $\sigma = \dfrac{6M}{bh^2} = \dfrac{6 \times 400}{1 \times 2^2} = 600$ psi,   tension

The net tension of 424 psi is shown acting on the element in Fig. 3-16(f).

$$\frac{b}{c} = 2$$

From Table 3-4: $\alpha_2 = 0.309$

Shear stress, Eq. (29): $\tau = \dfrac{T}{\alpha_2 bc^2} = \dfrac{366.7}{0.309 \times 2 \times 1^2} = 593$ psi

This stress, properly directed, is also shown on the element in Fig. 3-16(f).
Element at S:

As before, direct stress: $\sigma = 176$ psi, compression

Bending: $\sigma = \dfrac{6M}{bh^2}\dfrac{6 \times 483.3}{2 \times 1^2} = 1{,}450$ psi, tension

The net tension of 1,274 psi is shown on the element in Fig. 3-16(g).

$$\frac{b}{c} = 2$$

From Table 3-4: $\alpha_1 = 0.246$

Because of torque: $\tau = \dfrac{T}{\alpha_1 bc^2} = \dfrac{366.7}{0.246 \times 2 \times 1^2} = 745$ psi

Because of transverse shear: $\tau = \dfrac{3V}{2A} = \dfrac{3 \times 266.7}{2 \times 2} = 200$ psi

These shear stresses have opposite signs. The resultant, 545 psi, properly directed, is shown in Fig. 3-16(g).

19. Composite Sections

The membrane analogy indicates that the torsional moment carried by a cross section consisting of a number of areas joined together is equal to the sum of the torques of the separate parts. The angle θ_1 applies to each of the parts as well as to the whole section. Therefore, the total torque T for the cross section of Fig. 3-25 is equal to the sum of the torques T_1, T_2, and T_3 for the separate parts 1, 2, and 3, respectively. Hence by Eq. (30),

Torque carried by part 1: $\quad T_1 = \theta_1 G \beta' b_1 c_1^3$

Torque carried by part 2: $\quad T_2 = \theta_1 G \beta'' b_2 c_2^3$ $\hspace{3cm}$ (a)

Torque carried by part 3: $\quad T_3 = \theta_1 G \beta''' b_3 c_3^3$

Adding: $\quad T = \theta_1 G(\beta' b_1 c_1^3 + \beta'' b_2 c_2^3 + \beta''' b_3 c_3^3)$ $\hspace{1cm}$ (b)

Here β', β'', and β''' are the β values for parts 1, 2, and 3, respectively. It should be noted that the right-hand side of Eq. (b) will contain as many terms as there are rectangles in the cross section under consideration.

FIGURE 3-25 Composite section loaded in torsion.

The maximum value of the shearing stress occurs in the bar of greatest width. Let this bar be No. 1 in Fig. 3-25. Equation (28), after substitution for T_1 from Eq. (a), becomes

$$\tau_1 = \frac{T_1}{\alpha_1 b_1 c_1^2} = \frac{\theta_1 G \beta' c_1}{\alpha_1} \hspace{2cm} (c)$$

Elimination of $\theta_1 G$ between Eqs. (b) and (c) gives

$$\tau_1 = \frac{T \beta' c_1}{\alpha_1(\beta' b_1 c_1^3 + \beta'' b_2 c_2^3 + \beta''' b_3 c_3^3 + \cdots)} \hspace{1cm} (31)$$

The angular deformation per inch of length is found from Eq. (b).

$$\theta_1 = \frac{T}{G(\beta' b_1 c_1^3 + \beta'' b_2 c_2^3 + \beta''' b_3 c_3^3 + \cdots)} \hspace{1cm} (32)$$

As was previously mentioned, a concentration of stress exists at the reentrant corners of a composite section. This factor depends upon the ratio between the radius of the fillet and the thickness of the member. Values[11]

---

[11] See Huth, J. H., "Torsional Stress Concentration in Angle and Square Tube Fillets," *Trans. ASME*, **72**, 1950, *J. Appl. Mech.*, p. 388. See also p. 131 of Reference 5, end of chapter.

are shown in Table 3-5 for the structural angle, Fig. 3-26(a), and for the thin-walled square tube in sketch (b).

TABLE 3-5   Stress Concentration Factor K_t for Structural Angle and Thin-Walled Square Tube

r/c	0.125	0.25	0.50	0.75	1.00	1.25	1.50
Angle	2.72	2.00	1.63	1.57	1.56	1.57	1.60
Square Tube	2.46	1.70	1.40	1.25	1.14	1.25	1.07

FIGURE 3-26   Structural angle and square tube of Table 3-5.

Example 21. Find the torque that a 30-in. piece of $2 \times 2 \times \frac{3}{8}$-in. angle iron can carry if the maximum shearing stress on the fillet is to be 12,000 psi. Radius of the fillet is $\frac{1}{4}$ in. Find the angular deformation sustained with such loading.

Solution. $$\frac{r}{c} = \frac{0.25}{0.375} = 0.67$$

By Table 3-5: $$K_t = 1.59$$

Developed center line: $$b = 3\frac{5}{8} \text{ in.}$$

$$\frac{b}{c} = \frac{3.625}{0.375} = 9.7$$

By Table 3-4: $$\alpha_1 = 0.312$$
$$\beta = 0.312$$

By Eq. (28) at center of 2-in. leg:

$$\tau = \frac{T}{\alpha_1 bc^2} = \frac{T}{0.312 \times 3.625 \times 0.375^2} = \frac{T}{0.159}$$

On fillet: $$\tau = 1.59\frac{T}{0.159} = 12,000$$

From which: $$T = \frac{0.159 \times 12,000}{1.59} = 1,200 \text{ in. lb}$$

By Eq. (30): $\theta_1 = \dfrac{T}{\beta Gbc^3} = \dfrac{1{,}200}{0.312 \times 11{,}500{,}000 \times 3.625 \times 0.375^3}$

$$= 0.0017495 \text{ rad/axial in.}$$

Total deformation $\theta = 30 \times 0.0017495 = 0.052485 \text{ rad} = 3.007°$

20. Thin-Walled Tube

Thin-walled tubes are frequently loaded in torsion. It can be shown[12] that the shear stress in the wall is given by the following equation

$$\tau = \frac{T}{2Ac} \tag{33}$$

Here A is the area enclosed by the line running through the center of the wall section, and c is the width of the narrowest wall. The length of the center line, around the wall, is designated by a.

The deformation per unit axial length of the tube is given by

$$\theta_1 = \frac{T}{4A^2G} \sum \frac{a}{c} \tag{34}$$

The summation is to extend around the entire circumference of the tube.

Example 22. (a) Find the torque which the steel hollow cross section of Fig. 3-27(a) can carry at a shearing stress of 10,000 psi. Find the angular deformation θ_1 per inch of axial length.

(b) Do the same for the cross section of Fig. 3-27(b), where the wall is not continuous.

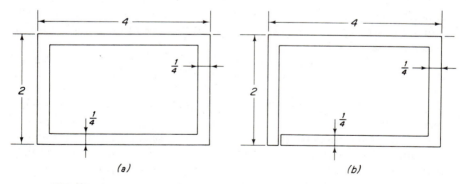

(a)   *(b)*

FIGURE 3-27 **Composite cross sections loaded in torsion. Example 22.**

[12]See p. 278 of Reference 8, end of chapter.

Solution. (a)   Area, $A = 1.75 \times 3.75 = 6.562$ in.^{2}

By Eq. (33): $T = 2Ac\tau_{max} = 2 \times 6.562 \times \frac{1}{4} \times 10,000 = 32,810$ in. lb

Length of center line, $a = 2 \times 1.75 + 2 \times 3.75 = 11.0$ in.

By Eq. (34): $\theta_1 = \dfrac{T}{4A^2G} \dfrac{a}{c} = \dfrac{32,810}{4 \times 6.565^2 \times 11,500,000} \times \dfrac{11}{\frac{1}{4}}$

$\qquad\qquad = 0.000\ 729$ radian/in. $= 0.0418$ deg/in.

(b) Length of center line, $b = 2 \times 1.75 + 2 \times 3.75 = 11.0$ in.

By Eq. (25): $T = 0.333bc^2\tau_{max} = 0.333 \times 11.0 \times (\frac{1}{4})^2 \times 10,000$

$\qquad\qquad = 2,290$ in. lb

By Eq. (26): $\theta_1 = \dfrac{T}{0.333Gbc^3} = \dfrac{2,290}{0.333 \times 11,500,000 \times 11.0 \times (\frac{1}{4})^3}$

$\qquad\qquad = 0.00348$ radian/in. $= 0.199$ deg/in.

The open cross section of sketch (b) has stresses and deflections like the wide rectangular bar of Fig. 3-24(a). It is much weaker in torsion and has a larger angular deflection than the closed section of sketch (a). When the section is closed and continuous a deflection like Fig. 3-23(d) cannot occur.

A stress concentration occurs at the reentrant corners of tubes. For square tubes of uniform wall thickness, values of the concentration factor may be had from Table 3-5. These results are valid only when the length of the sides of the tube is at least 15 or 20 times as great as the thickness.

21. Materials Used for Shafting

When service requirements are not too severe, the least expensive shaft material is hot-rolled, plain-carbon steel. For maximum machinability, a normalizing or annealing treatment may be necessary to improve the grain structure and to secure uniformity. Since hot-rolled bars as received from the mill are usually covered with scale, the shaft must be machined all over if a smooth surface is desired.

Cold-drawn bars, in contrast, have a smooth, bright finish and have diameters held to tolerances of a few thousandths of an inch. This material is sometimes erroneously called cold-rolled shafting. It is available in both plain-carbon and alloy compositions, and is in wide use in the field of general power transmission, since the amount of machining required is minimal. Cold drawing improves the physical properties; it raises the values for tensile strength and the yield point. When greater accuracy is required, shafting that has been turned and ground can be secured from the steel warehouses.

If greater strength is needed than can be secured by the use of a low-carbon steel in the as-rolled condition, a steel of somewhat higher-carbon content can be used. After the machining has been completed, the tensile and yield strengths and hardness can be increased by a quenching and tempering heat treatment. To respond to quenching, the carbon content must be about 0.30% or more. For forged shafts, such as are used in internal combustion engines and railroad cars, the carbon content is usually 0.45 or 0.50%. A widely used steel for such service is plain-carbon steel 1045.

When service conditions are more severe, or when certain desirable physical properties are to be obtained, an alloy steel can be used. As a rule, such steels are not used unless the part is to be heat treated, since full advantage of the expensive alloying elements can be secured only in this way. When heat treated to high strength and hardness, alloy steels are tougher, more ductile, and better adapted to shock and impact loads than are plain-carbon steels. The effect of the quenching penetrates deeper in alloy steels than in carbon steels, and a greater volume of the part is strengthened than if the hardening were confined to a shallow zone over the surface. Alloy steels warp and distort less in heat treatment, have less tendency to crack, and have smaller residual stresses than have carbon steels. Although practically all the alloy steels find application in the field of shafting, chromium-molybdenum steel 4140 and chromium-nickel-molybdenum steels 4340 and A8640 are in wide use as general-purpose alloy steels.

For equal hardness, alloy steels are superior in machining qualities. Where considerable machining is required, shop costs can sometimes be reduced by use of a free-cutting steel, such as 1137. This material is high in manganese and has a relatively high machinability rating for a heat-treating alloy steel.

If the service requirements demand resistance to wear rather than extreme strength, it is customary to harden only the surface of a shaft. The case-hardening or carburizing process is in wide use. Carbon is absorbed by a relatively thin layer while the part is held at a red heat in the furnace. Low-carbon alloy steels, such as 4320, 4820, and A8620, are frequently used for carburizing. The cyaniding and nitriding processes are also used to produce a hard surface. Sometimes it is necessary to localize the wear to a relatively small area. The hardening treatment is applied to those surfaces requiring it; the remainder of the shaft is left in its original condition.

ASME Code B17c-1927 recommends that the working stress τ_{max} in shear be taken at 8,000 psi for "Commercial Shafting" but without any definite specifications for physical and chemical properties of the material. The ultimate strength of such steels may range from 45,000 to 70,000 psi. The corresponding elastic limits would be from 22,500 to 55,000 psi. When there is a keyway at the section for which the stress calculations are made, the working stress is to be reduced to 75% of the value for a solid circular shaft. This reduction can be considered as making allowance for both loss of section and stress concentration.

REFERENCES

1. *ASME Annual Catalog and Directory.*

2. BORESI, A. P., O. SIDEBOTTOM, F. B. SEELY, and J. D. SMITH, *Advanced Mechanics of Materials*, 3d ed., New York: John Wiley & Sons, Inc., 1978.

3. FAUPEL, J. H., and F. E. FISHER, *Engineering Design*, 2d ed., New York: John Wiley & Sons, Inc., 1980.

4. LESSELS, J. M., *Strength and Resistance of Metals*, New York: John Wiley & Sons, Inc., 1954.

5. PETERSON, R. E., *Stress Concentration Factors in Design*, New York: John Wiley & Sons, Inc., 1974.

6. ROARK, R. J., and W. C. YOUNG, *Formulas for Stress and Strain*, 5th ed., New York: McGraw-Hill Book Company, 1976.

7. SPOTTS, M. F., *Mechanical Design Analysis*, Englewood Cliffs, N.J.: Prentice-Hall, Inc., 1964.

8. TIMOSHENKO, S., *Strength of Materials*, 3d ed., Part II, New York: Van Nostrand Reinhold Company, 1956.

9. TIMOSHENKO, S., and J. N. GOODIER, *Theory of Elasticity*, 3d ed., New York: McGraw-Hill Book Company, 1970.

PROBLEMS

1. A shaft carries a steady torque of 30,000 in. lb at a shearing stress of 8,000 psi. What is the diameter of the shaft?   *Ans.   d = 2.673 in.*

2. The torsional deformation of a steel shaft is to be 1° in a length of 600 mm when the shearing stress is equal to 69 MPa. Find the diameter of the shaft.

Ans.   d = 59.8 mm.

3. Suppose it is specified that the deflection at the center of a simply supported shaft under its own weight should not exceed 0.010 in. per foot of span.

(a) Find the maximum permissible span for a 3-in.-diam steel shaft.

(b) Find the stress caused by its own weight when the span length is determined as in part (a).   *Ans.   l = 13 ft; σ = 2,300 psi.*

4. The bent bar *ABCDE* in Fig. 3-28 lies in the horizontal plane and is simply supported at *A* and *E*. Angles at *B* and *D* are 90° each. Draw a view of the element

FIGURE 3-28   Problem 4.

lying on the top surface midway between C and D and show the values of the stresses. Mark the CD direction on your sketch. Diam $= 50$ mm.

Ans. $\sigma = 11.46$ MPa; $\tau = 27.5$ MPa.

5. Make a free-body diagram for the unit consisting of shaft AB and disk C (see Fig. 3-29), and show all forces and torques necessary for equilibrium. Coefficient of friction $\mu = 0.3$. Sufficient force is exerted between the disks to develop the full coefficient of friction. The entire torque is carried by the torque reaction at A. Force reactions are applied at the bearings as shown.

FIGURE 3-29 Problem 5.

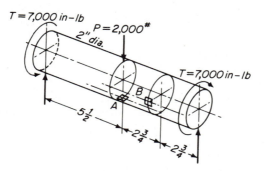

FIGURE 3-30 Problem 6.

6. Find the value of dimension b in Fig. 3-30 that causes the slope of the shaft to be zero at the bearings.   *Ans.* $b = 1,000$ mm.

7. (a) Apply the equation for combined stress and find the value of the maximum shearing stress for the element at A in Fig. 3-2.

(b) If this shaft rotates, find the value of the maximum shearing stress by the ASME Code.   *Ans.* (a) 6,150 psi; (b) 6,630 psi.

8. The element at A in Fig. 3-31 is located on the bottom of the shaft; the element at B is at the elevation of the shaft axis.

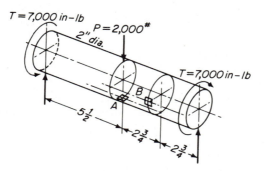

FIGURE 3-31 Problem 8.

(a) Compute the value of the bending and shear stresses in directions parallel and perpendicular to the shaft axis for the element at A. Draw a view of the element showing stress arrows and mark the value of each. Compute the value of the maximum shearing stress for this location on the shaft. Loads are steady and the shaft does not rotate.

(b) Repeat part (a) for the element at B. Take into account the transverse shear for this location.

(c) Find the value of the maximum shear stress for the element at A for steady loads and a rotating shaft by the ASME Code.   *Ans.*   (c) 6,890 psi.

9. Find the deflection at the center for the shaft of Fig. 3-30. Take $b = 1,000$ mm and diameter $d = 50$ mm.   *Ans.* $y = 3.33$ mm, up.

10. The shaft of Fig. 3-32 does not rotate, and is simply supported at A and B. The element at C is on the top surface; the element at D is at the elevation of the shaft axis.

(a) Draw a view of the element at C with sides parallel and perpendicular to the shaft axis; show arrows representing stresses, together with numerical values.

(b) Draw a view for the element at C, properly oriented with respect to the shaft axis, which has the maximum shearing stress. Show arrows and numerical values for all stresses acting.

(c) Work part (b) for the element at D.   *Ans.*   (b) $\tau_{max} = 9,180$ psi.

FIGURE 3-32   Problem 10.

11. The shaft of Fig. 3-33 is simply supported at A and B and is keyed against rotation at A.

(a) Draw a view of the element on the top surface of the shaft at B with sides parallel to the x- and y-axes. Show arrows and numerical values for all stresses acting.

(b) Draw the element at B properly oriented to give the maximum shearing stresses. Show arrows and numerical values for all stresses.

          Ans.   (b) $\tau_{max} = 96.1$ MPa.

12. Resistance-wire strain gages giving deformations at 45° with the shaft axis are attached at A and B in Fig. 3-34. Lead wires are carried from the gages through slip rings and brushes to electric instruments, which permit the deformations to be determined. The shaft carries torque only. Find its value if the elongation at A in x'-direction is 0.0006 mm/mm positive, and if elongation at B in y'-direction is the same amount but negative. Draw a view of the element with sides parallel to

FIGURE 3-33 Problem 11.

FIGURE 3-34 Problem 12.

the x- and y-axes and show stresses acting. Also draw a view for the element with sides parallel to x'- and y'-axes. $\mu = \frac{1}{3}$; $E = 200{,}000$ MPa.

Ans. $T = 2{,}209{,}000$ Nmm.

13. Find the permissible weight of the flywheel in Fig. 3-35 if the value of the maximum shearing stress in the shaft is to be 9,000 psi. The shaft rotates and carries a steady torque of 40,000 in. lb. Use the ASME Code. *Ans.* $W = 1{,}600$ lb.

FIGURE 3-35 Problem 13.   **FIGURE 3-36 Problem 14.**

14. How much torque will the shaft of Fig. 3-36 carry if the maximum shearing stress is not to exceed 55 MPa? Loads are steady and the shaft rotates. Use the ASME Code. *Ans.* $T = 746{,}000$ Nmm.

FIGURE 3-37   Problem 15.   FIGURE 3-38   Problem 16.

15. The shaft of Fig. 3-37 carries a torque of 20,000 in. lb. What must be the distance x in order to make the maximum shearing stress 8,000 psi? The shaft rotates and loads are steady. Use the ASME Code.   *Ans.* $x = 7.84$ in.

16. The shaft of Fig. 3-38 carries a torque of 6,000,000 Nmm. How large a hole may be drilled through the shaft so that the maximum shearing stress does not exceed 80 MPa? Loads are steady and the shaft rotates. Use the ASME Code.
                                    Ans.   49.5 mm.

17. A hollow shaft has a hole of diameter λd, where d is the outside diameter of the shaft, and λ is the appropriate constant. Show that the equation for the shearing stress by the ASME Code will be the same as Eq. (14) except that factor $(1 - \lambda^4)$ will appear in the denominator on the right-hand side.

18. What diameter hollow shaft is required to carry a bending moment of 1,800,000 Nmm together with a torque of 4,500,000 Nmm if the diameter of the hole is equal to 0.6 of the outside diameter of the shaft? Loads are steady and the shaft rotates. Maximum shearing stress equals 70 MPa.   *Ans.* $d = 76$ mm.

19. Determine the required diameter for the hollow shaft of Fig. 3-39 having a hole diameter 0.6 as great as the shaft diameter. The maximum shear stress is to be 12,000 psi. The shaft rotates and loads are steady. Use the ASME Code.
                                    Ans. $d = 1.982$ in.

FIGURE 3-39   Problem 19.

20. A hollow shaft has a hole of diameter λd_o, where d_o is the outside diameter and λ is a constant. Find the value of the ratio of d_o to the diameter d of a solid shaft for equal shearing stresses caused by the same torques for values of λ of 0.5, 0.55, and 0.6.   *Ans.* $d_o/d = 1.022, 1.033,$ and 1.047.

21. What is the ratio of the weight of the hollow shaft per unit length to the weight of the solid for the foregoing values of λ? *Ans.* 0.783, .0744, and 0.702.

22. Find the required shaft diameter (a) by the maximum shear theory, and (b) by the Mises–Hencky theory for the following conditions. The torque varies from 0 to 1,350,000 Nmm. The bending moment varies from 680,000 Nmm to 1,130,000 Nmm. Stress concentration due to fillet for both bending and torque is equal to 2.5. The shaft does not rotate. The material tests $\sigma_{ult} = 400$ MPa and $\sigma_{yp} = 270$ MPa. Endurance limit, $\sigma_e = 160$ MPa. The factor of safety equals 2 based on yield point. *Ans.* (a) 74.2 mm; (b) 71.4 mm.

23. The belt tensions for the pulleys of Fig. 3-40 fluctuate from the values shown in sketch (a) to those given in (b). Details of shaft and hubs are shown in sketch (c). Consider the stress concentration as being due to fillets only. Material tests $\sigma_{yp} = 69,000$ psi, $\sigma_{ult} = 104,000$ psi, and $\sigma_e = 39,000$ psi. Let $D/d = 1.20$ and $r/d = 0.125$. The factor of safety is 1.9 based on yield point.

Draw and dimension the bending moment diagram. Based on the moment just left of the right hub, find the value of d (a) by the maximum shear theory, and (b) by the Mises–Hencky theory. Include the dead weight of the pulleys.
Ans. (a) 2.561 in.; (b) 2.547 in.

(a) Minimum belt forces

(b) Maximum belt forces

(c) Shaft details

FIGURE 3-40 Problem 23.

24. In Fig. 3-35 let the shaft rotate and carry a steady torque of 40,000 in. lb. Material tests $\sigma_{ult} = 60,000$ psi, $\sigma_{yp} = 45,000$ psi, and $\sigma_e = 20,000$ psi. Stress concentration factor for bending and torsion is 1.35. For a maximum torsional stress of 9,000 psi find the value of W by the maximum shear theory of failure.
Ans. $W = 595$ lb.

25. A $2\frac{1}{2}$-in.-diam shaft is made of steel with $\sigma_{ult} = 120,000$ psi, $\sigma_{yp} = 90,000$ psi, and $\sigma_e = 44,000$ psi, machined surface. The shaft rotates and carries a steady 1,500-lb load at the center of a 40-in. simply supported span. Average torque is

20,000 in. lb. Assume $T_r = 0.1T_{av}$. Let $K = K_t = 1.7$. On the basis of (a) the maximum shear theory and (b) the Mises–Hencky theory, find the F_s for this shaft.

Ans.   (a) $F_s = 1.83$; (b) $F_s = 1.86$.

26. A 63.5-mm-diam shaft has a key 16 × 16 mm. The shaft material tests 400 MPa at yield point. Let $\tau_{yp} = 0.5\sigma_{yp}$. The factor of safety equals 2. The shaft fits into a cast-iron hub for which the working stress in compression is 125 MPa. What length of key in the hub material will be required to carry the torque of the solid shaft? The key material is assumed to be amply strong.   *Ans.* 158 mm.

27. A 3-in.-diam shaft of material with a yield point value of 50,000 psi has a 0.75 × 0.75 × 5-in. key. What must the minimum yield point value be for the material in the key in order to transmit the torque of the shaft? The factor of safety equals 2. $\tau_{yp} = 0.5\sigma_{yp}$.   *Ans.* 47,100 psi.

28. A square key has a breadth equal to one-fourth of the shaft diameter. The shaft and key are of materials that are equally strong with a yield point value in shear equal to one-half the yield point value in tension. Find the required length of the key in terms of shaft diameter necessary to transmit the shaft torque.

Ans. $l = 1.57d$.

29. A 75-mm-diam shaft is transmitting 300 kW at 600 rpm. A solid coupling similar to that shown in Fig. 3-13 has 6 bolts, each 18 mm in diameter. Find the required diameter of the bolt circle based on an average shearing stress of 27.5 MPa in the bolts.   *Ans.*   Diam = 227.4 mm.

30. Make the horizontal and vertical load and moment diagrams for the shaft shown in Fig. 3-41. Find the location and value of the minimum bending moment in the shaft for the portion lying between the left pulley and the right bearing.

Ans.   Min $M = 1,918,000$ Nmm for $x = 197.1$ mm.

FIGURE 3-41   Problem 30.

31. Repeat Problem 30 for the shaft in Fig. 3-42. The minimum moment should be for the portion between the two pulleys.

Ans.   Min $M = 10,540$ in. lb for $x = 19.96$ in.

FIGURE 3-42 **Problem 31.**

32. The shaft shown in Fig. 3-43 is fixed at the ends. Find the reactions at the ends and the stresses in each portion of the shaft. Draw views of the elements for each portion and show the stresses acting. *Ans.* $\tau = 125.7$ and 69.8 MPa.

FIGURE 3-43 **Problem 32.** **FIGURE 3-44** **Problem 33.**

33. The shaft in Fig. 3-44 is simply supported at *A* and *D* but is keyed against rotation at both points.

(a) Find reactions at the ends, and draw a view of the element at *B* on the top surface of the shaft. Show all stresses acting and their numerical values.

(b) Compute the value of the maximum shearing stress at *B* and the angle at which it acts. Make a view of the element properly oriented showing maximum shearing stresses acting as well as the normal stresses on all faces.

Ans. (a) $\sigma_x = 13,240$ psi; $\tau_{xy} = 3,060$ psi.

34. The shaft in Fig. 3-45 is simply supported at *A* and *C*, but is keyed against rotation. Draw and dimension the bending moment diagram, and find all reactions at the ends.

Find the resultant stress from all causes for the elements on the top surface of the shaft at *D* and *F*. Do the same for the elements at *E* and *G* at the elevation of the center line. Draw sketches for the elements with arrows properly directed for the

stress and show numerical values.   *Ans.*   At D, $\tau = 22.0$ MPa, $\sigma = 66.0$ MPa;
At E, $\tau = 23.2$ MPa, $\sigma = 0$.

FIGURE 3-45   Problem 34.

35. The shaft of Fig. 3-46 is built in at A and D. Find the value of the torque
reactions at the ends.   *Ans.*   39,230 in. lb; 60,770 in. lb.

FIGURE 3-46   Problem 35.

36. The shaft of Fig. 3-47 is built in at A and D. Find the value of the torque
reactions at A and D.   *Ans.*   2,058,000 Nmm; 7,942,000 Nmm.

FIGURE 3-47   Problem 36.

37. (a) Find the reactions, and draw and dimension the bending moment diagram for the shaft of Fig. 3-48. All bearings are on immovable supports at the same elevation. Include the effect of the dead load of the shaft. For the shaft, $E = 30{,}000{,}000$ psi: $\gamma = 0.283$ lb/in.^{3}.

(b) Find the values of the reactions if the bearing at C is $\frac{1}{8}$ in. lower than the others. Include the effect of the dead load of the shaft.

(c) Let the center support consist of a 6-in. I-beam, 12.5 lb per ft ($I = 21.8$ in.^{4}), 12 ft long, simply supported, with bearing C at its center. Find the reactions for the shaft, and draw and dimension the bending moment diagram. Neglect effects of the dead loads.

(d) What must the moment of inertia be for a beam supporting the bearing at C if the value of the bending moments, as caused by the 1,000-lb loads, at points B, C, and D, are to be numerically equal? What will be the deflection of point C? Neglect effects of the dead loads.

$\quad$ *Ans.* (a) $R_1 = 253$ lb, $R_2 = 1{,}734$ lb;
$\qquad\qquad$ (b) $R_1 = 460$ lb, $R_2 = 1{,}320$ lb;
$\qquad\qquad$ (c) $R_1 = 398$ lb, $R_2 = 1{,}204$ lb;
$\qquad\qquad$ (d) $\quad I = 123$ in.^{4}.

FIGURE 3-48 Problem 37.

38. Bearings A and B in Fig. 3-49 rest on immovable supports. Bearing at C is located at the center of a simply supported 8-in. I-beam 10 ft long. Let a fill be placed under this bearing so that all bearings are at the same elevation before load is applied. Neglect dead load of shaft.

(a) Find reactions at A, B, and C.

(b) If the bearing at C is resting on an unyielding support, find the three reactions.

FIGURE 3-49 Problem 38.

(c) What change in elevation of bearing at C of part (b) must be made if the bending moments in the shaft are to be numerically equal at the load and at bearing C?

(d) With the bearing at C supported by the beam, find the fill under the bearing required to make the bearing load equal to 1,000 lb.

Ans. (a) $R_c = 765.4$ lb; (b) $R_c = 875$ lb;
(c) fill $= 0.0184$ in.; (d) fill $= 0.0395$ in.

39. Bearings A and B in Fig. 3-50 rest on immovable supports. Bearing at C is located at the center of a simply supported 6-in. I-beam 12 ft long. Let a fill be placed under this bearing so that all bearings are at the same elevation before load is applied. Neglect dead load of the shaft.

(a) Find reactions at A, B, and C.

(b) If the bearing at C is resting on an unyielding support, find the three reactions.

(c) What change in elevation of bearing at C of part (b) must be made if the bending moments in the shaft are to be numerically equal at the load and at bearing C?

(d) With the bearing at C supported by the beam, find the fill under the bearing required to make the bearing load equal to 750 lb.

Ans. (a) $R_c = 533$ lb; (b) $R_c = 656.3$ lb;
(c) fill $= 0.039$ in.; (d) fill $= 0.110$ in.

FIGURE 3-50 Problem 39.

40. A rotating shaft has the 5,000-N centrifugal forces acting as shown in Fig. 3-51.

(a) Draw the bending moment diagram for the shaft. Note that large stresses are possible, even though bearing reactions are equal to zero.

FIGURE 3-51 Problem 40.

(b) Suppose an additional bearing is placed in the center of the shaft. Find the reactions, and draw and dimension the bending moment diagram. Ignore the effect of the dead-load deflection. Note how the presence of the central bearing can affect the bearing loads of an engine crankshaft.

Ans. (b) Reactions: 2,134 N downward; 4,268 N upward; 2,134 N downward.

41. Find the lowest critical speed for the steel shaft shown in Fig. 3-52.

Ans. n_{cr} = 1,700 rpm.

FIGURE 3-52　Problem 41.　　　　　　　FIGURE 3-53　Problem 42.

42. The static deflection at the center of the steel shaft in Fig. 3-53 is equal to 0.35 mm. Find the value of the critical speed. Shaft is steel.

Ans. n_{cr} = 1,810 rpm.

43. Find the value of the critical speed for the steel shaft of Fig. 3-54.

Ans. n_{cr} = 617 rpm.

FIGURE 3-54　Problem 43.　　　　　FIGURE 3-55　Problem 44.

44. Find the value of the critical speed for the steel shaft in Fig. 3-55.

Ans. n_{cr} = 402 rpm.

45. Find the deflection at the load and the slope at the end for the steel shaft shown in Fig. 3-56.

Ans. y_c = 0.0593 in.; θ = 0.132°.

FIGURE 3-56　Problem 45.　　　　　FIGURE 3-57　Problem 46.

46. Find the deflection at the load and the slope at the left end of the steel shaft in Fig. 3-57.

Ans. y = 0.345 mm; θ = 0.051°.

47. In Fig. 3-58, find the value of load P if the slope at the left bearing is 0.25°. Steel shaft.   *Ans.* $P = 776$ lb.

FIGURE 3-58   Problem 47.

48. Find the torque that a long piece of $75 \times 75 \times 12$-mm angle iron can carry if the maximum shearing stress at the fillet is 80 MPa. The radius of the fillet is 12 mm.   *Ans.* $T = 320,000$ Nmm.

49. (a) What percentage more torque will a square shaft carry than a round shaft of the same diameter if both have the same shear stress?

(b) If the cost per pound is the same, what will be the percentage increase in cost of the square over the round shaft?   *Ans.*   (a) 5.9%; (b) 27.3%.

50. Two pieces of shafting have the cross sections shown in Fig. 3-59. If the shear stresses in the two shafts are equal, find the ratio between the torques that the shafts are carrying. Approximate the stress concentration factor for the shaft of sketch (a) from Table 3-5.

What will be the ratio between the angular displacements if the applied torques are equal?   *Ans.*   6.27; 8.04.

FIGURE 3-59   Problem 50.

FIGURE 3-60   Problem 51.

51. The rectangular shaft in Fig. 3-60 is fixed at the wall.

(a) Cut the shaft at a point 2 in. from the wall, remove the portion to the right, and show all forces and moments on the end surface of the part that remains.

(b) Find the stresses from all causes on an element at A at the center of the top surface of the shaft at the cut. Take sides of the element parallel to the coordinate axes.

    (c) Repeat part (b) for an element at B at the center of the vertical side of the cut on the near side.   *Ans.* At A, $\tau = 1{,}550$ psi; $\sigma = 3{,}330$ psi.

    52. Let the shaft in Fig. 3-61 be fixed at the wall and have a total length of 300 mm. Answer the same questions as for Problem 51 but for a point 50 mm from the wall. Load is 5,000 N.   *Ans.* At B, $\tau = 16.16$ MPa; $\sigma = 20.0$ MPa.

FIGURE 3-61 Problem 52.   **FIGURE 3-62 Problem 53.**

    53. If the torque applied to a long bar having the cross section shown in Fig. 3-62 is equal to 5,000 in. lb, find the value of the maximum shearing stress.
                                             Ans. $\tau = 4{,}890$ psi.

    54. The beam shown in Fig. 3-63 is simply supported but is keyed at the ends to prevent rotation.

    (a) Find all reactions at the ends.

    (b) Find the value of the maximum shearing stress due to torque.

    (c) Find the angular rotation of the bracket in degrees.
                                     Ans. $\tau = 59.5$ MPa; $\varphi = 0.429°$.

FIGURE 3-63 Problem 54.

55. (a) Make an isometric drawing of the crankshaft shown in Fig. 3-64, and place all reactions thereon that are necessary for equilibrium.

(b) Cheek CD is rectangular in cross section, 2 in. wide in the x-direction, and 3 in. deep in the z-direction. Cut the cheek midway between C and D, and show all forces and moments acting on the cut surface.

(c) Draw the element at R lying at the midpoint of the top surface of the cheek with sides parallel to the coordinate axes, and show all stresses acting. Mark directions of the axes on the sketch.

(d) Repeat part (c) for the element at S at the center of the near-side vertical surface of the cheek.

(e) Suppose the loading consists of a force of 3,000 lb in the xy-plane acting to the left at A parallel to the y-axis. Find the bearing reactions for the shaft. Cut the cheek midway between C and D, and show all forces and moments acting on the cut surface.

(f) Draw the element at R showing all stresses acting.

(g) Draw the element at S showing all stresses acting.

Ans. (c) $\sigma = 4{,}667$ psi, $\tau = 3{,}717$ psi;
(d) $\sigma = 0, \tau = 5{,}329$ psi.

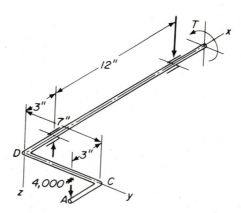

FIGURE 3-64   Problem 55.

56. The stroke of the air compressor shown in Fig. 3-65 is 75 mm, and the length of the connecting rod is 125 mm. Crank cheek CD is circular in cross section and is 25 mm in diameter. If the torsional shearing stress in the cheek is equal to 80 MPa, find the value of the gas force on the piston.

Ans. Gas force $= 15{,}420$ N.

57. (a) The bore of the air compressor of Fig. 3-66 is 4 in., and the stroke is 6 in. The value of the air pressure is 300 psi gage for a crank angle of 45°. The length of the rod is 11 in. Axes x, y, and z are mutually perpendicular. The shaft is turned by a pure torque applied at the right bearing as shown. Make sketches similar to Fig. 3-16(b), (d), and (e) for this problem, and place the value of all necessary forces and moments thereon.

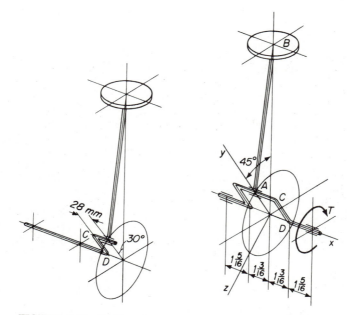

FIGURE 3-65   Problem 56.   **FIGURE 3-66   Problem 57.**

(b) The cheek is $\frac{7}{8}$ in. wide in the x-direction and $1\frac{3}{4}$ in. deep in the z-direction. Make an enlarged sketch for the element with sides parallel to the x- and y-axes lying at the center of the top surface of the cheek. Show all stresses acting.

(c) Repeat part (b) for the element lying at the center of the right vertical side of the cheek.   *Ans.*   (b) $\sigma = 15{,}370$ psi, $\tau = 5{,}060$ psi;
(c) $\sigma = 6{,}990$ psi, $\tau = 7{,}910$ psi.

58. Find the deflection of the weight in Fig. 3-67. The shaft and beam lie in horizontal planes, and are built in at the walls. Bending in crank and shaft is negligible, as is the extension in link connecting the members together. All joints are frictionless. The material is steel.   *Ans.* $\delta = 0.244$ in.

FIGURE 3-67   Problem 58.

FIGURE 3-68   Problem 59.   **FIGURE 3-69   Problem 61.**

59. A belt-driven air compressor has the crank mechanism shown in Fig. 3-68. The axes x, y, and z are mutually perpendicular. The bore is 100 mm; stroke is 125 mm; length of the rod is 225 mm. For a crank angle of 30° the air pressure is 700 kPa gage. The belts are perpendicular to the axis OB of the cylinder. The tight-sied tension is three times that of the slackside.

Make sketches similar to Fig. 3-16(b), (c), (d), and (e) for this problem, and place all necessary forces and moments thereon.

Ans.   Axial force, $F_y = 1,425$ N;
Transverse shear, $F_z = 3,030$ N;
Moment, $M_x = 118,870$ Nmm;
Torque, $M_y = 85,650$ Nmm;
Moment, $M_z = 170,050$ Nmm.

60. An air compressor has a crank arrangement similar to that shown in Fig. 3-16. The stroke is 4 in. and length of the rod is 7 in. The crank cheek is circular in cross section, and is 1 in. in diameter. Distance AC is equal to $1\frac{1}{8}$ in. If the torsional shearing stress at the midlength of the cheek is equal to 10,000 psi for a crank angle of 30°, find the value of the gas force on the piston.

Ans.   Gas force = 2,792 lb.

61. Find the deflection in Fig. 3-69 of the point where the load is applied. The bending moments are negligible since bearings are located close to the crank arms. Shafts and cranks lie in a horizontal plane. Ignore the effect of bending in cranks. The material is steel.   *Ans.* $\delta = 4.36$ mm.

62. Work Problem 61 but with the connection between the end of the crank and the beam replaced by a flexible member with a spring rate equal to 16,000 lb/in.
Ans. $\delta = 0.309$ in.

63. Find the three reactions for the shaft in Fig. 3-70. Assume the effect of the dead-load deflection of the beam to be negligible. *Ans.* $13wl/16$; $33wl/16$; $wl/8$.

64. If reactions R_1, R_2, and R_3 of Fig. 3-70 have values of $15wl/16$, $27wl/16$, and $3wl/8$, respectively, find the amount that support R_2 is lower than R_1 and R_3. Ignore the effect of dead-load deflection of the shaft. *Ans.* $\delta = wl^4/6EI$.

FIGURE 3-70   Problems 63 and 64.   FIGURE 3-71   Problem 65.

65. The three supports in Fig. 3-71 are all at the same elevation. Ignore the dead load of the shaft. Find the values of the three reactions.
Ans. $R_1 = 175$ lb, down; $R_2 = 680$ lb, up; $R_3 = 495$ lb, up.

66. The three supports in Fig. 3-72 are all at the same elevation. Ignore the dead load of the shaft. Find the values of the three reactions.
Ans. $R_1 = 2{,}110$ N, down; $R_2 = 5{,}310$ N, up; $R_3 = 1{,}800$ N, up.

FIGURE 3-72   Problem 66.

67. The center bearing in Fig. 3-73 rests on an immovable support. End bearings rest on structural beams with the spring rates shown. For the shaft, $EI = 7{,}200{,}000$ lb in.[2] Find the load carried by an end bearing. *Ans.* 391 lb.

FIGURE 3-73   Problem 67.

FIGURE 3-74   Problem 68.

68. Bearings *A* and *B* in Fig. 3-74 rest on immovable supports. The bearing at *C* rests on an I-beam with a spring rate of 875 N/mm. The shaft is steel. Find the values of the bearing reactions.   *Ans.* $R_3 = 1{,}479$ N.

69. Work Problem 19 except with the 10-in. dimension as 12 in. and the 18-in. radius as 15 in.

70. Make a figure like that in Fig. 3-45 except that the 300 mm length becomes 400 mm, and the 200 mm length becomes 175 mm. Mark numerical values for all the reactions on the drawing. Draw and dimension the bending moment diagram for the shaft.

71. Work Problem 37(c) but with only one pulley, on the right, present. Find the values of the bearing loads.

72. Show that the deflection at the center of the shaft of Fig. 3-75 is given by the following equation.

$$y = \frac{Pl^3}{162E}\left(\frac{1}{I_1} + \frac{19}{8I_2}\right)$$

FIGURE 3-75   Problem 72.   **FIGURE 3-76   Problem 73.**

73. Show that the deflection at the center of the shaft of Fig. 3-76 is given by the following equation.

$$y = \frac{Pl^3}{384E}\left(\frac{1}{I_1} + \frac{7}{I_2}\right)$$

74. Find the deflection at point *A* in Fig. 3-77. Consider the bearings as simple supports.   *Ans.* $y_a = 0.086$ mm, down.

FIGURE 3-77   Problem 74.   **FIGURE 3-78   Problem 75.**

75. In Fig. 3-78 the end bearings rest on immovable supports. The steel shaft $I = 4$ in.[4] The beam is 9 ft long, simply supported, with the bearing at the center, $I = 18$ in.[4] Find the bearing load at A.

76. The numerical values for the bending moments at B, C, and D in Fig. 3-79 are equal. Find the difference in elevation between the bearing at C and those at A and E.   *Ans.* $y = Pl^3/144EI$.

FIGURE 3-79   Problem 76.

77. The shaft of Fig. 3-80 has the bearing reactions shown. Find the change in elevation of the center bearing with respect to those at the ends.

FIGURE 3-80   Problem 77.

78. Show that the deflection at the center of the shaft of Fig. 3-81 is given by the following equation.

$$y = \frac{P}{24E}\left[\frac{4a^3}{I_1} + \frac{b}{I_2}(8a^2 + 4ab + l^2)\right]$$

FIGURE 3-81   Problem 78.

79. A square steel tube is 200 mm on the side with wall thickness 5 mm. Find the shearing stress in the wall if a torque of 10,000,000 Nmm is applied to the tube. Find the angular deformation per axial inch.   *Ans.* $\theta_1 = 0.0002$ deg/mm.

80. The shaft in Fig. 3-82 rotates 860 rpm. Its spring constant at the center is 2,000 lb/in. Weight W is 16 lb with center 2 in. from the shaft center. Find the deflection y at the center during rotation. Neglect weight of shaft.

Ans. $y = 0.404$ in.

FIGURE 3-82   Problems 80, 81, 82, 83.

81. Let the weight in Fig. 3-82 be 10 lb with center located 2 in. from the shaft center. The spring constant for the shaft at the center is 3,000 lb/in. Speed of rotation 860 rpm. Find deflection y of the shaft while rotating. Neglect weight of shaft.

Ans. $y = 0.212$ in.

82. Let the shaft in Fig. 3-82 be steel 1 in. in diameter and 24 in. long. The weight is 8 lb with center 3 in. from shaft center. Rotation is 1,200 rpm. Find the steady-state shaft deflection y when rotating. Neglect weight of shaft.

Ans. $y = 0.205$ in.

83. The shaft in Fig. 3-82 has a spring constant at the center of 700 N/mm. The mass is 5 kg with center located 80 mm from the shaft center. Find the deflection y at the center under a steady rotation of 1,750 rpm. Neglect weight of shaft.

Ans. $y = 2.01$ mm.

84. The bearings for the continuous shaft of Fig. 3-83 are all at the same elevation.

(a) Find the value of reaction R_2.

(b) Suppose the center bearing in Fig. 3-83 is 0.050 in. lower than the others. Find the value of R_2.

Note that a small change in the configuration of an indeterminate structure can make a large change in the reactive forces.

(c) Suppose the center support in Fig. 3-83 is replaced by a spring whose constant k is equal to 20,000 lb/in. Find the value of the center reaction.

Ans.   (a) 170.4 lb; (b) 180.0 lb; (c) 160.4 lb.

FIGURE 3-83   Problem 84.

85. The bearings for the continuous shaft of Fig. 3-84 are all at the same elevation.

(a) Find the value of reaction R_2.

(b) Suppose the two central bearings in Fig. 3-84 are resting on springs whose constants are 20,000 lb/in. each. Find the value of R_2.

Ans.   (a) $R_2 = 778.6$ lb; (b) $R_2 = 681.2$ lb.

FIGURE 3-84   Problem 85.

86. In Fig. 3-85 the three bearings are at the same elevation.
(a) When all bearings are on solid supports find the value of R_2.
(b) When the center bearing is 0.040 in. lower than the other two find the value of R_2.
(c) Suppose the center bearing is resting on an elastic support that could be represented by a spring of constant equal to 30,000 lb/in. Find R_2.

> *Ans.*   (a) $R_2 = 1,957$ lb; (b) $R_2 = 1,631.5$ lb;
> (c) $R_2 = 1,539.4$ lb.

FIGURE 3-85   Problem 86.

87. In Fig. 3-86 the three bearings are all at the same elevation.
(a) When all bearings are on solid supports find the value of R_2.
(b) When the center bearing is 1 mm lower than the other two find the value of R_2.
(c) Suppose the center bearing is resting on an elastic support that could be represented by a spring of constant equal to 5,200 N/mm. Find R_2.

> *Ans.*   (a) $R_2 = 8,480$ N; (b) $R_2 = 7,036$ N;
> (c) $R_2 = 6,637$ N.

FIGURE 3-86   Problem 87.

88. (a) If the shaft of Fig. 3-15(a) is of steel, find the bending stress in the material at the center support for Example 14.

(b) Find the bending stress for the shaft of Fig. 3-15(b), for Example 16.

Ans. (a) 4,326 psi; (b) 1,776 psi.

This problem illustrates the fact that the introduction of some flexibility in an indeterminate system can be used to reduce the stress in an important part.

89. (a) Find the deflection at the load for a uniform shaft on simple supports carrying the load P as shown in Fig. 1-14 sketch 7.

(b) Now work the same problem by the geometric method of Section 14. Make a complete drawing labeling all lengths and moments. *Ans.* $y = Pa^2b^2/3EIl$.

90. (a) Find the deflection at the load for the shaft shown in Fig. 1-15 sketch 14.

(b) Now work the problem by the geometric method of Section 14. Make a complete drawing labeling all lengths and moments. *Ans.* $y_{max} = Pb^2l/3EI$.

91. Let No. 7 of Fig. 1-14 represent a shaft of 36 in. span carrying a load of 90 lb located 24 in. from the left reaction. Find the deflection by the equation in terms of EI at a point 12 in. from the left reaction. Solve the same problem by the geometric method of Section 14. Make a diagram and dimension all lengths and moments. *Ans.* $y = 60,480/EI$.

92. Let No. 7 of Fig. 1-14 represent a shaft of 50 in. span carrying a load of 120 lb located 30 in. from the left reaction. Find the deflection by the equation in terms of EI at a point 10 in. from the left reaction. Solve the same problem by the geometric method of Section 14. Make a diagram and dimension all lengths and moments. *Ans.* $y = 160,000/EI$.

93. Let No. 13 of Fig. 1-15 represent a shaft of 48 in. span carrying two loads of 30 lb each with $a = 12$ in. Find the deflection at the center by the equation. Solve the same problem by the geometric method of Section 14. Make a diagram and dimension all lengths and moments. *Ans.* $y = 95,040/EI$.

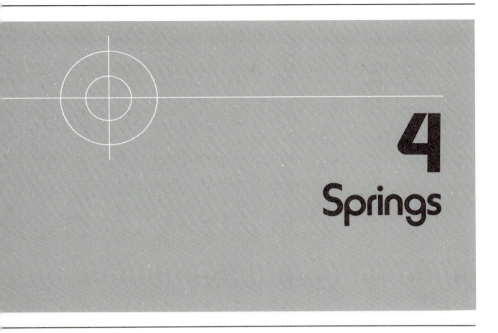

4
Springs

When flexibility or deflection in a mechanical system is specifically desired, some form of spring can be used. Otherwise, the elastic deformation of an engineering body is usually a disadvantage. Springs are employed to exert forces or torques in a mechanism or to absorb the energy of suddenly applied loads. Springs frequently operate with high values for the working stresses, and with loads that are continuously varying.

Helical and leaf springs are in widest use. A number of other types, such as Belleville, disk, spiral, ring, and volute springs, are shown in Fig. 4-1.

A,	area	k,	spring rate, force/unit deflection
c,	spring index		
d,	diameter of wire	kg,	kilogram, 2.2046 lb (mass)
E,	modulus of elasticity, normal stress	K_c,	stress concentration factor due to curvature (Wahl factor)
F_s,	factor of safety	K_s,	stress factor for transverse shear
G,	modulus of elasticity, shear stress		
in.,	inch, 25.4 mm	lb,	pound; 1 lb (force) = 4.4482 N 1 lb (mass) = 0.4536 kg
in. lb,	inch pound, 113 Nmm		
I,	moment of inertia	m,	meter, 39.370 in.
J,	joule, newton-meter, Nm	m,	milli (prefix), one thousandth
J,	polar moment of inertia	mm,	millimeter, 0.039 37 in.
k,	kilo (prefix), 1,000	M,	mega (prefix), 1 million

MPa,	megapascal, 1 N/mm² or 145 psi	δ,	(delta) deflection of helical spring
N,	newton, 0.2248 lb (force)	σ_{ult},	(sigma) ultimate tensile stress
N_c	number of active coils	σ_{yp},	yield point stress in tension
P,	load or force	τ	(tau) shearing stress
Q,	number of inactive end coils	τ'_e,	endurance limit in shear, zero to maximum stress
R,	mean radius of helix		
T,	torque, 1 in. lb = 113 Nmm	τ_{yp},	yield point stress in shear

Initially coned or Belleville

Radially tapered disk

Spiral   Ring   Volute

FIGURE 4-1   Various types of springs.

1. Helical Springs

The equations for stress and deformation of a closely coiled helical spring are derived directly from the corresponding equations for the torsion of a round bar, as shown in Fig. 4-2(a). The bar of length l and diameter d is fitted with brackets at each end of length R, and is in equilibrium under the action of the loads P. Assume that the straight bar is bent into the helix of N coils of radius R as shown in Fig. 4-2(b). The coiled bar is in equilibrium under the action of the two equal and opposite forces P.

The stress in the straight bar is shear caused by a torque equal to PR. The major stress in the helix is also torsional shearing stress. From Eq. (2) of Chapter 3, the stress in Fig. 4-2 is

$$\text{torsional shearing stress} = \frac{Tr}{J} = \frac{16PR}{\pi d^3} \tag{a}$$

FIGURE 4-2 Helical spring formed from round bar.

Let the spring index c be defined as the ratio between the helix diameter $2R$ and the wire diameter d or

$$c = \frac{2R}{d} \tag{1}$$

After being coiled into the helix, the cross sections have an additional stress from the transverse shear. An exact analysis shows that this stress at the midheight has the value $1.23P/A$. Then

$$\text{transverse shearing stress} = 1.23\frac{P}{A} = 1.23\frac{P4}{\pi d^2}\frac{c}{c} = 1.23\frac{4P}{\pi d^2}\frac{2R}{cd}\frac{2}{2}$$

$$= \frac{16PR}{\pi d^3} \times \frac{0.615}{c} \tag{b}$$

The total shearing stress τ on the inside of the coil at the midheight from static load P is given by the sum of Eqs. (a) and (b).

$$\tau = \frac{16PR}{\pi d^3}\left(1 + \frac{0.615}{c}\right) = K_s\frac{16PR}{\pi d^3} \tag{2}$$

where

$$K_s = 1 + \frac{0.615}{c} \tag{3}$$

Substitution of Eq. (1) into Eq. (2) yields two additional forms for the stress equations.

$$\tau = K_s\frac{8Pc}{\pi d^2} = K_s\frac{2Pc^3}{\pi R^2} \tag{4}, (5)$$

When Eq. (3) is substituted into Eq. (4) and the result solved for c, the following useful equation results:

$$c = \frac{\pi d^2 \tau}{8P} - 0.615 \tag{6}$$

Let Eq. (1) for c be substituted into Eq. (6).

$$R = \frac{d}{2}\left(\frac{\pi d^2 \tau}{8P} - 0.615\right) \tag{6a}$$

The deflection of the spring can be found by considering the rotation of the cross sections with respect to each other caused by the torque PR. Assume temporarily that element $ABCD$ in Fig. 4-2(b) is flexible, but that the remainder of the spring is rigid. Thus, from Eq. (5), Chapter 3, the rotation $d\varphi$ of section CD with respect to the adjacent cross section AB is equal to

$$d\varphi = \frac{PR\ dl}{JG}$$

This rotation of the differential length of the spring causes point E, located 90° away, to be moved an amount equal to

$$d\delta = R\ d\varphi = \frac{PR^2\ dl}{JG}$$

The total deflection caused by the torque when the entire spring is elastic is found by integration of this equation over the entire length of the spring.

Then
$$\delta = \frac{64PR^3 N_c}{Gd^4} = \frac{8Pc^3 N_c}{Gd} = \frac{4Pc^4 N_c}{GR} \tag{7}, (8), (9)$$

In these equations, the active length of spring is taken as $2\pi R N_c$ and polar moment of inertia J for the wire as $\pi d^4/32$. Equations (8) and (9) result from substitution of Eq. (1).

The average value of G for steel used for springs is equal to 11,500,000 psi—79,300 MPa in SI units.

If Q is the number of inactive end turns for the spring, its total length is $2\pi R(N_c + Q)$. The cross-sectional area for the wire is $\pi d^2/4$. The volume V of material in the spring is the product of these two terms.

$$V = \tfrac{1}{2}\pi^2 d^2 R(N_c + Q) \tag{10}$$

An equation for the spring rate k, or the force required for a unit deflection, can be had from Eqs. (7), (8), and (9) by replacing P by k and δ by unity.

$$k = \frac{Gd^4}{64R^3 N_c} = \frac{Gd}{8c^3 N_c} = \frac{GR}{4c^4 N_c} \tag{11}$$

Another equation for the spring rate k, can be had by considering Fig. 4-2(c).

$$k = \frac{P_1}{\delta_1} = \frac{P_2}{\delta_2} = \frac{P_2 - P_1}{\delta_2 - \delta_1} \tag{12}$$

Example 1. A helical compression spring is made from No. 8 wire and is to carry a 100-lb static load at a deflection of 1 in. Maximum shearing stress to be

80,000 psi, and number of inactive end coils Q equals 2. Find required values for mean radius R, number of active coils N_c, and volume V of spring material.

Solution. By Table 4-1, wire diameter $d = 0.1620$ in.

By Eq. (6a): $R = \dfrac{d}{2}\left(\dfrac{\pi d^2 \tau}{8P} - 0.615\right) = \dfrac{0.1620}{2}\left(\dfrac{\pi 0.1620^2 \times 80,000}{8 \times 100} - 0.615\right)$

$$= 0.0810(8.2448 - 0.615) = 0.618 \text{ in.}$$

By Eq. (1): $c = \dfrac{2R}{d} = \dfrac{2 \times 0.618}{0.162} = 7.63$

By Eq. (9): $N_c = \dfrac{\delta RG}{4Pc^4} = \dfrac{1 \times 0.618 \times 11,500,000}{4 \times 100 \times 7.63^4} = 5.24$ active coils

By Eq. (10): $V = \tfrac{1}{2}\pi^2 d^2 R(N_c + Q)$

$$= \tfrac{1}{2}\pi^2 0.1620^2 \times 0.618(5.24 + 2) = 0.5797 \text{ in.}^3$$

TABLE 4-1   Steel Spring Wire. Minimum Tensile Strengths, σ_{ult}, and ASTM Designations

W & M Gage* No.	Diameter d, in.	Hard-Drawn, Class 1, A227-64	Music Wire† A228-63T	Oil-Tempered Class 1, A229-64	302 Stainless Steel,† 18-8 A313-67
25	0.0204	283,000	349,000	293,000	299,000
24	0.0230	279,000	343,000	289,000	294,000
23	0.0258	275,000	337,000	286,000	289,000
22	0.0286	271,000	332,000	283,000	285,000
21	0.0317	266,000	327,000	280,000	278,000
20	0.0348	261,000	323,000	274,000	274,000
19	0.0410	255,000	314,000	266,000	269,000
18	0.0475	248,000	306,000	259,000	262,000
17	0.0540	243,000	301,000	253,000	260,000
16	0.0625	237,000	293,000	247,000	255,000
15	0.0720	232,000	287,000	241,000	250,000
14	0.0800	227,000	282,000	235,000	245,000
13	0.0915	220,000	275,000	230,000	240,000
12	0.1055	216,000	269,000	225,000	232,000
11	0.1205	210,000	263,000	220,000	225,000
10	0.1350	206,000	258,000	215,000	217,000
9	0.1483	203,000	253,000	210,000	210,000
8	0.1620	200,000	249,000	205,000	205,000
7	0.1770	195,000	245,000	200,000	195,000
6	0.1920	192,000	241,000	195,000	190,000
5	0.2070	190,000	238,000	190,000	185,000
4	0.2253	186,000	235,000	188,000	180,000
1/4 in.	0.2500	182,000	230,000	185,000	175,000
5/16 in.	0.3125	174,000		183,000	160,000
3/8 in.	0.3750	167,000		180,000	140,000

*Washburn and Moen.
†Values interpolated to nearest gage size.

No allowance is made in the design for stress concentration because of curvature for loads that are static or loads that may fluctuate a relatively small number of times during the expected life of the spring.

Example 1A. A steel helical compression spring is made from wire 4 mm in diameter and is to carry a load of 450 N at a deflection of 25 mm. Maximum shearing stress to be 550 MPa and number of inactive end coils Q equals 2. Find required values for mean radius R, number of active coils N_c, and volume V of spring material.

Solution. By Eq. (6a): $R = \dfrac{d}{2}\left(\dfrac{\pi d^2 \tau}{8P} - 0.615\right) = \dfrac{4}{2}\left(\dfrac{\pi 4^2 \times 550}{8 \times 450} - 0.615\right)$

$$= 2(7.68 - 0.615) = 14.14 \text{ mm}$$

By Eq. (1): $c = \dfrac{2R}{d} = \dfrac{2 \times 14.14}{4} = 7.07$

By Eq. (9): $N_c = \dfrac{\delta RG}{4Pc^4} = \dfrac{25 \times 14.14 \times 79{,}300}{4 \times 450 \times 7.07^4} = 6.23$ active coils

By Eq. (10): $V = \tfrac{1}{2}\pi^2 d^2 R(N_c + Q)$

$$= \tfrac{1}{2}\pi^2 4^2 \times 14.14(6.23 + 2) = 9{,}188 \text{ mm}^3$$

Torsion bar springs similar to Fig. 4-2(a) can sometimes be conveniently arranged for use in motor vehicles.

If a helical compression spring is relatively long as compared with its diameter, danger of column action or lateral buckling may exist at loads smaller than the desired working load. Sometimes it is practical to prevent the buckling of a long compression spring by placing it over a loosely fitting bar or in a tube, which serves as a guide.[1]

2. Properties of Spring Materials

Helical springs are either cold formed or hot formed depending on the size of the wire. Small sizes are wound cold, but when the bar has a diameter larger than about $\frac{3}{8}$ in., the spring is wound from a heated bar.

Numerous materials are used for cold-formed springs. Sizes and strength properties for several materials are shown in Table 4-1. These are cold-drawn wires and the smaller sizes are stronger because of the greater penetration of the hardening from the drawing. Hard-drawn wire is the cheapest and is commonly used for general-purpose springs where cost is an important factor. It is used where the stresses are low and where only static loading is present.

---

[1]See p. 43 of Reference 3, end of chapter. See also Burdick, W. E., F. S. Chaplin, and W. L. Sheppard, "Deflection of Helical Springs under Transverse Loadings," *Trans. ASME*, **61**, 1939, p. 623.

Springs of this material should not be used at temperatures above 250°F nor for subzero applications.

Music wire is a high-quality high-carbon steel wire that is widely used in the smaller diameters. The temperature restrictions are the same as for hard-drawn wire.

Oil-tempered wire is cold drawn to size and then quenched and tempered. It can be used in many places where the cost of music wire is prohibitive, but is ordinarily not considered suitable where long fatigue life is required.

Stainless steel 302 (18 percent chromium, 8 percent nickel) has high tensile strength and corrosion resistance. It can be used up to about 550°F and at subzero temperatures. It has superior creep resistant properties at the higher temperatures.

The properties for alloy spring steel wires are given by Table 4-2. These materials are used where the service conditions are severe, and where shock and impact loads may be present. For severe forming, the wire should be in the annealed condition. When wound from annealed stock, the spring must be heat treated after coiling. These materials are also available in the oil-tempered condition.

Valve spring wire, Table 4-3, has been developed especially for valve springs and other applications requiring high fatigue properties.

TABLE 4-2 Alloy Steel Spring Wire, Minimum Tensile Strengths, σ_{ult}

Diameter, in.	Cr-V A231-63T	Cr-Si A401-63T	Diameter, in.	Cr-V A231-63T	Cr-Si A401-63T
0.020	300,000	300,000	0.177		260,000
0.032	290,000		0.192	220,000	260,000
0.041	280,000	298,000	0.218		255,000
0.054	270,000	292,000	0.244	210,000	
0.062	265,000	290,000	0.250		250,000
0.080	255,000	285,000	0.283	205,000	
0.092		280,000	0.312	203,000	245,000
0.105	245,000		0.375	200,000	240,000
0.120		275,000	0.437	195,000	
0.135	235,000	270,000	0.500	190,000	
0.162	225,000	265,000			

TABLE 4-3 Carbon Steel Value Spring Wire. Minimum Tensile Strengths, σ_{ult} (ASTM Spec. A230-63T)

Diameter, in.	Class 1
0.093–0.128	210,000
0.129–0.162	205,000
0.163–0.192	200,000
0.193–0.250	195,000
	Class 2
0.093–0.128	230,000
0.129–0.162	225,000
0.163–0.192	220,000
0.193–0.225	215,000
0.226–0.250	210,000

Phosphor bronze, Table 4-4, has good corrosion resistance and electrical conductivity, and is frequently used for contact fingers in switches. Spring brass, Table 4-5, has similar properties and generally costs less than phosphor bronze. It should not be used above 150°F, but is suitable for subzero applications.

The maximum strength values for these materials are somewhat higher than the minimum values shown in the tables.

Tables 4-1 to 4-5 pertain only to the tensile strength of spring materials, while designs are usually based on the yield strength in shear. Experimental results indicate that the relationship between the tensile strength, σ_{ult}, and the yield strength in shear, τ_{yp}, is as given[2] in Table 4-6.

TABLE 4-4   Phosphor Bronze Wire. Copper Alloy No. 510, Spring Temper (ASTM Spec. B159-66)

Diameter, in., or Distance Between Parallel Surfaces	Minimum Tensile Strength
0.025 and under	145,000
Over 0.025 to 0.0625	135,000
Over 0.0625 to 0.125	130,000
Over 0.125 to 0.250	125,000
Over 0.250 to 0.375	120,000
Over 0.375 to 0.500	105,000

Sn, 4.2–5.8%; Ph 0.03–0.35%; Cu + Sn + Ph, 99.5% min

TABLE 4-5   Brass Wire. Minimum Tensile Strength, σ_{ult}, for 0.20-in. Diam and Larger (or Distance between Parallel Surfaces) (ASTM Spec. B134-66)

Copper Alloy No.	Copper %	Total Other Elements	Zinc	Minimum Tensile Strength
210	94.0–96.0	0.20	Remainder	72,000
220	89.0–91.0	0.20	Remainder	84,000
230	84.0–86.0	0.25	Remainder	100,000
240	78.5–81.5	0.25	Remainder	116,000
260	68.5–71.5	0.27	Remainder	120,000
268	63.0–68.5	0.30	Remainder	120,000
274	61.0–64.0	0.35	Remainder	120,000

TABLE 4-6   Ratio of Yield Strength in Shear, τ_{yp}, and Endurance Limit in Shear (Zero to Maximum), τ_e', to Ultimate Strength, σ_{ult}

Type	$\dfrac{\tau_{yp}}{\sigma_{ult}}$	$\dfrac{\tau_e'}{\sigma_{ult}}$
Hard-drawn wire	0.42	0.21
Music wire	0.40	0.23
Oil-tempered wire	0.45	0.22
302 stainless steel wire, 18-8	0.46	0.20
Cr-V and Cr-Si alloy wire	0.51	0.20

3. Hot-Formed Springs

Helical springs made of $\frac{3}{8}$-in. bar and larger are usually hot wound to avoid the high residual stresses that would be induced by cold forming. After forming, the spring is heat treated by quenching and tempering to produce the desired physical properties. Both plain-carbon and alloy steels are used for hot-formed springs. The materials in widest use[3] are shown in Table 4-7. Compositions for these steels are given in Chapter 14.

[2]See Elmendorf, H. J., "Ratio of Spring Torsional Elastic Limit to Wire Tensile Strength," *Metal Progr.*, **73**, Apr. 1958, p. 84. See also p. 165 of Reference 3, end of chapter.

[3]See Keysor, H. C., "Carbon and Alloy Materials for Hot-Formed Springs," *Prod. Eng.*, **17**, Nov. 1946, p. 86. See also Bittner, E. T., "Alloy Spring Steels." *Trans. ASM*, **40**, 1948, p. 263.

TABLE 4-7 Tensile Strength of Heat-Treated Steels for Hot-Formed Springs. 1-5/8-Diam Specimens Quenched and Drawn at Temperatures Shown*

Draw Temp. F	1095 Plain Carbon		6150 Chromium-Vanadium		8660 Chromium-Nickel-Molybdenum		9262 Silicon-Manganese	
	Ultimate	*Yield*	*Ultimate*	*Yield*	*Ultimate*	*Yield*	*Ultimate*	*Yield*
850	192,000	128,000	220,000	203,000	206,000	193,000	243,000	212,000
950	188,000	120,000	198,000	185,000	190,000	170,000	214,000	182,000
1050	172,000	107,000	180,000	168,000	171,000	150,000	188,000	156,000
1150	151,000	92,000	162,000	152,000	150,000	129,000	167,000	137,000

For torsion, use 0.60 of corresponding tensile value.

Plain-carbon steel 1095 is in wide use because of its availability and relatively low cost. This material, however, is shallow hardening, and sections larger than $\frac{3}{8}$ in. will not harden completely through. To obtain depth of hardening and a material with a higher yield point, an alloy steel must be used, even though the cost is greater. Silicon-manganese steel 9262 has been widely used as a low-priced alloy spring material. It has the disadvantages of being subject to decarburization in heat treatment and of being inclined to have excessive quantities of nonmetallic inclusions as well as a poor surface. Because of its higher cost, chromium-vanadium steel 6150 is being supplanted by other steels. Use of chromium-nickel-molybdenum steel 8660 is increasing because of its many desirable qualities such as good impact resistance, freedom from decarburization and surface defects, and relatively low cost. Alloy steels in general undergo less permanent set in service than plain-carbon steels, and are better suited to impact conditions, especially for service conditions where the surface of the spring becomes scratched and pitted.

4. Helical Spring of Minimum Volume of Material—Static Load

Helical spring design is dependent on numerous variables, and it is difficult by trial-and-error methods to find the most efficient design. For a statically loaded spring, let deflection δ, modulus G, number of inactive coils at both ends Q, load P, and stress τ be grouped together as shown here and the result[4] called B.

$$\bar{B} = \frac{\delta G}{Q\sqrt{8P\pi\tau}} \tag{13}$$

It can then be shown that if B is determined by the following equation, in

[4]See page 68 of Reference 5, end of chapter.

terms of spring index c, that the resulting statically loaded spring will contain the smallest possible amount of spring material.

$$B = \frac{c^3(5c + 1.23)}{2.46\sqrt{c + 0.615}} \qquad (13a)$$

Factor K_s has been included in obtaining this result.

Quantity B has been calculated for different values of c and the results shown in Table 4-8. After c is found, then wire diameter d, mean helix radius R, number of active coils N_c, and the volume V can be calculated by the equations previously given. Use of Eq. (13) is illustrated by the following example.

TABLE 4-8   Design Constants for Most Efficient Helical Compression Spring with Static Load

c	B	c	B	c	B	c	B	c	B
3.0	94	4.0	257	5.0	563	6.0	1,066	7.0	1,830
3.1	105	4.1	280	5.1	603	6.1	1,130	7.1	1,924
3.2	118	4.2	305	5.2	646	6.2	1,196	7.2	2,020
3.3	131	4.3	331	5.3	690	6.3	1,346	7.3	2,121
3.4	145	4.4	359	5.4	737	6.4	1,337	7.4	2,224
3.5	161	4.5	387	5.5	783	6.5	1,412	7.5	2,331
3.6	178	4.6	420	5.6	837	6.6	1,489	7.6	2,443
3.7	196	4.7	453	5.7	891	6.7	1,570	7.7	2,557
3.8	215	4.8	487	5.8	947	6.8	1,654	7.8	2,675
3.9	235	4.9	524	5.9	1,005	6.9	1,741	7.9	2,798
4.0	257	5.0	563	6.0	1,066	7.0	1,830	8.0	2,924

Example 2. A steel helical spring is to carry a static load of 100 lb at a deflection of 1 in. Stress τ is to be 80,000 psi. End turns Q equal 2. Find a suitable wire diameter d for the most efficient spring. Choose the nearest available standard wire size for the actual spring and find the corresponding values of R and N_c. Find the volume of material in the spring.

Solution. By Eq. (13): $\quad B = \dfrac{\delta G}{Q\sqrt{8P\pi\tau}} = \dfrac{1 \times 11,500,000}{2\sqrt{8 \times 100\pi 80,000}} = 406$

By Table 4-8: $\qquad c = 4.56, \quad$ tentative

By Eq. (3): $\qquad K_s = 1 + \dfrac{0.615}{c} = 1 + \dfrac{0.615}{4.56} = 1.135$

By Eq. (4): $\qquad d = \left(\dfrac{8K_sPc}{\pi\tau}\right)^{1/2} = \left(\dfrac{8 \times 1.135 \times 100 \times 4.56}{\pi 80,000}\right)^{1/2}$

$\qquad\qquad\quad = 0.1284$ in.

By Table 4-1, No. 10, $d = 0.1350$ in. is the nearest standard size wire to the above d.

By Eq. (6):
$$c = \frac{\pi d^2 \tau}{8P} - 0.615 = \frac{\pi 0.1350^2 \times 80,000}{8 \times 100} - 0.615 = 5.110,$$
exact value

By Eq. (1):
$$R = \frac{cd}{2} = \frac{5.110 \times 0.1350}{2} = 0.3450 \text{ in.}$$

By Eq. (9):
$$N_c = \frac{\delta R G}{4P c^4} = \frac{1 \times 0.3450 \times 11,500,000}{4 \times 100 \times 5.110^4} = 14.54 \text{ active coils}$$

By Eq. (10): $V_{min} = \frac{1}{2}\pi^2 d^2 R(N_c + Q)$
$$= \frac{1}{2}\pi^2 0.1350^2 \times 0.3450(14.54 + 2) = 0.5131 \text{ in.}^3$$

Data for Examples 1 and 2 are the same except in Example 1 the wire diameter was arbitrarily chosen while in Example 2 the most efficient size was first determined. It is noted that the spring of the arbitrarily chosen wire contains 13 percent more material than the spring of the most efficient design.

Example 2A. A steel helical spring is to carry a static load of 450 N at a deflection of 25 mm. Stress τ is to be 550 MPa. End turns Q equal 2. Find a suitable wire diameter d for the most efficient spring. Choose the nearest available standard wire size for the actual spring and find the corresponding values of R and N_c. Find the volume of material in the spring.

Solution. By Eq. (13):
$$B = \frac{\delta G}{Q\sqrt{8P\pi\tau}} = \frac{25 \times 79,300}{2\sqrt{8 \times 450\pi550}} = 397$$

By Table 4-8:
$$c = 4.53$$

By Eq. (3):
$$K_s = 1 + \frac{0.615}{c} = 1 + \frac{0.615}{4.53} = 1.136$$

By Eq. (4):
$$d = \left(\frac{8K_s Pc}{\pi\tau}\right)^{1/2} = \left(\frac{8 \times 1.136 \times 450 \times 4.53}{\pi 550}\right)^{1/2}$$
$$= 3.27 \text{ mm} = 0.1289 \text{ in.}$$

Use No. 10 wire for spring. $d = 0.1350$ in. $= 3.43$ mm by Table 4-1.

By Eq. (6):
$$c = \frac{\pi d^2 \tau}{8P} - 0.615 = \frac{\pi 3.43^2 \times 550}{8 \times 450} - 0.615 = 5.032$$

By Eq. (1):
$$R = \frac{cd}{2} = \frac{5.032 \times 3.43}{2} = 8.63 \text{ mm}$$

By Eq. (9):
$$N_c = \frac{\delta R G}{4P c^4} = \frac{25 \times 8.63 \times 79,300}{4 \times 450 \times 5.032^4} = 14.83 \text{ active coils}$$

By Eq. (10): $V_{min} = \frac{1}{2}\pi^2 d^2 R(N_c + Q)$
$$= \frac{1}{2}\pi^2 3.43^2 \times 8.63(14.83 + 2) = 8,432 \text{ mm}^3$$

Cold-formed springs must be wound to a smaller diameter than the desired size because of the spring back[5] or expansion that occurs after coiling. Cold-formed springs must sometimes be given a stress-relieving heat treatment before being put in service.

Cold-formed springs can also be wound from plain-carbon or alloy steel wire in the annealed condition. Afterwards the spring is heat treated to develop suitable strength values.

Compression springs are sometimes wound with considerably greater free height and pitch of coil than is desired. They are then compressed solid several times, which permanently sets the final height at the desired position. This operation is known as presetting or cold setting, and reduces the tendency of the spring to take a permanent set in service. The yield point is exceeded, and residual stresses are retained in the material with a sign opposite to those produced by normal operation. Such a residual stress permits a spring to carry larger loads than one in which the material was originally stress-free.

5. *Optimum Design of Helical Spring*

For many applications, a spring functions through an operating range and must be designed to fulfill the following conditions.

(1) At the most extended condition, the spring must be capable of exerting a given force P_1.

(2) At its most compressed condition, the torsional shear stress must not exceed a specified value τ_2.

FIGURE 4-3   Detent spring.

The detent spring of Fig. 4-3 is an example that fulfills these requirements. The parts rotate with respect to each other about a common center. When the ball is in the notch, the spring must exert sufficient force to hold the parts in position with respect to one another. After relative motion has taken place, and the ball is out of the notch, the stress in the spring must not exceed a safe value.

If only the torsional shear stress and active coils are considered, (K_s and Q neglected) it is easy to show[6] that if the spring is designed so that the mini-

[5]Gardiner, F. J., and H. C. R. Carlson, "The Spring-back of Coil Springs," *Mech. Eng.*, **80**, 1958, p. 74.

[6]See *Trans. ASME*, **75**, 1955, p. 435. See also p. 72 of Reference 5, end of chapter.

mum load, stress, and deflection are exactly one-half the maximum load, stress, and deflection, the spring will contain the least possible amount of material. This will be demonstrated in the following example.

Example 3. Suppose a steel spring must be capable of exerting a force P_1 of 30 lb in its most extended condition. After being compressed an additional amount ($\delta_2 - \delta_1$ of 0.375 in.) the torsional shearing stress τ_2 must not exceed the value 75,000 psi. Space limitations indicate that the mean helix radius should be about 0.40 in.

(a) Find the wire diameter, number of active coils, and the volume of material in the active coils for the optimum spring.

(b) Suppose the designer arbitrarily decides to make the spring of No. 9 wire. Find the volume of material contained in the active coils and compare with the volume of part (a). Retain the value for R of 0.40 in.

Solution. Neglect the transverse shear stress.

(a) For the optimum spring: $\quad \tau_1 = 0.5 \times 75,000 = 37,500$ psi

By Eq. (2), for $K_s = 1$:
$$d^3 = \frac{16P_1R}{\pi\tau_1} = \frac{16 \times 30 \times 0.4}{37,500\pi}$$

$$= 0.001630$$

$$d = 0.1177 \text{ in.}$$

Use No. 11 wire:
$$d = 0.1205 \text{ in.}$$

By Eq. (2):
$$R = \frac{\pi\tau_1 d^3}{16P_1} = \frac{\pi 37,500 \times 0.1205^3}{16 \times 30}$$

$$= 0.4294 \text{ in. adjusted for actual wire diameter}$$

By Eq. (7):
$$N_c = \frac{d^4 G\delta_1}{64P_1 R^3} = \frac{0.1205^4 \times 11,500,000 \times 0.375}{64 \times 30 \times 0.4294^3}$$

$$= 5.98 \text{ coils}$$

Volume, Eq. (10):
$$V_{min} = \tfrac{1}{2}\pi^2 d^2 R N_c$$

$$= \tfrac{1}{2}\pi^2 \times 0.1205^2 \times 0.4294 \times 5.98$$

$$= 0.1840 \text{ in.}^3$$

(b) Using No. 9 wire:
$$d = 0.1483 \text{ in.}$$

By Eq. (2):
$$P_2 = \frac{\pi d^3 \tau_2}{16R} = \frac{\pi 0.1483^3 \times 75,000}{16 \times 0.4}$$

$$= 120.1 \text{ lb}$$

Because of the change in $P_2 - P_1$, the number of active coils becomes

By Eq. (7):
$$N_c = \frac{d^4 G(\delta_2 - \delta_1)}{64(P_2 - P_1)R^3} = \frac{0.1483^4 \times 11,500,000 \times 0.375}{64(120.1 - 30) \times 0.4^3}$$

$$= 5.65 \text{ coils}$$

Volume, Eq. (10):
$$V = \tfrac{1}{2}\pi^2 d^2 R N_c$$
$$= \tfrac{1}{2}\pi^2 0.1483^2 \times 0.4 \times 5.65$$
$$= 0.2455 \text{ in.}^3$$

Volume ratio:
$$\frac{V}{V_{min}} = \frac{0.2455}{0.1840} = 1.33$$

Hence the arbitrarily selected wire diameter turned out to be a poor choice as the resulting spring contains 33 percent more material than the spring of minimum volume.

Example 3A. Suppose a steel spring must be capable of exerting a force P_1 of 135 N in its most extended condition. After being compressed an additional amount $\delta_2 - \delta_1$ of 9.4 mm, the torsional shearing stress τ_2 must not exceed the value 500 MPa. Space limitations indicate that the mean helix radius should be about 10 mm.

(a) Find the wire diameter, number of active coils, and the volume of material for the optimum spring.

(b) Suppose the designer arbitrarily decides to make the spring of No. 9 wire. Find the volume of material contained in the active coils and compare with the volume of part (a). Retain the value of R of 10 mm.

Solution. Neglect the transverse shear stress.

(a) For the optimum spring: $\tau_1 = 0.5 \times 500 = 250$ MPa.

By Eq. (2):
$$d^3 = \frac{16 P_1 R}{\pi \tau_1} = \frac{16 \times 135 \times 10}{\pi 250} = 27.5$$
$$d = 3.185 \text{ mm} = 0.125 \text{ in.}$$

Use No. 11 wire: $d = 0.1205$ in. $= 3.06$ mm

By Eq. (2):
$$R = \frac{\pi \tau_1 d^3}{16 P_1} = \frac{\pi 250 \times 3.06^3}{16 \times 135}$$
$$= 10.42 \text{ mm adjusted value}$$

By Eq. (7):
$$N_c = \frac{d^4 G \delta_1}{64 P_1 R^3} = \frac{3.06^4 \times 77{,}000 \times 9.4}{64 \times 135 \times 10.42^3}$$
$$= 6.49 \text{ active coils}$$

Volume, Eq. (10):
$$V_{min} = \tfrac{1}{2}\pi^2 d^2 R N_c$$
$$= \tfrac{1}{2}\pi^2 3.06^2 \times 10.42 \times 6.49$$
$$= 3{,}125 \text{ mm}^3$$

(b) No. 9 wire: $d = 0.1483$ in. $= 3.767$ mm

By Eq. (2):
$$P_2 = \frac{\pi d^3 \tau_2}{16 R} = \frac{\pi 3.767^3 \times 500}{16 \times 10} = 525 \text{ N}$$

Because of the change in $P_2 - P_1$, the number of active coils becomes

By Eq. (7):
$$N_c = \frac{d^4 G(\delta_2 - \delta_1)}{64(P_2 - P_1)R^3} = \frac{3.767^4 \times 77,000 \times 9.4}{64(525 - 135) \times 10^3}$$

$$= 5.84 \text{ active coils}$$

Volume, Eq. (10):
$$V = \tfrac{1}{2}\pi^2 d^2 R N_c$$

$$= \tfrac{1}{2}\pi^2 3.767^2 \times 10 \times 5.84$$

$$= 4,089 \text{ mm}^3$$

Volume ratio:
$$\frac{V}{V_{min}} = \frac{4,089}{3,125} = 1.31$$

The spring of part (b) with the arbitrarily chosen wire size thus contains 31% more material than the most efficient spring.

6. Fatigue of Springs

Since most failures are caused by fatigue, a poor surface is the worst handicap of hot-formed springs. Figure 4-4 shows a typical fatigue failure of a helical spring. A fatigue crack usually starts at a surface imperfection in a region of stress concentration. The endurance-limit stress for steel bars in the as-rolled condition may be from 30,000 to 45,000 psi for both plain-carbon and low-alloy steels. If the surface is badly pitted, the endurance limit may be as low as 18,000–20,000 psi. These values for actual springs are thus seen to be much lower than the endurance limit for the same material when polished specimens are tested in the laboratory. A layer of decarburized material on the surface, resulting from the heat treatment, is also a source of weakness since the endurance limit for the surface may then be less than the working stress for the spring. Decarburization can be avoided if the heat treatment is con-

FIGURE 4-4 Fatigue failure of helical spring. (From Wahl, Mechanical Springs, Penton Publishing Co.)

ducted in a controlled atmosphere. Corrosion, even in a mild form, greatly reduces the fatigue strength. Cadmium plating offers some degree of relief against corrosion.

Shot peening, which leaves the surface in compression, has proved to be very successful in raising the fatigue life of springs. A good surface can also be secured by the use of ground stock and controlled atmospheres, but the costs are relatively high. Such a surface, however, will be spoiled if the spring is subjected to rough usage.

If the spring operates under conditions of elevated temperature, there is a danger of creep or permanent deformation unless very low stress values are used. Such effects become noticeable above 350°F, and the ordinary spring steels cannot be used at temperatures above 400°F. Stainless steel of the 18-8 type resists high temperature better than other spring steels. For temperatures of 500–800°F, high-speed steel (18W, 4Cr, 1V) must be used.

For low-temperature service, the alloy steels, listed in Table 4-7, are in use. Failure under impact loads at low temperatures can frequently be guarded against by providing a stop to limit the deflection to a safe value.

Many special-type springs for appliances and other products are stamped from flat stock. For high stresses and severe service conditions, the sheared edges of flat springs must be polished to prevent the formation of fatigue cracks.

7. Design for Fluctuating Loads

If the loading on the spring is continuously fluctuating, due allowance must be made in the design for fatigue and stress concentration. The working-stress triangle, as explained in Chapter 2, should be modified, as shown in Fig. 4-5, when the spring material is tested in pulsating shear zero to maximum stress τ'_e. The relationship between τ'_e and the ultimate tensile strength, σ_{ult}, is given[7] in Table 4-6. For such test loading, the range stress is equal to

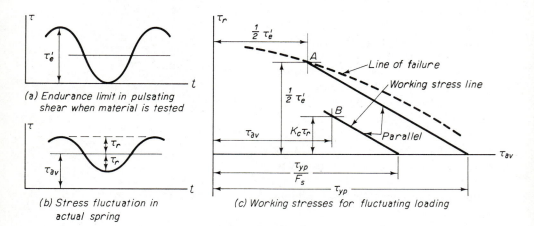

(a) Endurance limit in pulsating shear when material is tested

(b) Stress fluctuation in actual spring

(c) Working stresses for fluctuating loading

FIGURE 4-5 Working stress diagram for springs.

[7]See p. 164 of Reference 3, end of chapter.

the average stress or $\frac{1}{2}\tau_e'$. Thus the straight line approximating the line of failure can be drawn from point A in Fig. 4-5(c). The line giving actual working stresses is parallel to this line and can be drawn after dividing the yield point stress in shear τ_{yp} by the factor of safety F_s.

The stress concentration or Wahl factor K_c for curvature can be calculated by the following equation.

$$K_c = \frac{4c - 1}{4c - 4} \tag{14}$$

The range stress τ_r as determined by Eq. (2), (4), or (5) for range load P_r is multiplied by K_c before plotting in Fig. 4-5(c).

As is customary when designing for fluctuating loading, stress concentration is ignored when the average stress τ_{av}, as determined by the average load P_{av}, is plotted. It is also assumed that the residual stresses from coiling and from eccentric application of the load are small enough to be neglected.

By making a proportion of corresponding sides of similar triangles in Fig. 4-5, the following equation can be written.

$$\frac{K_c \tau_r}{\dfrac{\tau_{yp}}{F_s} - \tau_{av}} = \frac{\frac{1}{2}\tau_e'}{\tau_{yp} - \frac{1}{2}\tau_e'} \tag{15}$$

This is the basic equation for the design of springs with continuously fluctuating loading.

Example 4. A helical compression spring, made of No. 4 music wire, carries a fluctuating load. The spring index is 6, and the factor of safety is 1.5. If the average load on the spring is 120 lb, find the permissible values for the maximum and minimum loads.

Solution. From Eq. (1): $R = \dfrac{cd}{2} = \dfrac{6 \times 0.2253}{2} = 0.676$

By Eq. (14): $K_c = \dfrac{4 \times 6 - 1}{4 \times 6 - 4} = 1.15$

By Eq. (2): $\tau_{av} = \dfrac{16 \times 120 \times 0.676}{\pi 0.2253^3}\left(1 + \dfrac{0.615}{6}\right)$

$= 39,820$ psi

By Table 4-1: $\sigma_{ult} = 235,000$ psi

By Table 4-6: $\tau_{yp} = 0.4 \times 235,000 = 94,000$ psi

$\tau_e' = 0.23 \times 235,000 = 54,050$ psi

By Eq. (15): $\dfrac{1.15\tau_r}{\dfrac{94,000}{1.5} - 39,820} = \dfrac{27,025}{94,000 - 27,025}$

$\tau_r = 8,020$ psi

Then:
$$P_r = \frac{\tau_r}{\tau_{av}} P_{av} = \frac{8,020}{39,820} \times 120 = 24.2 \text{ lb}$$

$$P_2 = 120 + 24.2 = 144.2 \text{ lb}$$

$$P_1 = 120 - 24.2 = 95.8 \text{ lb}$$

The reader should plot a figure similar to Fig. 4-5(c) for this problem and check the stress values above by measuring lengths on the figure.[8]

Example 4A. A helical compression spring, made of No. 4 music wire, carries a fluctuating load. The spring index is 6, and the factor of safety is 1.5. If the average load on the spring is 535 N, find the permissible values for the maximum and minimum loads.

Solution. By Table 4-1, for No. 4 music wire, $d = 0.2253$ in. $= 5.72$ mm.

By Eq. (1):
$$R = \frac{cd}{2} = \frac{6 \times 5.72}{2} = 17.17 \text{ mm}$$

By Eq. (14):
$$K_c = \frac{4 \times 6 - 1}{4 \times 6 - 4} = 1.15$$

By Eq. (2):
$$\tau_{av} = \frac{16 \times 535 \times 17.17}{\pi 5.72^3}\left(1 + \frac{0.615}{6}\right) = 275.3 \text{ MPa}$$

By Table 4-1:
$$\sigma_{ult} = 235,000 \text{ psi} = 1,620 \text{ MPa}$$

By Table 4-6:
$$\tau_{yp} = 0.4 \times 1,620 = 648 \text{ MPa}$$

$$\tau'_e = 0.23 \times 1,620 = 372.6 \text{ MPa}$$

By Eq. (15):
$$\frac{1.15\tau_r}{\frac{648}{1.5} - 275.3} = \frac{186.3}{648 - 186.3}$$

$$\tau_r = 55.0 \text{ MPa}$$

Then:
$$\frac{P_r}{\tau_r} = \frac{P_{av}}{\tau_{av}} \quad \text{or}$$

$$P_r = \frac{P_{av}\tau_r}{\tau_{av}} = \frac{535 \times 55.0}{275.3} = 107 \text{ N}$$

$$P_2 = 535 + 107 = 642 \text{ N maximum load}$$

$$P_1 = 535 - 107 = 428 \text{ N minimum load}$$

8. Buckling of Compression Springs

In compression springs it is important that the height be not so great that buckling or column action will occur. A spring with squared and ground ends compressed between rigid plates can be likened to a column with fixed ends.

---

[8]Other fatigue diagrams are in use for the design of springs. See p. 102 of Reference 7, and p. 164 of Reference 3, end of chapter.

For such a spring the buckling length l_0 can be taken as[9]

$$l_0 = 2\delta + \frac{22.3R^2}{\delta} \tag{16}$$

However, factors such as eccentricity of the loading, effects of end turns, and so on tend to give a buckling length somewhat less than that indicated above.

9. Vibration or Surging of Helical Springs

A sudden compression of the end of a helical spring may form a compression wave that travels along the spring and is reflected at the far end. The material in the compressed wave is subjected to higher stresses, which may cause early fatigue failure. The natural frequency for a round trip of the wave is given by the following equation.

$$f = \frac{d}{2\pi R^2 N_c} \sqrt{\frac{Gg}{32\gamma}} \text{ cycles/sec} \tag{17}$$

Here g represents the acceleration constant of gravity, 386 in./sec^2, and γ represents the weight in pounds per cubic inch for the material of the spring.

For a steel spring, $G = 11,500,000$ psi and $\gamma = 0.285$ lb/in.^{3}. The equation above then becomes

$$f = \frac{3,510d}{R^2 N_c} \text{ cycles/sec} \tag{18}$$

Example 5. Find the lowest natural frequency for a valve spring of No. 4 steel wire with 10 active coils and mean diameter of helix of 2 in.

Solution.　　　　$R = 1$ in.

By Table 4-1:　　　$d = 0.2253$ in.

In Eq. (18):　　　$f = \dfrac{3,510 \times 0.2253}{1^2 \times 10} = 79.1$ cycles/sec

　　　　　　　　　　$= 4,745$ cycles/min

A spring can exhibit higher modes of vibration whose frequencies are 2, 3, 4, . . . , times the value given by Eqs. (17) and (18).

10. Commercial Tolerances

Sizes of helical springs are not standardized. They must therefore be made to order. Production costs can be reduced if liberal tolerances for dimensions and loading are allowed. Tolerances for commercial grade springs are given[10] in Tables 4-9, 4-10, and 4-11.

[9]See Chapter 24 of Reference 7, end of chapter.
[10]See Reference 2, end of chapter.

TABLE 4-9   Standard Tolerances for Spring Wire*

Diameter, mm (in.)	Tolerance, mm (in.)	Maximum Out-of-Roundness, mm (in.)
0.51–0.71 (0.020–0.028) 0.71–2.00 (0.028–0.078)	±0.010 (0.0004) ±0.015 (0.0006)	0.010 (0.0004) 0.015 (0.0006)
2.00–3.00 (0.078–0.118) 3.00–6.00 (0.118–0.240)	±0.020 (0.0008) ±0.030 (0.00118)	0.020 (0.0008) 0.030 (0.0012)
6.00–9.00 (0.240–0.354) 9.50–16.00 (0.375–0.625)	±0.050 (0.00197) ±0.070 (0.00276)	0.050 (0.002) 0.070 (0.0028)

*Data courtesy Associated Spring-Barnes Group. Most spring wires can be purchased to tighter tolerances. Music wire and most nonferrous materials are regularly made to closer tolerances.

TABLE 4-10   Coil Diameter Tolerances of Helical Compression and Extension Springs*

Wire Dia., mm (in.)	Tolerances, ±mm (in.)						
	Spring Index, D/d						
	4	6	8	10	12	14	16
0.38 (0.015)	0.05 (0.002)	0.05 (0.002)	0.08 (0.003)	0.10 (0.004)	0.13 (0.005)	0.15 (0.006)	0.18 (0.007)
0.58 (0.023)	0.05 (0.002)	0.08 (0.003)	0.10 (0.004)	0.15 (0.006)	0.18 (0.007)	0.20 (0.008)	0.25 (0.010)
0.89 (0.035)	0.05 (0.002)	0.10 (0.004)	0.15 (0.006)	0.18 (0.007)	0.23 (0.009)	0.28 (0.011)	0.33 (0.013)
1.30 (0.051)	0.08 (0.003)	0.13 (0.005)	0.18 (0.007)	0.25 (0.010)	0.30 (0.012)	0.38 (0.015)	0.43 (0.017)
1.93 (0.076)	0.10 (0.004)	0.18 (0.007)	0.25 (0.010)	0.33 (0.013)	0.41 (0.016)	0.48 (0.019)	0.53 (0.021)
2.90 (0.114)	0.15 (0.006)	0.23 (0.009)	0.33 (0.013)	0.46 (0.018)	0.53 (0.021)	0.64 (0.025)	0.74 (0.029)
4.34 (0.171)	0.20 (0.008)	0.30 (0.012)	0.43 (0.017)	0.58 (0.023)	0.71 (0.028)	0.84 (0.033)	0.97 (0.038)

TABLE 4-10 (Cont.)

	4	6	8	10	12	14	16
6.35 (0.250)	0.28 (0.011)	0.38 (0.015)	0.53 (0.021)	0.71 (0.028)	0.90 (0.035)	1.07 (0.042)	1.24 (0.049)
9.53 (0.375)	0.41 (0.016)	0.51 (0.020)	0.66 (0.026)	0.94 (0.037)	1.17 (0.046)	1.37 (0.054)	1.63 (0.064)
12.70 (0.500)	0.53 (0.021)	0.76 (0.030)	1.02 (0.040)	1.57 (0.062)	2.03 (0.080)	2.54 (0.100)	3.18 (0.125)

*Data courtesy Associated Spring-Barnes Group.

TABLE 4-11 Free Length Tolerances of Squared and Ground Helical Compression Springs*

Number of Active Coils per mm (in.)	Tolerances, ±mm/mm (in./in.) of Free Length						
	Spring Index, D/d						
	4	6	8	10	12	14	16
0.02 (0.5)	0.010	0.011	0.012	0.013	0.015	0.016	0.016
0.04 (1)	0.011	0.013	0.015	0.016	0.017	0.018	0.019
0.08 (2)	0.013	0.015	0.017	0.019	0.020	0.022	0.023
0.2 (4)	0.016	0.018	0.021	0.023	0.024	0.026	0.027
0.3 (8)	0.019	0.022	0.024	0.026	0.028	0.030	0.032
0.5 (12)	0.021	0.024	0.027	0.030	0.032	0.034	0.036
0.6 (16)	0.022	0.026	0.029	0.032	0.034	0.036	0.038
0.8 (20)	0.023	0.027	0.031	0.034	0.036	0.038	0.040

*Data courtesy Associated Spring-Barnes Group. For springs less than 12.7 mm (0.500 in.) long, use the tolerances for 12.7 mm (0.500 in.). For closed ends not ground, multiply above values by 1.7.

11. Effect of End Turns for Compression Springs

Several different types of end turns used for helical compression springs are shown in Fig. 4-6. The equations are derived for springs assuming that the loading is axial—a condition difficult to secure in practice. The end coils produce an eccentric application of the load, increasing the stress on one side of the spring. Under certain conditions, especially where the number of coils is small, this effect must be taken into account. The nearest approach to an axial load is secured by the spring shown in Fig. 4-6(a), where the end turns are squared and then ground perpendicular to the helix axis.

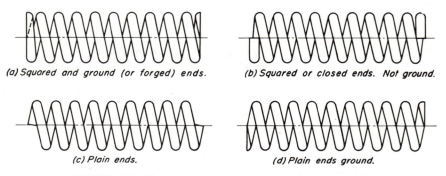

(a) Squared and ground (or forged) ends.   (b) Squared or closed ends. Not ground.

(c) Plain ends.   (d) Plain ends ground.

FIGURE 4-6   Types of end turns used for compression springs.

Equation (7) for the deflection requires the use of the proper number of active coils N_c. A deduction must be made from the total number of coils to take care of the turns at the ends, which do not affect the deflection. It is impossible to say definitely just how much this deduction should be. However, an average value for the number of active coils, based on experimental results, is found by deducting 1.75 turns from the total number, tip to tip, when ends are squared and ground as shown in Fig. 4-6(a). For plain ends, Fig. 4-6(c), the deduction from the total turns should be approximately one-half turn; and for plain ends ground, Fig. 4-6(d), the deduction should be one turn.

12. Helical Extension Springs

In helical tension springs, the shape of the hooks or end turns for applying the load must be designed so that the stress concentration effects caused by the presence of sharp bends are decreased as much as possible. In Fig. 4-7 the end of the tension spring has been formed by merely bending up a half-loop. If the radius of the bend is small, the stress concentration at cross section A will be large.

The most obvious method for avoiding these severe stress concentrations

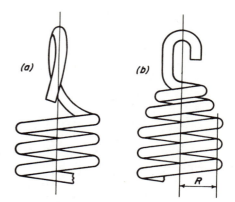

FIGURE 4-7 End of tension spring made by turning up half loop.

FIGURE 4-8 Types of end turns which reduce stress concentration.

is to make the radius r_1 of the bend larger. Figure 4-8(a) shows one method of doing this. Here the end has been formed by turning up a complete loop in a gradual sweeping curve; stress concentration has been greatly reduced. Another expedient is to reduce the radius of the end turns gradually from the maximum value R, as shown in Fig. 4-8(b). Although the curvature of the end turns has been increased, the moment arm for the force has been correspondingly reduced.

Special hooks or loops on the ends of tension springs, as well as the grinding of the ends of compression springs, add to the cost of manufacture and should be avoided whenever possible.

For computing the deflection of a tension spring, the end shown in Fig. 4-7 should be counted as about 0.1 turn for each of the hooks, and the full loop of Fig. 4-8(a) should count as 0.5 turn for each end so formed.

Another feature of helical tension springs is the initial tension which is induced at the time the spring is coiled. Such springs are wound solid and will not deflect until this initial tension is overcome. Table 4-12 gives average values of shear stress as caused by initial tension for springs produced on standard coiling machines.

Example 6. A helical extension spring is wound from 0.080-in.-diam wire with mean diameter of helix of 0.50 in. Find the approximate value of the load P that the spring can sustain before noticeable deflection occurs.

Solution. By Eq. (1): $c = \dfrac{2R}{d} = \dfrac{0.50}{0.08} = 6.25$

By Table 4-12: $\tau = 17{,}000$ psi

By Eq. (2): $17{,}000 = \dfrac{16P0.25}{\pi(0.08)^3}\left(1 + \dfrac{0.615}{6.25}\right)$

$P = 6.22$ lb

TABLE 4-12　Average Value of Shear Stress
Induced by Initial Tension in
Helical Extension Spring*

$c_1 = 2R/d$	Shear Stress τ, psi
4	22,500
5	20,400
6	18,100
7	16,400
8	15,100
9	13,500
10	12,200
11	11,000
12	10,000

*Data courtesy Spring Manufacturers Assn., Inc.

Working stresses for helical extension springs are usually about 75 percent as great as for corresponding compression springs.

13. Helical Springs of Rectangular Wire

When rectangular wire is used for helical springs, the value of the shearing stress can be found by use of the equations for rectangular shafts. For the springs of Fig. 4-9(a) and (b) the stresses at points A_1 and A_2 are found by

FIGURE 4-9　Helical spring of rectangular wire.

Eqs. (28) and (29), Chapter 3, which are as follows:

$$\tau = \frac{PR}{\alpha_1 bc_1^2}, \quad \text{for point } A_1 \tag{19}$$

$$\tau = \frac{PR}{\alpha_2 bc_1^2}, \quad \text{for point } A_2 \tag{20}$$

Values of α_1 and α_2 for various b/c_1 ratios are found in Table 3-4.

To these stresses must be added the transverse shearing stress of $1.5P/A$ to point A_1 in Fig. 4-9(a) and point A_2 in Fig. 4-9(b).

An equation for deflection may be derived from Eq. (30), Chapter 3 by substituting $\theta_1 = \delta/Rl$, and $l = 2\pi RN_c$. Hence

$$\delta = \frac{2\pi PR^3 N_c}{\beta Gbc_1^3} \tag{21}$$

Since a rectangular bar tends to become trapezoidal in cross section after it is wound into a spring, equations for the stresses and deflections for the true shape are more complicated. The equations above are only approximate since it is assumed that the cross section remains rectangular.

Example 7. Find the stresses at points A_1 and A_2 of the spring of Fig. 4-9(b) for $b = \frac{3}{4}$ in., $c_1 = \frac{1}{2}$ in., and $R = 1.2$ in. The static load on the spring is 2,500 lb. Find the deflection if the number of active coils is 6.

Solution.
$$c = \frac{2R}{b} = \frac{2 \times 1.2}{\frac{3}{4}} = 3.2$$

From Table 3-4: $b/c_1 = 1.50$; $\alpha_1 = 0.231$; $\alpha_2 = 0.269$; $\beta = 0.196$

By Eq. (19): $\tau = \dfrac{PR}{\alpha_1 bc_1^2} = \dfrac{2,500 \times 1.2}{0.231 \times 0.75 \times 0.5^2}$

$$= 69,270 \text{ psi}, \quad \text{for point } A_1$$

By Eq. (20): $\tau = \dfrac{PR}{\alpha_2 bc_1^2} = \dfrac{2,500 \times 1.2}{0.269 \times 0.75 \times 0.5^2}$

$$= 59,480 \text{ psi}, \quad \text{for point } A_2$$

Transverse shear stress at A_2:

$$\tau = 1.5\frac{P}{A} = 1.5\frac{2,500}{0.75 \times 0.5} = 10,000 \text{ psi}$$

Total shear stress at A_2:

$$\tau = 59,480 + 10,000 = 69,480 \text{ psi}$$

In Eq. (21): $\delta = \dfrac{2\pi \times 2,500 \times 1.2^3 \times 6}{0.196 \times 11,500,000 \times 0.75 \times 0.5^3} = 0.771 \text{ in.}$

For fluctuating loads, the design can be made as in Section 7, with the stress concentration factor K_c estimated by Eq. (14).

14. Helical Springs with Torsional Loading

A helical spring can be loaded by a torque about the axis of the helix. Such loading, as shown in Fig. 4-10(a), is similar to the torsional loading of a shaft. The torque about the axis of the helix acts as a bending moment on each section of the wire as shown in Fig. 4-10(b). The material is thus stressed in flexure, and the usual equation for bending stress can be used.

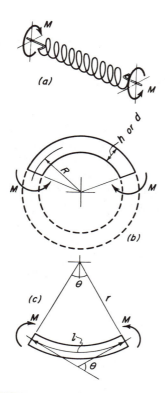

Because of the curved shape, the stress is larger on the inner edge of the coil and smaller on the outer edge. The adjustment can be made by multiplying the stress, as given by $\sigma = Mv/I$, Eq. (10), Chapter 1, by an appropriate stress concentration factor. For wire of rectangular cross section, these factors are

$$K_1 = \frac{3c^2 - c - 0.8}{3c(c-1)},$$

for inner edge   (22)

$$K_2 = \frac{3c^2 + c - 0.8}{3c(c+1)},$$

for outer edge   (23)

where $c = 2R/h$; h is the depth of section perpendicular to the axis.

For circular cross sections, the stress concentration factors are

$$K_3 = \frac{4c^2 - c - 1}{4c(c-1)},$$

for inner edge   (24)

$$K_4 = \frac{4c^2 + c - 1}{4c(c+1)},$$

for outer edge   (25)

FIGURE 4-10 Loading and angular deformation of torsional spring.

where $c = 2R/d$.

Cold forming leaves a residual stress of compression in the outer region of the cross section. The normal loading of the spring should therefore be such as to cause it to wind up and thus produce tensile stress in the outer portion of the wire. If the spring is mounted over an arbor, sufficient clearance must be provided to allow for the decrease in radius of the helix as the spring is wound. For fluctuating loads, care should be exercised in the design of the end turns or hooks to avoid stress concentration caused by sharp bends.

For commonly used values of the spring index, the curvature has no effect on the angular deformation. The deformation of the wire in the spring is the same as for a straight bar of the same length *l*. By No. 1 of Fig. 1-13 the total angular deformation θ between tangents drawn at the ends of the bar, as shown in Fig. 4-10(c), is

$$\theta = \frac{Ml}{EI} \qquad (26)$$

Angle θ in some cases may amount to a number of revolutions.

Example 8. A torsional window-shade spring is made from No. 17 music wire. The mean diameter of helix is 0.875 in. and the number of coils is 400. Assume $\sigma_{yp} = 0.6\sigma_{ult}$ and $F_s = 2$ based on the yield point. Compute stresses on the inside of the helix, taking into account the stress concentration due to curvature. Find the torque that the roller can exert after unwinding 12 revolutions from the most highly stressed condition.

Solution. By Table 4-1: $d = 0.0540$, $\sigma_{ult} = 301{,}000$ psi

$$c = \frac{2R}{d} = \frac{0.875}{0.054} = 16.2$$

By Eq. (24):

$$K_3 = \frac{4 \times 16.2^2 - 16.2 - 1}{4 \times 16.2(16.2 - 1)} = 1.048$$

$$\sigma_{yp} = 0.6 \times 301{,}000 = 180{,}600 \text{ psi}$$

$$K_3\sigma = \frac{\sigma_{yp}}{F_s} = \frac{180{,}600}{2} = 90{,}300 \text{ psi}$$

$$\sigma = \frac{90{,}300}{1.048} = 86{,}140 \text{ psi}$$

$$I = \frac{\pi d^4}{64} = \frac{\pi \times 0.054^4}{64} = 0.0000004174 \text{ in.}^4$$

At maximum stress:

$$M = \frac{\sigma I}{v} = \frac{86{,}140 \times 0.0000004174}{0.027} = 1.3317 \text{ in. lb}$$

$$l = 2\pi R N_c = \pi \times 0.875 \times 400 = 1{,}100 \text{ in.}$$

At maximum stress by Eq. (26):

$$\theta = \frac{1.3317 \times 1{,}100}{30{,}000{,}000 \times 0.0000004174} = 116.94 \text{ rad}$$

Deflection, after unwinding 12 revolutions:

$$\theta_2 = 116.94 - 12 \times 2\pi = 41.54 \text{ rad}$$

Remaining torque, by Eq. (26):

$$M = \frac{30{,}000{,}000 \times 0.0000004174 \times 41.54}{1{,}100}$$

$$= 0.473 \text{ in. lb}$$

15. Leaf Springs

The stresses and deformations of rectangular leaf springs for small deflections can be found by the appropriate equations for beams as given in Chapter 1. For cases where the width of the cross section is large compared with the thickness, it is necessary to multiply the deflection as given by the equation for a narrow beam by $(1 - \mu^2)$, where μ is Poisson's ratio, as explained in Section 13 of Chapter 1.

Example 9. Find the sidewise deflection of the spring-mounted table shown in Fig. 4-11. The steel springs are fixed at top and bottom so that tangents at the ends remain vertical at all times.

Solution. The springs are deformed into symmetrical curves with points of inflection at the midheight. A point of inflection has a zero value for the bending moment because the moment changes sign at such a point. If the springs are cut at the midheights, shear forces only are present as shown in the freebody diagram of Fig. 4-11(b). Hence

$$I = \frac{bh^3}{12} = \frac{2}{64 \times 12} = \frac{1}{384} \text{ in.}^4$$

$$y = (1 - \mu^2)\frac{Pl^3}{3EI} = \frac{(1 - 0.3^2) \times 25 \times 15^3 \times 384}{3 \times 30,000,000} = 0.3276 \text{ in.}$$

Total deflection of table $= 2 \times 0.3276 = 0.6552$ in.

FIGURE 4-11   Example 9.

FIGURE 4-12   Multiple-leaf spring approximately by trapezoid.

Multiple-leaf springs are in wide use, especially in motorcar and railway service. An exact analysis of this type of spring is complicated. An approximate solution can be obtained if the shorter leaves are tapered to a sharp point, and if it is assumed that the leaves remain in contact with one another throughout their length when the spring deflects. If these conditions are fulfilled, the curvature, and therefore the stress, may be obtained by replacing the actual spring shown in Fig. 4-12(a) by the trapezoid in Fig. 4-12(b). To form this trapezoid, it is assumed that each of the shorter leaves has been split along its center, that each half has been placed on either side of the longest leaf, and that all edges have then been welded together. The width b_0 at the support is equal to the width of each leaf multiplied by the number of leaves. When the trapezoid deflects under the load, the leaves on either side of a welded edge have the same deflection and curvature, and hence the same stress.

For small deflections, the equation for the cantilever trapezoid with a load on the end is as follows:

$$\delta = K_1 \frac{Pl^3}{3EI_0} \tag{27}$$

Factor K_1 depends on the ratio of the widths b/b_0, and is given by the curve of Fig. 4-13. The moment of inertia I_0 refers to the section at the wall and is equal to $b_0 h^3/12$, where h is the thickness of the leaf. If b is large in comparison with h, the previous remark relative to the deflection of wide beams should be observed.

FIGURE 4-13 K_1 **factor for deflection of trapezoidal spring.**

When the deflection is large compared to the thickness, the usual equations for small deflections of beams can no longer be used. The theory is complicated and beyond the scope of this book.[11]

---

[11]See Bisshopp, K. E., and D. C. Drucker, "Large Deflection of Cantilever Beam," *Quart. Appl. Math.*, 3, No. 3, Oct. 1945, p. 272. See also p. 90 of Reference 5, end of chapter.

Leaf springs frequently have a hole through the leaves for the tie bolt. This reduction of area, together with the stress concentration, usually occurs at the point of maximum bending stress and is frequently the cause of fatigue failures. The stress concentration factor for a hole, and also for semicircular notches, may be taken from Fig. 2-20 for a bar in tension. Another harmful effect is due to the clamping pressure used in attaching the spring. This compression also usually occurs at the location of maximum bending stress and causes a reduction in the fatigue strength.

Materials suitable for leaf springs are listed in Table 4-7. The previous remarks concerning condition of surface and decarburized surface layer apply equally well to leaf springs. If sharp bends occur in the region of high stress, the concentration must be taken into account, particularly for nonsteady loading.

16. Energy Storage by Springs

Springs are frequently used for storing energy in spring motors, or for absorbing the energy of shock or impact loads. When the force and deformation of a spring are proportional, the energy stored is equal to $\frac{1}{2}P\delta$, where P is the load and δ is the deformation. This equation assumes that the spring was carrying no load at the time P was applied.

The strain energy in a helical spring of round wire with a static load is

$$\text{energy} = \tfrac{1}{2}P\delta = \frac{P}{2} \times \frac{64PR^3N_c}{d^4G}$$

Let the following equality be formed from Eq. (a) of Section 1 and Eq. (10),

$$\frac{\tau^2}{4G} \times \text{volume} = \frac{1}{4G} \times \frac{256P^2R^2}{\pi^2 d^6} \times \frac{\pi^2 d^2 R N_c}{2}$$

The quantities on the right for the two expressions above are equal, so that

$$\text{energy} = \tfrac{1}{2}P\delta = \frac{\tau^2}{4G} \times \text{volume} \qquad (28)$$

For a given stress, some types of springs can store more energy, per unit volume of material, than can other types. For the purpose of comparing the helical spring with other kinds, Eq. (28) will be transformed into units of axial stress. Within the elastic range $\tau = 0.5\sigma$ and $G = E/2(1 + \mu)$, where $\mu = 0.3$. These substitutions are made in Eq. (28) to give

$$\text{energy} = \frac{\sigma^2}{6.15E} \times \text{volume} \qquad (29)$$

For a torsional spring, the energy is equal to $\frac{1}{2}\theta M$. Equation (26) is now substituted for θ; for M, its value $M = \sigma I/c$ is substituted. When the wire is rectangular, the further substitution, $I = bh^3/12$, is made. Hence,

after the terms have been rearranged,

$$\text{energy} = \frac{\sigma^2}{6E} \times \text{volume} \tag{30}$$

If the torsional spring is made of round wire, $I = \pi d^4/64$, and the equation for energy becomes

$$\text{energy} = \frac{\sigma^2}{8E} \times \text{volume} \tag{31}$$

The energy storage capacity is thus smaller when round wire is used.

For a rectangular cantilever with load P on the end, the deflection is $Pl^3/3EI$, and the expression for the energy in terms of the maximum stress, which occurs at the support, is

$$\text{energy} = \frac{\sigma^2}{18E} \times \text{volume} \tag{32}$$

If the cantilever is triangular ($b/b_0 = 0$) the deflection, from Eq. (21) and Fig. 4-13, is equal to $1.5Pl^3/3EI_0$, and the equation for the energy becomes

$$\text{energy} = \frac{\sigma^2}{6E} \times \text{volume} \tag{33}$$

Since the multiple-leaf spring may be approximated at the triangular cantilever, Eq. (33) indicates that such a spring is efficient from the standpoint of energy storage.

17. Belleville Spring

The initially coned or Belleville spring, Fig. 4-14, is characterized by the nonlinear relationship between load and deflection. The dimensions can be arranged so that a large deflection can be obtained with practically no change in load. This is sometimes a useful characteristic in machine construction.

The theory is exceedingly complex,[12] but simple expressions can be obtained for the spring with $h/t = 1.5$ and for loading such that deflection δ is equal to the height h. At this time the spring is compressed flat, but the

FIGURE 4-14 Cross section through initially coned or Belleville spring.

[12]See Chapter 13 of Reference 7, end of chapter. See also p. 80 of Reference 5, end of chapter.

load-deflection curve, Fig. 4-15, for the $h/t = 1.5$ spring, indicates that such a load will practically correspond to δ/t values between 1.0 and 2.0. The spring can thus undergo a large deflection range with practically no change in loading. Table 4-13 gives values by which calculations can be made. The following equation is also used.

$$\sigma_t = K_1 \frac{Et^2}{b^2} \qquad (34)$$

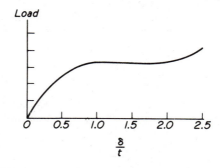

FIGURE 4-15   Load vs. deflection for $h/f = 1.5$.

TABLE 4-13   Constants for Steel Belleville Spring for $h/t = \delta/t = 1.5$

$\dfrac{b}{a}$	K_1	$\dfrac{b\sigma_t}{\sqrt{P}}$
1.25	−8.83	−22,090
1.50	−6.29	−19,430
1.75	−5.63	−19,050
2.00	−5.44	−19,350
2.50	−5.54	−20,630

High stresses are used in the design of Belleville springs. The maximum stress is compression on the upper surface and inner boundary in Fig. 4-14. For static loads, or loads that may be repeated but a few times, experience has shown that a maximum stress of 200,000 psi can be used even though the yield strength of the material may be considerably less. If the yield point is exceeded, a redistribution of the stresses occurs because of localized yielding. Residual stresses of the opposite sign are produced, which give lower values for the load stresses than the calculated values.

Example 10. Find the outside radius b and thickness t for a steel Belleville spring, $h/t = 1.5$, that carries a load of 1,000 lb at a maximum compressive stress of 200,000 psi (compressed flat). Let $b/a = 1.75$.

Solution.

From Table 4-13: $\dfrac{b\sigma_t}{\sqrt{P}} = -19{,}050$ and $K_1 = -5.63$

Then: $b = \dfrac{-19{,}050\sqrt{P}}{\sigma_t} = \dfrac{-19{,}050\sqrt{1{,}000}}{-200{,}000} = 3.012$ in.

$$a = \frac{b}{1.75} = \frac{3.012}{1.75} = 1.721$$

By Eq. (34): $t^2 = \dfrac{b^2\sigma_t}{K_1 E} = \dfrac{3.012^2(-200{,}000)}{-5.63 \times 30{,}000{,}000} = 0.01074$

or $t = 0.1036$ in.

and $h = 1.5t = 0.1554$ in.

The spring dimensions are thus completely determined.

18. Rubber Springs

Rubber springs and cushioning devices are finding an increasing range of application in industry. Knowledge of the behavior of rubber under stress is not clearly understood, and the results of calculations must be considered only approximate. The modulus of elasticity for rubber depends on the durometer hardness number, as shown by Fig. 4-16. Rubber does not follow

FIGURE 4-16 Relationship between durometer hardness number and modulus of elasticity for rubber.

Hooke's law, but becomes increasingly stiff as the deformation is increased. The relationship between load and deflection for a 1-in. cube of 55 durometer rubber in compression is shown by Fig. 4-17.

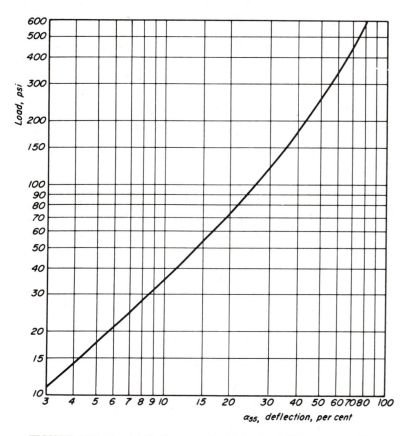

FIGURE 4-17   Load deflection curve for 1 in. cube of 55 durometer rubber between steel plates bonded to the rubber.

A rectangular block of rubber as shown in Fig. 4-18 is frequently used as a compression spring. Based on a considerable number of tests, an empirical equation[13] has been derived for finding the percentage of deflection α for such a spring.

$$\alpha = \frac{\alpha_{55} E_{55}}{E} \cdot \frac{(h\beta)^{2/3}}{\sqrt{A}} \tag{35}$$

[13]See Smith, J. F. D., "Rubber Mountings," *Trans. ASME,* **60,** 1938, p. A-13; **61,** 1939, p. A-159.

FIGURE 4-18 **Rubber block spring with compression loading.**

Here α_{55} represents the percentage of deformation of a 1-in. cube of 55 durometer rubber for the given compressive stress as shown by Fig. 4-17; E_{55} represents the modulus of elasticity, 313 psi, for 55 durometer rubber; modulus E refers to the durometer hardness of the slab under consideration, as given by Fig. 4-16. Thickness of the slab is h in., and β is the ratio of the length to the breadth of the slab. The area normal to the load is given by A.

Example 11. Find the percentage of deflection of a 3×6-in. slab of 65 durometer rubber 1 in. thick that carries a load of 5,000 lb.

Solution. Unit load: $\quad p = \dfrac{5,000}{3 \times 6} = 278$ psi

By Fig. 4-17: $\qquad \alpha_{55} = 54\%$

$$\beta = \tfrac{6}{3} = 2$$

By Fig. 4-16: $\qquad E = 3 \times 145 = 435$ psi for 65 durometer

In Eq. (35): $\qquad \alpha = \dfrac{54 \times 313(1 \times 2)^{2/3}}{435\sqrt{3 \times 6}} = 14.54\%$

A load-deflection diagram can be plotted for a rubber spring by finding the value of α for other loads. The foregoing discussion assumes that the rubber in Figs. 4-17 and 4-18 is bonded to metal plates top and bottom.

Figure 4-19 shows a simple shear spring of two blocks of rubber. Hooke's law holds reasonably well for rubber in shear, and the following equation can be used for the shearing deformation γ:

$$\gamma = \frac{\tau}{G} = \frac{P}{2AG} \qquad \text{(a)}$$

where A represents the area of one of the surfaces in shear.

FIGURE 4-19 **Simple rubber shear spring.**

Deformation δ is equal to $h \tan \gamma$, and for moderate values of γ the angle can be substituted for the tangent, with gives $\delta = h\gamma$. Substitution should now be made for γ in Eq. (a) to obtain an approximate equation for the deflection δ.

$$\delta = \frac{Ph}{2AG} \qquad (36)$$

A shear spring of cylindrical form is shown in Fig. 4-20. The rubber is bonded to a steel ring on the outside and a steel shaft in the center. Shear stress in the rubber at radius r is equal to

$$\tau = \gamma G = \frac{P}{2\pi rh} \qquad (b)$$

FIGURE 4-20   Cylindrical rubber spring with shear loading.

Deformation angle γ is then equal to

$$\gamma = \frac{P}{2\pi rhG} = \frac{b}{r} \qquad (c)$$

where

$$b = \frac{P}{2\pi hG} \qquad (d)$$

The slope of the deformed rubber at radius r is given by

$$\frac{dy}{dr} = \tan \gamma = -\tan \frac{b}{r} \qquad (e)$$

The minus sign is due to the fact that the slope is negative for the chosen coordinate system. The tangent term should now be expanded into a series and integrated term by term.

$$\frac{dy}{dr} = \frac{-b}{r} - \frac{b^3}{3r^3} - \cdots \qquad (f)$$

$$y = -b \log_e r + \frac{b^3}{6r^2} + \cdots + C \qquad (g)$$

When $\tan \gamma$ is less than about 0.4, only the $b \log_e r$ term need be retained. The constant of integration C can be evaluated from the condition that $y = 0$

when $r = r_o$. Then

$$C = b \log_e r_o \tag{h}$$

and

$$y = b \log_e \frac{r_o}{r} \tag{i}$$

The maximum value of y or δ occurs for $r = r_i$. Hence,

$$\delta = b \log_e \frac{r_o}{r_i} = \frac{P}{2\pi hG} \log_e \frac{r_o}{r_i} \tag{37}$$

A cylindrical spring loaded in torsional shear is shown in Fig. 4-21. At radius r the area in shear is $2\pi rh$. The shear force is this area multiplied by shearing stress τ, and the torque or moment is the force multiplied by the radius r.

$$M = 2\pi r^2 h\tau \quad \text{or} \quad \tau = \frac{M}{2\pi r^2 h} \tag{j}$$

The shear stress has a maximum value for the minimum value of r. Hence,

$$\tau_{max} = \frac{M}{2\pi r_i^2 h} \tag{38}$$

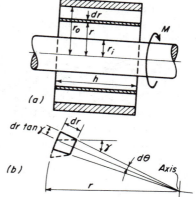

Figure 4-21(b) shows, to an enlarged scale, the deformation angle γ for the element at radius r from the axis. Since $r \, d\theta = dr \tan \gamma$, the contribution $d\theta$ to the total angular rotation of the shaft, given by the element at radius r, is

FIGURE 4-21 Cylindrical rubber spring with torsion loading.

$$d\theta = \frac{dr \tan \gamma}{r} \tag{k}$$

Angle γ can be found by the following equation.

$$\gamma = \frac{\tau}{G} = \frac{M}{2\pi r^2 hG} = \frac{c}{r^2} \tag{l}$$

where

$$c = \frac{M}{2\pi hG} \tag{m}$$

This value of γ should now be substituted in Eq. (k), and the tangent term expanded into a series. The resulting equation should be integrated term by term.

$$dθ = \frac{1}{r}\left(\tan\frac{c}{r^2}\right)dr = \frac{1}{r}\left(\frac{c}{r^2} + \frac{c^3}{3r^6} + \cdots\right)dr \tag{n}$$

$$θ = \left(-\frac{c}{2r^2} - \frac{c^3}{18r^6}\right)_{r_i}^{r_o} \tag{o}$$

This series converges rapidly, and the first term will usually be sufficient. When the limits for r are substituted, the following equation for $θ$ is obtained.

$$θ = \frac{M}{4\pi hG}\left(\frac{1}{r_i^2} - \frac{1}{r_o^2}\right) \tag{39}$$

The equations for many other types of rubber springs can be found in the literature.

Working stresses are generally limited to 25–50 psi in shear. For rubber compression springs, the deformation should not be greater than 10–20 percent of the free height to avoid excessive creep. Under fatigue loading even smaller values should be used. For satisfactory curing during vulcanization the maximum thickness should be limited to 2 in., with a preferred thickness of not over 1 in.

REFERENCES

1. *Design Handbook*, 1981, Bristol, Conn.: Associated Spring-Barnes Group.

2. *Manual on the Design and Application of Helical and Spiral Springs—SAE J 795, Handbook Supplement* **9**, New York: Society of Automotive Engineers, Inc., July 1962.

3. *Metals Handbook*, 8th ed., Vol. 1, Metals Park, Ohio: American Society for Metals, 1961.

4. ROTHBART, H. A., Ed., *Mechanical Design and Systems Handbook*, New York: McGraw-Hill Book Company, 1964, Section 33.

5. SPOTTS, M. F., *Mechanical Design Analysis*, Englewood Cliffs, N.J.: Prentice-Hall, Inc., 1964.

6. WAHL, A. M., "Analysis of Effect of Wire Curvature on the Allowable Stresses in Helical Springs," *Trans. ASME*, **61**, 1939, *J. Appl. Mech.*, p. A-25.

7. WAHL, A. M., *Mechanical Springs*, 2d ed., New York: McGraw-Hill Book Company, 1963.

PROBLEMS

Spring material for the problems is steel.

1. A steel helical spring must sustain a static load of 100 lb at a deflection of 2.3 in. The spring index is equal to 6. Find the value of the maximum stress if the spring is made of No. 8 wire. Find the required number of active coils.

Ans. $N_c = 25$.

2. A helical spring is made of No. 10 steel wire and has 20 active coils. The spring index is equal to 4.67. Loading is static. Use SI units and find the stress when the deflection is equal to 25 mm.   *Ans.*   477 MPa.

3. A helical spring is made of No. 8 wire with the spring index equal to 6. The load is steady and the spring rate is 42.5 lb/in. Maximum stress is 60,000 psi. Find the number of active coils.   *Ans.* $N_c = 25.4$ coils.

4. A helical spring of rate 10 N/mm is mounted on top of another spring of rate 7 N/mm. Find the force required to give a total deflection of 50 mm.
   Ans. $F = 206$ N.

5. The outer of two concentric helical springs has a rate of 2,400 lb/in. The inner spring has a rate of 1,750 lb/in. The outer spring is $\frac{1}{2}$ in. longer than the inner. If the total load is 8,000 lb, find the weight carried by each spring.
   Ans.   Outer $= 5,132.5$ lb; inner $= 2,867.5$ lb.

6. The larger of two concentric compression helical springs is made of 38-mm diam round bar stock, is 225-mm OD of helix, and has 6 active coils. The inner spring is of 25-mm bar, is 140-mm OD helix, and has 9 active coils. The free height of the outer spring is 19 mm more than the inner. Find the deflection of each spring for a load of 90,000 N. Take $G = 77,000$ MPa. Find the load carried by each spring.
   Ans. $P_o = 62,050$ N; $\delta_o = 117.7$ mm.

7. A compression spring is to be made of 0.120-in.-diam wire with 10 active coils, and 1-in. mean diameter of helix. Commercial limits for the wire diameter are 0.1185 and 0.1215 in. Helix diameter may vary from 0.980 to 1.020 in. The number of active coils may vary from $9\frac{3}{4}$ to $10\frac{1}{4}$. The modulus may vary from 11,400,000 to 11,800,000 psi. Find the spring rate when all variables tend to give the weakest spring. Find the spring rate when all variables tend to give the stiffest spring.
   Ans. $k = 25.8$ lb/in.; $k = 35.0$ lb/in.

8. A helical spring must be capable of exerting a force of 600 N after being released 10 mm from its most highly compressed position. Wire of 6 mm diameter is used; the spring index is equal to 6. The loading is static. The maximum stress is 400 MPa. Find the required number of active coils.   *Ans.* $N_c = 10.8$ coils.

9. A helical compression spring is made from No. 4 wire and carries a static load of 250 lb. The maximum shearing stress is to be 60,000 psi. Find the radius of the helix and the number of active coils for a deflection of $\frac{3}{4}$ in.
   Ans. $R = 0.470$ in.; $N_c = 13.4$ coils.

10. A helical spring of music wire is subjected to a continuously varying load. The length varies between 63.5 and 70 mm with corresponding loads of 510 N and 350 N. The spring is made of No. 7 wire and has a mean helix diameter of 25 mm. Find the value of the factor of safety and the number of active coils.
   Ans. $F_s = 1.32$; $N_c = 10.6$ coils.

11. A steel helical spring carries a compressive static load of 50 lb at a deflection of 0.25 in. Maximum stress is 75,000 psi. End turns Q equal 1.5 turns.

Find the value of c for the spring of minimum volume of material. Calculate the value of d and choose the nearest standard wire diameter for the actual spring. Calculate the corresponding R and the number of active coils N_c. Also find the volume of material in the spring.   *Ans.* $c = 3.7$.

12. A helical spring is subjected to a continuously varying load. No. 7 oil-tempered wire is used with a mean radius of helix of 13 mm. In the most compressed condition the force is 400 N. After 8 mm of release the minimum force is 260 N. Find the factor of safety and the number of active coils.

Ans. $F_s = 1.42$; $N_c = 13.2$ coils.

13. A helical spring is made from rectangular wire $\frac{1}{2} \times \frac{1}{4}$ in. with the narrow side parallel to the helix axis. Mean helix radius is 1.5 in. The static load is 150 lb.

(a) Find the value of the maximum stress, and the deflection, if the number of active coils is eight.

(b) Repeat part (a) with the spring wound with the long side of the cross section parallel to the axis.   *Ans.   (a) $\tau = 29{,}270$ psi; $\delta = 1.24$ in.*

14. A helical spring carries a static load of 1,110 N. There are 10 active coils, and the radius of the helix to the center of the wire is 25 mm.

(a) What size square wire must be used and what will be the stress if the deflection is to be 50 mm?

(b) If this spring is made of round wire, find the required diameter and the stress. What percentage of weight of material is saved by the use of round wire?

Ans.   (a) $c_1 = 6.64$ mm; $\tau = 494$ MPa.

15. A torsional helical spring for a 30-in. door (see Fig. 4-22) is to have a pull at the handle of 1 lb when the door is closed, and a pull of 3 lb after the door has rotated through 180°. No. 6 wire is to be used, and maximum stress is 150,000 psi.

Find the required diameter of the helix, the required initial angular deformation, and the number of active coils in the spring. Take into account the stress concentration.   *Ans.   Diam helix = 1.047 in.; $\theta_0 = 90°$; $N_c = 31.8$.*

FIGURE 4-22   Problem 15.

16. How many coils must there be in a window-shade spring that is to exert a pull on the shade of 13.5 N after being wound up 15 revolutions? The spring is made of 1.2-mm square wire. The helix diameter is 18.95 mm center to center of coils, and the roller diameter is 32 mm. What is the value of the maximum flexural stress?

Ans. $N_c = 262$; $\sigma = 783$ MPa.

17. A flexible coupling has an active element in the form of a torsional helical spring. Twenty-five horsepower are transmitted at 900 rpm. To what radius of helix must the spring be wound if made of $\frac{3}{4}$-in. square wire, and if the permissible stress is 30,000 psi? How many active coils must there be if the torsional deflection is not to exceed 5°?   *Ans. $R = 1.47$ in.; $N_c = 4.27$.*

18. A torsional spring is to be located within the hub of the reel for raising the weight shown in Fig. 4-23. The spring is wound from $\frac{5}{16}$-in. square wire. The OD of helix is $6\frac{1}{2}$ in. Ignore the effect of curvature on stress. The number of active coils is 20.

(a) Find the initial angular deformation required by the spring in order to support the load in the upper position.

(b) Find the additional force that must be applied to the weight in order to hold it in the lower position.

(c) Using Eq. (25), compute the strain energy stored in the spring because of lowering the weight; check the result by totaling the work done by the external forces.   *Ans.* $\theta_0 = 3.47$ rad; force $= 81.5$ lb.

FIGURE 4-23   Problem 18.   FIGURE 4-24   Problem 19.

19. (a) Find the force F necessary to move the top of the shaking table shown in Fig. 4-24 a distance of 1.75 in. to one side of the midposition. Tangents to the springs at the top and bottom remain vertical at all times. Find the value of the bending stress in the springs.

(b) Repeat part (a) except that the right spring is now $\frac{5}{16}$ in. thick. The left spring remains $\frac{1}{4}$ in. thick.   *Ans.*   (a) $F = 326$ lb; (b) $F = 481$ lb.

20. A cantilever, multiple-leaf spring similar to that in Fig. 4-12 is composed of five leaves 62 mm wide and 12 mm thick, and has a clear span to the point of application of the load of 500 mm. Assume that the shorter leaves are of the proper length and end taper to permit the spring to be approximated as a trapezoid. What static load will the spring support at a stress in the material of 400 MPa? What will be the deflection of the load?   *Ans.* $P = 5,950$ N; $\delta = 3.51$ mm.

21. A spring mount for a machine is built up of steel plates and rollers as shown in Fig. 4-25. Find the deflection caused by the load and the bending stress in the plates.   *Ans.* $y = 1.084$ in.; $\sigma = 11,850$ psi.

FIGURE 4-25   Problem 21.

22. Find the value of force *P* if the force between the contact points in Fig. 4-26 is to be 13 N. Find the deflection of point *B*. Leaves are steel, 0.75 mm thick, and 10 mm wide. Neglect effects of any friction.

Ans. $P = 21.7$ N; $y = 9.28$ mm.

FIGURE 4-26 Problem 22.

FIGURE 4-27 Problem 23.

23. Find the load carried by each spring in Fig. 4-27. The left spring is trapezoidal in plan, 2 in. wide at the hinge, and 10 in. wide at the wall. The right spring has a uniform width of 3 in. Assume that the hinges are frictionless and the connecting link is inextensible. Find the deflection of the load.

Ans.   Left $= 129$ lb; right 171 lb; $\delta = 0.2267$ in.

24. Forces *R* in Fig. 4-28 are just sufficient to bring the ends of the upper spring into contact with the lower spring. The material is steel. Find the value of *R*.

Ans. $R = 4,109$ N.

FIGURE 4-28 Problem 24.

FIGURE 4-29 Problem 25.

25. Find the deflection of the load in Fig. 4-29. The springs are of steel. Assume that the hinges are frictionless and the connecting link is inextensible.

Ans. $y = 0.1975$ in.

26. A cantilever spring has a load at the end. The spring is of uniform thickness, but is triangular in plan. Prove that the bending stress is the same for all cross sections.

27. Let the expression for the strain energy *U* for a simply supported leaf spring with load *P* in the center be given by $\frac{1}{2}Py$, where *y* is the deflection. Prove that stress $\sigma = c\sqrt{6EU/Il}$, where *c* is the distance from the neutral axis to the edge of the cross section. Use this relationship to prove that for a rectangular cross section the energy absorbed for a constant stress σ is the same regardless of whether the long or short dimension is vertical.

28. A helical spring is to carry a static load of 450 N at a stress of 480 MPa. The spring index equals 6. Find the theoretical diameter of the round wire.

29. A steel helical spring of 13 active coils is to have a deflection of 16 mm under a static load of 250 N. The spring index equals 5.25. Find the theoretical diameter of the wire.

30. A steel helical spring of index 4.64 has 20 active coils. The deflection is 38 mm when the stress is 480 MPa. Find the static load the spring is carrying and the mean radius of the helix.

31. A steel helical spring must exert a force of 2.5 lb after being released 0.10 in. from its most highly compressed condition. No. 19 wire is used with a spring index of 6. Find that static loading for a maximum stress of 60,000 psi. Find the number of active coils. *Ans.* $P_2 = 5.99$ lb; $N_c = 7.82$ coils.

32. The maximum and minimum static loads on a steel helical spring are 380 N and 320 N, respectively. These loads cause a change in the deflection of 6 mm. The spring index is 8, and the maximum stress is 480 MPa. Find the theoretical value of d. Also find R and N_c.

33. Find the F_s for a spring with the same data as Problem 12 except that R equals 15 mm.

34. An engine valve spring must exert a force of 60 lb when the valve is closed and 100 lb when the valve is open. The lift is $\frac{5}{16}$ in. Material tests $\tau_{yp} = 90,000$ psi and $\tau'_e = 48,000$ psi. $F_s = 1.5$. If the spring index is 6, find the theoretical wire diameter. Also find the number of active coils and the initial compression of the spring.

35. A helical spring is made from No. 10 hard-drawn wire. The load varies continuously from 25 to 50 lb. The radius of the helix is $\frac{5}{16}$ in. Find the value of the factor of safety.

36. Work Problem 22 but with the force between the contacts 9 N. The 40 mm dimension becomes 60 mm. *Ans.* $Y_b = 10.46$ mm.

37. In Fig. 4-30 each steel spring has 10 active coils of No. 12 wire. The mean radius of the helix is 0.3 in. The right spring has a free height 0.5 in. greater than the left spring. When the load is acting, the bar is level. Find the distance x.

FIGURE 4-30   Problem 37.   **FIGURE 4-31   Problem 38.**

38. In Fig. 4-31 assume the bar to be rigid and weightless, and the joint at the left to be frictionless. Each steel spring has 10 active coils of No. 10 wire wound with a spring index of 5. Springs are stress-free when the bar is level. Find the force carried by each spring and the deflection of the weight. *Ans.* $\delta = 2.07$ mm.

FIGURE 4-32   Problem 39.

39. The bar in Fig. 4-32 is made of steel. Find the force in the spring.

40. A steel helical spring carries a static load of 250 N at a deflection of 16 mm and a shear stress of 700 MPa. End turns equal 2. Find the wire diameter for the optimum spring. Select the nearest standard diameter and find the number of active coils and volume of material in the spring.   *Ans.* $N_c = 21.3$ coils.

41. Find the outside diameter b and thickness t for a Belleville spring, $h/t = 1.5$, that carries a load of 500 lb when compressed flat. Maximum compressive stress is 150,000 psi. Let $b/a = 2$.   *Ans.* $b = 2.88$ in.; $t = 0.087$ in.

42. A helical compression spring of music wire has a maximum load that is 4 lb greater than the minimum load. The deflection under the maximum load is 0.25 in. greater than the deflection under the minimum load. Assume tentatively that the number of active coils is 10. The factor of safety is 1.5. $R = 0.25$ in. Determine a suitable standard-size wire and find the exact number of active coils. Find the initial deflection of the spring.   *Ans.* $N_c = 11.0$; $\delta_1 = 0.620$ in.

43. Steel spring data: $P = 50$ lb static, $\delta = \frac{3}{4}$ in., $\tau = 95,000$ psi, and $Q = 2$. Find the wire size for the most efficient spring and the corresponding value for V.   *Ans.*   For No. 14 wire, $V = 0.3164$ in.^{3}

44. Steel spring data: $P = 50$ lb static, $\delta = \frac{3}{4}$ in., $\tau = 95,000$ psi, and $Q = 2$. Find the value for V if No. 11 wire is used.   *Ans.* $V = 0.1952$ in.^{3}

45. Steel spring data: $P = 75$ lb static, $\delta = 1$ in., $\tau = 100,000$ psi, and $Q = 2$. Find the wire size for the most efficient spring and the corresponding value of V.   *Ans.*   For No. 12 wire, $V = 0.2457$ in.^{3}

46. Steel spring data: $P = 75$ lb static, $\delta = 1$ in., $\tau = 100,000$ psi, and $Q = 2$. Find the value for V if No. 10 wire is used.   *Ans.* $V = 0.3055$ in.^{3}

47. Steel spring data: $P = 20$ lb static, $\delta = \frac{3}{8}$ in., $\tau = 90,000$ psi, and $Q = 1.75$. Find wire size for the most efficient spring and the corresponding value for V.   *Ans.*   For No. 17 wire, $V = 0.0306$ in.^{3}

48. Steel spring data: $P = 20$ lb static, $\delta = \frac{3}{8}$ in., $\tau = 90,000$ psi, and $Q = 1.75$. Find the value for V if No. 16 wire is used.   *Ans.* $V = 0.0323$ in.^{3}

49. A steel helical spring of the most efficient design must carry a load of 2.5 lb after being released 0.10 in. from its most highly compressed state where the maximum shearing stress is 60,000 psi. For a tentative value for c of 6, find a suitable standard wire diameter, the exact value for c, and the number of active coils. Compare the results with those of Problem 31.

Ans.   For No. 20 wire, $N_c = 10.76$ coils.

50. A steel helical spring of the most efficient design must carry a load of 8 lb after being released 0.25 in. from its most compressed condition when the shearing stress was 80,000 psi. For tentative value for c of 8 find a suitable standard wire size, the mean radius of the helix, and the number of active coils.

Ans. For No. 16 wire, $R = 0.240$ in.

51. A steel helical spring of the most efficient design must carry a load of 10 lb after being released 0.1875 in. from its most compressed condition when the shearing stress was 90,000 psi. For a tentative value for c of 6 find a suitable standard wire size, the mean radius of helix, and the number of active coils.

Ans. For No. 17 wire, $R = 0.139$ in.

52. A helical spring is wound from No. 12 music wire with mean radius of helix of 0.4 in. The load varies continuously between 25 and 15 lb. Find the value of the factor of safety. *Ans.* $F_s = 1.70$.

53. A helical spring is wound from No. 14 music wire with mean radius of helix of 0.3 in. The load varies continuously between 10 and 4 lb. Find the value of the factor of safety. *Ans.* $F_s = 2.28$.

54. A helical spring is wound from No. 16 music wire with a mean radius helix of 0.25 in. The load varies continuously between 8 and 2 lb. Find the value of the factor of safety. *Ans.* $F_s = 1.60$.

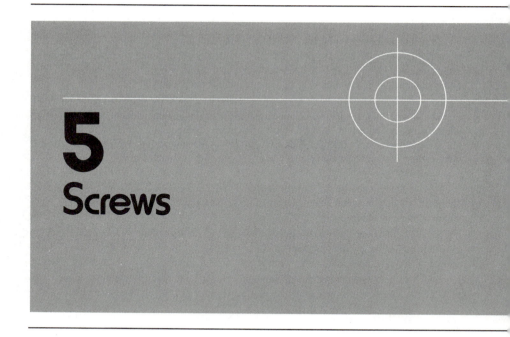

5
Screws

Bolts and screws are used to fasten the various parts of an assembly together. The designer should be familiar with the different kinds of threads in commercial use and with the method of specifying the desired tolerances for the fit between screw and nut. He should understand the reasons for the increase of fatigue strength obtained by the application of initial tension in the bolt. Power screws are employed in machines for obtaining motion of translation and also for exerting forces.

1. Kinds of Threads

In the inch system, the Unified thread is in general service for studs and bolts. This thread represents the agreement of standardization committees of Canada, Great Britain, and the United States. It is a replacement for the American National form of thread, which was formerly the standard in our country. Bolts and nuts of the two systems are interchangeable. The systems differ only in minor details, mainly in the arrangement of the tolerances. The included angle of the thread is 60°. The basic form of the thread is shown in Fig. 5-1(a), although actual threads are made with a rounded root. The crest is flat in the American National and may be either flat or rounded in the Unified thread. The Whitworth thread with 55° included angle, shown in Fig. 5-1(e), was formerly the standard in Great Britain.

(a) Unified, American National and SI threads

(b) Acme thread

(c) American National pipe threads

(d) Square thread

(e) Whitworth thread

FIGURE 5-1 **Basic forms of screw threads.**

For lead screws and power transmission the Acme screw shown in Fig. 5-1(b) is in wide use. It has an included angle of thread of 29°. The standard proportions of the American National pipe thread are given in Fig. 5-1(c). The taper, together with the smaller flat at crest and root, assists in producing a fluid-tight joint. Square threads shown in Fig. 5-1(d) are used to a limited extent for power transmission.

If an imaginary cylinder, coaxial with the screw, intersects the thread at the height that makes the width of thread equal to the width of space, the diameter of this cylinder is called the *pitch diameter* of the screw. See Fig. 5-1(a). The distance measured parallel to the axis from a point on one thread to the corresponding point on the adjacent thread is called the *pitch*. A screw made by cutting a single helical groove on the cylinder is called a *single-thread* screw. If the helix angle is somewhat steeper, and a second thread is cut in the space between the grooves of the first, a *double-thread* screw is formed, Fig. 5-2. For certain applications, triple and quadruple threads are in use. For multiple-thread screws the *lead* is the distance the nut advances in one revolution. The pitch is defined as for a single-thread screw. The helix may be cut either right hand or left hand. The screw shown in Fig. 5-2 is a right-hand screw.

FIGURE 5-2   Double-threaded screw.

Let the major diameter of the screw, or the diameter of the stock on which the thread is cut, be called d. The fundamental triangle with sides at 60° to the thread axis has a base equal to the pitch p. The height h perpendicular to the axis of the fundamental triangle then is

$$h = 0.86603p \tag{1}$$

The flats on the top and bottom of the fundamental thread shape have widths equal to $p/8$ or a height of $h/8$. The height of the thread perpendicular to the axis then is $\frac{3}{4}h$.

$$\text{height of Thread} = \tfrac{3}{4}h = \tfrac{3}{4} \times 0.86603p = 0.6495p \tag{2}$$

$$\text{pitch diameter} = d - \tfrac{3}{4}h$$

$$= d - 0.6495p \tag{3}$$

$$\text{root diameter} = d - 2 \times 0.6495p$$

$$= d - 1.2990p \tag{3a}$$

Example 1. Compute the pitch diameter, the helix angle, and the root area for a 1-in. coarse-thread series screw.

Solution. By Table 5-1, the 1-in. coarse-thread series has 8 threads per inch, or $p = \frac{1}{8}$ in.

By Eq. (3): $\qquad$ pitch diameter $= 1 - 0.6495 \times \frac{1}{8} = 0.9188$ in.

$$\text{pitch circumference} = 0.9188\pi = 2.88650 \text{ in.}$$

For helix angle α: $\quad \tan \alpha = \dfrac{0.125}{2.88650} = 0.04331$

$$\alpha = 2°29'$$

By Eq. (3a): $\quad$ root diameter $= d - 1.2990p$

$$= 1 - 1.2990 \times \frac{1}{8}$$

$$= 1 - 0.1624 = 0.8376 \text{ in.}$$

$$\text{root area} = \frac{\pi}{4} \times 0.8376^2 = 0.5510 \text{ in.}^2$$

The stress area, as given in Table 5-1 for this screw is 0.606 in.^{2} It is seen that the permissible rounding of the root has made the stress area larger. This increase is taken advantage of when figuring the capacity of a threaded rod to carry a tensile load.

2. Standardized Threads

Table 5-1 gives a summary of the various sizes and pitches for Unified and American National forms of threads. The size of a screw refers to the major diameter, or the size of the stock upon which the helix is cut. It should be noticed that sizes less than $\frac{1}{4}$ in. are designated by number. In general, there are two pitches, coarse and fine, that may be cut on each diameter. For certain sizes, there is also an extra-fine series of pitches. Complete information on sizes, pitches, tolerances, and allowances is given in *Unified and American Screw Threads*, ASA B1.1-1974, published by the American Society of Mechanical Engineers. There are three additional series of threads of 8, 12, and 16 threads per in., which may be cut on a wide variety of diameters.

Since it is impossible to manufacture parts exactly to a specified size, tolerances have been placed on the dimensions of the screw and nut. Such tolerance zones are illustrated in Figs. 5-3 and 5-4. When the measured dimension of a part falls within the tolerance zone, it is accepted as meeting the dimensional specifications.

The term *maximum metal* refers to the screw of largest pitch diameter and the nut of smallest pitch diameter permitted by the tolerance zones. Similarly

TABLE 5-1 **Dimension of Unified and American National Screw Threads**

Size	Outside or Major Diameter, in.	Coarse-Thread Series			Fine-Thread Series			Extra-Fine-Thread Series			Hex. Nut Width Across Flats
		Threads per in.	Basic Pitch Diameter, in.	Stress Area, in.²	Threads per in.	Basic Pitch Diameter, in.	Stress Area, in.²	Threads per in.	Basic Pitch Diameter, in.	Stress Area, in.²	
0	0.0600				80	0.0519	0.0018				
1	0.0730	64	0.0629	0.0026	72	0.0640	0.0028				
2	0.0860	56	0.0744	0.0037	64	0.0759	0.0039				
3	0.0990	48	0.0855	0.0049	56	0.0874	0.0052				
4	0.1120	40	0.0958	0.0060	48	0.0985	0.0066				
5	0.1250	40	0.1088	0.0080	44	0.1102	0.0083				
6	0.1380	32	0.1177	0.0091	40	0.1218	0.0102				
8	0.1640	32	0.1437	0.0140	36	0.1460	0.0147				
10	0.1900	24	0.1629	0.0175	32	0.1697	0.0200				
12	0.2160	24	0.1889	0.0242	28	0.1928	0.0258	32	0.1957	0.0270	
1/4	0.2500	20	0.2175	0.0318	28	0.2268	0.0364	32	0.2297	0.0379	7/16
5/16	0.3125	18	0.2764	0.0524	24	0.2854	0.0580	32	0.2922	0.0625	1/2
3/8	0.3750	16	0.3344	0.0775	24	0.3479	0.0878	32	0.3547	0.0932	9/16
7/16	0.4375	14	0.3911	0.1063	20	0.4050	0.1187	28	0.4143	0.1274	5/8
1/2	0.5000	13	0.4500	0.1419	20	0.4675	0.1599	28	0.4768	0.170	3/4
9/16	0.5625	12	0.5084	0.182	18	0.5264	0.203	24	0.5354	0.214	13/16
5/8	0.6250	11	0.5660	0.226	18	0.5889	0.256	24	0.5979	0.268	15/16
3/4	0.7500	10	0.6850	0.334	16	0.7094	0.373	20	0.7175	0.386	1-1/8
7/8	0.8750	9	0.8028	0.462	14	0.8286	0.509	20	0.8425	0.536	1-5/16
1	1.0000	8	0.9188	0.606	12	0.9459	0.663	20	0.9675	0.711	1-1/2
1-1/8	1.1250	7	1.0322	0.763	12	1.0709	0.856	18	1.0889	0.901	1-11/16
1-1/4	1.2500	7	1.1572	0.969	12	1.1959	1.073	18	1.2139	1.123	1-7/8
1-3/8	1.3750	6	1.2667	1.155	12	1.3209	1.315	18	1.3389	1.370	2-1/16
1-1/2	1.5000	6	1.3917	1.405	12	1.4459	1.581	18	1.4639	1.64	2-1/4
1-3/4	1.7500	5	1.6201	1.90				16	1.7094	2.24	2-5/8
2	2.0000	4-1/2	1.8557	2.50				16	1.9594	2.95	3
2-1/4	2.2500	4-1/2	2.1057	3.25							3-3/8
2-1/2	2.5000	4	2.3376	4.00							3-3/4
2-3/4	2.7500	4	2.5876	4.93							4-1/8
3	3.0000	4	2.8376	5.97							4-1/2
3-1/4	3.2500	4	3.0876	7.10							4-7/8
3-1/2	3.5000	4	3.3376	8.33							5-1/2
3-3/4	3.7500	4	3.5876	9.66							5-5/8
4	4.0000	4	3.8376	11.08							6

minimum metal refers to the screw of smallest pitch diameter and the nut of largest pitch diameter. *Allowance* is the clearance between the pitch diameters of screw and nut when both parts of the assembly are at the maximum metal condition.

In general, the quality of a product is highest when the tolerance zones are smallest. Manufacturing costs, however, rise as tolerance zones are made smaller.

Unified and American National threads are divided into various classes, which depend on the size of the tolerance zones. Figure 5-1(a) illustrates the manner in which the basic size is obtained for both screw and nut.

FIGURE 5-3 Tolerance zones and allowances for classes 1A, 1B, 2A, and 2B. Unified threads.

FIGURE 5-4 Tolerance zones for classes 3A and 3B. Unified threads and classes 2 and 3 American National threads.

3. Unified Threads

Unified threads are divided into classes 1A, 2A, and 3A for external threads and classes 1B, 2B, and 3B for internal threads.

Classes 1A and 1B are provided when bolts and nuts of the largest tolerances and the largest clearance after assembly are required. An assembly of such parts is illustrated in Fig. 5-3. The pitch diameter of the maximum metal nut is at the basic size, and the pitch diameter of the maximum metal screw is smaller than basic. Bar diagrams representing the tolerances and allowances for a typical screw are shown in Fig. 5-5.

FIGURE 5-5 Pitch diameter tolerances and allowances for different classes of fits for a 1 in. screw and nut of 8 threads per inch.

Classes 2A and 2B are similar to classes 1A and 1B but have smaller tolerances and a smaller clearance after assembly. These classes are well suited for the vast majority of bolts.

Classes 3A and 3B have the smallest tolerances and zero allowance for an assembly at the maximum metal condition at which both screw and nut have the basic pitch diameter. Such an assembly is illustrated in Fig. 5-4. These classes are intended for applications when closeness of fit and high accuracy are essential.

If desired, the screw can be of one class and the nut can be of another.

4. American National Threads

These threads are divided into Classes 2 and 3 for screws and nuts.

Class 2 has the larger tolerances for screw and nut but no allowance for a maximum metal assembly as shown in Fig. 5-4.

Class 3 has smaller tolerances and produces tighter fits upon assembly.

At the maximum metal condition there is no allowance between screw and nut as shown in Fig. 5-4.

It should be noted that Classes 2A and 2B Unified threads have an allowance but that Class 2 American National does not.

As illustrated in Fig. 5-6, a variation in the thread angle of a screw

FIGURE 5-6 Variation in pitch diameter caused by changes in thread angle.

FIGURE 5-7 Variation in pitch diameter caused by difference in lead.

causes a change in its pitch diameter. As shown in Fig. 5-7, a screw and nut, if made with slight differences in lead, may still be assembled if the extent of engagement is not too great, and if the two parts have a difference in pitch diameter.

The tolerance on the pitch diameter permitted by the standards must provide for errors from all causes. The effects of angle and lead variation must be included, as well as discrepancies in the pitch diameter itself.

5. Identification Symbols

Following are examples of approved identification symbols for use on drawings, tools, and specifications.

(1) Unified coarse-thread series, 1-in.-diam external thread, 13 threads per in., Class 2A fit:

$$\tfrac{1}{2}''\text{--}13\text{UNC--}2\text{A}$$

(2) American National coarse-thread series, $\frac{1}{2}$-in.-diam, 13 threads per in., Class 2 fit:

$$\tfrac{1}{2}''\text{--}13\text{NC--}2$$

(3) Unified fine-thread series, 1 in.-diam internal thread, 12 threads per in., Class 1B fit:

$$1''\text{--}12\text{UNF--}1\text{B}$$

For a left-hand thread the designations above should be followed by the letters LH.

6. SI Threads

The SI thread profile is similar to the Unified with a 60° included angle and a flat on the top equal to one-eighth of the pitch. The pitch, or distance from thread to thread, is given in millimeters. The basic outline is shown in Fig. 5-1(a). The flat at the bottom is equal to $p/4$. A selected list of SI threads is given in Table 5-2. Unfortunately, there is no interchangeability between SI and Unified threads.

TABLE 5-2   Selected List of ISO Screw Threads

Outside Diameter, mm	Pitch, mm	Pitch Diameter, mm	Stress Area, mm²	Outside Diameter, mm	Pitch, mm	Pitch Diameter, mm	Stress Area, mm²
1.6	0.35	1.373	1.27	20	2.5	18.376	245
2	0.4	1.740	2.07	24	3	22.051	353
2.5	0.45	2.208	3.39	30	3.5	27.727	561
3	0.5	2.675	5.03	36	4	33.402	817
3.5	0.6	3.110	6.78	42	4.5	39.077	1,120
4	0.7	3.545	8.78	48	5	44.752	1,470
5	0.8	4.480	14.2	56	5.5	52.428	2,030
6	1	5.350	20.2	64	6	60.103	2,680
8	1.25	7.188	36.6	72	6	68.103	3,460
10	1.5	9.026	58.0	80	6	76.103	4,340
12	1.75	10.863	84.3	90	6	86.103	5,590
14	2	12.701	115	100	6	96.103	6,990
16	2	14.701	157				

Metric threads are designated by diameter and pitch, as M10 × 1.5.

Example 1A. An SI thread has a diameter of 14 mm and a pitch of 2 mm. Calculate the pitch diameter, root diameter, and root area.

Solution. By Eq. (3):   pitch diameter $= d - 0.6495p$

$$= 14 - 0.6495 \times 2 = 12.701 \text{ mm}$$

The flat at the base of the fundamental SI thread form has a width $\frac{1}{4}p$ or a height $\frac{1}{4}h$. The perpendicular distance from the pitch line to the flat then is $\frac{1}{4}h$.

$$\text{root diameter} = \text{pitch diameter} = 2 \times \tfrac{1}{4}h$$

$$= 12.701 - \tfrac{1}{2} \times 0.86603p$$

$$= 12.701 - \tfrac{1}{2} \times 0.86603 \times 2 = 11.835 \text{ mm}$$

$$\text{root area} = \frac{\pi}{4} \times 11.835^2 = 110.01 \text{ mm}^2$$

By Table 5-2 it is seen that the rounding of the root has given a stress area slightly larger than the root area.

7. Effect of Initial Stress

Figure 5-8 shows a load P carried by the part that is attached to the support by means of a bolt. The bolt center is located at the center of gravity of the cross section of the part. Let it be assumed that the nut is merely made snug against the part but with no initial tension in the bolt. Also let it be temporarily assumed in Fig. 5-8 that the part is welded to the support, making it possible for both bolt and part to carry tensile loads.

When load P is placed on the assembly, the part immediately stretches, and since it is in contact with the nut, the bolt is stretched also. Both part and bolt have tension stresses, because load P divides and is carried partly by the bolt and partly by the part. In finding the portion of load carried by each, it is convenient to work with the so-called "spring constants" of these

FIGURE 5-8   Bolted part carrying load.

members. The spring constant k is the value of the force required to give a deformation of 1 in. It may be found by using the equation for deformation, $\delta = Pl/AE$, and letting P be equal to k when δ becomes equal to unity. When these substitutions are made, the equations for k_b for the bolt and k_p for the part become

$$k_b = \frac{A_b E_b}{l_b} \quad \text{and} \quad k_p = \frac{A_p E_p}{l_p} \tag{4}$$

where subscripts b and p refer to bolt and part, respectively, and where A is the cross-sectional area, E is the modulus of elasticity, and l is the length in the direction of the force. If the thread stops immediately above the nut, the gross area of the bolt must be used in computing k_b, since it is the unthreaded portion that is stretched by the load.[1]

Let P_b be the portion of load P that is carried by the bolt, and P_p be the remainder that is taken by the part. The deformation of the bolt is equal to P_b/k_b, and of the part, P_p/k_p. Because of the arrangement of the members in Fig. 5-8, these deformations are equal, and the following equation can be written.

$$\frac{P_b}{k_b} = \frac{P_p}{k_p} \tag{a}$$

---

[1] When a greater length of bolt is threaded, suitable allowance must be made for the increased flexibility that thereby results.

Substitution of $P - P_b$ for P_p gives

$$P_b = \frac{k_b}{k_b + k_p}P \qquad \text{tension in bolt} \qquad \text{(b)}$$

Similarly, $\qquad P_p = \frac{k_p}{k_b + k_p}P \qquad \text{tension in part} \qquad \text{(c)}$

Suppose now that the nut is tightened to the extent that an additional tensile load equal to F_o is placed in the bolt. The total tension in the bolt is now

$$F_b = \frac{k_b}{k_b + k_p}P + F_o \qquad (5)$$

At the same time that the bolt was receiving the additional tension F_o from the tightening, the part was receiving an equal and opposite force of compression F_o. The resultant force in the part now is

$$F_p = \frac{k_p}{k_b + k_p}P - F_o \qquad (6)$$

It is usual practice to induce force F_o initially upon assembly of the parts before load P is applied. Equations (5) and (6) are, of course, unaffected. Initial force F_o is substituted as a positive number in Eq. (6), and is always made sufficiently large to keep force F_p for the part negative or compression, and the assumed welding is thus not required. Should P be so great as to remove all the initial compression from the part, then neither Eq. (5) nor (6) can be used, for the bolt will then be carrying the total load P.

The maximum bolt force results when load P is a maximum, and the minimum bolt force results when the load is a minimum. The average bolt force $F_{b\ av}$ is equal to half the sum of $F_{b\ max}$ and $F_{b\ min}$. Thus

$$F_{b\ av} = \frac{k_b}{k_b + k_p}P_{av} + F_o \qquad (7)$$

The range of the bolt force is equal to one-half the difference between $F_{b\ max}$ and $F_{b\ min}$. Then

$$F_{br} = \frac{k_b}{k_b + k_p}P_r \qquad (8)$$

The minimum compressive force in the part occurs when load P is a maximum.

$$F_{p\ min} = \frac{k_p}{k_b + k_p}P_{max} - F_o \qquad (9)$$

The use of sufficient initial tension is very advantageous in reducing the fatigue effects in the bolt if load P is not steady but fluctuating. For varying load, the maximum bolt force is found by substituting the maximum value of P in Eq. (5), and the minimum bolt force is obtained by use of the minimum value of P. Whereas the maximum value of the bolt force is somewhat larger than if no initial force were used, the presence of F_o gives a considerably larger value of the minimum bolt force than if no tightening-up tension were used. The resultant effect is to give a much smaller range of variation[2] in the bolt force F_b than takes place in the applied load P, and the fatigue effects, which depend largely on the extent of the variation of the stress, are correspondingly reduced.

Average and range forces F_{bav} and F_{br} are divided by the stress area of the threads to give average and range stresses σ_{av} and σ_r. Since the stresses are continuously varying σ_{av} and $K\sigma_r$ are plotted on a diagram similar to Fig. 2-25 and the design made on the basis of such a diagram. Application is made in the following example using Fig. 5-9.

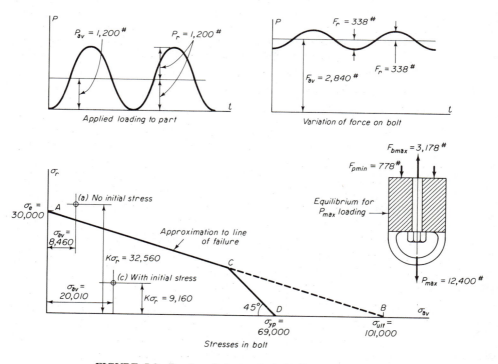

FIGURE 5-9   Loads and stresses on bolt: Example 2.

Example 2. Let the bolt in Fig. 5-8 be $\frac{1}{2}$ in. $\times$ 13UNC. Bolt and part are of the same length; the threads stop immediately above the nut. The bolt material has ground threads and a yield strength of 69,000 psi and ultimate of 101,000 psi. Take the stress concentration factor for the threads as 3.85. The steel part has a net area

[2] See Almen, J. O., *Machine Design*, **15**, Aug. 1943, p. 133; **16**, Feb. 1944, p. 158.

of 0.5 in.^{2} The load fluctuates continuously between 0 and 2,400 lb. Take σ_e as 30,000 psi.

(a) Find the F_s for the bolt when no initial force is present.

(b) Find the minimum required value of F_o to prevent loss of compression in the part.

(c) Find the F_s for the bolt when F_o is taken as 2,500 lb.

(d) Find the minimum force in the part for the given loading and F_o equal to 2,500 lb.

Solution. (a) By Table 5-1:   Stress area $= 0.1419$ in.^{2}

For loosely held part, when F_o is 0, load on bolt equals load on part.

$$P_{av} = 1,200 \text{ lb}; \qquad \sigma_{av} = \frac{1,200}{0.1419} = 8,460 \text{ psi}$$

$$P_r = 1,200 \text{ lb}; \qquad K\sigma_r = 3.85 \times \frac{1,200}{0.1419} = 32,560 \text{ psi}$$

When σ_{av} and $K\sigma_r$ are plotted as in Fig. 5-9, it is seen that the bolt loading is unsafe.

(b) For bodies of equal lengths and moduli, the spring constants are proportional to the cross-sectional areas.

$$\text{Gross area of bolt} = \frac{\pi}{4}\left(\frac{1}{2}\right)^2 = 0.1964 \text{ in.}^2$$

In Eq. (9), when $F_{p\ min}$ is equal to zero,

$$F_o = \frac{k_p}{k_b + k_p}P_{max} = \frac{0.5}{0.6964} \times 2,400 = 1,720 \text{ lb}$$

This is the theoretical minimum value for F_o.

(c) By Eq. (7): $F_{b\ av} = \dfrac{k_b}{k_b + k_p}P_{av} + F_o$

$$= \frac{0.1964}{0.6964} \times 1,200 + 2,500 = 2,840 \text{ lb}$$

$$\sigma_{av} = \frac{2,840}{0.1419} = 20,010 \text{ psi}$$

By Eq. (8): $F_{br} = \dfrac{k_b}{k_b + k_p}P_r = \dfrac{0.1964}{0.6964} \times 1,200 = 338 \text{ lb}$

$$\sigma_r = \frac{338}{0.1419} = 2,380 \text{ psi}$$

Eq. (11a), Chapter 2: $\dfrac{\sigma_{av}}{\sigma_{ult}} + \dfrac{K\sigma_r}{\sigma_e} = \dfrac{1}{F_s}$

$$\frac{20,010}{101,000} + \frac{9,160}{30,000} = \frac{1}{F_s}$$

$$0.1981 + 0.3053 = \frac{1}{F_s}$$

$$F_s = 1.99$$

(d) By Eq. (9): $F_{p\ min} = \dfrac{k_p}{k_b + k_p} P_{max} - F_o$

$$= \frac{0.5}{0.6964} \times 2{,}400 - 2{,}500 = -780 \text{ lb}$$

It should be noted that the use of initial stress has changed the situation to the extent that the bolt now has a reasonable value for the factor of safety.

8. Additional Design Considerations

When bolts are used to clamp large area plates together, the clamping effect is confined to a region adjacent to the bolts. As shown in Fig. 5-10(a), the region affected by the clamping force can be approximated as the fustums of two cones surrounding the bolt.[3] These cones have elements at about 45° from the nut and head of the bolt. The spring constant k_p is then taken as that of a hollow cylinder of outer diameter d_2 located at the midpoint of the sloping lines and inner diameter d_1.

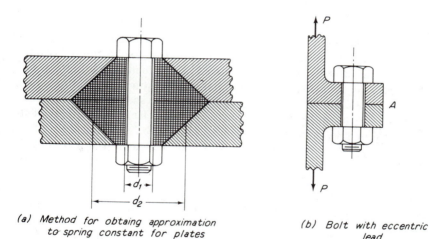

(a) Method for obtaing approximation   (b) Bolt with eccentric
    to spring constant for plates   lead

FIGURE 5-10   Additional factors to be considered in the design of bolted assemblies.

When the bolt is located eccentrically with respect to the force as shown in Fig. 5-10(b), a bending effect will be present, and the force in the bolt will be larger than P. Its value can be found by taking moments about A.

When two rigid parts are bolted together, Fig. 5-11(a), the lengthening

[3]See Radzimovsky, E. L., "Bolt Design for Repeated Loading," *Machine Design* **24**, Nov. 1952, p. 135.

FIGURE 5-11   Various types of bolted assemblies.

of the bolt required to induce force F_o is a very small quantity. Should the loading cause any creep or recession at the high spots of the contact surfaces, the initial stretch of the bolt may be lost and force F_o will disappear. A decrease in thickness of the part may also be caused by corrosion, wear, or displacement of platings and coatings.

The situation can be improved by the use of a spring washer under the nut, as shown in Fig. 5-11(b). The deformation sustained by the washer from force F_o may be many times as great as the stretch of the bolt. Any small shortening of the part during service then has but small effect in reducing the value of F_o, and the initial force can be expected to be retained. It is assumed that the washer is acting in its elastic range and is not compressed solid. Ordinary lock washers, which are compressed solid upon assembly, are of no help in reducing the fluctuations of the load in the bolt. In fact, such devices may constitute an additional hazard from loss of initial stress, caused by plastic flow at high spots or burrs on the surfaces.

A gasket is shown between the parts in Fig. 5-11(c). The stiffness of the gasket, hard or soft, influences the overall value of k_p. A soft gasket reduces the value of k_p and causes a larger proportion of the load P to be taken by the bolt. It should be kept in mind that any set or permanent deformation of the gasket may cause a loss of the initial force F_o. A periodic tightening of the nut to maintain the initial force may be required.

For structural applications, the Bethlehem Steel Company has developed a hardened steel washer with protrusions pressed out of the flat surface. These are flattened when the nut is tightened and give indication of the magnitude of the initial force in the bolt.

9. Power Screws

A power screw can be used for raising weights or exerting forces in machines. The weight W shown in Fig. 5-12, into which the supporting screw is threaded, can be raised or lowered by the rotation of the screw. It is, of course, assumed that W is prevented from turning when the screw rotates. An expression will now be derived for finding the value of the torque needed to raise the load.

FIGURE 5-12 Power screw.

The total force on the threads can be represented by a single force F_n, Fig. 5-13(a), which is normal to the thread surface. Force F_n is the diagonal of the parallelopiped. Side *ABEO* is an axial section through the bolt. The projection of F_n into this plane is inclined at θ or half the included thread angle. Side *ACHO* lies in the plane tangent to the pitch cylinder. The projection of F_n into this plane is inclined at the helix angle α calculated at the pitch radius of the screw. The length of this component is $F_n \cos \theta_n$ as indicated in sketch (c). It has the vertical and horizontal components shown.

The developed helix of a screw becomes merely an inclined plane with angle α equal to the helix angle of the screw. The lower triangular block thus represents the thread, which raises the weight when it is pushed to the left by force F.

The total friction force along the thread is $\mu_1 F_n$, where μ_1 is the coefficient of thread friction. Force $\mu_1 F_n$ has the vertical and horizontal components shown.

The upward reaction at the base is equal to weight W. During motion there is a resisting friction force $\mu_2 W$, where μ_2 is the coefficient of friction for the base or collar.

Let force F be just sufficient to cause motion to the left. Summation of the vertical forces in sketch (c) gives

$$F_n \cos \theta_n \cos \alpha = \mu_1 F_n \sin \alpha + W$$

or

$$F_n = \frac{W}{\cos \theta_n \cos \alpha - \mu_1 \sin \alpha} \qquad (10)$$

All thread forces, as well as force F, act at the pitch radius of the thread

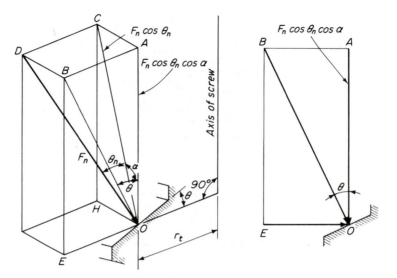

(a) Force parallelepiped

(b) Axial section

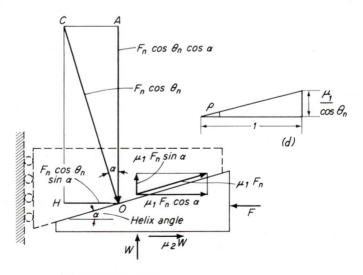

(c) Tangential plane

(d)

FIGURE 5-13   Force acting on screw when raising weight W.

r_t. The force of the collar friction acts at radius r_c to the midpoint of the collar surface. The torque required to raise the load is found by multiplying the horizontal forces by the appropriate radii. Hence,

$$T = Fr_t = r_t(F_n \cos \theta_n \sin \alpha + \mu_1 F_n \cos \alpha) + r_c \mu_2 W$$

Substitution of the above value for F_n gives

$$T_1 = W \left(r_t \frac{\cos \theta_n \sin \alpha + \mu_1 \cos \alpha}{\cos \theta_n \cos \alpha - \mu_1 \sin \alpha} + r_c \mu_2 \right) \tag{11}$$

$$= Wr_t \left(\frac{\cos \theta_n \tan \alpha + \mu_1}{\cos \theta_n - \mu_1 \tan \alpha} + \frac{r_c}{r_t} \mu_2 \right) \tag{12}$$

The relationship between θ_n and thread angle θ is obtained as follows. In sketch (a),

$$\tan \theta_n = \frac{CD}{OC} = \frac{AB}{OC} \tag{a}$$

$$AB = AO \tan \theta \qquad \text{and} \qquad CO = \frac{OA}{\cos \alpha}$$

These should now be substituted into Eq. (a) to give

$$\tan \theta_n = \tan \theta \cos \alpha \tag{13}$$

Equations (11) and (12) give the value of the torque required to raise the load when friction for both thread and collar are included. Sometimes the collar consists of an antifriction bearing, in which case μ_2 may be sufficiently small to be neglected. The equations then contain the μ_1 terms only.

Example 3. A quadruple Acme thread, 1-in. OD, has a pitch of 0.200 in. Collar radii are 1.5 and 0.5 in. Find the required torque to raise a load of 1,000 lb. Let μ_1 and $\mu_2 = 0.12$.

Solution.
$$\theta = \tfrac{1}{2} \times 29° = 14°30'$$

By Fig. 5-1:   Pitch diameter $= 1.000 - 0.5p = 0.900$ in.; $r_t = 0.45$ in.

Helix angle:
$$\tan \alpha = \frac{4 \times 0.200}{0.900\pi} = 0.282\ 94$$

$$\alpha = 15°48'$$

By Eq. (13):
$$\tan \theta_n = 0.258\ 62 \times 0.962\ 22 = 0.248\ 85$$
$$\theta_n = 13°58.5'$$

By Eq. (12): $$T_1 = 1,000 \times 0.450 \left(\frac{0.97040 \times 0.28294 + 0.12}{0.97040 - 0.12 \times 0.28294} + \frac{1.00}{0.45} \times 0.12 \right)$$

$$= 450 \left(\frac{0.39456}{0.93645} + 0.26667 \right) = 309.6 \text{ in. lb}$$

For standard screws that have small values for the helix angle α, Eq. (13) indicates that θ_n has almost the same value as θ. When this is so, θ_n can be replaced by one-half the included thread angle in the foregoing equations.

If the weight is being lowered by the application of force F acting to the right in Fig. 5-13, the sign of F, together with the signs of all the friction terms, are reversed. By a development similar to that of the foregoing, the torque required to lower the load is

$$T_2 = Wr_t \left(-\frac{\cos \theta_n \tan \alpha - \mu_1}{\cos \theta_n - \mu_1 \tan \alpha} + \frac{r}{r_t} \mu_2 \right) \tag{14}$$

Example 4. Find the torque required to lower the load for the Acme screw of Example 3.

Solution. By Eq. (14):

$$T = 1{,}000 \times 0.450 \left(-\frac{0.97040 \times 0.28294 - 0.12}{0.97040 - 0.12 \times 0.28294} + \frac{1.00}{0.45} \times 0.12 \right)$$

$$= 450 \left(-\frac{0.15459}{1.00436} + 0.26667 \right) = 50.7 \text{ in. lb}$$

If the angle is sufficiently great, the screw will overhaul, or the weight will revolve the screw. The inclined plane in Fig. 5-13(a) will then move to the right and force F must act to the left to preserve uniform motion. The torque equation for the overhauling screw can be shown to be

$$T_3 = Wr_t \left(\frac{\cos \theta_n \tan \alpha - \mu_1}{\cos \theta_n + \mu_1 \tan \alpha} - \frac{r_c}{r_t} \mu_2 \right) \tag{15}$$

Example 5. A single-pitch square thread screw has a pitch diameter of 1.75 in. and pitch of 0.875 in. Coefficient of friction $\mu_1 = 0.10$. Collar friction is negligible. Find the horsepower required for the screw to raise a weight of 3,000 lb at the rate of 20 ft/min.

Solution.
$$r_t = \tfrac{1}{2} \times 1.75 = 0.875 \text{ in.}$$

$$\cos \theta = 1$$

$$\tan \alpha = \frac{p}{\pi \, dp} = \frac{0.875}{1.75\pi} = 0.15916$$

By Eq. (12):
$$T_1 = 3{,}000 \times 0.875 \frac{1 \times 0.15916 + 0.10}{1 - 0.1 \times 0.15916}$$

$$= 2{,}625 \times \frac{0.25916}{0.98408} = 691.3 \text{ in. lb}$$

Rpm of screw:
$$n = \frac{20 \times 12}{0.875} = 274.3 \text{ rpm}$$

By Eq. (8), Chapter 3:
$$hp = \frac{Tn}{63{,}025} = \frac{691.3 \times 274.3}{63{,}025} = 3.01$$

If all friction could be eliminated in both screw and collar, Eq. (12) shows that the torque required to raise the load would be

$$T' = r_t W \tan \alpha \qquad (16)$$

The efficiency of a power screw with collar friction when it is raising the load is equal to the ratio of the torques of Eqs. (12) and (16). Hence for both screw and collar,

$$\text{efficiency} = \frac{T'}{T_1} \qquad (17)$$

If the collar friction is negligible, the following equation for the efficiency of the screw alone results.

$$\text{Efficiency} = \frac{\cos \theta_n - \mu_1 \tan \alpha}{\cos \theta_n + \mu_1 \cot \alpha} \qquad (18)$$

Efficiencies, as given by Eq. (18), have been plotted in Fig. 5-14 for several values of μ_1. It should be noted that the power screw has very low mechanical efficiency when the helix angle is in the neighborhood of either 0° or 90°.

FIGURE 5-14 Efficiency of power screw when collar friction is negligible.

The friction between the threads and between the nut and the abutment depends on a variety of factors, such as surface finish, degree of lubrication, alignment, materials, plating, burrs, and the like. Calculations for the induced force in the bolt will therefore vary as these factors affect the coefficient of friction. The torque wrench, although widely used, is not considered a very reliable means for obtaining an accurate value of the induced force.

A better way is to measure the bolt elongation when this is possible. The angle of twist of the nut can also be used. The nut is first firmly tightened to seat the parts on each other. It is then loosened and made finger-tight. The nut is then turned through a specified angle, which has been calculated to give the desired axial force in the bolt.

10. Torque Required to Produce Desired Clamping Force

Standard bolts with 60° threads are widely used to impart a clamping force. The force W can be determined by the power screw equation (12). For bolts $\frac{1}{2}$ in. diameter and larger, $\theta = 30°$, $\mu_1 = \mu_2$, and data from Table 5-1, the average value of the tightening torque T can be shown to be

$$\text{Coarse thread, } \mu = 0.15 \qquad T = 0.195dW \qquad (19)$$

$$\text{Coarse thread, } \mu = 0.10 \qquad T = 0.135dW \qquad (20)$$

$$\text{Fine thread, } \mu = 0.15 \qquad T = 0.189dW \qquad (21)$$

$$\text{Fine thread, } \mu = 0.10 \qquad T = 0.130dW \qquad (22)$$

Diameter d in these equations is the major or nominal diameter of the bolt. The radius for the collar was taken as the mean radius between the bolt diameter and the distance across the flats for a hex nut.

Example 6. Determine the torque required to produce a clamping force of 1,500 lb in a $\frac{1}{2}$–13 bolt with 60° threads. Take the collar radius r_c as the mean between bolt radius and radius across the flats for the nut $\mu_1 = \mu_2 = 0.15$.

Work by the exact equation (12) and check by approximate equation (19).

Solution. By Table 5-1:   Pitch diameter, $pd = 0.4500$ in.

$$r_t = 0.2250 \text{ in.}$$

$$r_c = \tfrac{1}{4}(0.50 + 0.75) = 0.3125 \text{ in.}$$

Helix angle:
$$\tan \alpha = \frac{p}{\pi \, dp} = \frac{1}{13\pi 0.450} = 0.05441$$

$$\alpha = 3°7'$$

By Eq. (13):
$$\tan \theta_n = \tan \theta \cos \alpha$$

$$= 0.57735 \times 0.99852 = 0.57650$$

$$\theta_n = 29°58'$$

By Eq. (12): $T = 1,500 \times 0.2250\left(\dfrac{0.86632 \times 0.05445 + 0.15}{0.86632 - 0.15 \times 0.05445} + \dfrac{0.3125}{0.2250} \times 0.15\right)$

$$= 337.5\left(\frac{0.04717 + 0.15}{0.86632 - 0.00817} + 0.2083\right)$$

$$= 337.5\left(\frac{0.19717}{0.85815} + 0.20833\right)$$

$$= 337.5(0.2298 + 0.2083) = 147.9 \text{ in.lb}$$

By Eq. (19): $T = 0.195\,dW = 0.195 \times \tfrac{1}{2} \times 1,500 = 146 \text{ in. lb}$

Use of the approximate equation will thus save a great amount of labor when calculating the clamping force for standard bolts with 60° threads.

11. Friction of Screws

Published articles on experiments for screw-thread friction indicate that the following values would be a good estimate for the coefficient of friction.[4]

Mean value of $\mu = 0.15$

Range of variation $= \pm 33\%$

The above is to apply to threads lubricated with mineral oil but with no special control over surface finish other than that normally found under conditions of large quantity production.

12. Stress Concentration

A stress concentration is present in the screw where the load is transferred through the nut to the adjoining member. This arises because the force in the screw must shift outwardly to the region near the boundary as it is transferred from the screw to the nut. Under ideal conditions, the tension in the screw and the compression in the nut should be reduced uniformly starting from full load at the first contact between screw and nut. However, the tension increases the pitch in the screw, and the compression decreases the pitch in the nut so that correct compliance between the loaded parts is not maintained. The major portion of the load is transferred at the first pair of contacting threads and a large stress concentration is present here. Although the stress concentration is somewhat relieved by bending of the threads and expansion of the nut, most bolt failures occur at this point.

Stress concentration factors for threads with static loads are usually determined by photoelastic analysis. Three-dimensional tests[5] have indicated

[4]See Lambert, T. H., "Effects of Variations in the Screw-Thread Coefficient of Friction on the Clamping Force of Bolted Connections," *J. Mech. Eng. Sci.*, 4, 1962, p. 401.

[5]See Hetenyi, M., "A Photoelastic Study of Bolt and Nut Fastenings," *Trans. ASME*, 65, 1943, p. A-93.

a stress concentration factor of 3.85 at the root of the first engaged thread. Other investigators,[6] however, have found higher as well as lower values. Various methods are used for increasing the flexibility of the nut and thus increasing the area over which this transfer of force takes place. A tension nut, or a nut with a tapered lip, as shown in Fig. 5-15(a), has been successfully used in fatigue service. The reduced cross-sectional area of the lip deforms in tension with the bolt, and the load is accordingly spread over a greater number of threads. Another method is to cut the thread of the nut on a very small taper, thus reducing the contact area for the first few threads. Since these threads will bend and carry less load, additional threads must come into service.[7] This method is expensive because of the close dimensional tolerance required. Nuts made of a material with a modulus smaller than the bolt have also been successful in spreading the load over a larger area. The material in the nut, however, must have a sufficient reserve of ductility to deform without breaking.

FIGURE 5-15   Typical methods for increasing the strength of bolts.

An increase in the flexibility of the bolt is also beneficial under fatigue conditions. Accordingly, there should be a considerable length of free thread beyond that required for engagement of the nut, as shown in Fig. 5-15(b). A radius between shank and head is helpful in reducing the stress concentration at this point. The run-out angle for the thread should have a small value, as shown in Fig. 5-15(c), since the stress concentration is highest at the first threads. For the same reason, a stress-relieving groove is shown on the right end of this stud. It should have a diameter equal to that of the root of the thread or slightly less, and should be merged to the shank by radii of consider-

[6]See Brown, A. F. C., and V. M. Hickson, "A Photo-elastic Study of Stresses in Screw Threads," *Inst. Mech. Eng., Proc. (B)*, **1B**, 1952–53, p. 605.

[7]See Stoeckly, E. E., and H. J. Macks, "The Effect of Taper on Screw Thread Load Distribution," *Trans. ASME*, **74**, 1952, p. 103.

able size. A large radius causes less stress concentration in transferring the load from the larger to the smaller cross section. In addition, if the stud is bent because of misfits in assembly of the parts, the smaller diameter will produce a smaller bending stress in the material. Flexibility of the bolt is increased by reducing the entire length between the head and the threads to the root diameter. If the bolt is holding two or more members together, short lengths of the original diameter must be left at each junction to serve as guide surfaces for the bolted parts, as shown in Fig. 5-15(d).

Heat-treated alloy steel is usually more sensitive to changes of form than plain-carbon steel. When an actual thread is tested in fatigue loading, it is more correct to speak of a stress reduction factor. This factor is defined as the ratio of the endurance limit of the material for a plain specimen to the endurance limit for a specimen with threads.

It has been estimated that bolt failures are distributed about as follows: 15 percent under the head, 20 percent in the bolt at the end of the threads, and 65 percent in the bolt at the nut face.

13. Locknuts

Many different kinds of locknuts have been devised to prevent a nut from loosening in service. Several different forms are shown in Fig. 5-16. Spring and lock washers placed beneath an ordinary nut are also in service.

(a) Castellated nut, drilled bolt and cotter.

(b) Upper part of nut is tapered. Segments press against bolt.

(c) Threads in section above slot are deformed to provide friction grip.

(d) Nylon or fiber insert exerts friction grip on threads.

(e) Nylon plug staked in flat of nut applies friction to bolt threads.

(f) Upper nut contracts the slotted projection of lower nut against bolt.

FIGURE 5-16 Several forms of locknuts.

Liquid thread-locking agents are available to prevent loosening of nuts under conditions of shock and vibration. The material fills the space between mating threads and binds them together by hardening into a tough plastic. This occurs upon contact with metal in the absence of oxygen. These materials come in many compositions and physical properties to fit the particular conditions of the application.

The special nut of Fig. 5-17(a) is effective in ensuring that the full body of the bolt, rather than the threaded portion, will be in bearing against the thin plates. The same effect can be had by the use of washers of suitable thicknesses. The speed nut of Fig. 5-17(b) provides a single thread nut of spring steel. When tightened, the prongs press against the stud and act as an arched spring lock.

(a) Nut to insure gross area of bolt will bear   (b) Standard flat speed nut
on thin plate

FIGURE 5-17   Examples of special nuts.

Self-tapping and drive screws are available for fastner service in sheet metal, plastics, nonferrous metals, and other materials. For use in die castings and plastics the thread can be shaped so that it merely displaces the metal without forming chips, which may not be permissible in some applications.

14. Materials and Methods of Manufacture

Bolts and screws are nearly always made on automatic forging machines which give a finished product with practically no scrap loss. Heading is done cold for diameters $\frac{3}{4}$ in. or less, but for larger diameters the end of the bar must be suitably heated. The threads are formed by rolling between dies which depress part of the material to form the root and which force the remainder up to make the top of the thread. The outside diameter of the thread is thus larger than the stock on which it was rolled. If the threaded portion of the bolt is to have the same diameter as the unthreaded part, the rolling must be done on a reduced diameter. Forging machines form this portion of smaller cross section at the same time the head is upset. Rolled threads, because of favorable grain structure at the root of the thread, are stronger than cut threads in fatigue and impact.

The tensile strength of bolts can be conveniently specified by the grade designations shown in Table 5-3. An increasing number indicates increasing tensile strength. Grade designation markings can be shown on the bolt heads as shown in Fig. 5-18.

TABLE 5-3 Mechanical Properties for Materials for Bolts, Screws, and Studs*

Grade Desig- nation	Material Properties				Core Hardness, Rockwell		C (carbon), %	
	Yield Strength, min. psi	Tensile Strength, min. psi	Elon- gation, min. %	Red. of Area, min. %	Min.	Max.	Min.	Max.
1	36,000	60,000	18	35	B70	B100	—	0.55
2($\frac{1}{4}$ through $\frac{3}{4}$)	57,000	74,000	18	35	B80	B100	—	0.55
2($\frac{3}{4}$ to $1\frac{1}{2}$)	36,000	60,000	18	35	B70	B100	—	0.55
4	100,000	115,000	10	35	C22	C32	—	0.55
5($\frac{1}{4}$ through 1)	92,000	120,000	14	35	C25	C34	0.28	0.55†
5(1 to $1\frac{1}{2}$)	81,000	105,000	14	35	C19	C30	0.28	0.55†
5.2	92,000	120,000	14	35	C26	C36	0.15	0.25
7	115,000	133,000	12	35	C28	C34	0.28	0.55
8	130,000	150,000	12	35	C33	C39	0.28	0.55
8.1	130,000	150,000	10	35	C32	C38	0.28	0.55
8.2	130,000	150,000	10	35	C35	C42	0.15	0.25†

*Proof load, psi, on full-size bolt is about 10% less than for yield strength, psi. All grades permitted nominal amounts of P and S. See standard. Grades 5 to 8.2 inclusive are quenched and tempered.

†Plus 0.74 Mn.

No.5   No.5.1   No.5.2   No.7   No.8   No.8.2

FIGURE 5-18 Grade designation markings for bolt heads. SAE standard J429. Grades not shown have no markings. Reproduced with permission from SAE Handbook 1983, Society of Automotive Engineers, Inc.

Bolts can, of course, be made from other kinds of steel as well as non-ferrous metals. They can also be turned on automatic screw machines, using free cutting stock of the same size as the head. The process is very wasteful of material, however, and has been largely abandoned in favor of cold heading.

Threads can also be formed by grinding.

15. Stress Due to Impact Load

Bolts are sometimes subjected to suddenly applied or impact loads. The stress caused by such loads can be found from the energy of impact U that the bolt must absorb.

The force-deformation diagram for a bolt under a tensile load is a triangle, as shown in Fig. 5-19. The area represents the strain energy U stored in the bolt. Thus

$$U = \tfrac{1}{2}F\delta \qquad (23)$$

where F is the force caused by impact, and δ is the corresponding deformation. Deformation δ is equal to F/k, where k is the spring constant for the bolt, AE/l. Substitution for δ gives

FIGURE 5-19 Strain energy diagram for bolt.

$$U = \frac{F^2}{2k} \qquad (24)$$

The stress in the bolt is equal to force F divided by the smallest cross-sectional area. For a threaded bolt, this area is the stress area of the threaded portion.

Example 7. A $\tfrac{1}{2}$ in.-13UNC $\times$ 12 in. long steel bolt must carry an impact load of 40 in. lb.

(a) Find the stress in the stress area for a standard bolt.

(b) Find the stress if the entire body of the bolt between head and nut is reduced to that of the stress area.

Solution. (a) If the threads stop adjacent to the nut, the full $\tfrac{1}{2}$-in. diameter is stretched by the impact force. Hence,

$$A = \frac{\pi}{4}\left(\frac{1}{2}\right)^2 = 0.1964 \text{ in.}^2$$

$$k = \frac{AE}{l} = \frac{0.1964E}{12} = 0.01637E$$

In Eq. (24): $F = \sqrt{2kU} = \sqrt{2 \times 0.01637 \times 30,000,000 \times 40}$

$$= 6,267 \text{ lb}$$

By Table 5-1:   stress area $= 0.1419$ in.^{2}

In stress area: $\sigma = \dfrac{6,267}{0.1419} = 44,160$ psi

It is possible for stress concentration to cause a large increase in the stress above.

(b) The value of k now depends on the stress area; hence,

$$k = \frac{AE}{l} = \frac{0.1419E}{12} = 0.01182E$$

In Eq. (24):
$$F = \sqrt{2 \times 0.01182 \times 30,000,000 \times 40} = 5,320$$

$$\sigma = \frac{5,320}{0.1419} = 37,530 \text{ psi}$$

Example 7A. A steel bolt has a major diameter of 14 mm and pitch p of 2 mm. It is 300 mm long and must carry an impact load of 4,500 Nmm. Threads stop adjacent to nut. $E = 210,000$ MPa.

(a) Find the stress in the root area for a standard bolt.

(b) Find the stress if the entire body of the bolt between head and nut is reduced to that of the root area.

Solution. (a) For the full 14-mm diameter,

$$A = \frac{\pi}{4} \times 14^2 = 153.9 \text{ mm}^2$$

$$k = \frac{AE}{l} = \frac{153.9 \times 210,000}{300} = 107,760 \text{ N/mm}$$

By Eq. (24): $F = \sqrt{2kU} = \sqrt{2 \times 107,760 \times 4,500} = 31,140$ N

By Example 1A, the root area is 110.0 mm².

In the root area: $\sigma = \dfrac{31,140}{110.0} = 283 \dfrac{N}{mm^2}$ or MPa

(b) The value of k now depends on the root area; hence

$$k = \frac{AE}{l} = \frac{110.0 \times 210,000}{300} = 77,000 \text{ N/mm}$$

In Eq. (24):
$$F = \sqrt{2kU} = \sqrt{2 \times 77,000 \times 4,500} = 26,320 \text{ N}$$

In the root area: $\sigma = \dfrac{26,320}{110.0} = 239 \dfrac{N}{mm^2}$ or MPa

By careful attention to details, this reduced diameter can be gradually blended into the threaded portion and the stress concentration factor reduced practically to unity.

This example illustrates how a smaller impact stress results from increasing the flexibility by making the diameter equal to the root diameter of the thread. Flexibility can also be increased by using a longer bolt. If the bolt is made of a ductile material, and the yield point is exceeded by the impact force, some permanent stretching results.

Experiments have indicated that the yield strength of a material under impact conditions is somewhat higher than when the loading is static.[8]

[8]See Forkois, H. M., R. W. Conrad, and I. Vigness, "Properties of Bolts under Shock Loading," *Proc. Soc. Exp. Stress Anal.*, **10**, No. 1, 1952, p. 165. See also Davidenkoff, N. N., "Allowable Working Stresses under Impact," *Trans. ASME*, **56**, 1934, APM-56, p. 97.

16. Relaxation

Bolts that connect heavy parts operate under conditions of constant elongation. In high-temperature service the initial tension will diminish until, after a sufficiently long time, the joint will no longer be tight. This phenomenon is known as *relaxation*. The initial tension must be sufficiently great to maintain a tight joint at the end of the expected life of the assembly. Similar conditions prevail in shrink or press-fitted joints in high-temperature service.

REFERENCES

1. *Guide to World Screw Thread Standards*, Bedford, England: W.H.A. Robertson & Co. Ltd., Lyton Works.
2. ROTHBART, H. A., Ed., *Mechanical Design and Systems Handbook*, New York: McGraw-Hill Book Company, 1964, Sections 20, 21, and 26.
3. *Screw-Threads for Federal Services, Handbook H-28*, Washington, D.C.: U.S. Government Printing Office, 1957.
4. *Unified and American Screw Threads*, ASA B1.1-1974, New York: American Society of Mechanical Engineers.

PROBLEMS

In Problems 1–5, let it be assumed that the threads stop immediately above the nut.

1. In an assembly with loading arrangement equivalent to Fig. 5-8, the bolt is $\frac{3}{4}$ in.–16UNF, and the total load varies from 2,000 to 10,000 lb. The material in the bolt tests $\sigma_{ult} = 130,000$ psi and $\sigma_{yp} = 110,000$ psi, with stress concentration factor for the threads equal to 3.85. Let the factor of safety based on the yield point be 2, and let E for both part and bolt be equal to 30,000,000 psi. The threads are machined. The cross-sectional area of the part is equal to 1.25 in.^{2}

(a) Draw the working-stress triangle for the bolt material, and plot the stresses when there is no initial stress in the bolt.

(b) If the initial force in the bolt is 8,000 lb, plot the stress values for the bolt on the working-stress triangle.

It will be necessary to determine whether or not the part will have a compressive force when the maximum load is acting. Should all the initial compression be removed, the force in the bolt will have the same value as the load.

(c) Repeat part (b) with an initial bolt force of 13,000 lb. Note the effect of excessive initial stress.

(d) Suppose the bolt is made of steel, but the part is made of aluminum, $E = 10,000,000$ psi. Plot the stress values for the bolt if other data are the same as in part (b).

(e) Suppose the bolt were turned to the stress area of the thread over its entire length except for the nut. Plot the stress values for the bolt; other data are the same as in part (b).   *Ans.*   (b) $\sigma_{av} = 25{,}650$ psi; $K\sigma_r = 10{,}790$ psi.

2. The connecting rod bolt of Fig. 5-20 is 10 mm in diameter; the thread pitch is 1.25 mm. It is drawn up to an initial force of 16,000 N. The bolt has ground threads. Material tests $\sigma_{ult} = 760$ MPa, $\sigma_{yp} = 620$ MPa, and $\sigma_e = 350$ MPa. The average cross-sectional area of the bolted parts is equal to 320 mm² for each bolt. The stress concentration factor for the threads is equal to 3.85. The load for the part varies continuously from 0 to 11,000 N for each bolt. Find the value of the factor of safety for the bolt. Find the value of the minimum force in the part. The material of the part has the same modulus as the bolt.   *Ans.* $F_s = 1.54$.

FIGURE 5-20   Problem 2.

3. Let the bolt in Fig. 5-8 be a 1 in.–12UNF made from cold-drawn 2317 steel. The load on the part varies continuously from 10,000 to 20,000 lb. The threads are machined. $K = 3.85$. Area of the steel part is 1.2 in.² Find the maximum permissible value of F_o if bolt is to be safe for continuous operation with an F_s equal to 2.   *Ans.* $F_o = 3{,}060$ lb.

4. The bolt in Fig. 5-8 is 20 mm diameter and pitch 2.5 mm and is made from 1137 cold-drawn steel. Threads are machined. $F_s = 2.5$; $K = 3.85$. Area of the steel part is 700 mm². Initial force $F_o = 34{,}000$ N. If the average stress in the root area is 170 MPa, find the maximum and minimum values of the varying load P on the part.   *Ans.* $P_{max} = 31{,}860$ N; $P_{min} = 15{,}340$ N.

5. (a) In Fig. 5-8 let the part undergo a permanent shortening in length equal to ϵ after the bolt has been tightened to an initial force of F_o. Derive an equation for the new value F_o' for the initial force in the bolt.

(b) Find the value of F_p' for the bolt of Problem 1(b) if the part for some reason should shorten 0.0003 in. Bolt and part are 5.0 in. long.

Ans.   (a) $F_o' = F_o - \dfrac{\epsilon k_b k_p}{k_b + k_p}$; (b) $F_p' = -23$ lb.

6. Find the force of tension in an 8-mm-diameter and 1-mm-pitch bolt if it is tightened by a wrench with a torque equal to 27,000 Nmm. Assume the coefficient of friction for the screw and the collar to be the same, and make computations for coefficients of 0.10, 0.15, and 0.20. Take the outer collar diameter as being equal to 13 mm.   *Ans.*   For $\mu = 0.10$, $W = 24,290$ N.

7. A load of 10,000 lb is carried by a 2.5-in. single-thread Acme screw of standard proportions. The pitch is $\frac{1}{3}$ in. and the pitch diameter is $2\frac{1}{3}$ in. The OD of the collar is equal to 4 in., and the ID is equal to 1.25 in.

(a) For $\mu_1 = \mu_2 = 0.15$, find the horsepower required to rotate the screw if the weight is to be raised at the rate of 10 fpm.

(b) What is the efficiency when friction of both screw and collar is considered? What is the efficiency if the collar friction is made negligible by use of an antifriction bearing?

(c) Find the horsepower required to lower the load at the same rate.

(d) What horsepower will raise the load at the given rate when the collar is supported on a ball thrust bearing for which $\mu_2 = 0.003$? Let the collar radius be the same as for the plain bearing. What will now be the efficiency?

(e) Find the pitch of the thread at which overhauling will take place, using a ball thrust bearing. The pitch diameter is the same.

(f) Suppose the screw is made with pitch just sufficient to overhaul. What will be the efficiency for the screw alone?

(g) If the minimum major diameter of the screw is 2.483 in. and the maximum minor diameter of the nut is 2.183 in., find the minimum length of the nut that must be engaged if the compressive stress on the projected area of the threads is equal to 600 psi.

(h) If the minimum minor diameter of the screw is 2.130 in., find the average value of the compressive stress at the root of the thread. What is the bearing pressure for the collar?   *Ans.*   (a) hp = 24.7; (b) eff = 12.3 and 22.5%;
(e) $p = 1.16$ in.; (g) $l = 5.05$ in.

8. (a) Write the expression for tan α for a square thread with negligible collar friction at which the maximum efficiency will occur.

(b) What is the value of the maximum efficiency, and the angle at which it occurs, for $\mu_1 = 0.1$?   *Ans.*   tan $\alpha = -\mu_1 + (1 + \mu_1^2)^{1/2}$; eff = 81.9%.

9. What pitch must be provided on a square thread power screw to raise a 2,000-lb weight at 40 fpm with power consumption of 4 hp? The pitch diameter is 1.375 in., μ_1 is 0.15, and collar friction is negligible.   *Ans.* $p = 1.106$ in.

10. A square thread screw has an efficiency of 65 percent when raising a load. The coefficient of friction for the threads is 0.15 with collar friction negligible. Pitch diameter is 70 mm. When lowering a load, a uniform velocity is maintained by a brake mounted on the screw. If the load is equal to 9,000 kg, what torque must be exerted by the brake?   *Ans.* $T = 510,800$ Nmm.

11. A square thread screw has an efficiency of 70 percent when raising a weight. The coefficient of friction is $\mu_1 = 0.12$ with collar friction negligible. The load is 8,000 lb and the pitch diameter is 1.15 in. Find the torque that a brake mounted on the screw must exert when lowering the load at a uniform rate.

Ans. $T = 890$ in. lb.

12. A square thread screw, 23 mm pitch diam, has an efficiency of 70 percent when raising a 1,800-kg weight. The coefficient of friction for the threads is 0.10, with collar friction negligible. Find the torque that a brake mounted on the screw must exert when the load is being lowered at a constant rate.

Ans. $T = 30,680$ Nmm.

13. A square thread screw has a pitch diameter of 1.5 in. and a lead of 1 in. The screw consumes 4 hp when raising a 2,800-lb weight at the rate of 30 fpm. Collar friction is negligible. Find the coefficient of friction for the threads.

Ans. $\mu_1 = 0.113$.

14. A square thread screw is at the point of overhauling when at rest. When raising a 450-kg weight at 4.6 m/min, the input is 0.69 kW. Find the pitch of the screw if the pitch diameter is 20 mm. Collar friction is negligible.

Ans. $p = 8.75$ mm.

15. A screw of 9 threads per in. is to be cut on a $1\frac{3}{16}$-in.-diam bar. Proportions are similar to the Unified thread. Find the value of the basic pitch diameter. If the screw is made double thread, find the helix angle.

16. The minimum major diameter of a 1 in.-8UNC screw is 0.9755 in. The maximum minor diameter of the nut is 0.8797 in. Find the required length of nut that must be engaged with the screw if the tensile load is 8,000 lb and permissible bearing stress on the projected area of the threads is 10,000 psi.

17. The OD of the aluminum part in Fig. 5-8 is 25 mm. The bolt is steel, 12 mm in diameter. The nut is torqued to 68,000 Nmm by the approximate equation. Find the maximum load P that can be placed on the part without losing all the initial compression in the part. Threads stop immediately above the nut. Part and bolt are of equal length.   *Ans.* $P = 73,780$ N.

18. A square thread power screw has μ_1 equal to 0.14 and negligible collar friction. When holding a load, it is just at the point of overhauling. What is the efficiency for this screw when raising a load?

19. Work Problem 11, but with the efficiency at 75 percent.

20. Work Problem 14, but with the input horsepower of 0.95.

21. Work Problem 13, but with the horsepower of 4.5.

22. A square thread screw is at the point of overhauling. Its efficiency when raising a load is 0.49. What is the coefficient of friction for the threads if the collar friction is negligible?

23. A square thread screw has an efficiency of 60% when mounted with an antifriction collar. The helix angle is 12°. Find the torque required to exert a 2,000-lb force if the pitch diameter of the screw is 2 in.

24. The bolts in Fig. 5-21 are 20 mm in diameter, pitch 1.5 mm, made from cold-drawn 8740 steel. Machined threads; $K = 3.85$. The steel cylinder is 100 mm ID. Fluid pressure varies continuously from 0 to 14 MPa. Assume the end plates are rigid. Find the value of initial force F_o if the bolts are to have an F_s equal to 4.

Ans. $F_o = 37,680$ N.

FIGURE 5-21   Problem 24.

25. By trial determine the value of the lead angle α, to the nearest $\frac{1}{2}°$, that will place the quadruple Acme screw of Example 3 at the point of overhauling.

26. For $\mu_1 = \mu_2$ find the value of the coefficient of friction that will cause the quadruple Acme screw of Example 3 to be just at the point of overhauling.
Ans. $\mu = 0.086$.

27. For a 60° standard thread, with collar diameter as the mean between the bolt diameter and the distance across the flats for a hex nut, let the torque T required to produce a clamping force W be taken as $T = C\, dW$. Find the value of C for a $\frac{3}{4}"$–10UNC thread for $\mu_1 = \mu_2 = 0.15$.   *Ans.* $C = 0.1949$.

28. Same as Problem 27 except for a $\frac{3}{4}"$–10UNC thread and $\mu_1 = \mu_2 = 0.10$.
Ans. $C = 0.1368$.

29. Same as Problem 27 except for a $\frac{3}{4}"$–16UNF thread and $\mu_1 = \mu_2 = 0.015$.
Ans. $C = 0.1894$.

30. Same as Problem 27 except for a $\frac{3}{4}"$–16UNF thread and $\mu_1 = \mu_2 = 0.10$.
Ans. $C = 0.1306$.

31. The bolt in Fig. 5-8 is $\frac{1}{2}"$–13UNC, machined threads made of 1045 cold-drawn steel. Stress concentration factor 2. Initial torque 300 in. lb. Assume coefficient of friction to be 0.15. Load P varies continuously from 1,000 to 3,000 lb. Part is steel with area equal to 0.5 in.² Find the factor of safety for the bolt. Find the minimum force in the part.   *Ans.* $F_s = 2.85$; $F_{p\ min} = -1,354$ lb.

32. The bolt in Fig. 5-8 is $1"$–12UNF, machined threads, made of 8740 hot-rolled steel. Stress concentration factor 2. Initial torque 1,200 in. lb. Assume coefficient of friction to be 0.10. Load varies continuously from 5,000 to 10,000 lb. Part is of steel with area 2 in.² Find the factor of safety for the bolt. Find the minimum force in the part.   *Ans.* $F_s = 4.18$; $F_{p\ min} = -2,051$ lb.

33. The bolt in Fig. 5-8 is $\frac{3}{8}"$–16UNC, ground threads, made of 1137 cold-drawn steel. Stress concentration factor 2. Initial torque 175 in. lb. Assume coefficient of friction to be 0.10. Load P varies continuously from 0 to 2,000 lb. Part is steel with area equal to 0.3 in.² Find the factor of safety for the bolt. Find the minimum force in the part.   *Ans.* $F_s = 1.66$; $F_{p\ min} = -1,264$ lb.

34. The bolt in Fig. 5-8 is $\frac{5}{8}"$–11UNC, ground threads, made of 1045 cold-rolled steel. Stress concentration factor 2.5. Initial torque 500 in. lb. Assume coefficient of friction to be 0.10. Load varies continuously from 1,000 to 5,000 lb. Part is steel with area equal to 1.5 in.² Find the factor of safety for the bolt. Find the minimum force in the part.   *Ans.* $F_s = 2.81$; $F_{p\ min} = -1,775$ lb.

6

Belts, Clutches, Brakes, and Chains

Belts, clutches, brakes, and chains are examples of machine elements that employ friction as a useful agent. A belt provides a convenient means for transferring power from one shaft to another. Belts are frequently necessary to reduce the higher rotative speeds of electric motors to the lower values required by mechanical equipment. Clutches are required when shafts must be frequently connected and disconnected. The function of the brake is to turn mechanical energy into heat. The design of frictional devices is subject to uncertainties in the value of the coefficient of friction that must necessarily be used. Chains provide a convenient and effective means for transferring power between parallel shafts.

A,	area	in.,	inch, 25.4 mm
b,	width of lining	J,	joule, newton-meter, Nm
c,	center distance	k,	kilo (prefix), 1,000
CW,	clockwise	kg,	kilogram, 2.2046 lb (mass)
CCW,	counterclockwise	kW,	kilowatt, 1.341 hp
d,	diameter	l,	length of belt
$e =$	2.718, base of natural logarithms	lb,	pound; 1 lb (force) = 4.4482 N
fpm,	feet per minute		1 lb (mass) = 0.4536 kg
F,	force	m,	meter, 39.370 in.
F_n,	force normal to surface	m,	milli (prefix), one thousandth
g,	acceleration due to gravity		
hp,	horsepower, 0.7457 kW	mm,	milimeter, 0.039 37 in.

M,	mega (prefix), 1 million	P,	force or load
M,	moment, 1 in. lb = 113 Nmm	Pa,	pascal, 1 N/m²
		r,	radius
M_f,	moment due to friction force	R,	reactive force
		T,	torque
M_n,	moment due to normal force	T_1,	force in tight side
		T_2,	force in slack side
M_1, M_2,	force peaks to failure	V,	velocity
MPa,	megapascal, 1 N/mm² or 145 psi	W,	watt, 1 Nm/sec or 1 J/sec
		α,	(alpha) angle of contact
N,	newton, 0.2248 lb (force); 1 lb (force) = 4.448 N	δ,	(delta) wear
		$δ_n$,	wear normal to surface
N,	number	μ,	(mu) coefficient of friction
p,	pressure, pitch	ψ,	(psi) half-angle of contact for belt
p_n,	pressure normal to surface		

1. V-belts

The rayon and rubber V-belt is widely used for power transmission. Such belts are made in two series: the standard V-belt shown in Fig. 6-1(a) and the high-capacity V-belt of Fig. 6-1(b). The belts can be used with short center distances and are made endless so that difficulty with splicing devices is avoided. The lengths are standardized and are given in Table 6-1 for multiple V-belts, and in Table 6-2 for high-capacity V-belts.

(a) Cross sections of multiple V-belts.

FIGURE 6-1

(b) Cross sections of high-capacity V-belts (full size).

FIGURE 6-1

TABLE 6-1   Pitch Lengths of Multiple V-Belts

Cross Section							
A	*B*	*C*	*A*	*B*	*C*	*D*	*E*
27.3			97.3		98.9		
32.3				98.8			
36.3	36.8		106.3	106.8	107.9		
39.3	39.8		113.3	113.8	114.9		
43.3	43.8		121.3	121.8	122.9	123.3	
47.3	47.8		129.3	129.8	130.9	131.3	
52.3	52.8	53.9		145.8	146.9	147.3	
56.3	56.8			159.8	160.9	161.3	
61.3	61.8	62.9		174.8	175.9	176.3	
69.3	69.8	70.9		181.8	182.9	183.3	184.5
76.3	76.8	77.9		196.8	197.9	198.3	199.5
81.3				211.8	212.9	213.3	214.5
	82.8	83.9		240.3	240.9	240.8	241.0
86.3	86.8	87.9		270.3	270.9	270.8	271.0
91.3	91.8	92.9		300.3	300.9	300.8	301.0

First cost is low, and power output may be increased by operating several belts side by side. All belts in the drive should stretch at the same rate in order to keep the load equally divided among them. When one of the belts breaks, the entire group must usually be replaced. The drive may be inclined at any angle with tight side either top or bottom. Since belts can operate on relatively small pulleys, large reductions of speed in a single drive are possible.

The included angle for the belt groove is usually from 34 to 38°. The wedging action of the belt in the groove gives a large increase in the tractive force developed by the belt.

Pulleys may be made of cast iron, sheet steel, or die-cast metal. Sufficient

TABLE 6-2   Outside Circumferences of High-Capacity V-Belts

3V	3V, 5V	3V, 5V, 8V	5V, 8V
25	50	100	150
$26\frac{1}{2}$	53	106	160
28	56	112	170
30	60	118	180
$31\frac{1}{2}$	63	125	190
$33\frac{1}{2}$	67	132	200
$35\frac{1}{2}$	71	140	212
$37\frac{1}{2}$	75		224
40	80		236
$42\frac{1}{2}$	85		250
45	90		265
$47\frac{1}{2}$	95		280

clearance must be provided at the bottom of the groove to prevent the belt from bottoming as it becomes narrower from wear. Sometimes the larger pulley is not grooved when it is possible to develop the required tractive force by running on the inner surface of the belt. The cost of cutting the grooves is thereby eliminated. Pulleys are on the market that permit an adjustment in the width of the groove. The effective pitch diameter of the pulley is thus varied, and moderate changes in the speed ratio can be secured.

2. Center Distance for V-belt Drive

In general the designer will know the pitch diameter, d_1 and d_2, for the two sheaves or pulleys and the standardized length, l, for the belt that is to be used. It is necessary to calculate the center distance, c, between the pulleys. This determines the angle of wrap ψ, Fig. 6-2(a), of the belt on the pulley, which determines the load-carrying capacity of the belt.

Although center distance c can be calculated from the equation shown in Fig. 6-2(a), this equation is very difficult to handle since both c and ψ are unknown. To an excellent degree of approximation length $ABCD$ in Fig. 6-2(b) will be equal to the half length of the belt. Thus center distance c can be easily had by solving the equation

$$\frac{1}{2}l = \frac{\pi}{2}r_1 + \sqrt{c^2 + (r_2 - r_1)^2} + \frac{\pi}{2}r_2$$

or

$$c^2 = \frac{1}{4}[l - \pi(r_1 + r_2)]^2 - (r_2 - r_1)^2 \qquad (1)$$

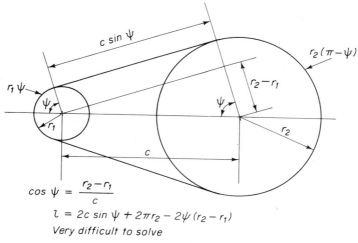

$$\cos \psi = \frac{r_2 - r_1}{c}$$

$$l = 2c \sin \psi + 2\pi r_2 - 2\psi (r_2 - r_1)$$

Very difficult to solve

(a)

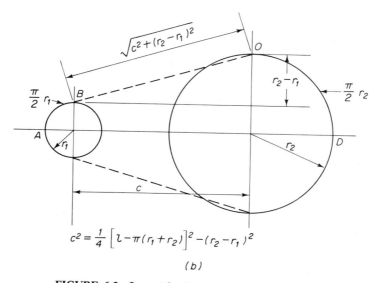

$$c^2 = \frac{1}{4} \left[l - \pi(r_1 + r_2) \right]^2 - (r_2 - r_1)^2$$

(b)

FIGURE 6-2   Layout for determining center distance c.

Example 1. A V-belt is 87.9 in. long and operates on sheaves of pitch diameters of 12 in. and 16 in. Find the center distance c.

Solution. By Eq. (1): $c^2 = \frac{1}{4}[l - \pi(r_1 + r_2)]^2 - (r_2 - r_1)^2$

$$= \frac{1}{4}(87.9 - \pi 14)^2 - 2^2 = \frac{1}{4}(87.9 - 43.98)^2 - 4$$

$$= 482.24 - 4 = 478.24$$

$$c = 21.9 \text{ in.}$$

By Fig. 6-2(a), angle ψ can be determined from

$$\cos \psi = \frac{r_2 - r_1}{c} \qquad (2)$$

For pulleys of equal diameter the tight-side tension T_1 is taken as five times the slack-side tension T_2. The angle of wrap on the smaller or drive pulley is less when the pulleys are of unequal diameter and the ratio T_1/T_2 is decreased. The relationship between $\cos \psi$ and T_1/T_2 is given by Fig. 6-3.

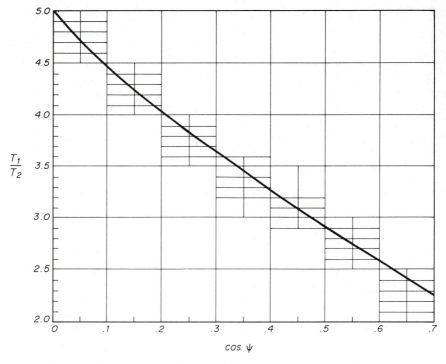

FIGURE 6-3   Relationship between angle of wrap ψ and tension ratio T_1/T_2.

3. Fatigue of V-belts

The velocity V of the belt, feet per minute, is given by the equation

$$V = \frac{\pi d n}{12} \qquad (3)$$

where d is the pulley diameter in inches and n is its speed, revolutions per minute.

If T_1 is the tension, pounds, in the tight side and T_2 the tension in the slack side, the nominal horsepower of the belt is

$$hp = \frac{(T_1 - T_2)V}{33,000} \qquad (4)$$

A service factor must usually be applied to the nominal horsepower requirements to take care of fluctuations in the loading and the prime mover. A recommended list of such factors is given in Table 6-3.

TABLE 6-3   Suggested Service Factors

Hours in daily service	AC Motors: Normal Torque, Squirrel Cage, Synchronous, Split Phase DC Motors: Shunt Wound Engines: Multicylinder Internal Combustion			AC Motors: High-Torque, High-Slip, Repulsion-Induction, Single-phase, Series Wound, Sliping DC Motors: Series Wound, Compound Wound Engines: Single-cylinder Internal Combustion Lineshafts: Clutches		
	3–5	8–10	16–24	3–5	8–10	16–24
Agitators for liquids Blowers and exhausters Centrifugal pumps and compressors Fans up to 10 hp Light-duty conveyors	1.0	1.1	1.2	1.1	1.2	1.3
Belt conveyors for sand, grain, etc. Dough mixers Fans over 10 hp Generators Lineshafts Laundry machinery Machine tools Punches, presses, shears Printing machinery Positive displacement rotary pumps Revolving and vibrating screens	1.1	1.2	1.3	1.2	1.3	1.4
Brick machinery Bucket elevators Exciters Piston compressors Conveyors (drag, pan, screw) Hammermills Papermill beaters Piston pumps Positive displacement blowers Pulverizers Sawmill and woodworking machinery Textile machinery	1.2	1.3	1.4	1.4	1.5	1.6
Crushers (gyratory, jaw, roll) Mills (ball, rod, tube) Hoists Rubber calenders, extruders, mills	1.3	1.4	1.5	1.5	1.6	1.8

The peak force in the belt is considered as made up of the tight-side tension T_1 plus a force T_b caused by bending around the pulley and a force T_c due to the centrifugal effects. Thus at point B in Fig. 6-4 the peak force F_1 is

$$F_1 = T_1 + T_{b1} + T_c \tag{5}$$

FIGURE 6-4   Force at different points along belt.

Similarly, at point E the peak force F_2 is equal to

$$F_2 = T_1 + T_{b2} + T_c \tag{5a}$$

The equation for the bending force T_b is

$$T_b = \frac{K_b}{d} \tag{6}$$

where K_b is a constant, depending on the size of the belt, and is given in Table 6-4.

The centrifugal force T_c is found by

$$T_c = K_c \left(\frac{V}{1,000}\right)^2 \tag{7}$$

Constant K_c is given by Table 6-4.

V-belts are designed on the assumption that a belt can sustain a certain number of force peaks before failure by fatigue will occur.[1] The fatigue effect

---

[1]Fatigue failure for a composite element such as a V-belt can occur in a number of different ways.

TABLE 6-4   Design Constants for V-belts*

Section	K_b	K_c
A	220	0.561
B	576	0.965
C	1,600	1.716
D	5,680	3.498
E	10,850	5.041
3V	230	0.425
5V	1,098	1.217
8V	4,830	3.288

Data courtesy Gates Rubber
Co., Denver, Colorado.

can be most easily calculated by an experimentally determined exponential equation. Thus suppose a belt can sustain M_1 applications of peak force F_1. Then

$$M_1 = \left(\frac{Q}{F_1}\right)^x \tag{8}$$

where constants Q and x are given in Table 6-5.

TABLE 6-5   Design Constants for V-belts

Section	10^8–10^9 Force Peaks		10^9–10^{10} Force Peaks		Minimum Sheave Diameter
	Q	x	Q	x	
A	674	11.089			3.0
B	1,193	10.924			5.0
C	2,038	11.173			8.5
D	4,208	11.105			13.0
E	6,061	11.100			21.6
3V	728	12.464	1,062	10.153	2.65
5V	1,654	12.593	2,394	10.283	7.1
8V	3,638	12.629	5,253	10.319	12.5

Similarly, for the slack side the belt is assumed able to sustain M_2 force peaks of magnitude F_2. Then

$$M_2 = \left(\frac{Q}{F_2}\right)^x \tag{8a}$$

For one belt passage or traverse both forces F_1 and F_2 occur and the combined effect of both forces must be determined. If the belt can sustain M_1 applications of force F_1 acting alone then, for one application, $1/M_1$ of the life is consumed. Similarly, if the belt can sustain M_2 applications of force F_2 alone then one application consumes $1/M_2$ of the belt life. One belt pass will consume $1/M_1$ plus $1/M_2$ of the life of the belt.

Let N' be the number of belt passes to failure. One belt pass consumes $1/N'$ of the belt life and the following equation results.[2]

$$\frac{1}{N'} = \frac{1}{M_1} + \frac{1}{M_2} \tag{9}$$

The belt velocity is $12V$ in./min where V is velocity feet/min. The time for one belt passage is $l/12V$ minutes where belt length l is in inches. The time or life, L, for N' belt passes is

$$L = \frac{N'l}{12V}\,\text{minutes}$$

$$= \frac{N'l}{12 \times 60V}\,\text{hours}$$

Assume that one month equals 20 days at 8 hours each or 160 hours. Then

$$L = \frac{N'l}{12 \times 60\,V \times 160} = \frac{N'l}{115{,}200\,V}\,\text{months} \tag{10}$$

As mentioned, length l is in inches.

Example 2. A C-section belt 87.9 in. long operates on pulleys of pitch diameter 12 in. and 16 in. Speed of smaller pulley is 1,160 rpm. Horsepower is 9 but a service factor of 1.6 must be used. Find the expected life.

Solution. By Eq. (3): $V = \dfrac{\pi\,dn}{12} = 3{,}644$ fpm

By Example 1: $c = 21.9$ in.

By Eq. (2): $\cos \psi = \dfrac{r_2 - r_1}{c} = \dfrac{8 - 6}{21.9} = 0.09145$

By Fig. 6-3: $\dfrac{T_1}{T_2} = 4.55 \qquad T_2 = \dfrac{T_1}{4.55} = 0.220T_1$

Design hp $= 9 \times 1.6 = 14.4$

By Eq. (4): $T_1 - T_2 = T_1 - 0.220T_1 = \dfrac{33{,}000\,\text{hp}}{V}$

$0.780T_1 = \dfrac{33{,}000 \times 14.4}{3{,}644} = 130.4$ lb

[2]This is known as Miner's equation. See Eq. (20), Chapter 2.

$$T_1 = \frac{130.4}{0.780} = 167.2 \text{ lb}$$

By Eq. (6): $$T_{b1} = \frac{K_b}{d} = \frac{1,600}{6} = 133.3 \text{ lb}$$

By Eq. (7): $$T_c = K_c \left(\frac{V}{1,000}\right)^2 = 1.716 \times 3,644 = 22.8$$

By Eq. (5): $$F_1 = 323.3 \text{ lb}$$

By Eq. (8): $$M_1 = \left(\frac{Q}{F_1}\right)^x = \left(\frac{2,038}{323.3}\right)^{11.173} = 6.304^{11.173}$$

$$\log M_1 = 11.173 \log 6.304 = 11.173 \times 0.79962 = 8.93415$$

$$M_1 = 859,000,000 \text{ force peaks}$$

$$T_1 = 167.2 \text{ lb}$$

By Eq. (6): $$T_{b2} = \frac{K_b}{d} = \frac{1600}{16} = 100.0 \text{ lb}$$

As above, by Eq. (5a): $$T_c = \frac{22.8}{290.0}$$

By Eq. (8a): $$M_2 = \left(\frac{Q}{F_2}\right)^x = \left(\frac{2,038}{290.0}\right)^{11.173} = 7.028^{11.173}$$

$$\log M_2 = 11.173 \log 7.028 = 11.173 \times 0.84683 = 9.46163$$

$$M_2 = 2,895,000,000 \text{ force peaks}$$

By Eq. (9): $$\frac{1}{N'} = \frac{1}{M_1} + \frac{1}{M_2} = \frac{1}{1,000,000,000}\left(\frac{1}{0.859} + \frac{1}{2.895}\right)$$

$$= \frac{1.164 + 0.345}{1,000,000,000}$$

$$N' = \frac{1,000,000,000}{1.509} = 662,700,000 \text{ belt passes}$$

By Eq. (10): $$= \frac{N'l}{115,200V} = \frac{662,700,000 \times 87.9}{115,200 \times 3,644}$$

$$= 139 \text{ months of 160 hr each}$$

It is seen that the larger pulley has but a small effect on the fatigue life. In fact, when d_2 is sufficiently large, its effect is so small that it can be neglected and fatigue effects depend only on the smaller pulley.

When calculations are made for the life at different values of the horsepower a curve similar to Fig. 6-5 results. The data for the example indicate that operation occurs at point A. It is noted that the expected life can be greatly increased by a reduction in the horsepower, but on the other hand can be markedly decreased when the horsepower is made larger.

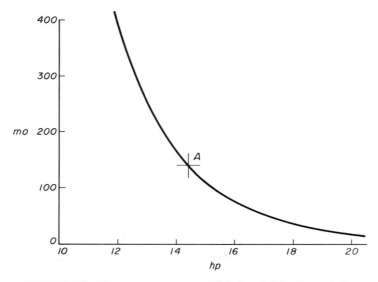

FIGURE 6-5   Horsepower vs. expected life (months) for Example 2.

Example 3. A 3V belt 106 in. long operates on two pulleys 11.7 in. diameter at 870 rpm. For an expected life of 20,000 hr find the design horsepower the belt can carry.

Solution. By Eq. (3): $V = \dfrac{\pi\, dn}{12} = \dfrac{\pi\, 11.7 \times 870}{12} = 2{,}665$ fpm

Life in months: $L = \dfrac{20{,}000}{160} = 125$ months at 160 hr each

By Eq. (10): $N' = \dfrac{115{,}200\,VL}{l} = \dfrac{115{,}200 \times 2{,}665 \times 125}{106}$

$$= 362{,}000{,}000 \text{ belt passes}$$

By Eq. (9): $\dfrac{1}{N'} = \dfrac{1}{M_1} + \dfrac{1}{M_2} = \dfrac{2}{M_1}$

$M_1 = 2N' = 2 \times 362{,}000{,}000 = 724{,}000{,}000$ force peaks

M_1 lies between 10^8 and 10^9 so $Q = 728$ and $x = 12.464$

By Eq. (8): $F_1^x = \dfrac{Q^x}{M_1}$

$12.464 \log F_1 = x \log Q - \log M_1$

$$= 12.464 \log 728 - \log 724{,}000{,}000$$

$$= 12.464 \times 2.86213 = 8.85974 = 26.81385$$

$$= 35.67359 - 8.85974 = 26.81385$$

$$\log F_1 = \frac{26.81385}{12.464} = 2.15130$$

$$F_1 = 141.7 \text{ lb}$$

By Eq. (6): $\qquad\qquad T_b = \dfrac{K_b}{d} = \dfrac{230}{11.7} = 19.7 \text{ lb}$

By Eq. (7): $\qquad\qquad T_c = K_c\left(\dfrac{V}{1,000}\right)^2 = 0.425 \times 2.665^2 = 3.0 \text{ lb}$

$$T_b + T_c = 22.7 \text{ lb}$$

By Eq. (5): $\qquad T_1 = F_1 - (T_b + T_c) = 141.7 - 22.7 = 119.0 \text{ lb}$

By Eq. (4): $\qquad \text{hp} = \dfrac{(T_1 - T_2)V}{33,000} = \dfrac{0.8T_1 V}{33,000} = \dfrac{0.8 \times 119.0 \times 2,665}{33,000}$

$$= 7.7$$

The design horsepower should be divided by a suitable service factor to obtain the nominal or operating horsepower.

Design constants for V-belts as appearing in the foregoing equations and tables are based on theory as well as the results of extensive testing. Lengths were in inches and forces in pounds. Corresponding values in SI units are not available. When data for a V-belt drive is given in SI units it is best, at present, to convert into U.S. customary units and follow the procedures outlined in the preceding sections.

4. Other Types of Belts

Other types of belts are available for power transmission purposes. Power bands, consisting of V-belts joined along their sides, as in Fig. 6-6(a), are of assistance in solving the length-matching problem when a number of V-belts must be run side by side.

The teeth of the so-called timing belt, Fig. 6-6(b), will keep shafts completely synchronized.

As the pulley size becomes smaller, the belt tension increases for a given horsepower output. This increased force on the overhanging pulley at the motor causes increased loads in the motor bearings. The recommended minimum sheave diameters for use with electric motors are shown in Table 6-6.

Allowance should be made in the design if the belt is to operate under adverse conditions such as may be caused by oil, high temperatures, or abrasive atmosphere.

Idler pulleys should be avoided because the additional flexing in passing around the pulley contributes to the fatigue effects. It is much better to main-

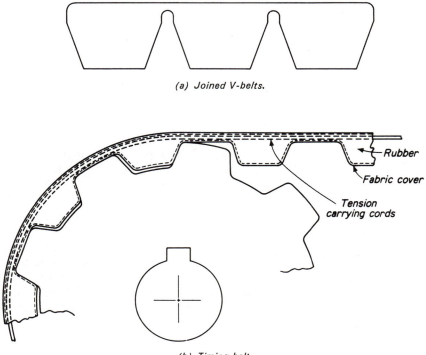

(a) Joined V-belts.

(b) Timing belt.

FIGURE 6-6

tain the tension as the belts stretch and wear by means of a take-up, which increases the center distance between the sheaves.

When designing a V-belt drive it is a good plan to calculate the cost of two or three different layouts of belts and pulleys to determine which has the smallest overall cost. The catalogs of the various manufacturers of V-belts contain much practical information.[3]

The V-belt is an important element in the field of power transmission. It is continually being improved by the various manufacturers and loading values are revised from time to time. The designer generally is guided by the current literature for the particular brand of belt he expects to use. The foregoing development should be considered a method that shows how the

[3]The following references should be consulted: Worley, W. S., "Design of V-Belt Drives for Mass Produced Machines," *Prod. Eng.*, **24**, Sept. 1953, p. 154; and Marco, S. M., W. L. Starkey, and K. G. Hornung, "A Quantitative Investigation of the Factors Which Influence the Fatigue Life of a V-Belt," *Trans. ASME*, **82**, Feb. 1960, *J. Eng. Ind.*, p. 47.

TABLE 6-6 Recommended Minimum Sheave Pitch Diameters, Inches, for General-Purpose Electric Motors*

Motor hp	Motor rpm					
	575	690	870	1,160	1,750	3,450
$1/2$			2.2			
$3/4$			2.4	2.2		
1	3.0	2.5	2.4	2.4	2.2	
$1^1/2$	3.0	3.0	2.4	2.4	2.4	2.2
2	3.8	3.0	3.0	2.4	2.4	2.4
3	4.5	3.8	3.0	3.0	2.4	2.4
5	4.5	4.5	3.8	3.0	3.0	2.4
$7^1/2$	5.2	4.5	4.4	3.8	3.0	3.0
10	6.0	5.2	4.4	4.4	3.8	3.0
15	6.8	6.0	5.2	4.4	4.4	3.8
20	8.2	6.8	6.0	5.2	4.4	4.4
25	9	8.2	6.8	6.0	4.4	4.4
30	10	9	6.8	6.8	5.2	
40	10	10	8.2	6.8	6.0	
50	11	10	8.4	8.2	6.8	

*See National Electrical Manufacturers Association Standard MG-1-14.43a and MG-1-14.43, Jan. 1968.

fatigue life of a belt is influenced by factors such as bending, centrifugal effects, and the horsepower transmitted.

5. Disk or Plate Clutch

The plate clutch,[4] shown diagrammatically in Fig. 6-7, is used in both automotive and industrial service. Let it be assumed that the parts are sufficiently rigid to give uniform wear over the lining, and that the wear is proportional

FIGURE 6-7 Plate clutch.

[4]See Jania, Z. J., "Friction-Clutch Transmissions," *Machine Design*, **30**, 1958, issues for Nov. 13, Nov. 27, Dec. 11, and Dec. 25.

to the product of velocity and the pressure. Since the velocity is proportional to the radius r to the element, the following equation can be written:

$$\delta = Kpr \tag{a}$$

where δ is the wear, and K is a constant. Since δ is constant for the entire face, the maximum pressure will occur at the smallest or inner radius r_i. Hence,

$$\delta = Kp_{max}r_i \tag{b}$$

Elimination of δ and K from the equations above gives

$$p = \frac{p_{max}r_i}{r} \tag{c}$$

The total normal force F_n, which must be exerted by the actuating spring, is found by multiplying the element of area $2\pi r\,dr$ by the pressure and integrating over the surface.

$$F_n = \int_{r_i}^{r_o} p\,dA = \int_{r_i}^{r_o} \frac{p_{max}r_i}{r} \times 2\pi r\,dr = 2\pi p_{max}r_i(r_o - r_i) \tag{11}$$

The torque is found by multiplying the force on the element by the coefficient of friction μ and the radius, and integrating over the area.

$$T = \int_{r_i}^{r_o} \mu p r\,dA = \int_{r_i}^{r_o} \mu\frac{p_{max}r_i}{r} \times 2\pi r^2\,dr$$

$$= \pi\mu p_{max}r_i(r_o^2 - r_i^2) = \tfrac{1}{2}\mu(r_o + r_i)F_n = \mu F_n r_{av} \tag{12}$$

Multiple-disk clutches have friction lining on facing sides of alternate plates. Since Eq. (12) gives the torque for a single face, this quantity must be multiplied by the number of faces to find the torque for the entire clutch.

Equations (11) and (12) are applicable only when the parts of the clutch are very rigid so that the wear is uniform over the lining. Suppose an alternate type of construction is used in which the clutch lining is attached to flexible plates so that the pressure p is uniform over the entire surface. The total axial force F_n required for operation is

$$F_n = \pi p(r_o^2 - r_i^2) \tag{13}$$

The torque exerted by the clutch is

$$T = \mu p \int r\,dA = 2\pi\mu p \int_{r_i}^{r_o} r^2\,dr = 2\pi\mu p\left(\frac{r^3}{3}\right)_{r_i}^{r_o}$$

$$= \frac{2}{3}\pi\mu p(r_o^3 - r_i^3) = \frac{2\mu(r_o^3 - r_i^3)F_n}{3(r_o^2 - r_i^2)} \tag{14}$$

Equations (13) and (14) are applicable when the assumption of uniform pressure can be made.

Example 4. A plate clutch with a single friction surface is 10 in. OD and 4 in. ID. $\mu = 0.2$.

(a) If the uniform wear theory is valid, find the required axial force for p_{max} = 100 psi. Find the torque for the clutch.

(b) Do the same for a similar clutch where the uniform pressure theory is valid for $p = 100$ psi.

(c) If the uniform wear theory is valid, find the torque the clutch will carry for $F_n = 5,000$ lb, and the value of p_{max}.

(d) Do the same for a clutch where the uniform pressure theory is valid.

Solution. (a) By Eq. (11):　$F_n = 2\pi 100 \times 2 \times 3 = 3,770$ lb

By Eq. (12):　$T = \pi 0.2 \times 100 \times 2(25 - 4) = 2,639$ in. lb

(b) By Eq. (13):　$F_n = \pi 100(25 - 4) = 6,597$ lb

By Eq. (14):　$T = \frac{2}{3}\pi 0.2 \times 100(125 - 8) = 4,900$ in. lb

(c) By Eq. (11):　$p_{max} = \dfrac{F_n}{2\pi r_i(r_o - r_i)} = \dfrac{5,000}{2\pi 2 \times 3} = 132.6$ psi

By Eq. (12):　$T = \dfrac{0.2 \times 7 \times 5,000}{2} = 3,500$ in. lb

(d) By Eq. (13):　$p = \dfrac{F_n}{\pi(r_o^2 - r_i^2)} = \dfrac{5,000}{\pi(25 - 4)} = 75.8$ psi

By Eq. (14):　$T = \dfrac{2 \times 0.2(125 - 8) \times 5,000}{3(25 - 4)} = 3,714$ in. lb

Example 4A. A plate clutch with a single friction surface is 250 mm OD and 100 mm ID. $\mu = 0.2$.

(a) If the uniform wear theory is valid, find the required axial force for p_{max} = 0.7 MPa. Find the torque for the clutch.

(b) Do the same for a similar clutch where the uniform pressure theory is valid for $p = 0.7$ MPa.

(c) If the uniform wear theory is valid, find the torque the clutch will carry for $F_n = 22,500$ N, and the value of p_{max}.

(d) Do the same for a clutch where the uniform pressure theory is valid.

Solution. (a) By Eq. (11):　$F_n = 2\pi 0.7 \times 50 \times 75 = 16,490$ N

By Eq. (12):　$T = \pi 0.2 \times 0.7 \times 50(125^2 - 50^2)$

　　　　　　　　$= 288,600$ Nmm

(b) By Eq. (13):　$F_n = \pi 0.7(125^2 - 50^2) = 28,860$ N

By Eq. (14):　$T = \frac{2}{3}\pi 0.2 \times 0.7(125^3 - 50^3) = 536,040$ Nmm

(c) By Eq. (11):　$p_{max} = \dfrac{F_n}{2\pi r_i(r_o - r_i)} = \dfrac{22,500}{2\pi 50 \times 75} = 0.955$ MPa

By Eq. (12):　$T = \dfrac{0.2(125 + 50) \times 22,500}{2} = 393,750$ Nmm

(d) By Eq. (13):　$p = \dfrac{F_n}{\pi(r_o^2 - r_i^2)} = \dfrac{22,500}{\pi(125^2 - 50^2)} = 0.546$ MPa

By Eq. (14):　$T = \dfrac{2 \times 0.2(125^3 - 50^3) \times 22,500}{3(125^2 - 50^2)}$

　　　　　　　　$= 417,850$ Nmm

6. Disk Brakes

Equations (11) and (12) for the disk clutch with rigid members can be adapted to the disk brake if the brake pad is shaped like a sector of a circle and the calculations made accordingly. A detailed analysis of the disk brake gives equations[5] which result in torques that are slightly larger than those resulting from the clutch equations. When the uncertain value of the coefficient of friction is considered, such results are satisfactory. Loads are balanced by locating pads on both sides of the disk. For better cooling, the disk is made hollow with an air stream flowing between. Servo action can be secured by the addition of several mechanical parts.

Example 5. A disk brake has two pads of included angle of 45° each. Outside radius 6 in. Inside radius 4 in. Coefficient of friction 0.4. Maximum lining pressure 100 psi. Find the force required to apply one shoe and find the torque capacity for both shoes.

Solution. By Eq. (11): $F_n = \dfrac{45}{360} \times 2\pi p_{max} r_i (r_o - r_i)$

$$= \frac{45}{360} \times 2\pi 100 \times 4(6 - 4) = 628.3 \text{ lb}$$

$$T = \mu F_n r_{av} = 0.4 \times 628.3 \times 5$$

$$= 1{,}256.6 \text{ in. lb, one shoe}$$

$$= 2{,}513.2 \text{ in. lb, both shoes}$$

7. Cone Clutch

The cone clutch utilizes the wedging action of the parts for increasing the normal force on the lining; thus an increase in the tangential friction force and the torque results. The assumption is again made that the normal wear is proportional to the product of the normal pressure and the radius.

$$\delta_n = K p_n r \qquad (a)$$

As shown in Fig. 6-8(b), the normal wear is uniform for all points on the cone. The maximum pressure thus occurs for the smallest value of r. Hence,

$$\delta_n = K p_{max} r_i \qquad (b)$$

Elimination of δ_n and K in the foregoing equations gives

$$p_n = \frac{p_{max} r_i}{r} \qquad (c)$$

---

[5]See Newcomb, T. P., and R. T. Spurr, *Braking of Road Vehicles*, Cambridge, Mass.: Robert Bentley, Inc., 1967, p. 111.

FIGURE 6-8 Cone clutch.

Let the radius r in Fig. 6-8(b) locate the elemental strip running around the cone. The differential area is then equal to $2\pi r\, dr/\sin \alpha$. The total normal force F_n is equal to the product of the normal pressure and the element of area integrated over the conical surface. Hence

$$F_n = \int_{r_i}^{r_o} p_n\, dA = \int \frac{p_{max} r_i}{r} \cdot \frac{2\pi r\, dr}{\sin \alpha} = \frac{2\pi p_{max} r_i (r_o - r_i)}{\sin \alpha} \qquad (15)$$

As shown by Fig. 6-8(a), the reaction R, which the actuating spring must be capable of exerting, is equal to

$$R = F_n \sin \alpha = 2\pi p_{max} r_i (r_o - r_i) \qquad (16)$$

This value of R is required for steady operation after the load has been put in motion. For engaging the clutch while picking up the load, a greater spring force than indicated by Eq. (16) is required.

The torque is given by

$$T = \int_{r_i}^{r_o} \mu p_n r\, dA = \int \frac{\mu p_{max} r_i}{r} \cdot \frac{2\pi r^2\, dr}{\sin \alpha}$$

$$= \frac{\pi \mu p_{max} r_i (r_o^2 - r_i^2)}{\sin \alpha} = \frac{\mu (r_o + r_i)}{2 \sin \alpha} R \qquad (17)$$

Values of the angle α range from a minimum of about $8°$ to larger values. Normal pressure on the lining may reach a value of 100 psi or more.

8. Band Brake

A simple band brake is shown schematically in Fig. 6-9. Sketch (b) shows, on an enlarged scale, the forces acting on an element of the lining lying between two axial planes a small angle $d\varphi$ apart. The tension on the tight side is equal to tension T of the slack side plus an increment dT. The drum is pressing outwardly on the element with a reactive force dR. The friction force μdR is equal to dT, the difference between the tangential forces in the band.

FIGURE 6-9   Simple band brake.

Thus

$$dT = \mu \, dR$$

Since angle $d\varphi$ is very small, the inward component from the tangential forces is $T \, d\varphi$, which is equal to the reaction dR. Then

$$dT = \mu T \, d\varphi$$

or

$$\frac{dT}{T} = \mu \, d\varphi$$

Band force T varies from T_2 where φ is zero, to the value T_1 where φ has the value α. The above equation should be integrated as follows.

$$\int_{T_2}^{T_1} \frac{dT}{T} = \int_0^\alpha \mu \, d\varphi$$

Then

$$\ln T \Big|_{T_2}^{T_1} = \mu \varphi \Big|_0^\alpha$$

$$\ln T_1 - \ln T_2 = \ln \frac{T_1}{T_2} = \mu \alpha$$

$$\frac{T_1}{T_2} = e^{\mu \alpha} \tag{18}$$

Values for the exponential function $e^{\mu \alpha}$ are given in Table 6-7.

TABLE 6-7 Values of Exponential, $e^{\mu\alpha}$

$\mu\alpha$	$e^{\mu\alpha}$	$\mu\alpha$	$e^{\mu\alpha}$	$\mu\alpha$	$e^{\mu\alpha}$	$\mu\alpha$	$e^{\mu\alpha}$	$\mu\alpha$	$e^{\mu\alpha}$	$\mu\alpha$	$e^{\mu\alpha}$	$\mu\alpha$	$e^{\mu\alpha}$
0.30	1.350	0.45	1.568	0.60	1.822	0.75	2.117	0.90	2.460	1.05	2.858	1.20	3.320
0.31	1.363	0.46	1.584	0.61	1.840	0.76	2.138	0.91	2.484	1.06	2.886	1.21	3.354
0.32	1.377	0.47	1.600	0.62	1.859	0.77	2.160	0.92	2.509	1.07	2.915	1.22	3.387
0.33	1.391	0.48	1.616	0.63	1.878	0.78	2.181	0.93	2.535	1.08	2.945	1.23	3.421
0.34	1.405	0.49	1.632	0.64	1.896	0.79	2.203	0.94	2.560	1.09	2.974	1.24	3.456
0.35	1.419	0.50	1.649	0.65	1.916	0.80	2.226	0.95	2.586	1.10	3.004	1.25	3.490
0.36	1.433	0.51	1.665	0.66	1.935	0.81	2.248	0.96	2.612	1.11	3.034	1.26	3.525
0.37	1.448	0.52	1.682	0.67	1.954	0.82	2.270	0.97	2.638	1.12	3.065	1.27	3.561
0.38	1.462	0.53	1.699	0.68	1.974	0.83	2.293	0.98	2.664	1.13	3.096	1.28	3.597
0.39	1.477	0.54	1.716	0.69	1.994	0.84	2.316	0.99	2.691	1.14	3.127	1.29	3.633
0.40	1.492	0.55	1.733	0.70	2.014	0.85	2.340	1.00	2.718	1.15	3.158	1.30	3.669
0.41	1.507	0.56	1.751	0.71	2.034	0.86	2.363	1.01	2.746	1.16	3.190	1.31	3.706
0.42	1.522	0.57	1.768	0.72	2.054	0.87	2.387	1.02	2.773	1.17	3.222	1.32	3.743
0.43	1.537	0.58	1.786	0.73	2.075	0.88	2.411	1.03	2.801	1.18	3.254	1.33	3.781
0.44	1.553	0.59	1.804	0.74	2.096	0.89	2.435	1.04	2.829	1.19	3.287	1.34	3.819

In Fig. 6-9, the brake is actuated by the application of force P at the end of the lever. A smaller force P is needed for operation when the tight side of the band is attached to the stationary support and the slack side is attached to the lever.

In the differential band brake, the friction force assists in applying the band. For the brake shown in Fig. 6-10 the following equation can be written:

$$Pa = T_2 m_2 - T_1 m_1 = T_2(m_2 - m_1 e^{\mu\alpha}) \tag{19}$$

Should the product $m_1 e^{\mu\alpha}$ be greater than m_2, the brake will grab or be self-locking. Sometimes this feature is desirable, and is utilized in back stop devices for preventing rotation in the reverse direction. For counterclockwise rotation, shown in Fig. 6-10, the friction tends to loosen the band, and the drum revolves freely. Should the rotation reverse and become clockwise, the friction will apply the brake, and if it is self-locking, the rotation will be stopped.

FIGURE 6-10 Differential band brake.

Let the analysis of Fig. 6-9(b) be applied to the band at the point of tangency for T_1. Reaction R is equal to $brp_{max}\,d\varphi$, where p_{max} is the pressure between drum and lining. The inward components of the band forces are equal to $T_1\,d\varphi$. These two forces are equal to each other, and give the following useful equation.

$$T_1 = brp_{max} \tag{20}$$

A similar equation can be written for the slack side.

Example 6. A band brake similar to Fig. 6-9 has a 16-in.-diam drum and a width of lining equal to 3 in. Speed is 200 rpm, $a = 10$ in., $m = 3$ in., $\alpha = 270°$, and $\mu = 0.2$. Find the torque and horsepower if the maximum lining pressure is 70 psi.

Solution. By Eq. (20): $T_1 = brp_{max} = 3 \times 8 \times 70 = 1{,}680$ lb

$$\mu\alpha = 0.2 \times 1.5\pi = 0.9425$$

By Table 6-7: $e^{\mu\alpha} = 2.566$

By Eq. (18): $T_2 = \dfrac{T_1}{e^{\mu\alpha}} = \dfrac{1{,}680}{2.566} = 655$ lb

Force: $P = \dfrac{T_2 m}{a} = \dfrac{655 \times 3}{10} = 196$ lb

Torque: $T = r(T_1 - T_2) = 8(1{,}680 - 655) = 8{,}200$ in. lb

Horsepower: $\text{hp} = \dfrac{Tn}{63{,}025} = \dfrac{8{,}200 \times 200}{63{,}025} = 26.0$

Example 6A. A band brake similar to Fig. 6-9 has a 400-mm-diam drum and a width of lining equal to 75 mm. Speed is 200 rpm, $a = 250$ mm, $m = 75$ mm, $\alpha = 270°$, and $\mu = 0.2$. Find the torque and horsepower if the maximum lining pressure is 0.5 MPa.

Solution. By Eq. (20): $T_1 = brp_{max} = 75 \times 200 \times 0.5$

$$= 7{,}500\text{-N tight-side force}$$

$$\mu\alpha = 0.2 \times 1.5\pi = 0.9425$$

By Table 6-7: $e^{\mu\alpha} = 2.566$

By Eq. (18): $T_2 = \dfrac{T_1}{e^{\mu\alpha}} = \dfrac{7{,}500}{2.566} = 2{,}293\text{-N slack-side force}$

Torque: $T = r(T_1 - T_2) = 200(7{,}500 - 2{,}923)$

$$= 915{,}400\text{ Nmm}$$

Absorbed power: $\text{kW} = \dfrac{Tn}{9{,}550{,}000} = \dfrac{915{,}400 \times 200}{9{,}550{,}000} = 19.2$

Band brakes are capable of exerting large torques. Careful attention must be paid to details so that the band seats properly on the drum when absorbing energy, and remains free and clear when not in use.

9. *Block Brake with Short Shoe*

The brake shown in Fig. 6-11 consists of a short block, which is pressed against the revolving drum by means of a lever. When the block is relatively short, it may be assumed that the normal force F_n between drum and shoe is concentrated at point B. If the lever is taken as the free body, a moment equation can be written about A as the center, and forces F_n and μF_n can be determined. It should be noted that if the direction of rotation of the drum is reversed, the direction of friction force μF_n is reversed also.

FIGURE 6-11   Block brake with short shoe.

When the rotation is as shown in Fig. 6-11, the moment of the friction force aids in applying the shoe to the drum. Care must be taken that the friction moment is not so large that the brake is applied without the assistance of force P; the brake will then seize or grab, and unsatisfactory or dangerous operation results. When the pivot is located on the other side of the line of action of μF_n, as shown by the dashed outline in Fig. 6-11, the friction force tends to unseat the shoe. When the shoe is long, equations that take account of the variation in the pressure along the shoe must be used.

Example 7. Consider the following dimensions for the (full outline) short shoe block brake of Fig. 6-11. Radius of drum 5 in., $a = 16$ in., $b = 2$ in., $c = 6$ in., $m = 1.5$ in., length of shoe 4 in., $p = 100$ psi, and $\mu = 0.4$. Find the value of the actuating force P.

Solution. Normal force, $F_n = 2 \times 4 \times 100 = 800$ lb

Friction force, $F_f = \mu F_n = 0.4 \times 800 = 320$ lb

Equilibrium of moments:

$$P_a + F_f m = F_n c$$

$$P = \frac{F_n c - F_f m}{a} = \frac{800 \times 6 - 320 \times 1.5}{16} = 270 \text{ lb}$$

10. Pivoted Block Brake with Long Shoe

When the shoe is relatively long,[6] as shown in Fig. 6-12(a), the normal pressure p_n varies with different values of the angle φ. Frictional wear in engineering equipment is usually assumed to be proportional to the product of the velocity and the pressure. Since the velocity is the same for all points of the shoe, the wear of the lining is proportional to the pressure. Hence,

$$\delta_n = Kp_n \tag{a}$$

where δ_n is the wear in the direction perpendicular to the lining, and K is a constant.

FIGURE 6-12   Block brake with long shoe.   FIGURE 6-13   Automotive type brake.

Figure 6-12(b) shows the motion δ of a point on the shoe as the lining wears and as the shoe rotates about point A through angle γ. Hence

$$\delta = \gamma m \tag{b}$$

The wear δ_n in the direction normal to the drum is then equal to

$$\delta_n = \delta \sin \theta = \gamma m \sin \theta \tag{c}$$

The perpendicular dropped from A to radius OB has a length equal to $r_1 \sin \varphi$ or $m \sin \theta$. Substitution in Eq. (c) gives

$$\delta_n = \gamma r_1 \sin \varphi \tag{d}$$

Substitution into Eq. (a) gives

$$Kp_n = \gamma r_1 \sin \varphi \tag{e}$$

[6]See Fazekas, G. A. G., "Graphical Shoe Brake Analysis," *Trans. ASME*, **79**, 1957, p. 1322. See also *J. Appl. Mech.*, **80**, 1958, p. 7.

One of the limiting factors in the design of a brake is the maximum pressure p_{max} between lining and drum. This occurs at the location having the maximum value of $\sin \varphi$ in Eq. (e). Hence

$$Kp_{max} = \gamma r_1 (\sin \varphi)_{max} \tag{f}$$

Elimination of $\gamma r_1/K$ between Eqs. (e) and (f) gives

$$p_n = \frac{p_{max}}{(\sin \varphi)_{max}} \sin \varphi \tag{g}$$

Should φ_2 in Fig. 6-12 be greater than $90°$, then $(\sin \varphi)_{max}$ is equal to unity. For smaller values of φ_2 use the value of $\sin \varphi_2$ for the denominator.

If b is the width of the lining, the area of a small element, cut by two radii an angle $d\varphi$ apart, is equal to $br\, d\varphi$. When this area is multiplied by the normal force p_n and the arm $r_1 \sin \varphi$, and integrated over the entire shoe, the moment M_n of the normal forces about point A results.

$$M_n = \int_{\varphi_1}^{\varphi_2} brr_1 p_n \sin \varphi \, d\varphi = \frac{brr_1 p_{max}}{(\sin \varphi)_{max}} \int_{\varphi_1}^{\varphi_2} \sin^2 \varphi \, d\varphi$$

$$= \frac{brr_1 p_{max}}{4(\sin \varphi)_{max}}(2\alpha - \sin 2\varphi_2 + \sin 2\varphi_1) \tag{21}$$

Normal force dF_n on the element of Fig. 6-12(a) causes a counterclockwise moment about point A. If numerical substitution into Eq. (21) for the total moment gives a positive result, moment M_n is therefore counterclockwise. A negative result means that M_n is clockwise about A.

The moment of the friction forces M_f about A is equal to

$$M_f = \int_{\varphi_1}^{\varphi_2} \mu p_n (r - r_1 \cos \varphi) br \, d\varphi$$

$$= \frac{\mu brp_{max}}{(\sin \varphi)_{max}} \int \left[r \sin \varphi - \frac{1}{2} r_1 \sin 2\varphi \right] d\varphi$$

$$= \frac{\mu brp_{max}}{4(\sin \varphi)_{max}}[r_1(\cos 2\varphi_2 - \cos 2\varphi_1) - 4r(\cos \varphi_2 - \cos \varphi_1)] \tag{22}$$

A positive value for M_f in Eq. (22) indicates a counterclockwise moment about A for the friction forces of Fig. 6-12, and a negative result indicates a clockwise moment. If the direction of rotation is reversed, these statements have the converse meanings for M_f.

In Fig. 6-12(a) the moment of the external force is equal to Pa. The shoe and lever are in equilibrium under the action of moments Pa, M_n, and M_f. A moment equation should be written for these terms with the proper signs. These signs depend on the location of the pivot with respect to the shoe and on the direction of rotation of the drum. When the friction moment assists in applying the shoe, the brake will be self-locking if M_f exceeds M_n. In practice, the ratio M_f/M_n should not be greater than about 0.7.

The torque exerted by the brake is found by taking the moment of the friction forces about the center of the drum O. Hence

$$T = \int_{\varphi_1}^{\varphi_2} \mu p_n br^2 \, d\varphi$$

$$= \frac{\mu br^2 p_{max}}{(\sin \varphi)_{max}} \int \sin \varphi \, d\varphi$$

$$= \frac{\mu br^2 p_{max}}{(\sin \varphi)_{max}} (\cos \varphi_1 - \cos \varphi_2) \tag{23}$$

Example 8. (a) Calculate the value of p_{max}, the torque, and the horsepower for the brake of Fig. 6-14. The coefficient of friction is equal to 0.2.

(b) Make the same calculations on the assumption that the total normal and friction forces are concentrated at point B and compare results with the more exact results of part (a).

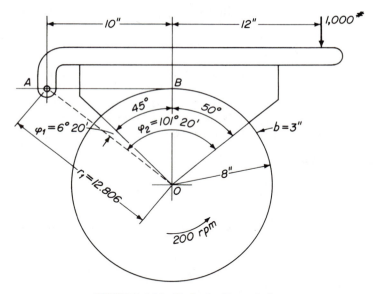

FIGURE 6-14   Brake for Example 8.

Solution. (a)   tan angle $AOB = \frac{10}{8} = 1.25$

angle $AOB = 51°20'$

Hence:

$\varphi_1 = 51°20' - 45°0' = 6°20'$, $2\varphi_1 = 12°40'$

$\varphi_2 = 6°20' + 95°0' = 101°20'$, $2\varphi_2 = 202°40'$

$\alpha = 95° = 1.65806$ rad

$r_1 = \sqrt{8^2 + 10^2} = 12.806$ in.

$$\cos \varphi_1 = 0.99390 \qquad\qquad \sin 2\varphi_1 = 0.21928$$
$$\cos \varphi_2 = -0.19652 \qquad\qquad \cos 2\varphi_1 = 0.97566$$
$$\sin 2\varphi_2 = -0.38537$$
$$\cos 2\varphi_2 = -0.92276$$

Since φ_2 is greater than $90°$: $(\sin \varphi)_{max} = 1$

By Eq. (21): $M_n = \frac{1}{4} \times 3 \times 8 \times 12.806 p_{max}(2 \times 1.65808 + 0.38357$
$$+ 0.21928) = 301.26 p_{max}$$

This moment is counterclockwise about point A.

By Eq. (22): $M_f = \frac{1}{4} \times 0.2 \times 3 \times 8 p_{max}[12.806(-0.92276 - 0.97566)$
$$- 4 \times 8(-0.19652 - 0.99390)] = 16.54 p_{max}$$

This moment is counterclockwise about point A.

Externally applied moment:

$$Pa = 1,000 \times 22 = 22,000 \text{ in. lb,} \qquad \text{clockwise}$$

Equilibrium of moments on shoe: $(301.26 + 16.54) p_{max} = 22,000$

$$p_{max} = 69.2 \text{ psi}$$

By Eq. (23): $T = 0.2 \times 3 \times 8^2 \times 69.2(0.99390 + 0.19652)$
$$= 3,164 \text{ in. lb}$$

$$\text{hp} = \frac{Tn}{63,000} = \frac{3,164 \times 200}{63,000} = 10.04$$

(b) Take moments about A.

Normal force at B: $\quad F_n = \dfrac{1,000 \times 22}{10} = 2,200 \text{ lb}$

Area of shoe: $\quad A = 2\pi 8 \times 3 \times \dfrac{95}{360} = 39.8 \text{ in.}^2$

Average pressure: $\quad p_{av} = \dfrac{2,200}{39.8} = 55.3 \text{ psi}$

Friction torque: $\quad T = \mu F_n r = 0.2 \times 2,200 \times 8 = 3,520 \text{ in. lb}$

$$\text{hp} = \frac{Tn}{63,000} = \frac{3,520 \times 200}{63,000} = 11.2$$

The short-shoe approximation thus indicates a smaller value for the lining pressure and a larger horsepower capacity than is given by the more exact solution of part (a).

Example 8A. Consider the long-shoe block brake of Fig. 6-14 but with the following changes in dimensions. Lengths 8, 10, and 12 in. become 200 mm, 250 mm, and 300 mm, respectively. Width b becomes 75 mm. Angles and speed remain as shown. The applied load is 4,500 N and the coefficient of friction is 0.2.
  (a) Find the maximum pressure p_{max}, the torque, and the power absorbed.
  (b) Make the same calculations on the assumption that the total normal and

friction forces are concentrated at point B and compare results with the more exact results of part (a).

Solution. (a)   tan angle $AOB = \dfrac{250}{200} = 1.25$

angle $AOB = 51°20'$

Hence:

$$\varphi_1 = 51°20' - 45°0' = 6°20' \qquad 2\varphi_1 = 12°40'$$

$$\varphi_2 = 6°20' + 95°0' = 101°20' \qquad 2\varphi_2 = 202°40'$$

$$\alpha = 95° = 1.658\ 06 \text{ rad}$$

$$r_1 = \sqrt{200^2 + 250^2} = 320.15 \text{ mm}$$

$$\cos \varphi_1 = 0.993\ 90 \qquad\qquad \sin 2\varphi_1 = 0.219\ 28$$

$$\cos \varphi_2 = -0.196\ 52 \qquad\quad \cos 2\varphi_1 = 0.975\ 66$$

$$\sin 2\varphi_2 = -0.385\ 37$$

$$\cos 2\varphi_2 = -0.922\ 76$$

Since φ_2 is greater than 90°, $(\sin \varphi_1)_{max} = 1$

By Eq. (21): $M_n = \frac{1}{4} \times 75 \times 200 \times 320.15 p_{max}(2 \times 1.658\ 06$
$$+ 0.385\ 37 + 0.219\ 28) = 4,707,100 p_{max}$$

This moment is counterclockwise about point A.

By Eq. (22): $M_f = \frac{1}{4} \times 0.2 \times 75$
$$\times 200 p_{max}[320.15(-0.922\ 76 - 0.975\ 66)$$
$$- 4 \times 200(-0.196\ 52 - 0.993\ 90)] = 258,400 p_{max}$$

This moment is counterclockwise about point A.

Externally applied moment:

$$Pa = 4,500 \times 550 = 2,475,000 \text{ Nmm clockwise about } A$$

Equilibrium of moments on shoe:

$$(4,707,100 + 258,400) p_{max} = 2,475,000$$

$$p_{max} = 0.4984 \frac{\text{N}}{\text{mm}^2} \text{ or MPa}$$

By Eq. (23): $T = 0.2 \times 75 \times 200^2 \times 0.4984(0.993\ 90 + 0.196\ 52)$
$$= 355,980 \text{ Nmm}$$

Power absorbed: $\text{kW} = \dfrac{Tn}{9,550,000} = \dfrac{355,980 \times 200}{9,550,000} = 7.46 \text{ kW}$

(b) Moments about A: $F_n = \dfrac{4,500 \times 550}{250} = 9,900 \text{ N}$

Area of shoe: $A = 2\pi 200 \times 75 \times \dfrac{95}{360} = 24,870 \text{ mm}^2$

Torque: $T = \mu F_n r = 0.2 \times 9,900 \times 200 = 396,000 \text{ Nmm}$

Power absorbed: $\text{kW} = \dfrac{Tn}{9,550,000} = \dfrac{396,000 \times 200}{9,550,000} = 8.29 \text{ kW}$

Figure 6-13 shows an internal brake of a type widely used in automotive service. The equation for the moment M_n of the normal forces on the shoe is the same as Eq. (21), but a positive result indicates a clockwise moment about A. The equations for friction moment M_f and torque T are the same as for the brake of Fig. 6-12. A positive or negative result for M_f should be interpreted in the same way as for a brake with an external shoe.

11. Brake with Pivoted Symmetrical Shoe

In Fig. 6-15 the brake shoe is supported by a symmetrically located pin at A. Let it be assumed that distance r_1 is of such magnitude that the friction forces dF_f have no resultant moment about A. The normal forces must then

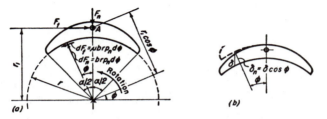

FIGURE 6-15 Block brake with pivoted symmetrical shoe.

be symmetrical with respect to the vertical center line, and also have zero moment about A. Thus, no rotation of the shoe on the pin can occur, and the vertical component of the wear is uniform for the entire shoe. Assume as before that normal wear is proportional to normal pressure.

$$\delta_n = Kp_n \qquad (a)$$

As shown in Fig. 6-15(b) the normal wear is equal to $\delta \cos \varphi$. Hence,

$$\delta \cos \varphi = Kp_n \qquad (b)$$

The maximum normal pressure occurs at the location for the maximum value of $\cos \varphi$, that is, for $\varphi = 0$, giving

$$\delta = Kp_{max}$$

Substitution into Eq. (b) gives: $p_n = p_{max} \cos \varphi \qquad (c)$

The equation for the moment of the friction forces about A is found by multiplying the tangential force on an element of area by the arm ($r_1 \cos \varphi - r$) and integrating over the entire shoe. By the foregoing assumption the result is equal to zero.

$$M_f = 2 \int_0^{\alpha/2} \mu b r p_n (r_1 \cos \varphi - r) \, d\varphi = 0$$

Substitution of the value for p_n from Eq. (c) gives

$$2 \mu b r p_{max} \int_0^{\alpha/2} (r_1 \cos^2 \varphi - r \cos \varphi) \, d\varphi = 0$$

Integrating and solving for r_1,

$$\left[r_1 \left(\frac{1}{2} \varphi + \frac{1}{4} \sin 2\varphi \right) - r \sin \varphi \right]_0^{\alpha/2} = 0$$

$$r_1 = \frac{4r \sin (\alpha/2)}{\alpha + \sin \alpha} \tag{24}$$

The torque of the friction forces about O as a center is given by the following equation:

$$T = 2 \int_0^{\alpha/2} \mu b r^2 p_n \, d\varphi$$

$$= 2 \mu b r^2 p_{max} \int_0^{\alpha/2} \cos \varphi \, d\varphi$$

$$= 2 \mu b r^2 p_{max} \sin \frac{\alpha}{2} \tag{25}$$

The pin reaction F_n can be found as the summation of the vertical components of the normal pressures.

$$F_n = 2 \int_0^{\alpha/2} b r p_n \cos \varphi \, d\varphi$$

$$= 2 b r p_{max} \int_0^{\alpha/2} \cos^2 \varphi \, d\varphi$$

$$= \tfrac{1}{2} b r p_{max}(\alpha + \sin \alpha) \tag{26}$$

The pin reaction F_t is found by summing the horizontal components of the frictional forces.

$$F_t = 2 \int_0^{\alpha/2} \mu b r p_n \cos \varphi \, d\varphi = \mu F_n \tag{27}$$

12. Lining Pressures

Wooden blocks are sometimes used as the friction elements for industrial brakes. Cast-iron shoes bearing on cast-iron or steel wheels are in wide use for brakes in railroad service. However, most brake linings depend upon asbestos as the basic friction material because of its ability to resist the effects of heat. Asbestos may be spun into yarn, woven and infused with a binder, and then consolidated by heat and pressure. Molded blocks and linings are also made directly from asbestos and binder without weaving.

Data on friction materials are shown in Table 6-8. The correct value to be used depends on the conditions of service for the particular application. The proper value should be obtained from the manufacturer's catalog for the particular type of lining considered. The coefficient of friction for brake materials may not remain constant in service but will vary because of conditions of temperature, or the presence of moisture and grease.[7]

TABLE 6-8   Typical Friction Materials for Brake and Clutch Service*

	Asbestos			Sintered Metal (Copper or Iron Base)
	Brake Blocks	Molded Linings	Woven Linings	
Coefficient of running friction	0.40–0.60	0.40–0.45	0.20–0.45	0.10–0.50
Max rubbing speed, fpm	7,500	5,000	3,000–7,500	
Max drum temperature, °F	750	500	500	Up to 1,000
Max working pressure, psi	150	100	100	Up to 500
Rate of wear †	E, G	G	G, M	E

*Properties can vary over a considerable range depending on composition and method of fabrication.

†E, excellent; G, good; M, moderate.

13. *Heating of Brakes*

The capacity of a brake or clutch to absorb power is largely determined by its ability to dispose of the frictional heat. The heat-dissipating quality is determined by factors such as the size, shape, and condition of the surface of the various parts. If the clutch or brake is not enclosed in a housing, or if the surrounding air is in motion, the brake can be more readily cooled.

The construction of the drum and adjacent parts, length of time the brake is applied, and the time between applications are also factors affecting the temperature. The horsepower per square inch can serve as an approximate guide to the rise in temperature of the drum.

Example 9. Suppose the drum of the short-shoe block brake of Example 7 is turning at 1,800 rpm. Find the hp/in.² for the drum surface.

[7]See Parker, R. C., "The Frictional Behavior of Engineering Materials," *Engineering*, 1949, pp. 193 and 217. See also Courtel, R., and L. M. Tichvinsky, "A Brief History of Friction," *Mech. Eng.*, **85**, Sept. 1963, p. 55; Oct., p. 33.

Solution. $V = \dfrac{\pi\, dn}{12} = \dfrac{\pi 10 \times 1,800}{12} = 4,712$ fpm

$$hp = \frac{F_f V}{33,000} = \frac{320 \times 4,712}{33,000} = 45.7 \text{ entire brake}$$

Drum area: $A = 2\pi rb = 2\pi 6 \times 2 = 62.8$ in.^{2}

$$\frac{hp}{\text{in.}^2} = \frac{45.7}{62.8} = 0.73$$

This value would probably be considered satisfactory. In industrial service, hp/in.^{2} is usually held to values less than unity. In automotive service, the hp/in.^{2} value is higher and may lie between 1.5 and 2.5.

14. Roller Chains

A roller chain provides a readily available and efficient method for transmitting power between parallel shafts. Details of construction are shown in Fig. 6-16. The rollers turn on bushings that are press-fitted to the inner link

FIGURE 6-16   **Construction of typical roller chain.**

plates. The pins are prevented from rotating in the outer link plates by the press-fit assembly. Dimensions for standard sizes are given in Table 6-9. The range in sizes is large so that chains can be used for both large and small amounts of power. Chains can be used for shaft spacings too great for gears. A large reduction in speed can be obtained if desired. The tolerances for a chain drive are greater than for gears, and the installation is relatively easy.

Chains present no fire hazard and are unaffected by relatively high temperatures and the presence of oil or grease. Chains are, however, more noisy than belts. A small adjustable idler sprocket should be provided to remove

TABLE 6-9   Dimensions of Standard Roller Transmission Chains

| Chain No. | Pitch, p, in. | Roller | | Pin Diameter, d, in. | Link Plate Thickness, a | Average Ultimate Strength, lb | Weight per Foot, lb |
		Diameter, H, in.	Width, E, in.				
25*	1/4	0.130*	1/8	0.0905	0.030	875	0.084
35*	3/8	0.200*	3/16	0.141	0.050	2,100	0.21
41†	1/2	0.306	1/4	0.141	0.050	2,000	0.28
40	1/2	5/16	5/16	0.156	0.060	3,700	0.41
50	5/8	0.400	3/8	0.200	0.080	6,100	0.68
60	3/4	15/32	1/2	0.234	0.094	8,500	1.00
80	1	5/8	5/8	0.312	0.125	14,500	1.69
100	1-1/4	3/4	3/4	0.375	0.156	24,000	2.49
120	1-1/2	7/8	1	0.437	0.187	34,000	3.67
140	1-3/4	1	1	0.500	0.219	46,000	4.93
160	2	1-1/8	1-1/4	0.562	0.250	58,000	6.43
180	2-1/4	1-13/32	1-13/32	0.687	0.281	76,000	8.70
200	2-1/2	1-9/16	1-1/2	0.781	0.312	95,000	10.51
240	3	1-7/8	1-7/8	0.937	0.375	130,000	16.90

*Without rollers.
†Lightweight chain.

excessive slack from the chain as it wears. It should be located on the outside of the chain on the slack side and near the smaller sprocket.

15. Horsepower Capacity of Roller Chains

At lower speeds, the horsepower capacity is determined by the fatigue life of the link plates. The following equation applies.[8]

$$\text{hp} = 0.004 N_1^{1.08} n_1^{0.9} p^{3.0-0.07p} \tag{28}$$

where N_1 = number of teeth in the smaller sprocket
  n_1 = speed, rpm, of the smaller sprocket
  p = chain pitch, inches

Constant 0.004 becomes 0.0022 for the No. 41 chain.
  At higher speeds the horsepower is determined by the roller bushing

[8]See *Design Manual, Roller and Silent Chain Drives*, Park Ridge, Ill.: American Sprocket Chain Manufacturers Assn., 1968, p. 46.

fatigue life. The equation is

$$\text{hp} = \frac{1,000 K N_1^{1.5} p^{0.8}}{n_1^{1.5}}$$ (29)

where $K = 29$ for chains Nos. 25 and 35
$\quad\quad = 3.4$ for chain No. 41
$\quad\quad = 17$ for chains Nos. 40 to 240

The data given in Tables 6-10 and 6-11 will be helpful in making numerical calculations. In many cases it is necessary to apply both Eqs. (28) and (29) and to use the smaller of the two results.

TABLE 6-10   Constants Used in Roller Chain Design

N_1	$N_1^{1.08}$	$N_1^{1.5}$	N_1	$N_1^{1.08}$	$N_1^{1.5}$	p	$p^{0.8}$	$p^{3.0-0.07p}$
11	13.33	36.48	23	29.55	110.3	1/4	0.3299	0.01601
12	14.64	41.57	24	30.95	117.6	3/8	0.4563	0.05411
13	15.96	46.87	25	32.34	125.0	1/2	0.5744	0.1281
14	17.29	52.38	28	36.55	148.2	5/8	0.6866	0.2492
15	18.63	58.10	30	39.38	164.3	3/4	0.7944	0.4283
16	19.97	64.00	32	42.22	181.0	1	1	1
17	21.32	70.09	35	46.52	207.1	1-1/4	1.195	1.915
18	22.68	76.37	40	53.73	253.0	1-1/2	1.383	3.234
19	24.05	82.82	45	61.02	301.9	1-3/4	1.565	5.004
20	25.42	89.44	50	68.37	353.6	2	1.741	7.260
21	26.79	96.23	55	75.78	407.9	2-1/4	1.913	10.025
22	28.17	103.2	60	83.25	464.8	2-1/2	2.081	13.310
						3	2.408	21.437

TABLE 6-11   Speed, n_1, RPM, of Small Sprocket Raised to Powers

n_1	$n_1^{0.9}$	$n_1^{1.5}$	n_1	$n_1^{0.9}$	$n_1^{1.5}$	n_1	$n_1^{0.9}$	$n_1^{1.5}$
5	4.257	11.18	175	104.4	2,315	1,200	590.6	41,570
10	7.943	31.62	200	117.7	2,828	1,400	678.4	52,380
15	11.44	58.10	250	143.9	3,953	1,600	765.1	64,000
20	14.82	89.44	300	169.6	5,196	1,800	850.6	76,370
25	18.12	125.0	350	194.8	6,548	2,000	935.3	89,440
30	21.35	164.3	400	219.7	8,000	2,400	1,102	117,600
40	27.66	253.0	450	244.3	9,546	2,800	1,266	148,200
50	33.81	353.6	500	268.6	11,180	3,200	1,428	181,000
60	39.84	464.8	600	316.5	14,700	3,600	1,587	216,000
80	51.62	715.6	700	363.6	18,520	4,000	1,745	253,000
100	63.10	1,000	800	410.0	22,630	5,000	2,133	353,600
125	77.13	1,398	900	455.8	27,000	6,000	2,514	464,800
150	90.88	1,837	1,000	501.2	31,620	7,000	2,888	585,700

Example 10. For a single-strand No. 60 chain, $p = \frac{3}{4}$ in., and the smaller sprocket has N_1 equal to 15 teeth. Smooth loading.

(a) Find the horsepower capacity at n_1 equal to 900 rpm for the smaller sprocket.

(b) Find the horsepower capacity at n_1 equal to 1,400 rpm for the smaller sprocket.

Solution. Numerical values should be substituted from Tables 6-10 and 6-11.

(a) Link plate fatigue:

$$\text{hp} = 0.004 \times 15^{1.08} \times 900^{0.9} \times 0.75^{3.0-0.07\times0.75}$$

$$= 0.004 \times 18.63 \times 455.8 \times 0.4283 = 14.60$$

Roller bushing fatigue:

$$\text{hp} = \frac{17{,}000 \times 15^{1.5} \times 0.75^{0.8}}{900^{1.5}} = \frac{17{,}000 \times 58.10 \times 0.7944}{27{,}000}$$

$$= 29.06$$

Link plate fatigue controls and the horsepower is 14.6.

(b) Link plate fatigue:

$$\text{hp} = 0.004 \times 15^{1.08} \times 1{,}400^{0.9} \times 0.75^{3.0-0.07\times0.75}$$

$$= 0.004 \times 18.63 \times 678.4 \times 0.4283 = 21.65$$

Roller bushing fatigue:

$$\text{hp} = \frac{17{,}000 \times 15^{1.5} \times 0.75^{0.8}}{1{,}400^{1.5}} = \frac{17{,}000 \times 58.10 \times 0.7944}{52{,}380}$$

$$= 14.98$$

At this speed, roller bushing fatigue controls, and the horsepower is 14.98.

The horsepower at different speeds for a single-strand No. 60 chain is plotted in Fig. 6-17. This figure illustrates graphically the need of using both Eqs. (28) and (29) when making calculations for the horsepower. The catalogs of the chain manufacturers contain extensive tables, which give the horsepower for different speeds and sprocket sizes.[9]

Chains have a considerable degree of elastic flexibility, which is helpful in prolonging their life. A service factor, however, must be applied to the nominal load under impact conditions to obtain the design horsepower carried by the chain. Factors are given by Table 6-12.

An increase in capacity can be obtained by the use of multiple-strand roller chains. These are essentially an assembly of single-strand chains placed side by side. The pins extend through the entire width of the chain and maintain the alignment of the different strands. Factors for multiplying the capacity of a single-strand chain for double, triple, and quadruple chains are given in Table 6-12. Overhanging shafts should be avoided when using multiple-strand chains unless the shaft is very rigid with a very small deflection.

[9] Loading tables are also given in the March 1966 supplement to ASA B29.1, same address as footnote 8.

FIGURE 6-17   Horsepower vs. speed curves for No. 60 single-strand roller chain, pitch, $p = 3/4$ in., operating on 15 tooth sprocket.

TABLE 6-12   Service Factors and Multiple-Strand Factors

Type of Driven Load	Type of Input Power			No. of Strands	Strand Factor
	I.C. Engine, Hydraulic Drive	Electric motor or Turbine	I.C. Engine, Mechanical Drive		
Smooth	1.0	1.0	1.2	2	1.7
Moderate shock	1.2	1.3	1.4	3	2.5
Heavy shock	1.4	1.5	1.7	4	3.3

16. Galling of Roller Chains

When the load is heavy and the speed is high, especially for pitches of 1 in. and greater, there is danger of galling or the welding together of the high spots of the contacting surfaces. Although such minute welds are immediately broken, rapid wear of the metal results. The maximum speed of operation to avoid galling for different chains and sprocket sizes is shown in Table 6-13. Operation at speeds higher than those shown in the table is permissible provided that a suitable reduction in the load is made. The calculations are

TABLE 6-13   Maximum Permissible Speed, RPM, to Avoid Galling

N_1	Chain Number and Pitch							
	No. 80, 1 in.	No. 100, 1-1/4 in.	No. 120, 1-1/2 in.	No. 140, 1-3/4 in.	No. 160, 2 in.	No. 180, 2-1/4 in.	No. 200, 2-1/2 in.	No. 240, 3 in.
11	2,800	1,800	1,500	1,100	800	600	300	175
12	2,800	1,800	1,500	1,100	800	400	300	175
13	2,800	1,800	1,500	1,100	800	400	300	150
14	2,800	1,800	1,400	1,100	800	400	300	150
15	2,700	1,800	1,400	1,000	750	400	300	150
16	2,700	1,800	1,400	1,000	700	400	300	150
17	2,600	1,600	1,400	1,000	700	400	250	150
18	2,600	1,600	1,300	1,000	700	400	250	150
19	2,600	1,600	1,300	1,000	700	400	250	150
20	2,400	1,600	1,300	1,000	650	400	250	150
21	2,400	1,600	1,300	900	600	400	250	125
22	2,400	1,600	1,200	900	600	400	250	125
23	2,400	1,600	1,200	900	500	350	250	125
24	2,400	1,600	1,200	900	450	350	250	125
25	2,200	1,400	1,200	800	450	350	250	125
28	2,200	1,400	1,100	800	400	300	250	125
30	2,200	1,400	1,100	700	400	300	250	125
32	2,000	1,400	1,000	600	350	300	250	100
35	2,000	1,200	1,000	600	350	300	200	100
40	1,600	1,200	700	550	350	300	200	100
45	1,600	1,000	700	550	350	250	200	100
50	1,400	900	600	550	350	250	200	80
55	1,400	900	600	550	350	250	150	80
60	1,200	800	600	550	350	250	150	80

involved and the maker of the chain should be consulted for a suitable recommendation.[10]

17. Lubrication of Roller Chains

Lubrication is a very important feature in the successful design of a roller chain drive. In general a chain should have a sheet metal casing for protection from atmospheric dust and to facilitate lubrication.

[10]See Horner, W., and W. Kilgard, "Speed Limits for Chain Drives," *Machine Design*, **38**, Apr. 14, 1966, p. 183.

There are four types of lubrication for chain drives. The maximum chain speed for the use of each type is indicated by Table 6-14.

A chain should be washed periodically in kerosene to remove gummed lubricant and wear particles. It should then be soaked in oil to restore the internal lubricant.

TABLE 6-14   Permissible Chain Speed, FPM, for Different Types of Lubrication

Type*	Number and Pitch of Chain, in.												
	25	35	40–41	50	60	80	100	120	140	160	180	200	240
	1/4	3/8	1/2	5/8	3/4	1	1-1/4	1-1/2	1-3/4	2	2-1/4	2-1/2	3
I	500	370	300	250	220	170	150	130	115	100	95	85	75
II	2,500	1,700	1,300	1,000	850	650	520	430	370	330	300	260	220
III	3,500	2,800	2,300	2,000	1,800	1,500	1,300	1,200	1,100	1,000	950	900	800
IV	Up to maximum permissible speed												

*Type   I, Oil supplied periodically with brush or spoutcan.
          II, Drip lubrication. Oil applied between link plate edges from a drip lubricator.
         III, Oil bath or slinger. Oil level maintained in casing at predetermined height.
         IV, Oil stream. Oil supplied by circulating pump inside chain loop on lower strand.

18. Polygonal Action

When the number of teeth in the sprockets is small, the driven shaft of a roller chain drive may be given a pulsating or jerky motion. In Fig. 6-18, the shaft centers are located not an integral number of pitches apart, but some number plus one-half pitches apart. If the driver is assumed to be rotating uniformly the driven shaft has the angular velocities indicated in the figures. In Fig. 6-18(a) the driven sprocket is rotating at a higher speed than the driver, whereas in (b) the driven shaft is moving slower. The equations for ω_2 indicate, for example, that for sprockets having 10 teeth, the velocity of the driven shaft will vary from 5 percent above to 5 percent below that of the driver. The chain velocity also varies. Sometimes long-link conveyor chains are operated on sprockets having as few as 5 or 6 teeth. The variation in velocity will then be most undesirable, and there will be an increase in the stresses in the chain and connected parts. [11]

19. Silent Chain

The silent or inverted tooth chain, shown in Fig. 6-19, is in wide use in the power transmission field. Different details of joint construction are used with the object of increasing chain life. Careful attention to lubrication is required.

---

[11]See Bouillon, G., and G. V. Tordion, "On Polygonal Action in Roller Chain Drives," *Trans. ASME*, **87**, 1965, *J. Eng. Ind.*, p. 243.

$$\omega_2 = \frac{\omega_1}{\cos\frac{180°}{N}}$$

(a)

Vel. chain $= r\omega_1$

$$\omega_2 = \omega_1 \cos\frac{180°}{N}$$

(b)

Vel. chain $= r\omega_1 \cos\frac{180°}{N}$

FIGURE 6-18 Variation in velocity of driven sprocket caused by shaft centers not being an integral number of pitches apart.

FIGURE 6-19 Silent chain.

Enclosures for the chain are usually necessary. Service factors should be employed in the design.

Standard pitches, or the distance center to center of pins, are $\frac{3}{8}$, $\frac{1}{2}$, $\frac{5}{8}$, $\frac{3}{4}$, 1, $1\frac{1}{4}$, $1\frac{1}{2}$, and 2 in. Sprockets may have 21–150 teeth. Center distance adjustment is required to compensate for wear. Greater power capacity can be obtained by using chains of greater width. Horsepower capacities are given in the manufacturer's catalog.

Data and design equations for roller chains are based on lengths in inches and forces in pounds. Equivalent SI information is not available. Should the data for a design be given in SI units, they can be changed into U.S. customary units and the foregoing procedure followed.

REFERENCES

1. BLACK, P., *Mechanics of Machines*, Elmsford, N.Y.: Pergamon Press, Inc., 1967, Chapter 3.

2. NAJI, M. R., and K. M. MARSHEK, "Experimental Determination of the Roller Chain Load Distribution," *Trans. ASME*, **105**, 1983, *J. Mech., Transmi., Autom. Design*, p. 331.

3. ROTHBART, H. S., Ed., *Mechanical Design and Systems Handbook*, New York: McGraw-Hill Book Company, 1964, Sections 28 and 29.

PROBLEMS

1. A 5V high-capacity V-belt carries a load horsepower of 12.5, but a service factor of 1.5 must be used. Pulley diameters are 9.75 in. and turn 1,750 rpm. Find the expected life in hours for belt lengths of 90, 95, and 100 in.
Ans.   18,800 hr; 19,900 hr; 20,900 hr.

2. A D-section V-belt has a length of 240.8 in. and operates on 16-in.-diam pulleys. Speed is 870 rpm. For an expected life of 20,000 hr, find the number of belts required to transmit 100 hp.   *Ans.*   4 belts.

3. An 8V high-capacity V-belt is 280 in. long and operates on 21.2-in.-diam pulleys. Speed is 870 rpm. For an expected life of 20,000 hr, find the number of belts required to transmit 200 hp.   *Ans.*   4 belts.

4. A C-section V-belt is 146.9 in. long and operates on pulleys 11 in. in diameter. Speed is 1,160 rpm. For a life of 20,000 hr, find the number of belts required to transmit 50 hp.   *Ans.*   4 belts.

5. A B-section V-belt is 129.8 in. long and operates on pulleys of diameters of 6 and 25 in. Speed of the small pulley is 1,750 rpm. Expected life is 10,000 hr. Find the horsepower transmitted by the belt. Assume that the large pulley does not contribute to the fatigue effects. Then show that this assumption is permissible.
Ans.   hp = 6.76.

6. A 5V high-capacity V-belt carries a load horsepower of 15, but a service factor of 1.5 must be used. Pulley diameters are 10.9 and 15 in. Speed of the small pulley is 1,750 rpm. Length of the belt is 100 in. Find the center distance and the expected life of the belt.   *Ans.*   c = 29.6 in.; life = 18,900 hr.

7. A 5V high-capacity V-belt is 71 in. long and carries a load horsepower of 9, but a service factor of 1.6 must be used. Pulley diameters are 10 and 14 in. Speed of the small pulley is 1,160 rpm. Find the expected life in hours.
Ans.   Life = 20,400 hr.

8. A 600-mm-OD plate clutch has a maximum lining pressure of 0.35 MPa. One hundred and thirty-five kW are to be carried at 400 rpm. If the coefficient of

friction is 0.30, find the inside diameter and the spring force required to keep the clutch engaged. *Ans.* $d_i = 274$ mm; $F_n = 49,100$ N and
$$d_i = 407 \text{ mm}; F_n = 42,330 \text{ N}.$$

9. Work Problem 8 but with the outside diameter 18 in. The maximum pressure is 75 psi, hp $= 150$, $\mu = 0.25$, and rpm $= 600$.
Ans. $d_i = 12.18$ in.; $F_n = 8,351$ lb and
$$d_i = 8.50 \text{ in.}; F_n = 9,513 \text{ lb}.$$

10. Find the torque that a cone clutch of mean raidus 150 mm and angle α of $8°$ can carry if the force R for steady running is 2,500 N; $\mu = 0.2$.

11. A cone clutch of inclination $10°$ is to carry 40 hp at 600 rpm. The width of the lining along an element of the cone is 2 in. Maximum lining pressure is to be 50 psi. The coefficient of friction is 0.2. Find suitable values for r_o and r_i.
Ans. $r_o = 6.04$ in.; $r_i = 5.70$ in.

12. A cone clutch has a mean radius of 200 mm and a cone angle of $8°$. Maximum lining pressure is 0.7 MPa and the coefficient of friction is 0.2. Find the torque the clutch can exert and the engaging force required for steady operation. The lining measures 75 mm along an element of the cone. What is the friction power for a speed of 600 rpm? *Ans.* $R = 8,944$ N; 161.5 kW.

13. Work Problem 12 but with the cone angle of $12°$.
Ans. $R = 13,190$ N; 159.4 kW.

14. Find the friction power that the brake of Fig. 6-20 can absorb, and the length of arm a. *Ans.* 4.31 kW; 121.2 mm.

FIGURE 6-20 Problem 14.

FIGURE 6-21 Problem 15.

15. The band brake in Fig. 6-21 is to absorb 6 hp at 150 rpm. The maximum pressure between lining and drum is 100 psi. The width of the band is 2 in. and $\mu = 0.12$. Find the angle of wrap α and the distance a.
Ans. $\alpha = 205.7°$; $a = 7.94$ in.

16. Find the horsepower that the brake of Fig. 6-22 can absorb, and the length of arm *a*.   *Ans.*   9.89 kW; *a* = 36.6 mm.

17. Find the value of the torque that the backstop in Fig. 6-23 can resist if the maximum pressure between lining and drum is 200 psi. What must be the minimum value of the coefficient of friction to ensure the brake's holding the load?
Ans.   *T* = 9,310 in. lb; μ = 0.174.

FIGURE 6-22   Problem 16.   FIGURE 6-23   Problem 17.

18. In Fig. 6-24 the maximum lining pressure for any point on the brake is to be limited to 0.7 MPa; μ = 0.2. Find the torque capacity of the brake and the value of force *P*.   *Ans.*   *T* = 984,800 Nmm; *P* = 2,113 N.

FIGURE 6-24   Problem 18.

19. In Fig. 6-25, the maximum lining pressure for any point on the brake is limited to 100 psi; $\mu = 0.2$. Find the torque capacity of the brake.

Ans. $T = 9,820$ in. lb.

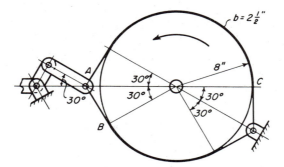

FIGURE 6-25 Problem 19.

20. (a) Write the expression for torque and the maximum lining pressure in terms of T_2 for the brake shown in Fig. 6-26(a); $\mu = 0.3$.

(b) **Do** the same for the brake in sketch (b). Note that for the same value of the actuating force T_2, the brake in (b) can exert approximately 3 times the torque as the brake in (a), but the lining pressure is almost twice as great.

Ans. (a) $T = 0.87T_2r, p = 1.874T_2/br$;
(b) $T = 2.51T_2r, p = 3.51T_2/br$.

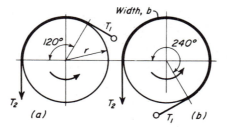

FIGURE 6-26 Problem 20.

21. A band brake exerts a torque of 15,000 in. lb. The drum is 2 in. wide and 10 in. in radius. If the maximum pressure between the lining and drum is 100 psi and the coefficient of friction is 0.3, find the angle of contact between lining and drum.

Ans. $\alpha = 264°46'$.

22. A band brake exerts a torque of 1,700,000 Nmm. The drum is 62 mm wide and the radius is 250 mm. If the maximum pressure between the lining and the drum is 0.7 MPa and the coefficient of friction is 0.32, find the angle of contact between lining and drum.

Ans. $\alpha = 176°$.

23. Work Problem 21 with a torque of 13,000 in. lb and μ equal to 0.25.

24. Find the value of dimension c in Fig. 6-27 that will cause the friction forces neither to help nor resist the application of the shoe to the drum.

Ans. $c = 1.414r$.

FIGURE 6-27   Problem 24.

FIGURE 6-28   Problem 25.

25. (a) For $\alpha/2 = 45°$, find the value of r_1 in Fig. 6-28 that will cause the shoe to be free from moment about its pivot.

(b) Work part (a) using $\alpha/2 = 60°$.   *Ans.   (a) $r_1 = 1.10r$; (b) $r_1 = 1.17r$.*

26. If the permissible value for p_{max} in Fig. 6-29 is 0.70 MPa, find (a) the corresponding value of P, and (b) the friction power that the brake will absorb; $\mu = 0.25$.

Ans.   (a) $P = 3,738$ N; (b) 20.6 kW.

FIGURE 6-29   Problem 26.

FIGURE 6-30   Problem 27.

27. (a) Find the resultant vertical component of the normal and friction forces for the brake of Fig. 6-30.

(b) Find the resultant horizontal component of the normal and friction forces.

$$\textit{Ans.}\quad (a)\ F_v = \frac{\sqrt{2}}{8} brp_{max}[2(1-\mu) + \pi(1+\mu)].$$

$$(b)\ F_h = \frac{\sqrt{2}}{8} brp_{max}[2(1+\mu) - \pi(1-\mu)].$$

28. Find the force P required if the maxmum pressure of the lining in Fig. 6-31 is equal to 0.7 MPa; $\mu = 0.25$. The width of the lining is 50 mm. Find the torque about center O.   *Ans. $P = 3,160$ N; $T = 167,440$ Nmm.*

29. (a) Make an exploded view of the double-block brake in Fig. 6-32. Draw each part, including the drum, separately, and show all forces and moments necessary for equilibrium. Assume that the normal and friction forces for the shoes are concentrated at points B.

(b) Compute the value of the maximum shoe pressure and the horsepower absorbed by the brake. Find the horsepower absorbed per unit area, and the pV value. *Ans.* $p_{max} = 93.4$ psi; hp $= 20.5$; hp/in.$^2 = 0.34$; $pV = 58,700$.

FIGURE 6-31   Problem 28.   FIGURE 6-32   Problem 29.

30. (a) Make an exploded view of the double-block brake of Fig. 6-33 and show all forces and moments required for equilibrium of each part. $\mu = 0.2$; width $b = 4$ in.

(b) Find the value of p_{max} for each shoe and the value of the torque exerted by the brake. Also find the absorbed power.
Ans. (b) $p_{max} = 0.187$ MPa and 0.166 MPa; 39.1 kW.

FIGURE 6-33   Problem 30.

FIGURE 6-34   Problem 31.   FIGURE 6-35   Problem 32.

31. Make an exploded view of the brake in Fig. 6-34 showing the forces required for the equilibrium of each part. Write the expression for the torque in terms of μ, P_1, r, a_1, and a_2.   *Ans.* $T = 4.4\mu P_1 r a_1 / a_2$.

32. A crane brake is lowering the load shown in Fig. 6-35 at a speed of 5.5 m/min. The motor shaft is connected to the cable drum shaft by a gear train with a 130:1 ratio. Assume that the gears and bearings are frictionless. The brake drum rotates at one-half motor speed and is 400 mm in diameter; the lining is 75 mm wide; μ is equal to 0.2. The shoes and levers are arranged as in Problem 31. Find the value of p_{max} and the value of P_1 for $a_1 = 1$ and $a_2 = \frac{1}{3}$.

Ans. $p_{max} = 0.1814$ MPa; $P_1 = 569$ N.

33. Assume the brake in Fig. 6-36 has short shoes. The coefficient of friction is 0.3. Find the value of distance a that will cause the wear to be the same for both shoes.   *Ans.* $a = 1.74$ in.

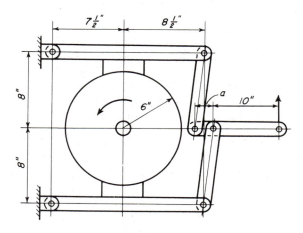

FIGURE 6-36   Problem 33.

34. A block brake with a short shoe is to be designed for a pV value of 55,000; μ is equal to 0.2; the area of the shoe is 18 in.²; and the diameter of the brake drum is 16 in. A cable drum 12 in. in diameter is connected to the brake drum by means of gearing. The brake drum revolves 3 times as fast as the cable drum. See Fig. 6-37. Find the uniform velocity at which a 1,000-lb weight at the end of the cable is being lowered. Find the value of the pressure of the shoe.

Ans. $V = 791$ fpm; $p = 69.5$ psi.

FIGURE 6-37 Problem 34.   **FIGURE 6-38 Problem 36.**

35. A 3V high-capacity V-belt carries a load horsepower of 1.25, but a service factor of 1.5 must be used. Pulley diameters are 3.15 and 6.00 in. Length of the belt is 63 in. Speed of the small pulley is 1,160 rpm. Find the center distance and the expected life in hours.   *Ans.* $c = 24.3$ in.; life $= 19,600$ hr.

36. The weight in Fig. 6-38 must be raised at the rate of 10 fps. Motor pulley is 9.25 in. in diameter, and turns 870 rpm. Length of the belt is 125 in. Expected life is 10,000 hr. Neglect fatigue effects at large pulley. Assume a service factor of 1.6 will care for impact effects and friction of the system. Find the center distance, diameter of the large pulley, and number of belts required.

Ans. $c = 31.8$ in. No. of belts $= 5$.

37. Work Problem 14 but for an actuating force of 1,350 N.

38. Find the power absorbed and force P in Fig. 6-39 if the maximum lining pressure is 0.7 MPa. $\mu = 0.2$.

FIGURE 6-39 Problem 38.

39. In Fig. 6-40 the horsepower absorbed is 15; $\mu = 0.3$. Find the value of the actuating force P.

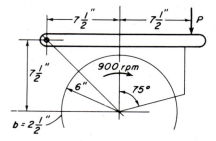

FIGURE 6-40   Problem 39.

40. Find the value of force P for the brake in Fig. 6-41. $p_{max} = 0.7$ MPa. $\mu = 0.25$.

FIGURE 6-41   Problem 40.

41. Work Problem 33 but with the dimensions of $7\frac{1}{2}$, 8, and $8\frac{1}{2}$ in. in Fig. 6-36 each increased by $\frac{1}{2}$ in.

42. Work Problem 30 but with short shoes rigidly attached to the two uprights.

43. An automobile of weight W with four-wheel brakes has a wheel base of l inches. The center of gravity is c in. forward of the rear axle, and is h in. above the ground. If the coefficient of friction between the pavement and the tires is μ, show that the load carried by the two front tires during braking with the motor disconnected is equal to $W(c + \mu h)/l$.

44. At the time the clutch is engaged in Fig. 6-42, each shaft is rotating at the velocity shown. If Ω_1 is greater than Ω_2, the system on the right will be accelerated when the two shafts are clutched together. Upon engagement assume that the clutch slips, but applies a constant torque of T in. lb. Show that the time required for the two shafts to be rotating at the same velocity is given by $(\Omega_1 - \Omega_2)I_1 I_2 / T(I_1 + I_2)$. Neglect any effects due to torsional vibration.

FIGURE 6-42   Problem 44.

45. The body of weight W_1 in Fig. 6-43 has a downward velocity of v_o when a constant retarding torque T from a brake is applied. Total mass moment of inertia of the rotating body is I lb in. $\sec^2$. Neglect friction and the inertia of the rope and lower pulley. Show that the velocity of W after t sec have elapsed is given by $v = v_o - (T - \frac{1}{2}Wr)t/[(2I/r) + (Wr/2g)]$.

FIGURE 6-43   Problem 45.

46. (a) Find the horsepower capacity for a No. 40 chain at a speed of 900 rpm for the small sprocket, which has 12 teeth.

(b) Find the horsepower for a speed of 2,400 rpm.

Ans.   (a) hp $= 3.42$; (b) hp $= 3.45$.

47. (a) Find the horsepower capacity for a No. 120 chain at a speed of 500 rpm for the small sprocket, which has 20 teeth.

(b) Find the horsepower for a speed of 900 rpm.

Ans.   (a) hp $= 88.3$; (b) hp $= 77.9$.

48. A No. 60 chain operates with a small sprocket of 15 teeth. Calculate and plot the curves for horsepower vs. n_1 for link plate fatigue and roller bushing fatigue and compare with Fig. 6-17.

49. A No. 80 roller chain operates on a small sprocket of 20 teeth. Calculate and plot the curves for horsepower vs. speed from 0 to 2,400 rpm.

50. A No. 120 roller chain operates on a small sprocket of 12 teeth. Calculate and plot the curves for horsepower vs. speed from 0 to 1,400 rpm.

51. A No. 60 roller chain operates on a small sprocket of 16 teeth. Calculate and plot the curves for horsepower vs. speed from 0 to 4, 000 rpm.

52. A C-section belt has a pitch length of 87.9 in. and carries a load horsepower, but a service factor of 1.6 must be used. Pitch diameter of both pulleys is 12 in. operating at a speed of 1,160 rpm. Find the expected life of the belt.

Ans.   101 months of 160 hr each.

53. A D-section V-belt has a length of 240.8 in. and operates on 16-in.-diameter pulleys. Speed is 870 rpm. For an expected life of 20,000 hr find the number of belts required to transmit 100 hp.   *Ans.*   hp = 26.5 per belt. Use 4.

54. An 8V belt is 280 in. long and operates on 21.2-in.-diameter pulleys. Speed is 870 rpm. For an expected life of 20,000 hr, find the number of belts required to transmit 200 hp.   *Ans.*   hp . 51.6 per belt. Use 4.

55. A C-section belt is 146.9 in. long and operates on pulleys 11 in. in diameter. Speed is 1,160 rpm. For a life of 20,000 hr, find the number of belts required to transmit 50 hp.   *Ans.*   hp = 13.5 per belt. Use 4.

56. A B-section belt is 129.8 in. long and operates on pulleys of diameters 6 in. and 25 in. Speed of small pulley is 1,750 rpm. Expected life is 240 months of 160 hr each. Assume that the large pulley does not contribute to the fatigue effects and then show that this assumption is justified.   *Ans.*   hp = 5.24.

57. A 5V belt carries a nominal horsepower of 15 but a service factor of 1.5 must be used. Pulley diameters are 10.9 in. and 15 in. Speed of small pulley is 1,750 rpm. Length of belt is 100 in. Find the center distance and the expected life of the belt.   *Ans.* c = 29.6 in.; L = 118 months of 160 hr each.

58. A 5V belt 71 in. long operates on pulleys of pitch diameters of 10 in. and 14 in. Horsepower is 9 but a service factor of 1.6 must be used. Smaller pulley turns 1,160 rpm. Find the expected life of the belt.   *Ans.*   128 months of 160 hr each.

59. A C-section belt has a pitch length of 87.9 in. and operates on pulleys of 12 in. pitch diameter at 1,160 rpm. Find the expected life in months of 160 hr each for transmitted horsepower of 12, 14, 16, 18, and 20. Make solution by tabular method. Plot the results. Note the rapid decrease in expected life as the transmitted power is increased even by nominal amounts.

Ans.   282, 122, 56, 27.2, and 13.7 months of 160 hr each.

60. Same data and requirements as for Problem 59 except the pulley diameters are now 12 and 16 in.   *Ans.*   390, 164, 73, 41.3, and 17.0 months of 160 hr each.

61. A D-section belt has a length of 240.8 in. and operates on 16-in.-diameter pulleys at 870 rpm. Find the horsepower the belt can transmit for expected lives of 5,000, 10,000, 15,000, 20,000, and 25,000 hr.

Make solution by tabular method and plot the results.

Ans.   hp = 34.8, 30.5, 28.1, 26.5, and 25.3.

62. A V-belt operates on pulleys of diameter 20 in. and 35 in. on a center distance of 56.5 in. Calculate the exact length of the belt by the equation of Fig. 6-2(a). Using the length of belt thus found and the given pulley diameters, calculate the center distance by the approximate equation of Fig. 6-2(b).

63. A V-belt operates on pulleys of diameters 400 mm and 740 mm on a center distance of 1,450 mm. Calculate the exact length of the belt by the equation of Fig.

6-2(a). Using the length of belt thus found and the given pulley diameters, calculate the center distance by the approximate equation of Fig. 6-2(b).

64. Consider the internal brake of Fig. 6-13. Pivot for shoe is located 3.5 in. below the center and 1.5 in. to the left of the vertical through the center. Radius of drum 5 in. Width of lining 2 in. Arc of lining is 90° starting 15° CW from shoe pivot. Maximum lining pressure 100 psi. Coefficient of friction 0.25. Horizontal actuating force is 3.5 in. above the center. Drum rotates CW.

Find value of actuating force P and torque exerted by the brake.

Ans. $P = 723$ lb; $T = 1,531$ in. lb.

65. Consider the internal shoe brake of Fig. 6-13. Pivot for shoe is located 4 in. below the center and 1.75 in. to the left of the vertical through the center. Radius of drum 5.5 in. Width of lining 2.25 in. Arc of lining is 90° starting at 18° CW from the shoe pivot. Maximum lining pressure 90 psi. Coefficient of friction 0.30. Horizontal actuating force is 4 in. above the center. Drum rotates CW.

Find value of actuating force P and torque exerted by the brake.

Ans. $P = 872$ lb; $T = 2,316$ in. lb.

66. Consider the internal shoe brake of Fig. 6-13. Pivot for shoe is located 3.75 in. below the center and 1.6 in. to the left of the vertical through the center. Radius of drum 5.25 in. Width of lining 2 in. Arc of lining is 90° starting at 12° from shoe pivot. Maximum lining pressure 100 psi. Coefficient of friction 0.25. Horizontal actuating force is 3.75 in. above center. Drum rotates CW.

Find the value of actuating force P and torque exerted by the brake.

Ans. $P = 768$ lb; $T = 1,635$ in. lb.

67. Consider the internal shoe brake of Fig. 6-13. Pivot for shoe is located 3.90 in. below the center and 1.80 in. to the left of the vertical through the center. Radius of drum 5.00 in. Width of lining is 2.40 in. Arc of lining is 90° starting at 20° from shoe pivot. Maximum lining pressure 90 psi. Coefficient of friction 0.30. Horizontal actuating force is 3.90 in. above center. Drum rotates CW.

Find the value of the actuating force P and the torque exerted by the brake.

Ans. $P = 856$ lb; $T = 2,076$ in. lb.

7
Welded and Riveted Connections

A weld is a union between metal surfaces brought about by the localized application of heat. Welding has assumed an important place in speeding the manufacture of component parts and in the assembly of these parts into engineering structures. Forge or pressure welding of wrought iron has been practiced for centuries; methods such as torch, arc, or resistance welding have appeared only in recent years. The field of usefulness of these latter methods is being rapidly expanded.

Rivets, also, can be used for forming the joints and connections between the parts of a structure. Although welding has replaced riveting to a considerable extent, rivets are customarily employed for certain types of joints. Long experience with this method of fastening has given confidence in the reliability of riveted joints.

A,	throat area		1 lb (mass) = 0.4536 kg
F_s,	factor of safety	K,	stress concentration factor
FS	far side	l,	length of weld
h,	size of weld leg	lb,	pound;
in.,	inch, 25.4 mm		1 lb (force) = 4.4482 N
I,	moment of inertia		1 lb (mass) = 0.4536 kg
J,	joule, newton-meter, Nm	m,	meter, 39.370 in.
J,	polar moment of inertia	mm,	millimeter, 0.039 37 in.;
k,	kilo (prefix), 1,000		1 in. = 25.4 mm
kg,	kilogram, 2.2046 lb (mass);	M,	mega (prefix), 1 million

340

M,	moment, 1 in. lb = 113 Nmm	T,	torque
MPa,	megapascal, 1 N/mm² or 145 psi	σ,	(sigma), stress, tension or compression
N,	newton, 0.2248 lb (force)	σ_{av},	average stress
NS,	near side	σ_e,	endurance limit stress
p,	pressure, force/unit area	σ_r,	range stress
P,	load or force	σ_{ult},	ultimate stress
Pa,	pascal, 1 N/m²	σ_{yp},	yield point stress
r,	radius	τ,	(tau) shearing stress
r_1,	distance, center of gravity of joint to center of weld or rivet	τ_{yp},	yield point stress in shear

1. Fabrication by Welding

Because of lower initial cost, many structural parts of machinery formerly made by casting are now fabricated by welding. The components can be sheared or flame cut from hot-rolled steel plate and then welded together. Figure 7-1 shows a number of typical welded assemblies. Sometimes the intricate portion of the body can be cast or stamped. The flat areas, made of plates, then can be attached by welding.

FIGURE 7-1 **Machine parts fabricated by fusion welding.**

Welded assemblies usually provide greater strength at a reduction in weight—an important advantage for moving parts of machines and transport equipment. In a welded design it is usually necessary to do a smaller amount of machining than for an equivalent casting. The design must provide accessibility to the welds so they can be properly made and inspected.

2. Fusion Welding

In the fusion process, heat is obtained from an oxyacetylene flame or from an electric arc passing between an electrode and the work. The edges of the parts are heated to the fusion temperature and joined together with the addition of molten filler material from a welding rod.

In metallic arc welding, the electrode is composed of suitable filler material, which is melted and fed into the joint as the weld is progressively formed. Shielded arc welding uses an electrode with a heavy coating of fluxing materials. These are consumed as the rod is melted, and perform the usual functions of a flux, as shown in Fig. 7-2. When the oxyacetylene torch is used, the molten metal is protected from the atmosphere by the outer envelope of the flame. The flame is generally adjusted until it is neutral or slightly reducing. A flux is used in the gas welding of some metals to float out any impurities that may be present and thus aid in forming a sound weld.

FIGURE 7-2   Arc welding with coated electrode.

Either direct or alternating current can be used with the metallic arc process. When the weld is larger than about $\frac{3}{8}$ in. minimum thickness, it is usually made in successive layers. Deposited weld metal frequently has the coarse structure characteristic of cast metals.

3. Strength of Fusion Welds

Several different types of welds with the equations for the stresses arising from the given loadings are shown in Fig. 7-3. The height h for a butt weld does not include the bulge or reinforcement a. Plates that are $\frac{1}{4}$ in. thick and heavier should be beveled before welding, as shown.

The fillet weld loaded parallel to its length is shown in Fig. 7-3(c). Stresses are calculated for the throat or the minimum cross section located at 45° to the legs h. The throat area is $0.707hl$, where h is the length of the leg.

(a) Butt weld in tension

Stress, $\sigma = \dfrac{P}{hl}$

(b) Butt weld in shear

Av. stress, $\tau = \dfrac{P}{hl}$

(c) Fillet weld in shear

(d) Fillet weld in tension perpendicular to its length

Av. stress, $\tau = \dfrac{P}{.707hl} = 1.414 \dfrac{P}{hl}$

FIGURE 7-3 Various types of welds and loadings.

The average shearing stress on the throat area is

$$\tau = \frac{P}{0.707hl} = 1.414\frac{P}{hl} \tag{1}$$

The usual equation for the factor of safety for static loads can be employed.

$$F_s = \frac{\tau_{yp}}{\tau} = \frac{0.5\sigma_{yp}}{\tau} \tag{2}$$

Example 1. A $\frac{1}{4}$-in. fillet weld 2 in. long carries a steady load of 3,000 lb lying along the weld. The weld metal has a yield strength of 50,000 psi. Find the value of the factor of safety.

Solution. By Eq. (1): $\tau = 1.1414 \times \dfrac{3,000}{(1/4) \times 2} = 8,490$ psi

By Eq. (2): $F_s = \dfrac{25,000}{8,490} = 2.95$

Example 1A. A 6-mm fillet weld 50 mm long carries a steady load of 13,000 N lying along the weld. The weld metal has a yield strength of 360 MPa. Find the value of the factor of safety.

Solution. By Eq. (1): $\tau = 1.414\dfrac{P}{hl} = \dfrac{1.414 \times 13,000}{6 \times 50} = 61.3$ MPa

By Eq. (2): $F_s = \dfrac{0.5\sigma_{yp}}{\tau} = \dfrac{0.5 \times 360}{61.3} = 2.94$

If the working stress in shear is taken as 11,300 psi, a $\frac{1}{8}$-in. fillet weld, 1 in. long, by Eq. (1), can carry the following load.

$$P = 0.707hl\tau = 0.707 \times \tfrac{1}{8} \times 1 \times 11,300 = 1,000 \text{ lb}$$

This provides a convenient rule of thumb for figuring the capacity of a fillet weld, loaded parallel to its length as calculations can be made mentally for welds of different widths and lengths.

The fillet weld loaded in a direction perpendicular to its length, Fig. 7-3(d), has the advantage that precise fit-up of the plates is not required. The stress situation on the throat area is very complex, and no exact design equations are available. In addition to the forces, the throat is subjected to a bending moment which adds to the complications.[1]

It is common practice to consider the stress as being the force divided by the throat area and to consider it as a shear stress which results again in Eq. (1).

4. Eccentrically Loaded Welds—Static Loads

When the load on a welded joint is applied eccentrically, the effect of the torque or moment must be taken into account as well as the direct load. The state of stress in such a joint is complicated, and it is necessary to make simplifying assumptions.

When a joint consists of a number of welds, it is customary to assume that the moment stress at any point is proportional to the distance from the

FIGURE 7-4   Stress on element of eccentrically loaded welded joint.

[1]See Jennings, C. H., "Welding Design," *Trans. ASME*, **58**, 1936, p. 497; **59**, 1937, p. 462.

center of gravity of the group of welds. Let the weld shown in Fig. 7-4 be one of a group forming a joint with the center of gravity of all the weld areas at O. The moment stress τ acts perpendicularly to radius r on element dA of the weld. The external moment or torque T is equal to the moment from stress τ integrated over all the welds of the joint.

$$T = \int \tau \, dA = \int \frac{\tau}{r} r^2 \, dA = \frac{\tau}{r} \int r^2 \, dA = \frac{\tau J}{r}$$

$$\tau = \frac{Tr}{J} \tag{3}$$

Ratio τ/r is a constant since the stress is assumed to vary directly with r. The integral $\int r^2 \, dA$ in Eq. (3) has been replaced by J, the polar moment of inertia about O for the group of welds. For the maximum torsional stress, the value of r to the point furthest removed from the center of gravity O must be used. The stress from the direct load must be added vectorially to the moment stress in order to obtain the resultant stress. For static loads, it is usual practice to assume that the direct stress in a weld is uniformly distributed throughout its area.

The parallel axis equation can be used for finding the value for J for a weld about an axis through O perpendicular to the plane of the weld. For the weld in Fig. 7-4 this equation would be written

$$J = J_o + A r_1^2 \tag{4}$$

The area A in this equation refers to the throat area of the weld. The fact that the throat for a fillet weld is inclined at 45° to the plane of the joint has no effect on the value of J. Radius r_1 extends from the center O_1 of the weld to the center of gravity O for the group. Symbol J_o represents the moment of inertia of the single-weld area about its own center O_1. This value can be found from the following equation.

$$J_o = \frac{A l^2}{12} \tag{5}$$

where A is again the throat area and l is the length of the weld.

When Eq. (5) is substituted into Eq. (4), the result is

$$J = A\left(\frac{l^2}{12} + r_1^2\right) \tag{6}$$

The value of J for each weld about O should be computed by Eq. (6); the results are added to obtain the moment of inertia of the entire joint.

Example 2. An eccentrically loaded bracket is welded to its support as shown in Fig. 7-5. If the load is steady, find the value of the maximum stress in the weld. Find the factor of safety if the yield strength of the weld metal is 50,000 psi.

FIGURE 7-5   Example 2.

Solution. Consider the bracket as the free body. By symmetry the center of gravity of the welds in the x-direction is halfway between the vertical welds. The center of gravity in the vertical direction is found by taking moments about the top weld.

$$22\bar{y} = 2 \times 8 \times 4$$
$$\bar{y} = 2.909 \text{ in. from top}$$

Width of throat: $b = 0.707h = 0.707 \times \frac{1}{4} = 0.1768 \text{ in.}$

For vertical weld: $A = 0.177 \times 8 = 1.414 \text{ in.}^2$

$$r_1 = \sqrt{1.091^2 + 3^2} = \sqrt{10.1903} = 3.192 \text{ in.}$$

By Eq. (6): $J = 1.4142\left(\frac{8^2}{12} + 3.192^2\right) = 21.953 \text{ in.}^4$

For top weld: $A = 0.1768 \times 6 = 1.061 \text{ in.}^2$

By Eq. (6): $J = 1.061\left(\frac{6^2}{12} + 2.909^2\right) = 12.159 \text{ in.}^4$

Total J: $J = 2 \times 21.953 + 12.159 = 56.065 \text{ in.}^4$

Total area: $A = 2 \times 1.414 + 1.061 = 3.889 \text{ in.}^2$

Direct stress: $\tau = \dfrac{7,500}{3.889} = 1,930 \text{ psi}$

The maximum stress will occur at *A*.

Radius *OA*: $r = \sqrt{5.091^2 + 3^2} = 5.909 \text{ in.}$

Torsional stress: $\tau = \dfrac{Tr}{J} = \dfrac{7,500 \times 9 \times 5.909}{56.065} = 7,110 \text{ psi}$

This stress is directed perpendicular to the radius from the center of gravity to A. It is now resolved into the components shown. The total vertical component is $3,610 + 1,930$ or $5,540$ psi.

Resultant stress: $\qquad \tau = \sqrt{5,540^2 + 6,130^2} = 8,260$ psi

By Eq. (2): $\qquad F_s = \dfrac{0.5\sigma_{yp}}{\tau} = \dfrac{25,000}{8,260} = 3.03$

The foregoing method assumes that failure would take place by shear on the weld throats. This is correct only when the weld pattern is a circle lying about the center of gravity. The results for a pattern like Fig. 7-5 will only be approximate.

Example 3. Find the value of static force P in Fig. 7-6 if electrode E6010 is used at a factor of safety equal to 2.
(a) All welds are $\frac{1}{4}$-in. fillets.
(b) Welds on the left side are $\frac{1}{4}$ in. and on the right side are $\frac{1}{2}$ in.

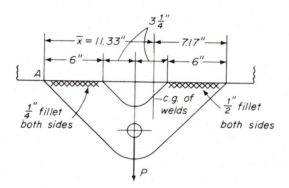

FIGURE 7-6   Example 3.

Solution. (a) By Table 7-3, for rod E6010, $\sigma_{yp} = 50,000$ psi.

Throat area of all welds: $\qquad A = 4 \times \frac{1}{4} \times 6 \times 0.707 = 4.242$ in.^{2}

Then: $\qquad P = \tau A = \dfrac{25,000}{2} \times 4.242 = 53,020$ lb

(b) Left, throat area: $\qquad A = 2 \times \frac{1}{4} \times 6 \times 0.707 = 2.121$ in.^{2}
$\qquad\qquad\qquad\qquad\qquad\qquad$ (both welds)

Right, throat area: $\qquad A = 2 \times \frac{1}{2} \times 6 \times 0.707 = 4.242$ in.^{2}
$\qquad\qquad\qquad\qquad\qquad\qquad$ (both welds)

Total area: $\qquad A = 6.363$ in.^{2}

Take moments at left end: $\qquad \bar{x} = \dfrac{2.121 \times 3 + 4.242 \times 15.5}{6.363}$

$\qquad\qquad\qquad\qquad = 11.33$ in.

For left welds: Eq. (6): $J = 2.121\left(\dfrac{6^2}{12} + 8.33^2\right) = 153.65$ in.⁴

For right welds: Eq. (6): $J = 4.242\left(\dfrac{6^2}{12} + 4.17^2\right) = 86.37$ in.⁴

Total: $J = 153.65 + 86.37 = 240.02$ in.⁴

Direct stress: $\tau = \dfrac{P}{6.363} = 0.1572P$

Moment arm of the load is $11.33 - 9.25$ or 2.08 in.

Moment stress at A: $\tau = \dfrac{Tr}{J} = \dfrac{P2.08 \times 11.33}{240.02} = 0.0984P$

Total stress at A: $\tau = 0.1572P + 0.0984P = 0.2556P$

By Eq. (2): $0.2556P = \dfrac{25,000}{2}$ or $P = 48,900$ lb

It should be noted that, although the welds of part (b) are larger, because of the eccentricity, the carrying capacity of the joint has been actually reduced. It is generally advantageous to maintain symmetry in the design of a welded joint.

Example 3A. In Fig. 7-6, let the successive dimensions across the top be 150 mm, 82 mm, 82 mm, and 150 mm. The welds on the left are 6 mm and on the right are 12 mm. Find the value of the static load P if electrode E6010 is used at a factor of safety of 2.

Solution. By Table 7-3, for rod E6010, $\sigma_{yp} = 50,000$ psi $= 34.5$ MPa.

Left, throat area: $A = 2 \times 6 \times 150 \times 0.707 = 1,273$ mm² both welds

Right, throat area: $A = 2 \times 12 \times 150 \times 0.707 = 2,546$ mm² both welds

Total area, $A = 3,819$ mm²

Moments at left: $\bar{x} = \dfrac{1,273 \times 75 + 2,546 \times 389}{3,819} = 284.3$ mm from left

Distance, c.g. to center: left $= 284.3 - 75 = 209.3$ mm

right $= 464 - (284.3 + 75) = 104.7$ mm

By Eq. (6), left welds: $J = 1,273\left(\dfrac{150^2}{12} + 209.3^2\right) = 58,152,000$ mm⁴

right welds: $J = 2,546\left(\dfrac{150^2}{12} + 104.7^2\right) = 32,683,000$ mm⁴

Total: $J = 58,152,500 + 32,683,000 = 90,835,000$ mm⁴

Direct stress: $\tau = \dfrac{P}{3,819} = 0.000\ 261\ 8P$

Moment arm of load $= 284.3 - (150 + 82) = 52.3$ mm

Torsion stress at A: $\qquad \tau = \dfrac{Tr}{J} = \dfrac{52.3P\,284.3}{90,835,500} = 0.000\ 163\ 7P$

Total stress at A: $\qquad\qquad \tau = (0.000\ 261\ 8 + 0.000\ 163\ 7)P$

$$= 0.000\ 425\ 5P$$

By Eq. (2): $\qquad 0.000\ 425\ 5P = \dfrac{17.25}{2} \qquad$ or $\qquad P = 202,700$ N

The method explained in this section cannot be considered an exact analysis of weld stresses, but should be looked upon simply as a reasonable effort to take into account the fact that capacity of a joint to resist moment loads is increased by locating the welds further from the center. The theory also assumes that the weld stresses are within the elastic limit of the weld material, and that any effects from the deflection of the welded parts can be neglected.

5. Stress Concentration in Welds

Since abrupt changes in form occur in welds, stress concentrations are present as the force passes from one portion of the assembly to the other. Stress concentration effects are usually ignored for static or steady loads, and are applied only to the variable or range component in accordance with general design practice for fluctuating loads.

Care must be exercised that the weld metal and the plates at the base of a butt weld be thoroughly fused together. With insufficient fusion, sharp-cornered notches extend inward, as at A in Fig. 7-7(a), and a serious stress concentration results. Stress concentration also occurs at points B, where the force spreads into the reinforcement. A fillet weld has concentrations at the toe and heel, points A and B in Fig. 7-7(b), where the force is passing from one plate to the other through the weld. Values of stress concentration factors are given in Table 7-1.

A fillet weld loaded by parallel forces as in Fig. 7-3(c) has stress concentrations at each end caused by the unequal elongations of the plates. The upper plate at point B has the maximum elongation, because here it is

(a) Butt weld in tension   (b) Fillet weld in tension

FIGURE 7-7 Stress concentration in welds.

TABLE 7-1   Stress Concentration Factors, *K*, for Welds

Location	K
Reinforced butt weld	1.2
Toe of transverse fillet weld	1.5
End of parallel fillet weld	2.7
T-butt joint with sharp corners	2.0

carrying the entire load *P*. The lower plate at *A* has a small elongation, because it is carrying very little load at this point. Since the weld joins the two plates together, it is subjected to a greater deformation than the average for the weld as a whole, and an increase or concentration of stress occurs at the end of the weld. Similar reasoning applied to the other end, where the lower plate at *C* has a large elongation and the upper plate at D has a small elongation. Similar stress concentration effects are present at the ends of the butt weld in shear, shown in Fig. 7-3(b).

6. Residual Stress—Weldability

Residual stresses of considerable magnitude can result from contraction of the weld metal upon cooling. Such stresses are usually greatest in the transverse direction and are thus more harmful when the weld is subjected to tensile loads. The presence of residual stress is particularly dangerous if the weld is subjected to repeated loading or impact. Removal of residual stresses can be accomplished by annealing, or by application of an overload, which stresses the entire weld to the yield point value.

It is important that both the weld metal and adjacent parent metal be ductile and free from brittleness. The properties of the weld metal depend on the composition of the welding rod. However, for steel parent metal with a carbon content higher than about 0.15%, there is danger of air hardening upon rapid cooling from the welding temperature. The quenching effect of the cold metal surrounding a weld can be reduced if the parts are preheated to 600°–1500°F before the welding is done. An annealing treatment after welding may be required to restore the original ductility of the parent metal. The fusion welding characteristics of some commonly used metals and alloys are given in Table 7-2.

Residual welding stresses in gray and alloy cast iron can be eliminated by preheating before welding, followed by slow cooling after welding. Plain-carbon and alloy steel castings are ordinarily welded by the same procedure used for rolled steel of similar composition. Air hardening occurs if the carbon or alloy content is sufficiently high.

TABLE 7-2   Weldability of Various Metals and Alloys*

Metal or Alloy	Gas	Arc	Metal or Alloy	Gas	Arc
Carbon steels			Magnesium alloys	A	No
1. Low- and medium-carbon	A	A	Copper and copper alloys		
2. High-carbon	B	A	1. Deoxidized copper	A	B
3. Tool steel	B	B	2. Pitch, electrolytic, and lake	B	A
Cast steel, plain-carbon	A	A	3. Commercial bronze, red brass, low brass, and ounce metal	A	B
Gray and alloy cast iron	A	B	4. Spring, admiralty, yellow, and commercial brass	A	B
Malleable iron	B	B			
Low-alloy, high-tensile steels			5. Muntz metal, tobin bronze, naval brass, and managenese bronze	A	B
1. Ni-Cr-Mo and Ni-Mo	B	B	6. Nickel silver	A	B
2. All other usual compositions	A	A	7. Phosphor bronze, bell metal, and bearing bronze	A	A
Stainless steels			8. Aluminum bronze	B	A
1. Chromium	B	A	9. Beryllium copper	—	A
2. Chromium-Nickel	A	A	Nickel and nickel alloys	A	A
Aluminum			Lead	A	No
1. Commercially pure Al	A	A			
2. Al-Mn alloy	A	A			
3. Al-Mg-Mn and Al-Si-Mg alloy	B	A			
4. Al-Cu-Mg-Mn alloys	No	B			

*A, commonly used; B, occasionally used under favorable conditions.

For the copper and aluminum alloys, strips of the parent metal are frequently used as filler material where an exact color match is required.

Fluxes are necessary in welding aluminum to remove the oxide coating from both the parent metal and filler rods. When the welding is completed, the parts must be thoroughly cleaned of the flux to avoid corrosion. Since the welding temperature of aluminum and magnesium is below the visible light range, it is difficult for the operator to determine when the welding temperature is being approached. At high temperatures these alloys are very weak, and there is a tendency for the member to collapse unless positive support is provided during the welding operation.[2]

7. Welding Electrodes

Many different types of electrodes have been standardized to fit the various conditions found in the welding of machinery and structures. The strength and ductility properties of several of these are given in Table 7-3.

[2]For the welding of aluminum, see "How to Weld Aluminum," *Iron Age*, **157**, June 20, 1946, p. 27; **158**, July 4, 1946, p. 11.

TABLE 7-3   Tensile Strength, Yield Point, and Elongation Requirements
for All-Weld-Metal Tension Test in the As-Welded Condition
(ASTM Designation A233-64T)

E60 Series				E70 Series			
AWS-ASTM Classification	Tensile Strength, min. psi	Yield Point, min. psi	Elongation in 2 in., min. percent	AWS-ASTM Classification	Tensile Strength, min. psi	Yield Point, min. psi	Elongation in 2 in., min. percent
E6010	62,000	50,000	22	E7014	72,000	60,000	17
E6011	62,000	50,000	22	E7015	72,000	60,000	22
E6012	67,000	55,000	17	E7016	72,000	60,000	22
E6013	67,000	55,000	17	E7018	72,000	60,000	22
E6020	62,000	50,000	25	E7024	72,000	60,000	17
E6027	62,000	50,000	25	E7028	72,000	60,000	22

The numbering system is based on the use of an "E" prefix followed by four digits. The last digit indicates a group of welding technique variables, such as current supply and application. The next to the last digit indicates a welding position number as 1 for all positions—flat, horizontal, vertical, and overhead; 2 for flat and horizontal fillet welding; and 3 for flat only. The two digits on the left indicate the approximate tensile strength in thousands of pounds per square inch. Welding electrodes are available in diameters from $\frac{1}{16}$ to $\frac{5}{16}$ in.

The application properties for a number of electrodes are shown in Table 7-4. The choice of the proper electrode for a particular application should be left to a man with long experience in the field of welding.

TABLE 7-4   Electrode Classification

Classification	Penetration	Basic Application
E6010 E6011	Deep	Good mechanical properties, especially where multipass welds are employed, as in buildings, bridges, pressure vessels, and piping
E6012	Medium	Good for single-pass high-speed horizontal fillet welds easy to handle; especially adapted to cases of poor fit-up
E6013	Medium	Designed for obtaining good-quality welds in thin metal
E6020	Medium Deep	A high-production electrode for horizontal fillet welds in heavy sections
E6027	Medium	Iron powder electrode; fast and easy to handle

8. Design for Fluctuating Loads

When the loading fluctuates, design calculations can be made by the methods explained in Chapter 2, as illustrated by the following examples.

Example 4. The load on a butt weld Fig. 7-3(a) fluctuates continuously between 10,000 and 40,000 lbs. Plates are 1 in. thick. Factor of safety equals 2.5. Use an E6010 welding rod. Let the endurance limit for the weld be equivalent to that of an as-forged surface. Find the required length of weld.

Solution. By Table 7-3, for E6010:

$$\sigma_{ult} = 62{,}000 \text{ psi}$$

By Fig. 2-11: $\quad\quad\quad \sigma_e = 16{,}000 \text{ psi}$

By Table 7-1: $\quad\quad\quad K = 1.2$

Given loading: $\quad\quad P_{av} = 25{,}000 \text{ lb}; \quad \sigma_{av} = \dfrac{25{,}000}{1 \times l}$

$$P_r = 15{,}000 \text{ lb}; \quad \sigma_r = \dfrac{15{,}000}{1 \times l}$$

By Eq. (11), Chapter 2: $\quad\quad \dfrac{\sigma_{av}}{\sigma_{ult}} + \dfrac{K\sigma_r}{\sigma_e} = \dfrac{1}{F_s}$

$$\dfrac{25{,}000}{62{,}000l} + \dfrac{1.2 \times 15{,}000}{16{,}000l} = \dfrac{1}{2.5}$$

$$\dfrac{l}{1}(0.403 + 1.125) = 0.40$$

$$l = \dfrac{1{,}528}{0.40} = 3.82 \text{ in.}$$

In the examples so far considered, the torque or moment, if any, lay in the plane of the welds. Sometimes, however, the moment may lie in the plane perpendicular to the welds. In this case moment of inertia I is computated rather than J, and the stress determined by the ordinary bending equation.

Example 5. A short length of heavy-wall tubing is welded to the support in Fig. 7-8. The loading fluctuates continuously between the values shown. Let an E6010

FIGURE 7-8   Example 5.

electrode be used. Factor of safety is 4. Let the endurance limit for the weld material be equivalent to that of an as-forged surface. Find the required diameter d to carry the load.

Solution. Width of throat, $b = 0.707h = 0.707 \times 0.5 = 0.354$ in.

Area of element: $dA = br\, d\theta$

$$I = 2 \int_0^{\pi} (r \sin \theta)^2\, dA = 2r^2 \int \sin^2 \theta\, br\, d\theta$$

$$= 2br^3 \int \frac{1}{2}(1 - \cos 2\theta)\, d\theta = br^3 \left(\theta - \frac{1}{2}\sin\theta\right)_0^{\pi}$$

$$= \pi br^3 = 0.354\pi r^3 = 1.112r^3$$

$$M_{av} = 150{,}000 \text{ in. lb}, \qquad M_r = 50{,}000 \text{ in. lb}$$

$$\sigma_{av} = \frac{M_{av}r}{I} = \frac{150{,}000r}{1.112r^3} = \frac{134{,}900}{r^2}$$

$$\sigma_r = \frac{1}{3}\sigma_{av} = \frac{44{,}960}{r^2}$$

By Table 7-1: $K = 1.5$

By Table 7-3: $\sigma_{ult} = 62{,}000$ psi

By Fig. 2-11: $\sigma_e = 16{,}000$ psi

By Eq. (11), Chapter 2: $$\frac{\sigma_{av}}{\sigma_{ult}} + \frac{K\sigma_r}{\sigma_e} = \frac{1}{F}$$

$$\frac{134{,}900}{62{,}000r^2} + \frac{1.5 \times 44{,}960}{16{,}000r^2} = \frac{1}{4}$$

$$\frac{1}{r^2}(2.176 + 4.215) = 0.25$$

$$r^2 = \frac{6{,}391}{0.25} = 25.56$$

$$r = 5.05 \text{ in.}, \qquad d = 10.10 \text{ in.}$$

9. Resistance Welding

In resistance welding, a heavy electric current is passed through the parts at the place where the weld is desired. The resistance of the metals to the current causes the temperature to rise quickly and the metal to become plastic. The weld is completed by mechanical pressure from the electrodes. This pressure brings the parts into complete union. Alternating current is used from a suitable transformer, which must be capable of supplying a large current at low voltage. The copper electrodes are faced with harder alloys at the contact end. To confine the heat in the proper region, the resistance between electrode and part must be less than the resistance between the surfaces to be joined.

Spot welding is a type of resistance welding in which cylindrical electrodes are used that have a contact area approximately equal to the size of the desired weld. As shown in Fig. 7-9, a chamber is provided for cooling water, which prevents overheating and prolongs the life of the electrode. The squeezing pressure must be properly adjusted to the thickness and strength of the parts being welded. A chart showing the spot welding possibilities of various combinations of metals and alloys is given in Fig. 7-10. The metal surfaces to be welded must be clean if sound welds are expected. It is especially necessary that the oxide coating on aluminum be removed by mechanical or chemical means before spot welding.

FIGURE 7-9 Schematic diagram of spot welding.

In the course of spot welding coated materials, the coatings frequently dissolve in the other metals present or burn away.

FIGURE 7-10 Spot welding chart for various combinations of metals.

Seam welding is similar to spot welding except that copper alloy disks about 6–9 in. in diameter are used. The work is rolled between the electrodes, and evenly spaced spot welds are obtained by periodic application of the current. It is possible to overlap the welds by close spacing and thus form a fluid-tight joint.

10. Other Kinds of Welding

(a) *Cold Welding.* Some metals, notably aluminum and copper, can be welded cold[3] if they are subjected to very high localized pressure. There must be sufficient lateral displacement to break down the normal oxide coating and to bring pure metals into contact with each other.

(b) *Atomic Hydrogen Welding.* When molecular hydrogen is passed through the welding arc it is dissociated into atomic hydrogen. When the gas in this condition strikes the colder work it recombines into molecular hydrogen with a large evolution of heat, which is used in forming the weld.

(c) *Electron Beam Welding.*[4] Here a beam of high-speed electrons moving at about one-half the velocity of light is focused on a small spot on the workpiece. The workpiece is quickly brought to the melting temperature and the parts joined together. The process must be conducted in a vacuum.

(d) *Ultrasonic Welding.* Here the workpieces are clamped together and given a high-frequency vibration in the direction of the interface. Local plastic deformation occurs; oxide and other surface films are broken and dispersed and nascent metallic contact is achieved.[5]

(e) *Friction Welding.* A cylindrical workpiece is attached to the end of a flywheel shaft and rotated at high speed and the power is shut off. The other part is stationary but is pressed firmly against the moving member. The friction raises the interface to the welding temperature as the flywheel coasts down. Upon cooling, the parts are welded together.[6]

(f) *Hard Facing.* This is the process of depositing by welding of an alloy material on metallic parts to build up a worn area or to form a protective surface to resist abrasion, impact, heat, or corrosion.

[3]See p. 21 of Reference 9, end of chapter.
[4]See Solomon, J. L., "Electron Beam Welding," *Mech. Eng.*, **87**, Jan. 1965, p. 28. See also Miller, K. J., and L. M. Reese, "Electron Beam Welding," *Machine Design*, **36**, Apr. 23, 1964, p. 218; May 7, 1964, p. 165.
[5]See *Machine Design*, **36**, Apr. 9, 1964, p. 130.
[6]See Alm, G. V., "Grown Together Joints Grow Up," *Machine Design*, **40**, Jan. 4, 1968, p. 100. See also Kiwalle, J., "Designing for Inertia Welding," *Machine Design*, **40**, Nov. 7, 1968, p. 161.

11. Soldering and Brazing

Many cast and wrought metals can be united by soldering and brazing. Soft solders are tin-lead alloys of low melting points; hard solders comprise the silver solders and brazing alloys of different compositions and melting points. In any soldering or brazing process, the parts are heated above the melting point of the solder but below that of the parts. The wetting action of the solder brings it into intimate contact with the surfaces to be joined. The solder, after cooling, serves as an adhesive to bind the two parts together, even though in some cases the solder forms an alloy with the metal of the part. In general, the strength of the joint is improved as the wetting action becomes more perfect.

The surfaces must be clean and covered with a flux, which is liquid at the soldering temperature. The flux dissolves any oxides present as well as preventing oxidation while the parts are being heated. The flux is drawn into the joint by capillary attraction. When the soldering temperature is reached, the solder is also drawn in by capillarity and displaces the flux. For this action, the surfaces must be closely spaced with respect to each other. The capillarity is greater for a tightly fitted joint, but higher temperatures are required to secure sufficient fluidity of the solder. In practice, clearances of 0.003 or 0.004 in. have been found satisfactory.

The strength of a soldered joint depends on many factors, such as the quality of the solder, thickness of the joint, smoothness of the surfaces, kind of materials soldered, soldering temperature, and duration of contact between solder and part at the soldering temperature. Optimum stress values in shear for soft-soldered joints can be taken as 6,000 psi for copper, 5,000 psi for mild steel, and 4,000 psi for brass. These are ultimate values.

The properties of soft-solder metals are given in Table 7-5. Melting of the alloy begins at the melting point shown, and is completed at the temperature of the flow point.

The silver solders have higher melting temperatures, and the soldering operation is carried on while the parts are at a red heat. Cast-iron, wrought-iron, and carbon steels can be joined to each other or to brass, copper, nickel, silver, monel, and other nonferrous alloys. Silver solders are malleable and ductile and are suitable for vibration and impact loads. They are also resistant to corrosion. The tensile strength of cast-silver solder varies from 40,000 to 60,000 psi. Joint preparation and clearances are the same as for soft solder. Heating can be done with a torch or with a high-frequency induction heating coil.

Brazing is a form of soldering at high temperatures that uses alloys of copper and zinc. The properties of several brazing alloys are given in Table 7-5. Tobin bronze is in wide use as a welding rod for general repair service

TABLE 7-5   Compositions and Uses for Solders and Brazing Alloys

Soft Solders

Sn	Pb	Melting Point, °F	Flow Point, °F	Uses
60	40	361	372	High-grade solder; has low flow-point temperature
50	50	361	421	Widely used general-purpose solder
40	60	361	453	For wiped joints; for automobile radiators and heating units
30	70	361	486 ⎫	Low-grade solder; for filling dents and seams
20	80	361	523 ⎭	in automobile bodies

Silver Solders

Ag	Cu	Zn	Melting Point, °F	Flow Point, °F	Uses
10	52	38	1,510	1,600	Solders of low silver give best results in stainless steel; application of pressure aids in forming good joint
20	45	35	1,430	1,500	Solder of minimum silver for good joints in brass
50	15.5	16.5	1,160	1,175	Cd 18, general-purpose solder; suitable for joining unlike metals.*
65	20	15	1,280	1,325	Solders of high silver are preferred for copper, especially for thin sections

Miscellaneous Brazing Alloys

Alloy	Zn	Sn	Cu	Melting Point, °F	Uses
Brazing alloy	Rem		50–53	1,595–1,620	General-purpose brazing alloy
Tobin bronze	38–43	0.50–1.50	Rem	1,625	Strong oxyacetylene welds on steel, cast iron, copper, and nickel alloys
Manganese bronze	38–43	0–1.5	Rem	1,600	Mn 0.50–0.75; hard and wear resistant
Cu-Ni-Zn alloy	43		47	1,700	Ni 10; strong, general-purpose alloy

*Proprietary alloy, "Easy-Flo," Handy and Harman.

and for building up defects in iron castings. The tensile strength is approximately 50,000 psi. Manganese bronze is similarly used, especially where the deposited metal is called upon to resist abrasion. Brazing alloys for aluminum are available with a melting point below that of the aluminum parts. In fusion welding, the high temperature permits impurities to be floated away by the flux. However, for brazing, the oxides must be removed and the joint cleaned by mechanical means.

12. Furnace Brazing

Furnace or copper brazing[7] is a very satisfactory method for joining parts with pure copper. The component parts are assembled togethei, a ring of pure copper wire is placed adjacent to the joint as shown in Fig. 7-11, and the assembly is passed through a furnace. The copper melts at 1980°F and is drawn into the joint by capillary attraction. The strength of the assembly increases with the tightness of the fit. For steel, the clearance should run from about 0.001 in. loose to a light press fit.

FIGURE 7-11 Typical joints fabricated by furnace welding.

Castings, forgings, and parts machined from solid stock can sometimes be redesigned and fabricated from stampings, screw machine parts, and pieces of tubing brazed together. Machining operations can be minimized and substantial economies effected. Parts must be held in their proper relationship with each other in their trip through the furnace. It is advantageous to have self-locking joints and thus avoid the use of holding fixtures. When the heating is done in an electric furnace with a reducing atmosphere, no flux is required, and the parts are delivered with smooth bright surfaces.

[7]See Webber, H. M., "Furnace Brazing of Machine Parts," *Mech. Eng.*, **72**, 1950, pp. 863 and 969. See also *Iron Age*, **142**, 1938, Sept. 8, 15, and 22, Nov. 3, 10, and 24, and Dec. 8 and 29; **143**, 1939, Feb. 2, Mar. 16, and Apr. 6.

13. *Riveted Joint with Central Load*

Rivets are in wide use as fastenings for joints in buildings, bridges, boilers, tanks, ships, and miscellaneous frameworks. For centrally applied loads, it is customary to assume that all the rivets in a joint are equally stressed.[8] This assumption, however, is only approximately valid. As a simple example, consider the two bars in Fig. 7-12, which are held together by a single row of rivets. The material at B and C has large elongations, because it is carrying almost the entire load P. The material at A and D has small elongations because the bars at these points are carrying only small loads. Because of these unequal elongations, the end rivets carry excessive loads. The loads are smaller for the next rivets toward the center, until at the middle of the joint the rivets will be carrying less than the average load. For ductile materials, a more uniform distribution of the loading among the rivets occurs if the joint is first subjected to an overload, which causes the material in the highly stressed regions to yield in plastic flow.

FIGURE 7-12   **Bars carrying tensile forces joined by single row of rivets.**

In structural steel work, it is standard practice to use punched holes, which are $\frac{1}{16}$ in. larger than the diameter of the rivet. Although the rivet after driving may completely fill the hole, computations are made on the basis of the original rivet diameter. Since punching injures the metal around the circumference of the hole, work of higher quality is secured by subpunching the hole to a smaller diameter and then reaming it until it is large enough to admit the rivet. Boiler codes required the holes to be either drilled or subpunched and reamed. In boiler work, computations are made on the basis of the hole diameter, since the rivets must be tightly driven and must completely fill the holes.

Rivets must be spaced neither too close nor too far apart. The minimum rivet spacing, center to center, for structural steel work is usually taken as three rivet diameters. A somewhat closer spacing is often used in boilers. Rivets should not be spaced too far apart or buckling of the plates will take place. The maximum spacing is usually taken as 16 times the thickness of the outside plate. The edge distance, or distance from the rivet center to the

[8]See Jenkins, E. S., "Rational Design of Fastenings," *Soc. Auto. Eng. J.*, **52**, Sept. 1944, p. 421. See also *Proc. Soc. Exp. Stress Anal.*, **2**, No. 2, 1944, p. 28; **7**, No. 2, 1949, p. 17.

edge of the plate, must not be less than a specified amount or there is danger of failure, as shown in Fig. 7-13(d) and (e).

14. Stresses in Rivets

Power-driven hot rivets contract upon cooling and draw the plates tightly together so that the friction between the parts assists in the transfer of the load. Under certain conditions, the entire load might be carried in this way. The friction in a joint breaks down as the loading increases, and the joint can then fail in a number of different ways, as shown in Fig. 7-13. For fluctuating loads, it is especially desirable to prevent slip in the joint. The loading should therefore be conservative and the rivets tightly driven.

| (a) Shearing of rivet | (b) Tension in plate between rivets | (c) Failure in compression | (d) Shearing of edge of plate | (e) Tearing of edge |

FIGURE 7-13 Types of failure for riveted joints.

Despite the assistance given by the friction, it is customary to compute the strength of a riveted joint from the strength of the rivets in shear, the tension in the plates between the rivet holes, or the strength in compression or bearing for the rivets on the plates. Power-driven rivets are rivets driven by a riveting machine or by a hand-operated air hammer. When turned bolts are used, they must fit the holes with not over $\frac{1}{50}$-in. diametral clearance. When rivets are loaded in single shear, as shown in Fig. 7-14, the stress situation is complicated by the effects of the eccentricity. When loaded in single shear, lower working-stress values must be used for shear and compression than when the rivet is in double shear. Proper provision must be made in riveted joints for the effects of fluctuating loads, impact, and extremes of temperature.

FIGURE 7-14 Eccentric load on rivet of single lap joint.

15. Stresses in a Cylindrical Shell

The stresses in the walls of a thin cylindrical shell are generally computed on the assumption that the stress is uniform throughout the wall thickness. Let the shell of Fig. 7-15 be loaded by the internal pressure p. If the length

(a) Hoop stress in shell   (b) Longitudinal stress in shell   (c) Shell with eliptical end closure

FIGURE 7-15   Stresses in thin cylindrical shell.

of the section perpendicular to the paper is taken as l, an element of area on the drum surface is equal to $rl\,d\theta$. The horizontal component of the force on the element is $prl\cos\theta\,d\theta$. When this force is integrated over a quarter-circle, the result is equal to the hoop force F. Thus,

$$F = \int_0^{\pi/2} prl\cos\theta\,d\theta = prl\sin\theta \Big|_0^{\pi/2} = prl \qquad (7)$$

The hoop or tangential stress σ_1 is found by dividing force F by the area of the cross section tl, giving

$$\sigma_1 = \frac{F}{tl} = \frac{pr}{t} \qquad (8)$$

In the longitudinal direction, the stress σ_2 in the material must resist the force of the internal pressure acting on the end surface of the drum. Hence,

$$2\pi rt\sigma_2 = \pi r^2 p$$

$$\sigma_2 = \frac{pr}{2t} \qquad (9)$$

The hoop stress is seen to be twice as great as the longitudinal stress. Equations (8) and (9) are valid only when thickness t is much less than radius r. From Eq. (7) the conclusion can be drawn that the bearing or compressive force on a rivet, Fig. 7-13(c), is equal to the stress in bearing times the projected area, or the diameter times the thickness of the plate.

Stresses at discontinuities in shells increase rapidly when the plate thickness is such that bending must be taken into account.[9] The theory becomes very complex.

A riveted boiler joint is weaker than the plate composing the shell because of the area lost in the rivet holes. Computations can be made for various assumed types of failure that will free the main plate of the joint in order to make a comparison with the strength of the unpunched plate and thus arrive at a value for the efficiency of the joint. The process is illustrated by the following example.

Example 6. Find the working force carried by an $8\frac{1}{4}$-in. width of the triple-riveted butt joint in Fig. 7-16. Make computations for the assumed types of failure

FIGURE 7-16 Triple riveted butt joint. Example 6.

shown in Table 7-6, and compute the efficiency of the joint. If this joint is used on a 60-in.-diam cylindrical tank, find the permissible internal pressure. Use the following stress values.

Tension: $\sigma_t = 55,000$ psi ult for steel plate

Compression: $\sigma_c = 95,000$ psi ult for steel plate

Shear: $\tau = 44,000$ psi ult for steel rivets

Factor of safety: $F_s = 5$

[9]See Timoshenko, S., *Strength of Material*, 3d ed., Part II, New York: Van Nostrand Reinhold Company, 1956, p. 164. See also Timoshenko, S., and S. Woinowsky-Krieger, *Theory of Plates and Shells*, 2d ed., New York: McGraw-Hill Book Company, 1959, p. 481.

TABLE 7-6   Loads for Joint of Fig. 7-16

Type of Failure		Working Force, lb
(a)	Tension at A-A	·58,270
(b)	Shear in rivets	87,710
(c)	Bearing on rivets	78,970
(d)	Tension at B-B Shear at A-A	58,220
(e)	Tension at B-B Bearing at A-A	59,750
(f)	Bearing at B-B and C-C Shear at A-A	77,430
(g)	Tension in gross plate	68,060
(h)	Efficiency	85.5%
(i)	Permissible pressure	235 psi

Solution. (a) Tension in main plate at section A-A.

$$F = \frac{55,000}{5}(8.25 - 1.1875) \times 0.75 = 58,270 \text{ lb}$$

(b) Shear in rivets. There are nine surfaces in shear: four on B-B, four on C-C, and one on A-A.

Area of rivet in shear: $A = \frac{\pi}{4} \times 1.1875^2 = 1.108 \text{ in.}^2$

$$F = \frac{44,000}{5} \times 9 \times 1.108 = 87,710 \text{ lb}$$

(c) Bearing on rivets. On B-B and C-C, failure occurs in the main plate. On A-A, failure occurs in the butt plate.

Area in bearing: $A = 4 \times 1.1875 \times 0.75 + 1.1875 \times 0.5 = 4.156 \text{ in.}^2$

$$F = \frac{95,000}{5} \times 4.156 = 78,970 \text{ lb}$$

(d) Tension at section B-B, and shear at section A-A.

Area in tension: $A = (8.25 - 2 \times 1.1875) \times 0.75 = 4.406 \text{ in.}^2$

$$F = \frac{55,000}{5} \times 4.406 + \frac{44,000}{5} \times 1.108$$

$$= 48,470 + 9,750 = 58,220 \text{ lb}$$

(e) Tension at section B-B and bearing at A-A.

$$F = 48,470 + \frac{95,000}{5} \times 1.1875 \times 0.5 = 59,750 \text{ lb}$$

(f) Bearing at sections B-B and C-C, and shear at section A-A.

$$F = \frac{95,000}{5} \times 4 \times 1.1875 \times 0.75 + 9,750 = 77,430 \text{ lb}$$

(g) Tension in unpunched plate.

$$F = \frac{55,000}{5} \times 8.25 \times 0.75 = 68,060 \text{ lb}$$

(h)   Efficiency $= \dfrac{58,220}{68,060} \times 100 = 85.5\%$

(i) By Eq. (7): $p = \dfrac{F}{rl} = \dfrac{58,220}{30 \times 8.25}$

$$= 235 \text{ psi maximum permissible internal pressure}$$

16. Riveted Joint with Eccentric Load

When the load is applied eccentrically to a group of rivets forming a joint, the effect of the torque or moment must be taken into account, as well as the direct load. A typical example is shown in Fig. 7-17, where the joint of

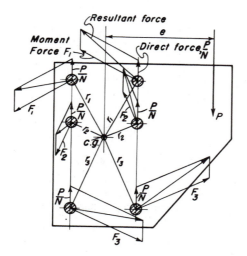

FIGURE 7-17   Riveted joint with eccentric load.

N rivets is subjected to a moment equal to Pe. Let it be assumed that the moment load on a rivet varies directly with the distance from the center of gravity of the group of rivets, and is directed perpendicular to the radius to the center of gravity. Equations for moment forces F_1, F_2, and F_3 in Fig. 7-17 may then be written

$$F_1 = Cr_1; \qquad F_2 = Cr_2; \qquad F_3 = Cr_3 \qquad \text{(a)}$$

where C is the constant of proportionality.

The externally applied moment is equal to the summation of the products of these forces and their arms to the center of gravity. Hence

$$T = Pe = N_1F_1r_1 + N_2F_2r_2 + N_3F_3r_3 + \cdots$$
$$= C(N_1r_1^2 + N_2r_2^2 + N_3r_3^3 + \cdots) \tag{10}$$

where N_1 is the number of rivets with radius r_1, N_2 is the number with radius r_2, and so on, until the entire joint has been taken care of. When the value of C has been determined from Eq. (10), the moment force for each rivet can be computed by multiplying by the appropriate r. It is customary to assume that the direct load P/N is the same for all rivets of the joint. The vectorial sum of moment force and direct force is the resultant force on the rivet.

Example 7. Find the value of the force carried by the most heavily loaded rivet for the joint of Fig. 7-18. Find the value of the shear stress for $\frac{3}{4}$-in. rivets and the value of the bearing stress if the plate is $\frac{5}{16}$ in. thick.

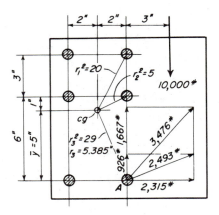

FIGURE 7-18   Example 7.

Solution. The plate is considered the free body.

Direct force on rivets:
$$F = \frac{10,000}{6} = 1,667 \text{ lb}$$

Take moments about the bottom row of rivets:
$$\hat{y} = \frac{2 \times 6 + 2 \times 9}{6} = 5 \text{ in.}$$

In Eq. (10):
$$10,000 \times 5 = C(2 \times 20 + 2 \times 5 + 2 \times 29)$$
$$C = 463$$

Rivet at A has the maximum force.

Moment force:
$$F_3 = 463 \times 5.385 = 2,493 \text{ lb}$$

This force has the horizontal and vertical components shown in Fig. 7-18.

Resultant force: $\qquad F_r = \sqrt{(926 + 1{,}667)^2 + 2{,}315^2} = 3{,}476$ lb

Area in shear: $\qquad A = \dfrac{\pi}{4} \times 0.75^2 = 0.4418$ in.^{2}

Shear stress in rivet: $\qquad \tau = \dfrac{3{,}476}{0.4418} = 7{,}870$ psi

Area in bearing: $\qquad A = \dfrac{5}{16} \times \dfrac{3}{4} = 0.2344$ in.^{2}

Bearing stress on rivet: $\qquad \sigma = \dfrac{3{,}476}{0.2344} = 14{,}830$ psi

Example 8. Find the value of P for the two joints shown in Fig. 7-19 based on a working stress of 15,000 psi in shear.

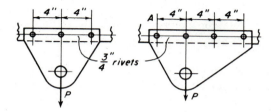

FIGURE 7-19 **Example 8.**

Solution. Area in shear $= \dfrac{\pi}{4}\left(\dfrac{3}{4}\right)^2 = 0.4418$ in.^{2}, each rivet.

Joint (a) $\qquad P = 3 \times 0.4418 \times 15{,}000 = 19{,}880$ lb

Joint (b) Moment arm of the load is 2 in.

In Eq. (10): $\qquad 2P = C(2 \times 2^2 + 2 \times 6^2), \qquad C = 0.025P$

On the rivet at A: moment force $= 6 \times 0.025P = 0.15P$

$$\text{direct force} = \frac{P}{4} = 0.25P$$

$$\text{total force} = 0.40P$$

$$\text{stress} = \frac{0.4P}{0.4418} = 15{,}000$$

$$P = 16{,}570 \text{ lb}$$

Thus, although joint (b) has more rivets and is more expensive to make, the carrying capacity is less than the symmetrical joint (a).

Example 8A. A gusset plate is supported by a horizontal row of 4 rivets 20 mm diameter and spaced 75 mm apart. Find the value of the downward load P located 100 mm to the right of the last rivet for a shearing stress of 70 MPa in the most heavily loaded rivet.

Solution. Consider the gusset as the free body.

Area of 1 rivet:
$$A = \frac{\pi}{4} d^2 = \frac{\pi}{4} \times 20^2 = 314.2 \text{ mm}^2$$

Force carried by most heavily loaded rivet $= 314.2 \times 70 = 21,990$ N

$$e = 37.5 + 75 + 100 = 212.5 \text{ mm}$$

Torque from load:
$$T = Pe = 212.5P$$

By Eq. (10):
$$212.5P = C(2 \times 112.5^2 + 2 \times 37.5^2) = 28,125C$$
$$C = 0.007\ 556P$$

moment force on last rivet on right $= 112.5 \times 0.007\ 556P = 0.850P$

direct force on this rivet $= 0.25P$

total force on rivet $= (0.25 + 0.85)P = 1.10P$

Then:
$$1.10P = 21,990 \quad \text{or} \quad P = 19,990 \text{ N}$$

When the load in a bolted connection is carried by shear stress, the calculations should be made either for the full body of the bolt or for the stress area of the threads, depending on which area is carrying the load.

17. Instantaneous Center of Rotation

As demonstrated in the preceding section, superposition indicates that the forces on a rivet are those due to the direct load plus the force resulting from the moment of the applied load. The total force is the geometric sum of these forces. The forces can be resolved into vertical and horizontal components which can also be used if desired when calculating moments.

The data for a simple joint of two rivets carrying an eccentric load are shown in Fig. 7-20. The direct force on each rivet is 500 lb. Calculations of the moment force of 450 lb on each rivet are shown on the figure. These have been resolved into vertical and horizontal components and combined with the direct forces, giving resultant forces F_1 and F_2.

By a principle of mechanics, when a body is in equilibrium under a set of forces, the moments will be in balance about every point in the plane. Let M be such an arbitrarily selected point. Calculations on the figure indicate that the moments are in equilibrium about point M.

The plate of Fig. 7-20, with the forces, is reproduced in Fig. 7-21. The direct force of 1,000 lb causes a small downward movement of the plate

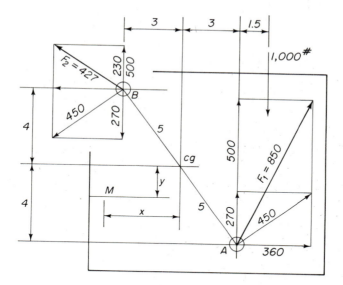

$$T = C(r_1^2 + r_2^2)$$
$$1000 \times 4.5 = C(5^2 + 5^2) \quad C = 90^{\#}/in.$$
$$\text{Moment force} = 5 \times 90 = 450^{\#}$$

Moments about cg :

$$1,000 \times 4.5 + 500 \times 3 = 500 \times 3 + 450 \times 5 + 450 \times 5$$

Moments about M :

$$1,000 (4.5 + x) = 770 (3 + x) + 360 (4 - y) + 230 (x - 3) + 360 (4 + y)$$
$$4,500 + 1,000x = 4,500 + 1,000x$$

FIGURE 7-20   Flat plate with eccentric load.

against the elastic resistance. The 4,500 in. lb moment causes a small clockwise rotation about the center of gravity of the rivets. Movement of points of the plate will be perpendicular to the radius from the center of gravity and proportional to the distance from the center of gravity. Thus all points lying on the horizontal radius to the left of the center of gravity will have an upward movement. There is one point O where upward movement from the moment is exactly equal to the downward movement from the direct load. This point, which is stationary, is called the "instantaenous center of rotation."

Let the two forces F_1 and F_2 be considered as producing moments about O. The center for each force will lie somewhere along the perpendicular to the force as shown in the figure. To lie on both radii, the point must be at their intersection O, which also lies on the horizontal from the center of gravity. Calculations for the moments are shown on the figure. Lengths required are easily found by simple trigonometry.

Moments about O:

$1,000 \times 10.055 = 850 \times 9.444 + 427 \times 4.747$

$10,055 = 8,028 + 2,027$

FIGURE 7-21   Additional details for the plate of Fig. 7-20.

18. Adhesives

A design technique of rapidly growing importance is metal-to-metal adhesive bonding. High-strength phenolic elastomers, vinyl-phenolics, and epoxy-resin adhesives are being used to an increasing extent for the structural joining of metals. Organic materials can be bonded as well. Adhesive bonding is in wide use in the automotive and aircraft industries.

The important matter of selecting the right adhesive for each type of service is not a simple procedure. There is a great variety of adhesive products, and there is little standardization of their properties. Adhesives seldom display their best properties when substituted directly for other fastenings. In general it is necessary to design the joint specifically for the use of adhesives. It is possible under certain conditions to form a successful joint by the use of adhesives that would not be possible by mechanical means.

The adhesive joint has the advantage of a larger contact area than other types of joints. Thick or thin materials of any shape can be joined. Any combination of similar or dissimilar materials can be employed. The adhesive layer minimizes or prevents electrochemical corrosion between dissimilar metals. Such joints in general show good resistance to fatigue or cyclic loading and aid in dampening vibration. The adhesive can sometimes be arranged to provide sealing for fluids and gases. The warping and residual stress that accompany welding are avoided.

The surfaces to be joined must be carefully cleaned and sometimes given special treatments. Heat and pressure may be required for curing and sometimes the curing time is a disadvantage. Some adhesives will harden only in the absence of air. Jigs and fixtures may also be required, and rigid process control may be necessary. Some adhesives may be subject to attack by bacteria, moisture, or certain solvents. The upper service temperature is limited to about 350°F. Some adhesives lose strength faster than others with increase of temperature. As a consequence of the differential strain between the parts, a severe stress concentration occurs at the ends of the overlap in an ordinary lap joint.

In mechanical bonding the adhesive infiltrates porous materials to give cementing of a mechanical nature. In chemical bonding the adhesive reacts with the parts or joins them largely through intermolecular attraction. For creep resistance the adhesive material should be rigid, but when the joint undergoes flexure in service, a more flexible material should be used.

Adhesive bonding is a rapidly growing field with excellent possibilities for future development. Simpler, cheaper, stronger, and more easy-to-use adhesives can be expected in the future.[10]

REFERENCES

1. FISH, G. D., *Arc Welded Steel Frame Structures*, New York: McGraw-Hill Book Company, 1941.

2. KOOPMAN, K. H., "Elements of Joint Design for Welding," *Welding J.*, **37**, 1958, p. 579.

3. *Metals Handbook*, 8th ed., Vol. 6, Metals Park, Ohio: American Society for Metals, 1971.

4. *Procedure Handbook of Arc Welding—Design and Practice*, Cleveland: The Lincoln Electric Co., 1957.

5. ROSSI, B. E., *Welding and Its Applications*, New York: McGraw-Hill Book Company, 1941.

6. ROTHBART, H. A., Ed., *Mechanical Design and Systems Handbook*, New York: McGraw-Hill Book Company, 1964, Section 25.

[10]See Goland, M., and E. Reissner, "The Stresses in Cemented Joints," *Trans. ASME*, **66**, 1944, p. A-17. See also Sharpe, L. H., "Assembling with Adhesives," *Machine Design*, **38**, Aug. 18, 1966, p. 178; Scott, R. C., "Guide to Joint Sealants," *Prod. Eng.*, **35**, Mar. 16, 1964, p. 73; Eickner, H. W., "Basic Shear Strength Properties of Metal-Bonding Adhesives . . . ," *Bull. 1850*, Madison, Wis.: U.S. Dept. of Agriculture, Forest Products Laboratory, Aug. 1955; and Kuenzi, E. W., "Determination of Mechanical Properties of Adhesives for Use in the Design of Bonded Joints." *Bull. 1851*, Madison, Wis.; U.S. Dept. of Agriculture, Forest Products Laboratory, Jan. 1956.

7. SKEIST, I., Ed., *Handbook of Adhesives*, New York: Van Nostrand Reinhold Company, 1962.

8. *Symposium on the Welding of Iron and Steel*, Vols. 1 and 2, London: The Iron and Steel Institute, 1935.

9. UDIN, H., E. R. FUNK, and J. WULFF, *Welding for Engineers*, New York: John Wiley & Sons, Inc., 1954.

10. *Welding Handbook*, 4th ed., New York: American Welding Society, 1960.

PROBLEMS

1. Find the value of the maximum stress in the weld of Fig. 7-22. Assume the direct stress to be uniformly distributed over the throat area.

Ans. $\tau = 11,500$ psi.

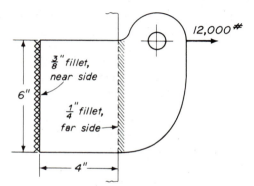

FIGURE 7-22   Problem 1.

2. Find the permissible static load P in Fig. 7-23 if an E6010 electrode is used and the factor of safety by the maximum shear theory of failure is to be equal to 2.

Ans. $P = 341,000$ N.

FIGURE 7-23   Problem 2.

3. Work Problem 2 for the joint of Fig. 7-24.   *Ans.* $P = 29,600$ lb.

FIGURE 7-24 Problem 3.   FIGURE 7-25 Problem 4.

4. Work Problem 2 for the joint of Fig. 7-25.   *Ans.* $P = 49,700$ lb.

5. The ends of the channel of Fig. 7-26 lie on immovable supports. If all welds are of the same size, find the length of leg h of the fillet welds required to carry the given load. Use E6010 electrode. Factor of safety is equal to 3. The beam is to be fixed at the ends by the welds.   *Ans.* $h = 0.38$ in.

FIGURE 7-26 Problem 5.

6. The I-beam in Fig. 7-27 is 300 mm deep with flange 162 mm wide and of a uniform thickness of 9 mm. The beam is welded to immovable supports by butt welds. Assume the bending moment is carried by the flange welds, and the shear forces are carried by the web welds. Flange welds are of the same size as the flange of the beam. Find the value of force P for a bending stress of 80 MPa in the flange welds.   *Ans.* $P = 109,800$ N.

FIGURE 7-27 Problem 6.

7. Find the required length of weld *l* in Fig. 7-28 if an E6010 electrode is used with an F_s of 2. *Ans. l = 263 mm.*

FIGURE 7-28   Problem 7.

8. Find the required length of weld *l* in Fig. 7-29 if an E6010 electrode is used with an F_s of 2. Let the endurance limit for the weld metal be equivalent to that of an as-forged surface. Stress concentration factor is equal to 1.5.

Ans. l = 18.9 in.

FIGURE 7-29   Problem 8.   **FIGURE 7-30   Problem 9.**

9. Find the value of P_{max} for the joint of Fig. 7-30 if P_{min} is equal to zero. Other data is the same as for Problem 8. *Ans. P_{max} = 7,460 lb.*

10. Find the factor of safety for the most highly stressed point of the welds of Fig. 7-31. The weld metal has a yield strength of 50,000 psi. *Ans. F_s = 2.95.*

FIGURE 7-31   Problem 10.

11. Find the value of the maximum stress in the joint of Fig. 7-32. The weld extends across the top and along the vertical edge. *Ans.* $\sigma = 74.8$ MPa.

FIGURE 7-32   Problem 11.

12. Derive an equation for the value of the maximum stress in the welds of Fig. 7-33. Let A be the area of one weld.

$$Ans. \quad \sigma = \frac{P}{2A}\sqrt{K^2 + (1 + aK)^2}, \text{ where } K = \frac{3(2e - a)}{1 + 3a^2}$$

FIGURE 7-33   Problem 12.   **FIGURE 7-34   Problem 13.**

13. Same as Problem 12 but for the joint of Fig. 7-34.

$$Ans. \quad \sigma = \frac{P}{2A}\sqrt{a^2 K_1^2 + (1 + K_1)^2}, \text{ where } K_1 = \frac{3(2e - 1)}{1 + 3a^2}$$

14. Compute and tabulate the working forces for a pitch length of 125 mm for the double-riveted butt joint of Fig. 7-35 for the following assumed methods of failure. Also find the efficiency of the joint. (a) Tension at section A-A. (b) Shear in the rivets. (c) Bearing on the rivets. (d) Tension at B-B and shear at A-A. (e) Tension at section B-B and bearing at A-A. (f) Tension in gross plate. Ultimate strength of steel plate in tension is 380 MPa, and in compression, 655 MPa. Ultimate shear strength for rivets is 300 MPa. Use a factor of safety of 5.

$\qquad$ *Ans.* Tension at A-A, 84,440 N; eff = 80.8 percent.

FIGURE 7-35   Problem 14.

15. Make computations similar to those of Example 6, but for the triple-riveted butt joint in Fig. 7-36. Tank diam = 60 in. Use stresses from Example 6.

$\qquad$ *Ans.* Tension at B-B and shear at A-A, 70,860 lb; eff = 84.1 percent.

FIGURE 7-36   Problem 15.   **FIGURE 7-37   Problem 16.**

16. Find the resultant force on each rivet of the joint in Fig. 7-37.

17. Repeat Problem 16 for the joint of Fig. 7-38.

FIGURE 7-38   Problem 17.

FIGURE 7-39   Problem 18.

18. Find the permissible load *P* for the riveted joint in Fig. 7-39 if the resultant shearing stress for the most highly stressed rivets is 100 MPa. Rivets are 25 mm in diameter.   *Ans.*   *P* = 170,270 N.

19. Find the shearing stress in the most heavily loaded rivet in the group shown in Fig. 7-40. Rivets are $\frac{3}{4}$ in. in diameter.   *Ans.* τ = 12,680 psi.

FIGURE 7-40   Problem 19.

FIGURE 7-41   Problem 20.

20. Find the permissible load *P* for the joint in Fig. 7-41 if the maximum value of the shearing stress is 100 MPa. Rivets are 20 mm in diameter.

Ans.   *P* = 145,460 N.

21. Find the distance *b* in Fig. 7-42 if the maximum shearing stress on the most heavily loaded rivet is to be 13,500 psi.   *Ans.*   *b* = 5.04 in.

FIGURE 7-42   Problem 21.

FIGURE 7-43   Problem 22.

22. Repeat Problem 21 for Fig. 7-43 for a maximum shearing stress of 93 MPa.
                                                        Ans.   *b* = 152.3 mm.

23. Find the value of the force on the most heavily loaded rivet in Fig. 7-44.

Ans.   1,270 lb.

FIGURE 7-44   Problem 23.   FIGURE 7-45   Problem 24.

24. Find the value of the force on the most heavily loaded rivet in Fig. 7-45.

Ans.   20,400N.

25. Find the value of the force on the most heavily loaded rivet in Fig. 7-46.

Ans.   4,300 lb.

FIGURE 7-46   Problem 25.   FIGURE 7-47   Problem 26.

26. Derive an equation that will give the value of the maximum force on the right-hand rivets in Fig. 7-47

Ans. $F = \dfrac{P}{4}\sqrt{\dfrac{(e + p_1)^2 + p_2^2}{p_1^2 + p_2^2}}.$

27. The lap joint of Fig. 7-48 is made of carbon steel plates and rivets. Find the efficiency of the joint based on shear in the rivets on *A-A* and tension in the plate on *B-B*. Use stress values from Example 6.

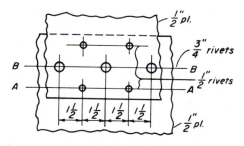

FIGURE 7-48   Problem 27.

28. Find the resultant force on the most heavily loaded rivet in Fig. 7-49.

Ans.   20,220 N.

FIGURE 7-49   Problem 28.

29. Find the value of *P* in Fig. 7-50 if the shearing stress on the most heavily loaded rivet is 12,000 psi.

FIGURE 7-50   Problem 29.   **FIGURE 7-51   Problem 30.**

30. Compute the value of the shearing stress at the center of rivets *A* and *B* in Fig. 7-51.

Ans.   At $A, \tau = 1,382$ psi.
At $B, \tau = 1,568$ psi.

31. Take the data of Example 7 and calculate and draw to scale the resultant force at each rivet. Draw a radius vector from each rivet perpendicular to the resultant force and note that they all intersect at a point called the instantaneous center or pole. Also note that each resultant force is proportional to its distance from the pole.

32. In Fig. 7-21 let the width between the two rivets become 8 in., and let the vertical height become 6 in. Find the location for the instantaneous center of rotation.   *Ans.*   4.546 left of center.

33. Consider the force system on the rivets of Example 7. Draw a perpendicular from rivet *A* and note where it intersects the horizontal through the center of gravity. Now find the resultant force on the rivet next above *A*. Draw the perpendicular to this force and note that it intersects the point as determined above for rivet *A*.

Ans.   3.600 in. to left of center of gravity.

34. Consider a welded joint similar to Fig. 7-31. Let the vertical lengths of the welds be 6 in. on the left and 8 in. on the right. Fillet welds have $\frac{3}{8}$-in. legs. Other dimensions unchanged. Find the value of the maximum stress.

Ans. $\tau = 5{,}256$ psi.

35. Consider a welded joint similar to Fig. 7-31. Let the vertical lengths of the welds be 5 in. on the left and 7 in. on the right. Fillet welds have $\frac{5}{16}$-in. legs. Other dimensions unchanged. Find the value of the maximum stress.

Ans. $\tau = 7{,}260$ psi.

36. Consider the welded joint of Fig. 7-32. Let vertical and horizontal fillet welds be $\frac{1}{4}$ in. Load is 5,000 lb. Horizontal dimensions are 6 in. and 5 in. Vertical dimension is 6 in. Find the value of the maximum stress.   *Ans.* $\tau = 10{,}680$ psi.

37. Consider a welded joint similar to Fig. 7-32. Let the vertical and horizontal fillet welds be $\frac{5}{16}$ in. Let the horizontal dimensions be 7 in. and 6 in. and the vertical dimension be 7 in. Load is 5,000 lb. Find the value of the maximum stress.

Ans. $\tau = 7{,}460$ psi.

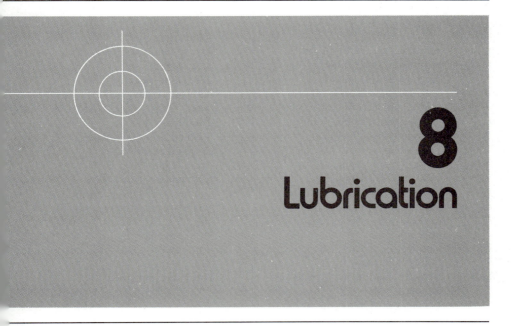

8

Lubrication

A lubricant is used to reduce the friction of bearings and sliding surfaces in machines and thus diminish the wear, heat, and possibility of seizure of the parts. Although a layer of oil will eliminate the excessive friction of metal-to-metal contact, the friction within the oil film must be taken into account. The study of lubrication and the design of bearings is therefore concerned mainly with phenomena related to the oil film between the moving parts. The literature on lubrication is very extensive, and a mathematical treatment of the subject is beyond the scope of this book. Fortunately it is possible to make design calculations from the graphs obtained by mathematical analysis.

A,	area		unit axial length
A_c,	cooling area of bearing housing	h,	film thickness
		h_0,	minimum film thickness
c,	radial clearance of journal bearing	hp,	horsepower, 0.7457 kW
		in.,	inch, 25.4 mm
C,	celsius, 1 deg C $= \frac{9}{5}$ deg Fahrenheit	J,	joule or newton-meter, Nm
		k,	kilo (prefix), 1,000
d,	diameter of journal bearing	kg,	kilogram, 2.2046 lb (mass);
e,	eccentricity of journal bearing		1 lb (mass) = 0.4536 kg
		kW,	kilowatt, 1.341 horsepower
f,	coefficient of friction	l,	length in axial direction
fpm,	feet per minute	lb,	pound;
F,	total friction force		1 lb (force) = 4.4482 N
F_1,	tangential friction force per		1 lb (mass) = 0.4536 kg

m,	meter, 39.370 in.	t,	temperature
m,	milli (prefix), one thousandth	ΔT_h,	rise in temperature of
mm,	millimeter, 0.039 37 in		bearing housing
M,	mega (prefix), 1 million	U,	tangential velocity
MPa,	megapascal, 1 N/mm²;	VI,	viscosity index
	1 MPa = 145 psi	W,	watt, 1 Nm/sec or 1 J/sec
n,	rpm, revolutions per minute	W,	total load
N,	newton, 0.2248 lb (force);	W_1,	load per unit axial length
	1 lb (force) = 4.448 newtons	Z,	viscosity, centipoises
Nmm,	newton millimeter;	$sp.gr_{60}$,	specific gravity at 60°F
	1 in. lb = 113 Nmm	$sp.gr_t$,	specific gravity at t°F
p,	oil pressure in film	γ,	(gamma) weight per cu in.
psi,	pounds per square inch	ϵ,	(epsilon) eccentricity ratio,
P,	load per unit of projected		e/c
	area	μ,	(mu) viscosity
Pa,	pascal, 1 N/m²	ρ,	(rho) density, mass per unit
r,	radius of journal		volume
SUS,	viscosity, Saybolt Universal	τ,	(tau) shearing stress
	Seconds		

1. Viscosity and Newton's Law

The plate in Fig. 8-1 is resting on top of an oil film of thickness h, and is being moved with velocity U under the action of the force F parallel to the stationary base. When the plate moves, it does not slide along on top of the film. The oil adheres tightly to the plate, and motion is accompanied by slip or shear between the oil particles throughout the entire height of the film. Thus, if the plate and contacting layer of oil move with velocity U, the velocity at intermediate heights is directly proportional to the distance from the fixed or bottom plate, as shown in Fig. 8-1.

FIGURE 8-1   Flat plate moving on oil film.

According to Newton, the shearing stress in the oil film varies directly with the velocity U and inversely with the film thickness h. Hence

$$\tau = \mu \frac{U}{h} = \frac{F}{A} \tag{1}$$

The factor of proportionality μ is called the coefficient of viscosity, or simply the viscosity. This equation assumes that the area A is very large as compared to h, in order that disturbances around the edges may be neglected.

Viscosity is the measure of the ability of the lubricant to resist shearing stress. It is molecular phenomenon, and the work done by force F is turned into heat, which raises the temperature of the oil and the surrounding parts.

By Eq. (1) the dimensions of μ in English units are

$$\mu = \frac{F(\text{lb})\,h(\text{in.})}{A(\text{in.}^2)U\left(\frac{\text{in.}}{\text{sec}}\right)} = \frac{FH}{AU}\,\frac{\text{lb sec}}{\text{in.}^2} \tag{2}$$

In SI units the equation becomes

$$\mu = \frac{Fh}{AU}\,\frac{\text{Nsec}}{\text{mm}^2} \tag{3}$$

The conversion from English to SI units can be made as follows.

$$1\,\frac{\text{lb sec}}{\text{in.}^2} = \frac{(4.448\ \text{N})(\text{sec})}{(25.4\ \text{mm})^2}$$

$$1\,\frac{\text{lb sec}}{\text{in.}^2} = \frac{1}{145}\,\frac{\text{Nsec}}{\text{mm}^2} \tag{4}$$

or

$$1\,\frac{\text{Nsec}}{\text{mm}^2} = 145\,\frac{\text{lb sec}}{\text{in.}^2} \tag{4a}$$

Numerically, the viscosity of lubricating oils is a small number. Average viscosities for a number of typical automobile oils are shown in Fig. 8-2. There are in units of lb sec/in.^{2}. For SI values, data from these curves can be divided by 145 to give Nsec/mm^2.

2. Petroff's Bearing Equation

Equation (1) for the flat plate can be easily adapted to the cylindrical or journal bearing, provided that the speed and viscosity are high and that the load is very light so that the journal is in a central position in the bearing. Thus the plate in Fig. 8-1 is assumed to be wrapped into the cylindrical shaft in Fig. 8-3. If $2r$ or d is the diameter of the journal, and l its length in the axial direction, the developed journal area A is $2\pi rl$.

The thickness h becomes the radial clearance c, or the difference between the radius of the bearing and the radius of the shaft. Substitution for A and h in Eq. (1) gives $F = 2\pi\mu Url/c$. However, if F_1 is taken as the tangential friction force per unit of axial length, then $F = F_1 l$, and the foregoing equation becomes

$$\frac{F_1}{\mu U}\left(\frac{c}{r}\right) = 2\pi \tag{5}$$

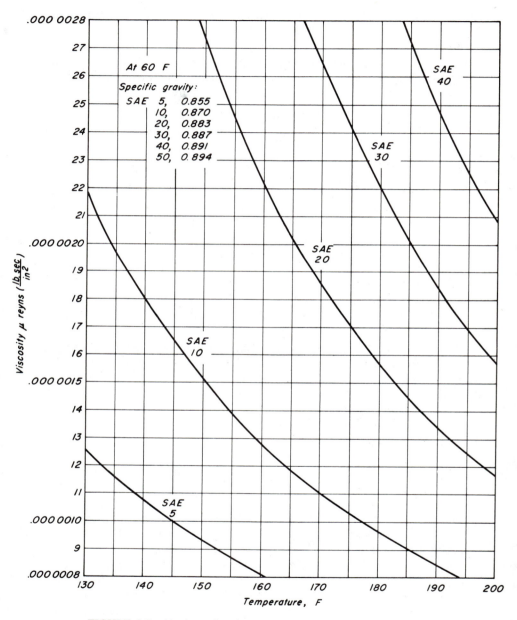

FIGURE 8-2 Absolute viscosity vs. temperature for the SAE oils of
Fig. 8-13.

FIGURE 8-3 **Very lightly loaded journal centered in bearing.**

This is known as Petroff's equation. It is valid only for the hypothetical case of zero load and centrally located journal.

The tangential velocity of the journal is given by

$$U = \frac{\pi \, dn}{60} \text{ units per sec} \tag{6}$$

where n represents the revolutions per minute.

$$\text{hp} = \frac{\text{ft lb/min}}{33,000} = \frac{\text{in. lb /min}}{33,000 \times 12} = \frac{\text{in. lb/sec}}{33,000 \times 12/60} = \frac{\text{in. lb/sec}}{6,600}$$

The power absorbed by the bearing is equal to the tangential force $F_1 l$ times the tangential velocity U. The horsepower of the bearing then is

$$\text{hp} = \frac{F_1 lU}{6,600} \tag{7}$$

The friction torque T is $F_1 lr$ or $F_1 l = T/r$. Let this, together with Eq. (6) for U, be substituted into Eq. (7).

$$\text{hp} = \frac{\dfrac{T}{r} \times \dfrac{2\pi rn}{60}}{6,600} \qquad \frac{Tn}{63,025} \qquad (T \text{ in in. lb}) \tag{8}$$

Example 1. Find the friction horsepower by Eqs. (7) and (8) for a very lightly loaded 360° journal bearing 4 in. in diameter and 6 in. long. Radical clearance c is 0.002 in., speed is 900 rpm. SAE 10 oil is used at 175°F.

Solution. By Fig. 8-2: $\quad \mu = 0.000\ 001\ 03 \text{ lb sec/in.}^2$

By Eq. (6):
$$U = \frac{\pi dn}{60} = \frac{\pi 4 \times 900}{60} = 188.5 \text{ in./sec}$$

By Eq. (5):
$$F_1 = \frac{2\pi \mu Ur}{c} = \frac{2\pi 0.000\ 001\ 03 \times 188.5 \times 2}{0.002}$$

$$= 1.22 \text{ lb/in.}$$

By Eq. (7):

$$hp = \frac{F_1 lU}{6,600} = \frac{1.22 \times 6 \times 188.5}{6,600} = 0.209$$

Also:
Torque:

$$T = F_1 lr = 1.22 \times 6 \times 2 = 14.64 \text{ in. lb}$$

By Eq. (8):

$$hp = \frac{Tn}{63,025} = \frac{14.64 \times 900}{63,025} = 0.209$$

In SI units F_1 is the force in newtons per axial mm. Total friction force for the bearing is $F_1 l$. Tangential velocity U has dimensions of mm/sec. The power loss is $F_1 lU$ in Nmm/sec or $F_1 lU/1,000$ in Nm/sec or watts. The power loss in kilowatts then is

$$kW = \frac{F_1 lU}{1,000,000} \tag{9}$$

The friction torque T, Nmm, is $F_1 lr$ or $F_1 l = T/r$. Let this, together with Eq. (6) for U, be substituted into Eq. (9).

$$kW = \frac{\dfrac{T}{r} \times \dfrac{2\pi rn}{60}}{1,000,000} = \frac{Tn}{9,550,000} \qquad (T \text{ in Nmm}) \tag{10}$$

Example 1A. Find the friction power loss for a Petroff bearing 100 mm in diameter and 150 mm long. Radial clearance c is 0.05 mm. Speed is 900 rpm. SAE oil 10 is used at 175°F.

Solution. By Fig. 8-2: $\mu = 0.000\ 001\ 03$ lb sec/in.^{2}

By Eq. (4):

$$\text{Viscosity} = \frac{0.000\ 001\ 03}{145} = 0.000\ 000\ 0071 \text{ Nsec/mm}^2$$

By Eq. (6):

$$U = \frac{\pi dn}{60} = \frac{\pi 100 \times 900}{60} = 4,712 \text{ mm/sec}$$

By Eq. (5):

$$F_1 = \frac{2\pi \mu Ur}{c} = \frac{2\pi 0.000\ 000\ 0071 \times 4,712 \times 50}{0.05}$$

$$= 0.21 \text{ N/mm}$$

By Eq. (9):

$$kW = \frac{F_1 lU}{1,000,000} = \frac{0.21 \times 150 \times 4,712}{1,000,000} = 0.148$$

Also:
Torque:

$$T = F_1 lr = 0.21 \times 150 \times 50 = 1,575 \text{ Nmm}$$

By Eq. (10):

$$kW = \frac{Tn}{9,550,000} = \frac{1,575 \times 900}{9,550,000} = 0.148$$

3. Load-Carrying Journal Bearing

The flat plate in Fig. 8-1 is incapable of supporting a vertical load because no pressure exists in the oil film. If a vertical load were placed on the plate, the oil would squeeze out around the edges, and the plate would approach the surface beneath, resulting in metal-to-metal contact. If, however, the plates are tipped at a small angle to each other, as shown in Fig. 8-4, the oil will be drawn into the wedge-shaped opening, and sufficient pressure will be generated in the film to support the load *W*. This fact is utilized in the thrust bearings for hydraulic turbines and propeller shafts of ships, as well as in the conventional journal bearing with flooded-oil lubrication.

FIGURE 8-4 Load on moving plate supported on wedge-shaped film.

Since the same quantity of oil is drawn past each cross section, the velocity distribution throughout the height of the film is no longer linear as it is for parallel plates. The velocity curve will be concave at the wider cross sections and convex at the narrower, as indicated in Fig. 8-4.

If the shaft in Fig. 8-3 carried a downward load, the journal would no longer remain central with the bearing. For the classical case of the infinitely long bearing, the shift of the journal center is not downward, but is theoretically to one side as shown in Fig. 8-5. The shift of the journal center is usually expressed as a proportion of the radial clearance, *c*. Thus in Fig. 8-5 the shaft has shifted ϵc inches, where the eccentricity ratio ϵ is a pure number less than unity. A copious supply of oil is assumed, which fills the entire clearance space. The oil clings to the journal and is drawn around into the region of decreasing film thickness, and thereby builds up sufficient pressure to support the load *W*. The bearing automatically adjusts itself to the imposed load. Thus, should *W* be increased, the shift to the left would become greater, and the resulting decrease of the film thickness would cause higher oil pressures, which would take care of the larger load.

FIGURE 8-5 Oil-film pressure in full journal bearing.

4. Load and Friction Curves for Journal Bearings

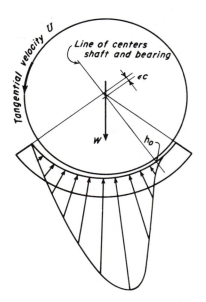

FIGURE 8-6 **Pressure distribution in partial bearing.**

In actual bearings, the full continuous film of Fig. 8-5 does not exist. The film ruptures and the load is carried by a partial film located beneath the journal. Even in partial bearings, as in Fig. 8-6, the ruptured film leaves a portion of the arc at the trailing edge with zero pressure.

A computer solution for journal bearings for different bearing arcs and length-diameter ratios has been made[1] and is given by the curves in Figs. 8-7, 8-8, and 8-9. The ordinates are the non-dimensional load variable $(W_1/\mu U)(c/r)^2$ and friction variable $(F_1/\mu U)(c/r)$, where W_1 is the load per unit axial length of bearing, and F_1 is the tangential friction force per unit axial length. The abscissa is taken as the eccentricity ratio ϵ.

It is seen from Fig. 8-6 that the minimum film thickness h_0 is

$$h_0 = c(1 - \epsilon) \tag{11}$$

Example 2. A 4-in.-diam $\times$ 6-in.-long 120° central partial bearing has a minimum film thickness of 0.001 in. Radial clearance is 0.002 in. SAE 10 oil like that in Fig. 8-2 is used. The bearing carries a load of 82 psi of projected journal area at 900 rpm. Find the temperature of the film and the friction horsepower.

Solution. By Eq. (6): $U = \dfrac{2\pi rn}{60} = \dfrac{2\pi 2 \times 900}{60} = 188.5$ in./sec

By Eq. (11): $\epsilon = 1 - \dfrac{h_0}{c} = 1 - \dfrac{0.001}{0.002} = 0.5$

$$\frac{c}{r} = \frac{0.002}{2} = \frac{1}{1,000}$$

From Fig. 8-7(a): $\dfrac{W_1}{\mu U}\left(\dfrac{c}{r}\right)^2 = 1.69$

From Fig. 8-7(b): $\dfrac{F_1}{\mu U}\left(\dfrac{c}{r}\right) = 3.60$

[1]See Raimondi, A. A., and J. Boyd, *A Solution for the Finite Journal Bearing and Its Application to Analysis and Design*, Report III, **1**, No. 1, American Society of Lubrication Engineers, Elmsford, N.Y.: Pergamon Press, Inc., p. 194.

FIGURE 8-7 Load and friction characteristics of 120° central partial bearing.

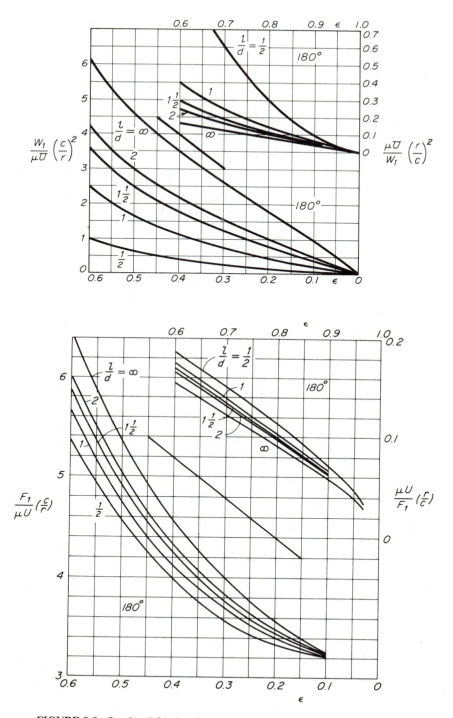

FIGURE 8-8 Load and friction characteristics of 180° central partial bearing.

FIGURE 8-9 Load and friction characteristics of 360° journal bearing.

$$\mu = \frac{W_1}{1.69U}\left(\frac{c}{r}\right)^2 = \frac{82 \times 4}{1.69 \times 188.5 \times 1,000^2}$$

$$= 0.000\ 001\ 028\ \text{lb sec/in.}^2$$

From Fig. 8-2, the oil film temperature is 175°F:

$$F_1 = 3.60\mu U\left(\frac{r}{c}\right)$$

$$= 3.60 \times 0.000\ 001\ 028 \times 188.5 \times 1,000$$

$$= 0.697\ \text{lb/in.}$$

By Eq. (7):   friction hp $= \dfrac{0.697 \times 6 \times 188.5}{6,600} = 0.120$

Other variables are in use for presenting lubrication data. Thus, $W_1 = Pd$, where P is the pressure, psi, on the projected area of the bearing, and $U = \pi dN'$, where N' is the speed of rotation, revolutions per second, rps. Then

$$\frac{W_1}{\mu U}\left(\frac{c}{r}\right)^2 = \frac{P}{\pi\mu N'}\left(\frac{c}{r}\right)^2 = \frac{1}{\pi S} \tag{12}$$

where

$$S = \frac{\mu N'}{P}\left(\frac{r}{c}\right)^2 \tag{13}$$

Here S is the so-called "Sommerfeld number."

For journal bearings, the coefficient of friction, f, is defined in the usual way as the ratio of the tangential force to the normal force. Then

$$f = \frac{F}{W} = \frac{F_1}{W_1} \tag{14}$$

Example 2A. A 100-mm-diameter × 150-mm-long 120° central partial bearing has a minimum film thickness of 0.025 mm. Radial clearance is 0.05 mm. SAE 10 oil like that in Fig. 8-2 is used. The bearing carries a load of 0.57 MPa of projected journal area at 900 rpm. Find the temperature of the film and the frictional power loss.

Solution. By Eq. (6): $U = \dfrac{2\pi rn}{60} = \dfrac{2\pi 50 \times 900}{60} = 4,712\ \text{mm/sec}$

By Eq. (11): $\epsilon = 1 - \dfrac{h_0}{c} = 1 - \dfrac{0.025}{0.050} = 0.5$

$$\frac{c}{r} = \frac{0.05}{50} = \frac{1}{1,000}$$

From Fig. 8-7(a): $\dfrac{W_1}{\mu U}\left(\dfrac{c}{r}\right)^2 = 1.69$

From Fig. 8-7(b): $\dfrac{F_1}{\mu U}\left(\dfrac{c}{r}\right) = 3.60$

Then: $\mu = \dfrac{W_1}{1.69U}\left(\dfrac{c}{r}\right)^2 = \dfrac{0.57 \times 100}{1.69 \times 4,712 \times 1,000^2}$

$$= 0.000\ 000\ 007\ 158\ \frac{\text{Nsec}}{\text{mm}^2}$$

The viscosity in U.S. customary units can be had by multiplying both sides of Eq. (4a) by 0.000 000 007 158.

$$0.000\ 000\ 007\ 158\ \frac{\text{Nsec}}{\text{mm}^2} = 0.000\ 000\ 007\ 158 \times 145$$

$$= 0.000\ 001\ 038\ \frac{\text{lb sec}}{\text{in.}^2}$$

By Fig. 8-2, oil film temperature is 174°F.

By Eq. (a):
$$F_1 = 3.60\mu U \left(\frac{r}{c}\right)$$

$$= 3.60 \times 0.000\ 000\ 007\ 158 \times 4{,}712 \times 1{,}000$$

$$= 0.1214\ \text{N/mm}$$

By Eq. (9):
$$\text{kW} = \frac{F_1 l U}{1{,}000{,}000}$$

$$= \frac{0.1214 \times 150 \times 4{,}712}{1{,}000{,}000} = 0.0858$$

5. Power Lost in Friction

The frictional loss of energy in a bearing is turned into heat, which raises the temperature of the lubricant and the adjacent parts.[2] The rate at which such heat is produced will now be determined. Let P be the load on the bearing per unit of projected journal area ld. Since the total load on the bearing is Pld, the tangential friction force is $fPld$, where f is the coefficient of friction. The friction moment is $\frac{1}{2}fPld^2$. The number of radians turned by the shaft per minute is $2\pi n$, where n is the revolutions per minute. The work of friction per minute, or power, H_1, then is $\frac{1}{2}fPld^2 \times 2\pi n$ or

$$H_1 = \pi f Pld^2 n \ \frac{\text{work}}{\text{min}} \text{ of friction} \tag{15}$$

In this equation let f be replaced by F_1/W_1 and P be replaced by W_1/d.

Then
$$H_1 = \pi F_1 l \ dn \ \frac{\text{work}}{\text{min}} \tag{15a}$$

Work, in Eqs. (15) and (15a), can be in units of in. lb or Nmm.

6. Power Lost by Self-Cooled Bearings

Under steady operating conditions, the heat is removed from a bearing at the same rate that it is produced. The heat, however, must be dissipated at a sufficiently rapid rate to maintain an operating temperature low enough

[2]For additional information on the thermal equilibrium of journal bearings, see p. 101 of Reference 9, p. 246 of Reference 6, and p. 210 of Reference 12, end of chapter.

so that no harm will come to the oil. Sometimes cooling coils are provided for large and important bearings. In forced-feed systems the excess heat is carried away by the lubricant, which is cooled before being returned to the bearing.

However, the majority of bearings in use depend upon the surrounding air as the cooling medium. Many bearings are constructed in the form of a self-contained unit with an oil reservoir in the base or pedestal,[3] as shown in Fig. 8-10 or 8-17(d). An oil ring or chain, rotating with the shaft, dips into the lubricant and carries a supply of oil to the top of the shaft, where it spreads over the surface.

FIGURE 8-10   Ring-oiled partial journal bearing.

For self-cooled bearings of moderate rises in temperature,[4] Newton's law of cooling is sufficiently accurate. This law states that the power lost is

$$H_2 = C_1 A_c \, \Delta T_h \frac{\text{work}}{\text{min}}$$

where C_1 is the cooling rate in appropriate dimensions, A_c is the area losing heat, and ΔT_h is the rise in temperature of the housing above the surrounding air. Area of cooling A_c can be taken as a constant C_2 times the developed journal area πdl. Then

$$H_2 = C_1 C_2 \pi \, dl \, \Delta T_h \frac{\text{work}}{\text{min}} \tag{16}$$

Under steady operating conditions, $H_1 = H_2$:

$$\pi f P l d^2 n = C_1 C_2 \pi dl \, \Delta T_h$$

or
$$fPdn = C_1 C_2 \, \Delta T_h \tag{17}$$

Let f be replaced by F_1/W_1 and P be replaced by W_1/d. Then

$$F_1 n = C_1 C_2 \, \Delta T_h \tag{18}$$

---

[3]See Chapter 5 of Reference 9, end of chapter.

[4]See Karelitz, G. B., "Heat Dissipation in Self-Contained Bearings," *Trans. ASME,* **64**, 1942, p. 463; **52** (1), 1930, p. APM 57. See also Lemmon, D. C., and E. R. Booser, "Bearing Oil-Ring Performance," *Trans. ASME,* **82**, 1960, *J. Basic Eng.,* p. 327.

When C_1 is given in terms of Btu/(hr ft² deg F), it must be transformed into units of minutes and inches. The Btu, or British thermal unit, represents the amount of energy required to raise the temperature of 1 pound of water 1 degree Fahrenheit. One Btu is equal to 778 ft lb or 778×12 in. lb. Then

$$1 \frac{\text{Btu}}{\text{hr ft}^2 \text{ deg F}} = \frac{778 \times 12 \text{ in. lb}}{(60 \text{ min})(12 \text{ in.})^2 \text{ deg F}} = 1.0805 \frac{\text{in. lb}}{\text{min in.}^2 \text{ deg F}} \qquad (19)$$

Example 3. Let the shaft diameter of a 120° central partial bearing be $3\frac{1}{2}$ in., the axial length $5\frac{1}{4}$ in., and the rpm 900. The load is 120 pounds times the projected journal area. The r/c ratio is 1,000, and the minimum film thickness 0.00075 in. The housing area is 8 times the developed journal area ($C_2 = 8$). Assume the rate of cooling C_1 is 2 Btu/(hr ft² deg F). Find the average rise in temperature of the housing above the surrounding air.

Solution. $c = \dfrac{r}{1,000} = 0.001 \ 75$ in. radial clearance.

By Eq. (11): eccentricity ratio, $\epsilon = 1 - \dfrac{h_0}{c} = 1 - \dfrac{0.000\ 75}{0.001\ 75} = 0.571$

By Fig. 8-7(a): $\dfrac{W_1}{\mu U} \left(\dfrac{c}{r} \right)^2 = 2.27$ \qquad\qquad (a)

By Fig. 8-7(b): $\dfrac{F_1}{\mu U} \left(\dfrac{c}{r} \right) = 4.13$ \qquad\qquad (b)

Divide Eq. (b) by Eq. (a): $f\left(\dfrac{c}{r} \right) = \dfrac{4.13}{2.27} = 1.82$ \qquad or \qquad $f = 0.001\ 82$

By Eq. (19): $C_1 = \dfrac{2 \text{ Btu}}{\text{hr ft}^2 \text{ deg F}} = \dfrac{2 \times 1.0805 \text{ lb}}{\text{min in. deg F}}$

By Eq. (17): $\Delta T_h = \dfrac{fP \, dn}{C_1 C_2} = \dfrac{0.001\ 82 \times 120 \left(\frac{\text{lb}}{\text{in.}^2}\right) 3.5 \text{ (in.) } 900 \left(\frac{\text{rev}}{\text{min}}\right)}{2.16 \left(\frac{\text{lb}}{\text{min in. deg F}}\right) 8}$

$$= 39.8 \text{ deg F}$$

If the average temperature rise of the housing is assumed to be one-half as much as the film, and if the surrounding air would be at 100°F, the film temperature would then be at about 180°F.

The maximum temperature at which the film should operate depends on the physical properties of the lubricant. For usual applications, an average film temperature in the range 160°–180°F is considered satisfactory.

The cooling rate in SI units is usually given in watts/(m² deg C). The relationship between this and U.S. customary units is found as follows.

$$1 \frac{\text{Btu}}{\text{hr ft}^2 \text{ deg F}} = \frac{778 \text{ ft lb}}{\text{hr ft}^2 \text{ deg F}} = \frac{(778)(0.3048 \text{ m})(4.448 \text{ N})}{(3,600 \text{ sec})(0.3048 \text{ m})^2(\frac{5}{9} \text{ deg C})}$$

$$= \frac{5.677 \text{ Nm}}{\text{sec m}^2 \text{ deg C}} = \frac{5.677 \text{ watts}}{\text{m}^2 \text{ deg C}} \qquad (20)$$

For use in solving problems, meters and seconds in the equation above should be replaced by minutes and millimeters as follows.

$$1 \frac{\text{watt}}{\text{m}^2 \text{ deg C}} = \frac{\text{Nm}}{\text{sec m}^2 \text{ deg C}} = \frac{\text{N}}{\text{sec m deg C}}$$

$$= \frac{\text{N}}{\left(\frac{\text{min}}{60}\right)(1,000 \text{ mm}) \text{ deg C}} = \frac{0.06 \text{ N}}{\text{min mm deg C}} \qquad (21)$$

Example 3A. A 120° central partial bearing has a diameter of 90 mm and axial length 135 mm. Speed is 900 rpm; load is 0.83 N/mm² of projected area; r/c = 1,000; h_0 = 0.019 mm. Cooling rate C_1 is 11.5 watts/(m² deg C); C_2 = 8. Find the expected rise in temperature of the housing.

Solution.

$$c = \frac{r}{1,000} = 0.045 \text{ mm radial clearance}$$

By Eq. (11):

$$\epsilon = 1 - \frac{h_0}{c} = 1 - \frac{0.019}{0.045} = 0.578$$

By Fig. 8-7(a):

$$\frac{W_1}{\mu U}\left(\frac{c}{r}\right)^2 = 2.27$$

By Fig. 8-7(b):

$$\frac{F_1}{\mu U}\left(\frac{c}{r}\right) = 4.13$$

Divide:

$$f\left(\frac{r}{c}\right) = \frac{4.13}{2.27} = 1.82 \quad \text{or} \quad f = 0.001\ 82$$

By Eq. (21):

$$C_1 = 11.5 \frac{\text{watts}}{\text{m}^2 \text{ deg C}} = \frac{11.5 \times 0.06 \text{ N}}{\text{min mm deg C}} = \frac{0.69 \text{ N}}{\text{min mm deg C}}$$

By Eq. (17):

$$\Delta T_h = \frac{fP\,dn}{C_1 C_2} = \frac{0.001\ 82 \times 0.83 \left(\frac{\text{N}}{\text{mm}^2}\right) 90 \text{ mm} \times 900 \frac{\text{rev}}{\text{min}}}{8 \times 0.69 \left(\frac{\text{N}}{\text{min mm deg C}}\right)}$$

$$= 22.2 \text{ deg C}$$

From this the expected film temperature can be calculated.

The calorie, the heat unit of the now obsolete centimeter-gram-second system, is the amount of energy required to raise 1 gram of water 1 degree Celsius. The Btu is the energy to raise 1 pound of water 1 degree Fahrenheit. A pound mass is equal to 0.4536 kg or 453.6 g.

$$1 \text{ Btu} = (1 \text{ lb}) \text{ raised } (1 \text{ deg F})$$

$$(456.3 \text{ g})(\tfrac{5}{9} \text{ deg C}) = 252 \text{ calories} \qquad (22)$$

$$1 \text{ calorie} = \frac{778 \times 12}{252} = 37.05 \text{ in. lb} \qquad (23)$$

Also:

$$1 \text{ Btu} = 778 \times 12 \text{ in. lb}$$

$$= 778 \times 12 \times 25.4 \text{ mm} \times 4.448 \text{ N}$$

$$= 1,055,000 \text{ Nmm} \qquad (23a)$$

7. Designing for Film Temperature and Minimum Film Thickness

The temperature of the oil film and the minimum film thickness at which the bearing operates are usually the important factors that the designer attempts to predict when he makes the computations for a new design. The curves of Figs. 8-7, 8-8, and 8-9 permit these quantities to be found, although not by direct calculation. The load and film thickness for a number of assumed values of the film temperature can be found and curves plotted. From such curves, the temperature and film thickness can be read for any given load that the bearing might be carrying. The procedure can be carried out conveniently in tabular form as shown by the following example.

Example 4. A 3-in.-diam × 3-in.-long 120° central partial oil ring bearing operates at 1,200 rpm with SAE 10 oil of Fig. 8-2. Room temperature is 80°F; assume that housing temperature rises one-half as much as the film temperature. Assume the cooling rate to be 2 Btu/(hr ft² deg F), and that the cooling area is 16 times the developed journal area. The r/c ratio is 1,000. Plot the curves for P vs. temperature and P vs. h_0 for the range 155°F–165°F at 2.5-deg intervals.

$$\text{Solution.} \qquad U = \frac{\pi \, dn}{60} = \frac{\pi 3 \times 1,200}{60} = 188.5 \text{ in./sec}$$

By Eq. (19), cooling rate: $C_1 = 2 \times 1.0805 \dfrac{\text{in. lb}}{\text{min in.}^2 \text{ deg F}}$

In Eq. (18):
$$F_1 = \frac{C_1 C_2 \, \Delta T_h}{n} = \frac{2 \times 1.0805 \times 16 \, \Delta T_h}{1,200}$$

$$= 0.0288 \, \Delta T_h$$

The following table should be constructed. The values in columns 2–5 can be found by computation. Column 6 is determined from Fig. 8-7(b), and column 7 from Fig. 8-7(a). The values in the remaining columns can then be calculated.

1	2	3	4	5	6	7	8	9	10
Film t_{av}	ΔT_h	$F_1 =$ $0.0288 \, \Delta T_h$	μ	$\dfrac{\mu U}{F_1}\left(\dfrac{r}{c}\right)$	ϵ	$\dfrac{\mu U}{W_1}\left(\dfrac{r}{c}\right)^2$	W_1	P	h_0
155	37.5	1.0805	0.000 001390	4.124*	0.585	1.85*	485	161.6	0.00062
157.5	38.75	1.1165	1332	0.225	0.622	0.460	546	182.0	0.00057
160	40	1.1525	1278	0.209	0.653	0.397	607	202.3	0.00052
162.5	41.25	1.1885	1228	0.195	0.680	0.347	667	222.4	0.00048
165	42.5	1.2246	1184	0.182	0.704	0.305	733	244.2	0.00044

*Indicates inverse of quantity at column heading.

The desired curves are plotted in Fig. 8-11.

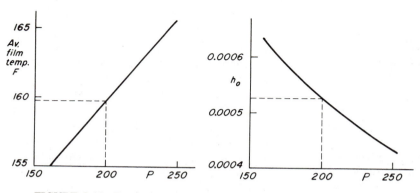

FIGURE 8-11 Example 4. Self-cooled ring-oiled bearing. Increase in load causes increase in film temperature and decrease in film thickness.

Now suppose the probable film temperature and minimum film thickness are desired when the bearing is carrying a load of 200 psi of projected journal area. Reference to Fig. 8-11 shows that the film temperature, for the assumed conditions, is 160°F, and the minimum film thickness is 0.000 53 in.

The permissible film thickness is largely dependent on the surface roughness of the parts, as discussed in Section 14 and Fig. 8-18.

8. Measurement of Viscosity

Other units for designating viscosity are in use. In the now-obsolete centimeter-gram-second system the unit is the poise, which has dimensions of dyne sec/cm². The dyne is the force that will accelerate a 1 gram mass at the rate of 1 centimeter per second squared. By Newton's second law of motion,

$$\text{force} = \text{mass} \times \text{acceleration}$$

$$1 \text{ dyne} = \text{gram} \times \frac{\text{cm}}{\text{sec}^2} = \left(\frac{\text{kg}}{1{,}000}\right)\frac{\left(\frac{\text{m}}{100}\right)}{\text{sec}^2}$$

$$= \frac{1}{100{,}000}\frac{\text{kg m}}{\text{sec}^2} = \frac{1}{100{,}000} \text{ N by Newton's second law} \quad (24)$$

The poise can be easily expressed in units of Nsec/mm² and lb sec/in.².

$$1 \text{ poise} = 1 \frac{\text{dyne sec}}{\text{cm}^2} = \left(\frac{\text{N}}{100{,}000}\right)\frac{\text{sec}}{(10 \text{ mm})^2}$$

$$= \frac{1}{10{,}000{,}000}\frac{\text{Nsec}}{\text{mm}^2} \quad (25)$$

or $\quad 1 \dfrac{\text{Nsec}}{\text{mm}^2} = 10{,}000{,}000 \text{ poises} = 1{,}000{,}000{,}000 \text{ centipoises} \quad (26)$

By Eq. (4):
$$1\,\frac{\text{lb sec}}{\text{in.}^2} = \frac{1}{145}\,\frac{\text{Nsec}}{\text{mm}^2} = \frac{10{,}000{,}000}{145}\,\text{poises}$$

The centipoise is $\frac{1}{100}$ of a poise. It is usually represented by Z.

$$1\,\frac{\text{lb sec}}{\text{in.}^2} = \frac{1{,}000{,}000{,}000}{145}Z = 6{,}900{,}000Z \tag{27}$$

or
$$1Z = 0.000\,000\,145\,\frac{\text{lb sec}}{\text{in.}^2} \tag{28}$$

Commercially, viscosity is usually specified by terms related to the instrument used for its measurement. One method in wide use is to specify viscosity in Saybolt Universal Seconds, SUS, at a given temperature. The Saybolt viscosity is the time in seconds required for 60 cm³ of the oil to flow through a calibrated capillary tube at a specified temperature F, Fig. 8-12.

FIGURE 8-12 Saybolt Universal viscosimeter.

The conversion from SUS seconds to centipoises can be made by the empirical equation

Centipoises:
$$Z = sp.gr._t\left(0.22\,\text{SUS} - \frac{180}{\text{SUS}}\right)\frac{\text{dyne sec}}{\text{cm}^2} \tag{29}$$

The specific gravity of the oil changes with the temperature, becoming smaller as the temperature rises. This property can be determined by means of the empirical equation

$$sp.gr_t = sp.gr_{60} - 0.000\ 35(t - 60) \tag{30}$$

where $sp.gr_{60}$ is the specific gravity at 60°F, and t is the temperature in °F.

The weight per cubic inch, γ, for oil is then the weight of water per cubic inch times the specific gravity. Taking water at 62.4 lb/ft³ the value of γ then is

$$\gamma = \frac{62.4}{1,728} \times sp.gr_t \tag{31}$$

Example 5. Find the weight per cubic inch, γ, for SAE20 oil at 167.5°F.

Solution. By Fig. 8-13: $sp.gr_{60} = 0.883$

By Eq. (30): $\quad sp.gr_{167.5} = 0.883 - 0.00035(167.5 - 60)$

$$= 0.883 - 0.038 = 0.845$$

By Eq. (31): $\quad \gamma = \dfrac{62.4}{1728} \times 0.845 = 0.0305 \text{ lb/in.}^3$

Viscosity is only one of the many properties of a satisfactory lubricating oil. It is, however, the only variable directly related to the lubricant that appears in the design equations. The SAE classification is widely used for specifying the viscosity of lubricating oils. The viscosity for any given designation can vary over a considerable range. To make calculations, precise information is needed concerning the viscosity of the oil that is to be used. Fair average values for the viscosities of SAE oils are shown by the curves of Fig. 8-13 plotted on ASTM Chart D341–39. It is unique that when the viscosities for two temperatures, usually 100°F and 210°F, are plotted on this chart, a straight line between them will give the viscosity at other desired temperatures. Since the designer requires the viscosity in absolute units, the viscosities of the oils of Fig. 8-13 are shown in lb sec/in.² units in Fig. 8-2. These curves are easily drawn by reading the SUS values from Fig. 8-13 and then by transforming to the absolute units of Fig. 8-2.

Example 6. An oil has a viscosity of 550 SUS at 100°F and a viscosity of 64 SUS at 210°F. Specific gravity at 60°F is 0.887. Find the viscosity of this oil in lb sec/in.² units at 180°F.

Solution. The given SUS values should be plotted on the ASTM chart D341–39. See Fig. 8-13. It is seen that the given oil is equivalent to SAE 30. At 180°F, viscosity, SUS at 180°F is equal to 91 sec from the chart.

By Eq. (30): $\quad sp.gr_{180} = sp.gr_{60} - 0.000\ 35(t - 60)$

$$= 0.887 - 0.000\ 35(180 - 60) = 0.845$$

By Eq. (29): $\quad Z = sp.gr_t \left(0.22\ \text{SUS} - \dfrac{180}{\text{SUS}} \right)$

$$= 0.845 \left(0.22 \times 91 - \dfrac{180}{91} \right) = 15.24 \text{ centipoises}$$

Let both sides of Eq. (28) be multiplied by 15.24.

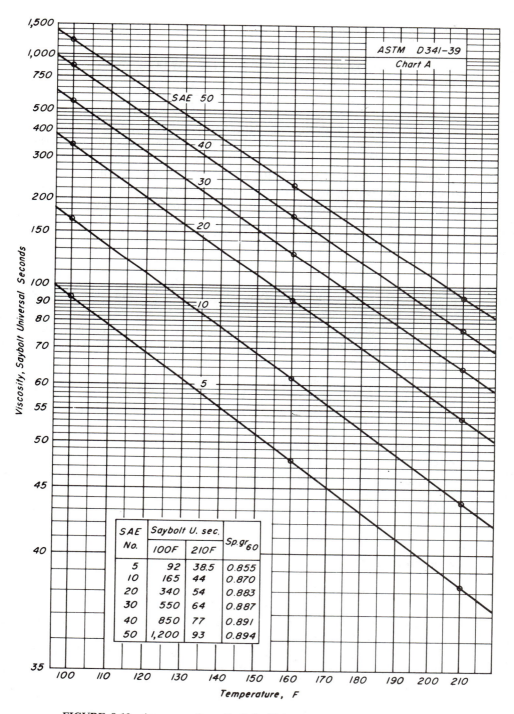

FIGURE 8-13 Average values, Saybolt Universal viscosities, seconds,
of some common SAE lubricating oils.

$$15.24 \text{ centipoises} = 15.24 \times 0.000\ 000\ 145\ \frac{\text{lb sec}}{\text{in.}^2}$$

$$= 0.000\ 002\ 21\ \frac{\text{lb sec}}{\text{in.}^2}$$

This value agrees with Fig. 8-2. Other values for SUS can be read for this oil on Fig. 8-13 and transformed into lb sec/in.² units and plotted to form the curve of Fig. 8-2.

Another metric unit in use is the kinematic viscosity, v, called the stoke and found by dividing the viscosity in poises by the density. It has the dimensions of cm²/sec, determined as follows. By Newton's second law,

$$\text{mass} = \frac{\text{force}}{\text{accel}} = \frac{\text{dyne sec}^2}{\text{cm}} \text{ grams}$$

$$\text{density} = \frac{\text{mass}}{\text{cm}^3} = \frac{\text{dyne sec}^2}{\text{cm}^4}$$

Then: $\text{stoke} = \dfrac{\text{poise}}{\text{density}} = \dfrac{\text{dyne sec}}{\text{cm}^2} \times \dfrac{\text{cm}^4}{\text{dyne sec}^2} = \dfrac{\text{cm}^2}{\text{sec}}$

In the metric system, the density of a substance has the same numerical value as its specific gravity. When Eq. (29) is divided by density, or specific gravity, the result will be centistokes represented by v.

Centistokes: $v = \dfrac{Z, \text{ centipoises}}{\text{density}} = \left(0.22 \text{ SUS} - \dfrac{180}{\text{SUS}}\right) \dfrac{\text{cm}^2}{\text{sec}}$ (32)

A centistoke is $\frac{1}{100}$ of a stoke.

Example 7. Find the kinematic viscosity for the oil of Example 6 at the given temperature.

Solution. By Example 6, SUS = 91 sec at temperature 180°F.

By Eq. (32): $v = 0.22 \times 91 - \dfrac{180}{91} = 18.04 \dfrac{\text{cm}^2}{\text{sec}}$ centistokes

9. *Viscosity Index*

A widely used method for specifying the rate of change of viscosity with temperature is known as the *viscosity index*.[5] The method was originally devised by testing samples of oils made from two extreme types of crudes. The series of oils having a small change of viscosity with temperature were given a viscosity index of 100, and the oils with a large change of viscosity with temperature were given a viscosity index of 0.

[5] See Dean, E. W., and G. H. B. Davis, "Viscosity Variation of Oils with Temperature," *Chem. Met. Eng.*, **36**, 1929, p. 618. See also *Ind. Eng. Chem.*, **32**, 1940, p. 102.

To find the viscosity index, *VI*, of a sample of oil, its SUS values are first determined for 210°F and 100°F and plotted on the ASTM chart as *x* and *y*, respectively, as shown in Fig. 8-14. Among the *VI* = 100 oils, there will be one whose SUS value at 210°F will be *x*, the same value as that of the sample. Let the SUS at 100°F for this oil be designated by *H*, as shown in the figure. Similarly, the viscosity at 100°F for the *VI* = 0 oil, which has the same SUS at 210°F as the sample, is represented by *L*.

FIGURE 8-14 **Determination of viscosity index.**

By definition, the viscosity index is found by means of the following equation:

$$VI = \frac{L-y}{L-H} \times 100\% \tag{33}$$

The viscosity index thus designates the rate of change of viscosity of an oil as compared to oils with very small and very large rates of change of viscosity. If should be noted that the *VI* tells nothing about the value of the viscosity at any given temperature. Values of *VI* below 0 and above 100 are recognized.

When the SUS value at 210°F for the sample lies between 50 and 350, the values of *H* and *L* can be determined from the following equations.

$$H = 0.0408x^2 + 12.568x - 475.4 \tag{34}$$

$$L = 0.2160x^2 + 12.070x - 721.2 \tag{35}$$

Example 8. Find the viscosity index for SAE 20 oil in Fig. 8-13.

Solution. $\qquad x = 54 \qquad$ and $\qquad y = 340$.

By Eq. (34): $\quad H = 0.0408 \times 54^2 + 12.568 \times 54 - 475.4 = 322.2$

By Eq. (35): $\quad L = 0.216 \times 54^2 + 12.07 \times 54 - 721.2 = 560.4$

By Eq. (33): $\quad VI = \frac{L-y}{L-H} \times 100 = \frac{560.4 - 340}{560.4 - 322.2} \times 100 = 92.5\%$

10. The Zn/P Curve

The phenomenon that takes place when a shaft is rotating in a bearing in the presence of a lubricant can be represented by the experimentally determined curve shown in Fig. 8-15. Here the abscissa is taken as the nondimensional group Zn/P, where Z is the viscosity in centipoises, n is the speed in revolutions per minute, and P is the load in pounds per square inch of the projected journal area. The ordinate is taken as the coefficient of friction f, where f is the ratio of the tangential friction force to the load carried by the bearing.

FIGURE 8-15   Plot of operating characteristic vs. coefficient of friction for journal bearing.

Nondimensional coordinates such as these are very convenient for engineering problems where many variables are present. Experiments can be made by changing the value of the nondimensional group without the need for changing each individual variable, and much labor can be saved. The curve can be divided into three more or less distinct parts.

(1) Operation in the AB portion occurs for very heavy loads and slow speeds, and generally with an insufficient supply of oil. The film thickness is smaller than the combined heights of the asperities on the opposing surfaces. Interlocking of the asperities may occur, and gross seizure is prevented by the coating of lubricant on the surfaces. This is called boundary lubrication.

(2) In the BC region the separation between the parts is greater and the load is carried partly on the asperities and partly on localized patches of lubricant, which form a load-supporting film. This is called mixed lubrication.

(3) From C to D the viscosity Z and the speed n are relatively large and

the unit load P is relatively small so that the film is relatively thick if a copious supply of lubricant is present. The friction depends on the viscosity of the lubricant and is independent of the smoothness of the surfaces and the kinds of materials used for the bearing. It is only in this region that the hydrodynamic theories, as discussed in the following sections, are valid. To avoid danger of seizure, the operating value of Zn/P should be at least 5 or 6 times that of the minimum point of the curve. It can be noted from the CD portion of the curve that the coefficient of friction for lubricated surfaces increases with increase of velocity. This is in contrast with dry surfaces, where the coefficient of friction is relatively independent of the velocity.

Example 9. Let the curve for Zn/P and f for a 3-in.-diam $\times$ 6-in.-long bearing be a straight line passing through points $Zn/P = 75, f = 0.002$, and $Zn/P = 375$, $f = 0.0065$. SAE 10 oil like that in Fig. 8-13 is used. The total load on the bearing is 1,350 lb, and the rpm $= 1,200$. What is the horsepower friction loss for this bearing if the oil film has a temperature of 170°F?

Solution. In Eq. (30): $\quad sp.gr_t = 0.870 - 0.00035(170 - 60)$

$$= 0.832 \text{ at } 170°\text{F}$$

By Fig. 8-13, at 170°F: $\qquad \text{SUS} = 56.3 \text{ sec}$

In Eq. (29):

$$Z = 0.832\left(0.22 \times 56.3 - \frac{180}{56.3}\right)$$

$$= 7.64 \text{ centipoises at } 170°\text{F}$$

$$P = \frac{1,350}{18} = 75 \text{ psi of projected journal area}$$

$$\frac{Zn}{P} = \frac{7.64 \times 1,200}{75} = 122$$

Draw the given Zn/P and f curve, and read $f = 0.0027$ for $Zn/P = 122$.

$$F = fW = 0.0027 \times 1,350$$

$$= 3.645 \text{ lb tangential friction force}$$

Friction: $\qquad \text{hp} = \dfrac{Tn}{63,000} = \dfrac{3.645 \times 1.5 \times 1,200}{63,000} = 0.104$

11. Bearing Materials

Some of the qualities required in a material for a sleeve bearing are load-carrying capacity, thermal conductivity, low coefficient of friction, smoothness of surface, and resistance to wear, fatigue, and corrosion. No one material can possess all the desired characteristics to a sufficient degree,[6] and

[6]See References 1 and 4, end of chapter.

it is necessary to compromise in most designs. The chemical compositions of a number of typical bearing materials are given in Tables 8-1 and 8-2.

TABLE 8-1   Percent Composition of Typical Bearing Alloys

	Babbitt		Bronze		Copper-Lead
	Tin Base, SAE 11	Lead Base, SAE 13	SAE 791	SAE 794	SAE 48
Copper, Cu	5.75		Rem	73.5	70
Tin, Sn	87.5	6	4	3.5	
Lead, Pb		Rem	4	23	30
Antimony, Sb	6.75	10	4		
Zinc, Zn			4		

TABLE 8-2   Maximum Design Pressures for Engine Bearings*

Bearing Material	Per Cent Composition of Material	Maximum Mean Bearing Pressure, psi
Lead-base babbitt	Pb (75–85), Sn (4–10), Sb (9–15)	600–800
Tin-base babbitt	Pb (0.35–0.6), Sn (86–90), Sb (4–9), Cu (4–6)	800–1,000
Cadmium-base alloy	Cu (0.4–0.75), Cd (97), Ni (1–1.5), Ag (0.5–1.0)	1,200–1,500
Copper-lead alloy	Pb (45), Cu (55)	2,000–3,000
Copper-lead alloy	Pb (25), Sn (3), Cu (72)	3,000–4,000
Silver	Ag (99), 0.5–1.0 lead on surface	5,000 up

*Pressures are based on fatigue life of 500 hr at a bearing temperature of 300°F. Bearing metal thicknesses range from 0.01 to 0.015 in. for lead, tin, and cadmium-base metals, and 0.025 in. for copper, lead, and silver, all on steel backs. At lower temperatures, the fatigue life will be greatly extended. See the reference works at the end of the chapter.

(a) *Tin- and Lead-Base Babbitts.* Babbitt bearings are in very wide use and are of two general types: tin base and lead base. They are quickly run in and assume very smooth surfaces. Application is usually made to a steel back as shown in Fig. 8-17(b) and (c). Thickness of the lining is usually about 0.015 in. Thin babbitt bearings with linings 0.002–0.005 in. thick are also used and can carry somewhat heavier loads. Babbitt bearings have good *conformability*, or the property of adjusting themselves to small misalignments or shaft deflections. They also make excellent bearings from the standpoint of *embeddability* because a reasonable amount of dirt or foreign matter in the lubricant can be absorbed by the soft bearing material and the shaft is thus protected against scoring. Journal material may be soft steel, hard steel, or cast iron.

(b) *Bronze Bearings.* Bronze bearings are suitable for high loads and slow speeds, but the alignment between shaft and bearing must be good. They are made in a great number of alloy compositions to obtain various physical properties. Standaidized bushings, as shown in Fig. 8-17(a), can be purchased ready for use.

(c) *Copper-Lead Bearings.* These bearings are used where the loads are higher than can be carried by babbit. The conformability, however, is less and consequently such bearings must be used where the shafts are rigid and the alignment is good.

(d) *Cast-Iron Bearings.* Cast iron is a bearing material widely used where the service is not too severe. Journals should have a hardness in the 150–250 Brinell range, and should have a smooth finish. Some of the good performance of cast iron is usually attributed to the graphite inclusions that are normally present. The alignment between journal and bearing, however, must be good. Cast iron runs well with phosphor bronze, and also with cast iron as evidenced by the cylinder block and pistons of internal combustion engines.

(e) *Porous Bearings.* The so-called "self-lubricating" or "porous" bearing is made by sintering powdered metal and then impregnating with oil. Various compositions of bronze are in wide use. Iron is used to a lesser extent. Figure 8-16 shows a self-aligning porous-metal bearing applied to a small electric motor. The felt washer can frequently be omitted. Such bearings are available from stock in both cylindrical and spherical form.

Porous bearings are sometimes designed to a specified value of PV, where P is the load per square inch of projected area, and V is the surface velocity of the shaft in feet per minute.[7]

FIGURE 8-16 Self-aligning porous bearing with oil-saturated felt washer.

Example 10. Find the required length for a porous bearing $\frac{1}{2}$ in. in diameter turning at a speed of 1,750 rpm. Load on the bearing is 50 lb. $PV = 30,000$.

Solution. Let W be the load on the bearing.

$$PV = \frac{W}{ld} \times \frac{\pi\, dn}{12} = \frac{\pi\, Wn}{12l}$$

Then:
$$l = \frac{\pi Wn}{12(PV)} = \frac{\pi 50 \times 1,750}{12 \times 30,000} = 0.76 \text{ in.}$$

[7]See ASTM Specifications B 438 and 439, Vol. 7.

(f) *Carbon and Plastic Bearings.* For high-temperature service, or where conventional types of lubrication cannot be used, bearings of pure carbon have been satisfactorily employed. Graphite is a layer-lattice solid, and the low friction of such bearings is due to the low resistance to slip between the lattice planes.

Teflon has an extremely low coefficient of friction, and requires no oil lubrication. Bearings can be operated at high loads combined with slow or oscillating speeds. Wear resistance is increased by filling with glass fiber, powdered carbon, bronze, or metallic oxides. Bearings can be used under conditions of corrosion, abrasion, or stickiness. Bearings are frequently formed merely by the use of liners cut from Teflon sheet or strips.

(g) *Laminated Phenolic Bearings.* These are plastic bearings formed by impregnating paper or fabric with the phenolic resin, forming into shape, and then curing by heat and pressure. Since this is a thermosetting material subsequent heating will have no effect. Such bearings show good resistance to corrosion, fatigue, and shock. They have been successfully applied to large sizes and heavy loads, such as in the steel mill industry. Water is sometimes used as the lubricant.

(h) *Rubber and Wood.* Soft rubber, lubricated with water, has been used for many years in pumps and general marine service. Lignum vitae and wood are also used for bearings to a limited extent.

(i) *Bonded Coatings.* Molybdenum disulfide, MoS_2, which resembles graphite in appearance, possesses very-low-friction characteristics. Like graphite, it has a laminar structure. To ensure retention of the lubricant on the surface where it is needed, it can be mixed with a suitable binder, usually a thermosetting plastic, spread on the surface, and then dried at a suitable temperature. Wear life in general is satisfactory. This lubricant has proved to be very effective under operating conditions not possible for fluid lubricants.[8]

Long experience has shown that certain combinations of materials will work well together.[9] Other combinations will not but will exhibit excessive wear.

Soft steel works well with cast iron, babbit, and soft brass, but works poorly with soft steel, bronze, nylon, and thermoset resins.

Hard steel works well with soft bronze, phosphor bronze, brass, cast iron, nylon, and the thermoset resins, but poorly with the harder heat-treated bronzes. Hardened nickel steel does not work well with hardened nickel steel.

[8]See Campbell, W. F., p. 197 of Reference 2, end of chapter. See also Reference 3, end of chapter.

[9]See Rothbart, H. A., *Cams*, New York: John Wiley & Sons, Inc., 1956, p. 277.

12. Bearing Loads

Bearing loads are determined mainly from experience with similar applications that have proved successful. The permissible loading is not a direct function of the particular material under consideration, but depends rather on the type of service. Table 8-3 gives a survey of current practice and can be used as a general guide when more specific information is lacking.[10] Careful attention should be paid to the reliability of the oil supply for small values of Zn/P, which occur when the speed is slow and the load is large.

TABLE 8-3   Current Practices in Bearing Design Pressures

Type of Bearing	Design Pressure, psi
Diesel engines	
Main bearings	800–1,500
Connecting rod bearings	1,000–2,000
Wrist pins	1,800–2,000
Electric motor bearings	100–200
Marine lineshaft bearings	25–35
Steam turbines and reduction gears	100–250
Automotive gasoline engine	
Main bearings	500–700
Connecting rod bearings	1,500–2,500
Aircraft engine connecting rod bearings	700–2,000
Centrifugal pumps	80–100
Roll neck bearings	1,500–2,500
Railway axle bearings	300–350
Light lineshaft bearings	15–25
Heavy lineshaft bearings	100–150

13. Construction of Bearings

A separate insert or bushing is usually provided to form the bearing surface for supporting the shaft in a sleeve bearing. It is made of a material that is known to have desirable qualities of high load capacity and low friction. The types and details of construction for sleeve bearings can vary over a considerable range.

Figure 8-17(a) shows a solid bronze bushing, a type of bushing that is in wide use. It can be cast, machined from bar stock, or formed from tubing or sheets. It is usually held in place by a light press fit. The bearings shown in

[10]See Fuller, D. D., "Design Analysis of Journal Bearings," *Machine Design*, **28**, Feb. 9, 1956, p. 119. See also p. 221 of Reference 6, end of chapter.

(a) Solid bronze bushing

(b) Flanged all around lined bearing

(c) Lined bearing split on shaft center

(d) Oil ring bearing

FIGURE 8-17   Types of plain sleeve bearings.

sketches (b) and (c) consist of a steel or bronze back with a thin lining of babbitt or one of the numerous other bearing alloys. Such bearings can be made either all around or split on the centerline of the shaft. A wide variety of types and sizes of such bearings is available from stock ready to use. They may be obtained solid, split longitudinally or flanged, and with many kinds of oil grooves.

The oil ring bearing[11] of Fig. 8-17(d) has a cast-iron back with a babbitt liner cast in place. To obtain a better oil supply in slow-speed service, the ring is sometimes replaced by an endless loop of chain hanging on the shaft.

14. Clearance and Oil Grooves

In addition to choosing the material, the designer must specify the shaft clearance and the oil grooving for the bearing.

The clearance depends to some extent on the desired quality. Small

[11]Karelitz, G. B., "Performance of Oil Ring Bearings." *Trans. ASME,* **52**(1), 1930, p. APM 57.

clearances can be maintained with high-grade workmanship, but when costs must be reduced, the clearances usually are made larger. Table 8-4 gives a summary of clearances used in industrial applications.

TABLE 8-4 Bearing Clearances in Industrial Applications*

	Running Clearance, Thousandths of an Inch, for Shaft Diameter under				
	1/2 in.	*1 in.*	*2 in.*	*3-1/2 in.*	*5-1/2 in.*
Precision spindle practice—hardened and ground spindle lapped into the bronze bushing. Below 500 fpm and 500 psi	0.00025 to 0.00075	0.00075 to 0.0015	0.0015 to 0.0025	0.0025 to 0.0035	0.0035 to 0.005
Precision spindle practice—hardened and ground spindle lapped into bronze bushing. Above 500 fpm and 500 psi	0.0005 to 0.001	0.001 to 0.002	0.002 to 0.003	0.003 to 0.0045	0.0045 to 0.0065
Electric motor and generator practice—ground journal in broached or reamed bronze bushing or reamed babbitt bushing	0.0005 to 0.0015	0.001 to 0.002	0.0015 to 0.0035	0.002 to 0.004	0.003 to 0.006
General machine practice (continuous rotating motion)—turned steel or cold-rolled steel journals in bored and reamed bronze or poured and reamed babbitt bushings	0.002 to 0.004	0.0025 to 0.0045	0.003 to 0.005	0.004 to 0.007	0.005 to 0.008
General machine practice (oscillating motion)—journal and bearing material as above	0.0025 to 0.0045	0.0025 to 0.0045	0.003 to 0.005	0.004 to 0.007	0.005 to 0.008
Rough machine practice—turned steel or cold-rolled steel journals in poured babbitt bearings	0.003 to 0.006	0.005 to 0.009	0.008 to 0.012	0.011 to 0.016	0.014 to 0.020

*Data courtesy Johnson Bronze Co.

For small clearances, the surface roughness of the shaft and bearing must be considered. These surfaces, when examined by a microscope, are usually far from being geometrically smooth. In many cases, the roughness consists of two orders of irregularities. Over most of the surface there are relatively small, low irregularities at close intervals. In addition, at relatively large intervals, there are high peaks and deep valleys. Roughness is usually

measured from the tops of the high peaks, as shown in Fig. 8-18. To accentuate the roughness, it is customary to prepare samples with a vertical scale
that is 25 times as great as the horizontal scale. Profilometers give readings
for either the root mean square or the arithmetic average of the roughness.
The predominant peak roughness can then be had by mutliplying by a suitable
factor. For example, for ground surfaces, the arithmetic average roughness
should be multiplied by 5 to obtain the predominant peak roughness.[12]

Predominant
peak roughness

Deepest
maximum
roughness

h_o

rms roughness

FIGURE 8-18   **Lower limit at which hydrodynamic lubrication is assumed
to be possible.**

Figure 8-18 shows the surfaces for journal and bearing located so that
the high peaks are just in contact. It has been proposed[13] that the condition
illustrated be taken as the limit at which hydrodynamic lubrication can be
assumed to exist. Thus if each surface has an arithmetic average roughness of
30μ in., the minimum permissible film thickness in Fig. 8-18 would be
$2 \times 30 \times 5$ or 0.0003 in. Surface finish should thus be specified and closely
controlled if the design calculations indicate that the bearing will operate
with a very thin film. Experience has indicated, however, that other factors,
such as misalignment, deformation, foreign matter in the oil, and fatigue
due to dynamic loads, are the cause of most bearing failures.

Oil grooves with carefully rounded edges should be provided to distribute
the oil over the entire surface. It is very important, however, that no grooves
be placed in the load-carrying portion of the lining. When so placed, such
grooves merely provide an easy exit for the oil and prevent the formation
of the film and the high pressures necessary to support the load. A good rule
is to use the least possible amount of grooving. Sometimes the chamfer
between the bearing and cap provides a channel for the lateral flow of the
lubricant. Figure 8-19 shows the reduction in load capacity (for equal minimum film thickness) due to an incorrectly located oil groove. If the load on
the bearing with the middle groove is not reduced, operation may take place

[12]See Tarasov, L. P., "Relation of Surface Roughness Readings to Actual Surface
Profile," *Trans. ASME,* **67**, 1945, p. 189. The rms average for a surface is usually about 10
percent greater than the arithmetic average roughness.

[13]See Ocvirk, F. W., and G. B. Dubois, "Surface Finish and Clearance Effects on
Journal-Bearing Load Capacity and Friction," *Trans. ASME,* **81**, 1959, *J. Basic Eng.,*
p. 245.

FIGURE 8-19 Loss of capacity from incorrect grooving.

FIGURE 8-20 Loss of capacity caused by oil collar.

with a dangerously thin oil film. If a belt pulley is adjacent to a bearing, care should be taken that the forces from the belt do not pull the shaft toward the side where oil grooves may be located and where a satisfactory oil film cannot be formed.

In a similar manner, a circumferential groove reduces the load-carrying qualities, as shown in Fig. 8-20. An oil collar is a very effective means of distributing the lubricant. However, an oil collar practically cuts the bearing in two, and the sum of the loads carried by each part is less than that carried by a single bearing of the total width. Forced-feed-lubricated bearings frequently are constructed without oil grooves.

The most important point to be considered when choosing the method of lubricating a bearing is the reliability of the oil supply. Because of low first cost, many bearings are lubricated by hand oilcans, drip or wick oilers, oil-soaked waste, or grease. As previously mentioned, such bearings usually operate in the region of boundary lubrication, with its attendant high friction and danger of seizure, should the scanty oil supply become temporarily deranged.

Safety lies in having a copious oil supply at all times. Such a supply can be provided by an oil ring and reservoir, or by piping the oil to the bearing under pressure from a central pumping station. Whatever method is used, close attention to the mechanical details is required if satisfactory operation is to be expected.

Means for preventing leakage and loss of oil at the ends of the bearing must often be provided. Felt gaskets or patented oil seals are in wide use. Sometimes the outer housing is enlarged to accommodate a small disk fastened to the shaft. The oil that has leaked past the side of the bearing is thrown off the disk by centrifugal force and caught by the housing. Often a circumferential groove at the end of the bearing is provided with drain holes to return the oil to the reservoir.

Grease is widely used for exposed locations because it is easier to retain in the bearing. Investigations have shown that the action can be hydrodynamic with a pressure distribution somewhat like that for oil.

15. Elastic Matching

When the bearings are some distance apart and the shaft carries a lateral load, the inclination of the shaft at the ends may be great enough to give rise to the misalignment shown in Fig. 8-21(a). Obviously such a bearing is subjected to rapid wear and is not able to form a satisfactory load-carrying film. The difficulty can be reduced by setting the bearings closer together and using a larger-diameter shaft or a smaller value for the l/d ratio. The spherical seat shown in sketch (b) gives self-alignment from all causes, but its expensive, and increases the size of the bearing.

(a) Rigid housing   (b) Spherical housing   (c) Elastic matching of
                                                              shaft and bearing

FIGURE 8-21   Types of bearing supports.

Sometimes the bearing can be located eccentrically on an elastic support in such a way that it tilts slightly under load and assumes the same inclination as the shaft. See Fig. 8-21(c) and Problem 116, Chapter 1. The oil film then has a uniform thickness throughout the axial length of the bearing, and the difficulty shown in Fig. 8-21(a) is overcome.[14]

16. Dry Friction

Dry friction or the friction between surfaces without intentional lubrication is a very complex phenomenon. Despite more than a century of study, a satisfactory explanation of the mechanism of dry friction does not as yet

[14]See Baker, J. G., "Elastic Matching Improves Bearing Life and Performance," *Machine Design*, **20**, May 1948, p. 106.

exist. The so-called "classical rules of dry friction" are usually stated as follows.

(1) The friction force is directly proportional to the force pressing the bodies together.

(2) The friction force is independent of the area of the contacting surfaces.

(3) The friction force is largely independent of the velocity of sliding.

(4) The friction force depends on the nature of the sliding surfaces.

The coefficient of static friction for bodies at rest, or with motion impending, is somewhat greater than for kinetic friction, which prevails after the sliding member is set in motion.

According to Coulomb, 1785, the friction force is due to the interlocking of the asperities of the two surfaces and the necessity of deforming the irregularities as the two bodies pass by each other. Within rather wide limits the magnitude of the friction force is independent of the degree of roughness of the surfaces. The asperities that are deformed are very minute so that it apparently makes no difference whether they are located on a smooth surface or consist of protuberances located upon larger irregularities.

For very rough surfaces, the protuberances may interlock to the extent that parts of them are torn off as the bodies move with respect to each other. This is a wear phenomenon and is usually not considered to be friction in the usual meaning of the term. Such action is especially likely to occur when one of the bodies is much harder than the other.

Although the nature of the sliding surfaces determines the friction force, surfaces of naked metal do not exist in everyday life. Exposure to the atmosphere will contaminate the surface with oxide films, moisture, absorbed gases, grease, dirt, and so on. These films cling very tightly to a surface and can be removed only with difficulty. Ordinary cleaning does not suffice. Even in the laboratory, unless the body is kept in an inert atmosphere, an oxide film quickly forms. Thus when two ordinary bodies are sliding together, it is the surface films that are touching each other and not the metals themselves.

Reference handbooks usually make no attempt to give exact specifications for the surfaces under consideration, and this may partially account for the wide variations in the published values for the coefficient of friction.[15]

When bodies with very smooth surfaces are brought into intimate contact, adhesive or cohesive forces come into action and hold the bodies tightly together. Thus when gage blocks or optical flats are wrung together, a relatively large force may be required for tangential motion even though the

[15] See Schmidt, A. O., and E. J. Weiter, "Coefficients of Flat Surface Friction," *Mech. Eng.*, **79**, 1957, p. 1130. See also Bikerman, J. J., "Surface Roughness and Sliding Friction," *Rev. Mod. Phys.*, **16**, 1944, p. 53.

normal force pressing them together is very small. The classical rules of friction do not apply under such conditions.

A surface of naked metal can be obtained by removal of the oxide and surface film and degasing in the laboratory. When two metallic surfaces of this kind are brought together in a vacuum or inert atmosphere, the metals weld together at the high spots where contact occurs between them. When one body is drawn along the other the welds will be made and broken continuously. The resulting motion is somewhat jerky or discontinuous and has been given the name of *stick slip*. The force required to produce motion under such conditions is relatively very large and may be as much as 20 times the value for corresponding contaminated surfaces. The force also depends upon the velocity of sliding and is thus not in accord with the classical rules of friction.

When two bodies with contaminated surfaces are brought together, actual contact occurs only at the high spots of the two surfaces. The actual area of touching may be only a very small fraction of the nominal area. The contact stress at the isolated points where the bodies touch reaches very high values. Such stresses are required to deform the asperities sufficiently to permit motion to take place. Under such conditions of extreme stress, the surface film may rupture and the welding phenomenon described above may take place. Such welding, together with any adhesive forces that may be present, serves to further complicate the situation.[16]

The wide scatter in the published values for the coefficient of friction is undoubtedly due to the lack of precise information on the conditions of the actual surfaces being tested. "... in view of the extremely high friction found with really clean metals it is fortunate, for engineering, that metals are not found with perfectly clean surfaces in practise."

17. Boundary or Thin-Film Lubrication

Machine parts coated with a lubricant and rubbing together without benefit of a fluid film are a common occurrence in machinery. Such operation can also occur in a journal bearing for low values of the viscosity or speed or for high values for the load as indicated by the left end of the curve in Fig. 8-15. Such action, known as boundary lubrication, can also occur during starting, stopping, or the reversing of journal bearings. It can also occur during the running-in period of a new machine. The separation of the surfaces is of molecular dimensions.

The rules previously stated for dry friction apply also to boundary lubrication with the following additions.

[16]For elastohydrodynamic lubrication see Section 14 of Chapter 9.

(5) The friction force depends on the composition of the lubricant and its reaction with the contacting surfaces.

(6) The friction force is influenced by the temperature and surface roughness of the bodies.

(7) The friction force is independent of the viscosity of the lubricant.

The ordinary rules of hydrodynamic lubrication do not apply to boundary lubrication, and design equations for this type of service are not available.[17]

Fatty acids, as found in certain animal and vegetable oils, can be added in small quantities to the lubricant to give a friction reducing quality distinct from the viscosity of the lubricant. These react with the surfaces to form very tough and durable coatings, and at the same time provide a plane of low shearing strength on which the relative motion between the parts takes place. Massive seizure between the parts is thus prevented. Should the film be ruptured when two peaks come together exposing pure metals, instantaneous welding will occur. Such welds of the high spots are immediately broken by the relative movement between the bodies resulting in wear and increased friction force.

Boundary lubrication breaks down at elevated temperatures resulting in a large increase in the coefficient of friction accompanied by wear and surface damage.

True boundary lubrication is difficult to achieve even in the laboratory. Usually there will be sufficient lubricant present to form localized patches of film, which carry part of the load while the remainder of the surfaces are rubbing without complete separation of the peaks of the two surfaces. The result is the so-called mixed lubrication.

18. Mixed or Semifluid Lubrication

Mixed lubrication, as shown by the *BC* section of Fig. 8-15, represents a partial breakdown of the full hydrodynamic film as might occur under heavy loads, slow speed, and low viscosity. An insufficient supply of lubricant as given by drip cups or wick oilers may be a contributing factor.[18] Bearing and journal should be made of materials that experience has shown will operate together satisfactorily. Mixed lubrication can also occur when starting a heavily loaded bearing that has been inactive for some time.

For larger values of Zn/P, operation is in the CD region. The lubrication is hydrodynamic and follows the laws given in the earlier sections of the

[17]See Reference 2, end of chapter.

[18]See Larsen, R. G., and G. L. Perry, "Investigation of Friction and Wear under Quasi-hydrodynamic Conditions," *Trans. ASME*, **67**, 1945, p. 45.

chapter. Such operation is largely independent of the smoothness of the surfaces or the kinds of materials used for the moving parts.

Many other means are available for achieving lubrication in mechanical equipment. Some of these are slider bearings, hydrostatic bearings, and gas bearings. Information can be obtained from the appended list of reference works.

REFERENCES

1. BASSETT, H. H., *Bearing Metals and Alloys*, New York: Longmans, Green & Co., Inc., 1937.

2. *Boundary Lubrication, An Appraisal of World Literature*, New York: American Society of Mechanical Engineers, 1969.

3. BRAITHWAITE, E. R., *Solid Lubricants and Surfaces*, Elmsford, N.Y.: Pergamon Press, Inc., 1963.

4. DAYTON, R. W., Ed., *Sleeve Bearing Materials*, Metals Park, Ohio: American Society for Metals, 1949.

5. FULLER, D. D., "Design Analysis of Journal Bearings," *Machine Design*, **28**, Feb. 9, 1956, p. 119.

6. FULLER, D. D., *Theory and Practice of Lubrication for Engineers*, New York: John Wiley & Sons, Inc., 1956.

7. *General Discussion on Lubrication*, London: Institution of Mechanical Engineers, 1938.

8. GROSS, W. A., *Gas Film Lubrication*, New York: John Wiley & Sons, Inc., 1962.

9. HERSEY, M. D., *Theory of Lubrication*, New York: John Wiley & Sons, Inc., 1938.

10. NORTON, A. E., *Lubrication*, New York: McGraw-Hill Book Company, 1942.

11. *Physical Properties of Lubricants*, 2nd ed., Chicago: American Society of Lubrication Engineers, 1951.

12. RADZIMOVSKY, E. I., *Lubrication of Bearings*, New York: The Ronald Press Company, 1959.

13. SHAW, M. C., and E. F. MACKS, *Analysis and Lubrication of Bearings*, New York: McGraw-Hill Book Company, 1949.

14. SLAYMAKER, R. R., *Bearing Lubrication Analysis*, New York: John Wiley & Sons, Inc., 1955.

15. SPOTTS, M. F., *Mechanical Design Analysis*, Englewood Cliffs, N.J.: Prentice-Hall, Inc., 1964.

16. TRUMPLER, P. R., *Design of Film Bearings*, New York: Macmillan Publishing Company, 1966.

PROBLEMS

1. A very lightly loaded 360° bearing, 6 in. in diameter and 9 in. long, consumes 2 hp in friction when running at 1,200 rpm. Radial clearance is 0.003 in. Find the temperature of the oil film using SAE 10 oil in Fig. 8-2.   *Ans.* $t = 145°F$.

2. A 2.5-in.-diam $\times$ 3.75-in.-long 360° bearing turns at 1,200 rpm and consumes 0.1 hp in friction. SAE 10 oil in Fig. 8-2 is used; $r/c = 1,000$. Find the total load for the bearing if the film is at a temperature of 175°F.

Ans. $W = 1,030$ lb.

3. A 360° journal bearing has a diameter of 60 mm and an axial length of 90 mm. Bearing turns at 1,200 rpm and consumes 0.137 kW in friction. $r/c = 1,000$. SAE 30 oil at 180°F is used. Find load bearing is carrying.

Ans. Load = 10,300 N.

4. A 75-mm-diam $\times$ 75-mm-long 360° bearing turns at 1,200 rpm; $c/r = 0.001$; $h_0 = 0.025$ mm. SAE 20 oil is used. The film temperature is 180°F. Find the total load W.   *Ans.* $W = 3,600$ N.

5. A 3-in.-diam $\times$ 6-in.-long 120° central partial bearing is operating at a value for ϵ equal to 0.6. Coefficient of friction is equal to 0.0025. Find the value of the minimum film thickness.   *Ans.* $h_0 = 0.00098$ in.

6. A bearing 250 mm in diameter and 250 mm long operates at 1,500 rpm, carrying 108,000 N. The c/r ratio is 0.0015. Let the viscosity of the lubricant be taken as 0.000 000 03 N sec/mm². Find the friction power loss in accordance with Petroff's equation, the 360° loaded bearing, and for the 120° central partial bearing.

Ans. 2.11; 14.36; and 7.38 kW.

7. Use data as for Example 2, except that the oil film temperature and cooling area are not known. For this bearing, P is 125 psi of projected journal area, and SAE 10 oil in Fig. 8-2 is used. Find the required cooling area of the housing.

Ans. 3.8 ft.²

8. A 120° central partial bearing 90 mm in diameter and 135 mm long carries 0.86 MPa projected journal area at 600 rpm. Room temperature is 38°C and housing is at 60°C. $C_1 = 11.5$ W/(m² deg C); $C_2 = 8$; $\epsilon = 0.52$. What is the minimum film thickness?   *Ans.* $h_0 = 0.028$ mm.

9. A 120° central partial bearing 4 in. in diameter and 5 in. long carries 90 psi of projected journal area. The r/c ratio is 1,000; h_0 is equal to 0.001 in. SAE 20 oil in Fig. 8-2 is used. Room temperature is 100°F. Rpm = 600. Assume the rate of cooling is 2.5 Btu/hr/ft²/°F. Assume housing temperature rises one-half as much as oil film. Find the oil film temperature and the required area of housing.

Ans. $t = 168°F$; $A_c = 2.44$ ft.²

10. A 120° central partial bearing 62 mm in diameter and 100 mm long has a minimum film thickness of 0.015 mm. SAE 20 oil in Fig. 8-2 is used; $n = 860$ rpm and $r/c = 1,000$. If the oil film is at 74°C, find the load the bearing is carrying.

Ans. $W = 7,460$ N.

11. Find the required area of housing in order to maintain the film temperature of a 76-mm-diam × 114-mm-long, 120° central partial bearing at 74°C; rpm = 900 and r/c = 1,000. Room temperature is 38°C. Assume housing temperature rises one-half as much as the oil film. Cooling rate is 11.5 W/(m² deg C). SAE 10 oil is used. Load is 0.686 MPa of projected journal area.   *Ans.* A_c = 0.212 m².

12. A 2-in.-diam × 4-in.-long 120° central partial bearing turns 1,350 rpm. SAE 10 oil is used. Film temperature is 175°F and room temperature is 100°F. Assume housing rises one-half as much as the film, and that the cooling rate is 2 Btu/hr/ft²/°F. Let the area of the housing be 8 times the developed journal area.

If the shaft and bearing are dimensioned as shown in Fig. 8-22, determine the load and minimum film thickness for an assembly at the loosest permissible fit, and also for an assembly at the tightest permissible fit.

Ans.   Loosest, W = 880 lb; tightest, W = 500 lb.

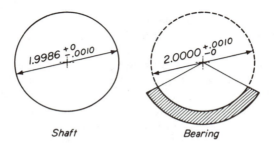

Shaft   Bearing

FIGURE 8-22   Problem 12.

13. A 120° central partial bearing is 2 in. in diameter and 3 in. long. r/c = 1,000; n = 900 rpm. SAE 10 oil is used. Bearing is self-cooled with C_1 = 2 Btu/hr/ft²/°F and C_2 = 8. Room temperature is 100°F. Assume the housing surface temperature rises one-half as much as the average film temperature.

(a) Plot the curve for average film temperature vs. P.

(b) Plot the curve for h_o vs. P.

Use temperature range from 145°F to 170°F inclusive at 5-deg intervals.

14. A 360° pressure-lubricated bearing with a central groove is 4 in. diam × 4 in. long (l' = 2 in.). r/c = 1,000; P = 300 psi projected area. n = 900 rpm. SAE 20 oil is used. Take C_3 = 0.5.

For an inlet temperature of 100°F and an outlet temperature of 220°F, find the value of the inlet pressure p_i if all the frictional heat is carried away by the oil.

Ans. p_i = 27.2 psi.

15. A 120° central partial bearing is 2 in. in diameter × 3 in. long with a radial clearance of 0.001 in. The shaft turns at 900 rpm. Loading is 150 psi of projected area. Room temperature is 100°F. SAE 10 oil is used. Heat transfer coefficient C_1 equals 2 Btu/hr/ft²/°F. For a minimum film thickness of 0.0004 in. find the average temperature in the film and the required value for constant C_2.

Ans. t_{av} = 162°F; C_2 = 6.86.

16. (a) Use the data in Example 4 and calculate the values of P for the given temperature range for values of r/c of 600, 800, 1,200, and 1,400. Calculations for r/c of 1,000 have already been made in Example 4. Plot the curves for P vs. film temperature.

(b) From the curves read and tabulate the temperature values for loads P of 180, 200, and 220 psi of projected area for each of the values above of r/c. Plot the curve for film temperature vs. r/c for $P = 180$ psi. Do the same for $P = 200$ psi and for $P = 220$ psi. Note that there is an optimum value for r/c at which the film temperature is a minimum.

(c) Plot the ϵ values vs. P for the above r/c ratios. From the curves read and tabulate the ϵ values for loads P of 180, 200, and 220 psi for each r/c curve. Calculate the corresponding h_0 values and plot vs. r/c for $P = 180$ psi. Do the same for $P = 200$ psi and for $P = 220$ psi. Note that there is an optimum r/c at which h_0 is a maximum for any given pressure P.

17. A journal bearing 3 in. in diameter and 6 in. long has the same Zn/P vs. f curve as for Example 9. Assume the specific gravity at the film temperature to be 0.84. Total loads is 1,800 lb, and rpm = 1,500. Find the SUS value at the operating temperature if the power loss for the bearing is 0.15 hp. *Ans.* SUS = 51 sec.

18. Same data as for Problem 17 except horsepower loss in the bearing is 0.12. *Ans.* SUS = 42.6 sec.

19. A journal bearing 3 in. in diam and 6 in. long has the same Zn/P vs. f curve as that in Example 9. The oil film is at 180°F. The oil has a SUS value of 52 sec at 210°F, a VI of 80 percent, and a $sp.gr_{60}$ of 0.9. Total load is 1,500 lb and rpm = 1,200. Find the horsepower loss for this bearing. *Ans.* hp = 0.133.

20. A 120° central partial bearing is 62 mm in diameter and 100 mm long. Rpm = 860, SAE 20 oil is used, and the r/c ratio is 1,000. For an oil temperature of 71°C and minimum film thickness of 0.015 mm, find the load on the bearing. *Ans.* $W = 8,160$ N.

21. Plot the curve for SAE 10 oil in Fig. 8-13 in coordinates of lb sec/in.² vs. temperature. For this type of problem it is convenient to carry out the work in tabular form. A separate column heading should be used for each successive step in passing from the SUS values of Fig. 8-13 to μ in. lb sec/in.² units.

22. An oil has a VI of 60 percent and a Saybolt viscosity at 100°F of 400 sec. Find the Saybolt viscosity for this oil at 180°F. *Ans.* SUS = 70.5 sec.

23. An oil has a VI of 90 percent and a Saybolt viscosity at 100°F of 300 sec. Find the Saybolt viscosity of this oil at 160°F. *Ans.* SUS = 84 sec.

24. Three bearings have respective film arcs of 120°, 180°, and 360°. The diameter is 62 mm and length is 100 mm. Rpm = 1,200, r/c ratio is 1,000, and load is 1.4 MPa of projected journal area. SAE 10 oil. Find the film temperature for each bearing if the minimum film thickness is 0.015 mm.

This problem illustrates the fact that, when the angle is reduced, the bearing must be operated at a lower temperature if the film thickness is to be maintained. *Ans.* 20.2°C; 24.3°C; 25.7°C.

25. Let the film arcs for a series of bearings be 120°, 180°, and 360°. The diameter is 62 mm and the length is 100 mm. Rpm = 1,200, r/c ratio is 1,000, and film temperature is 77°C. SAE 10 oil is used and minimum film thickness is 0.015 mm. Find the load P on the projected journal area that each bearing is carrying.

This problem illustrates the fact that the load must be reduced for smaller values of the bearing arc in order to maintain a given film thickness and oil temperature.   *Ans.*   0.91, 1.32, and 1.47 MPa.

26. Let the film arcs for three bearings be 120°, 180°, and 360°, respectively. Diameter is 90 mm and length is 113 mm. The load on the bearings is 8,000 N. Rpm = 900 and film temperature is 74°C. SAE 10 oil is used; r/c ratio is 800. Find the minimum film thickness and the friction power loss.

This problem shows that a reduction in film arc will give a reduction in the friction power loss, but that the film thickness will also be reduced.

Ans.   Power loss = 0.068, 0.080, and 0.113 kW.

27. A 2 in. diam × 2 in. long 360° bearing is lubricated with SAE 20 oil. Speed is 900 rpm. Load P is 40 psi projected area. $r/c = 500$. Cooling rate for housing is 2 Btu per hour per square foot per degree F. Take housing area in cooling as 8 times the developed journal area. Find rise in housing temperature ΔT_h if the film temperature is 175°F.   *Ans.* ΔT_h = 32.7 deg F.

28. Same as Problem 27 but film is at 180°F.   *Ans.* ΔT_h = 30.7 deg F.

29. Same as Problem 27 but film is at 185°F.   *Ans.* ΔT_h = 29.0 deg F.

30. Same as Problem 27 but film is at 190°F.   *Ans.* ΔT_h = 27.5 deg F.

31. A 3 in. diam × 3 in. long 360° bearing is lubricated with SAE 30 oil. Speed is 1,800 rpm. Load P is 500 psi projected journal area. $r/c = 500$. Cooling rate for housing is 2 Btu per hour per square foot per degree F. Take housing area in cooling as 8 times the developed journal area. Find the horsepower required to operate the bearing and the rise in temperature ΔT_h of the housing if the film temperature is 175°F.   *Ans.*   hp = 0.574; ΔT_h = 46.5 deg F.

32. Same as Problem 31 but film temperature is at 180°F.

Ans.   hp = 0.546; ΔT_h = 44.2 deg F.

33. Same as Problem 31 but film temperature is at 185°F.

Ans.   hp = 0.518; ΔT_h = 42.0 deg F.

34. Same as Problem 31 but film temperature is at 190°F.

Ans.   hp = 0.497; ΔT_h = 40.3 deg F.

9

Ball and Roller Bearings

Ball and roller bearings have been brought to their present state of perfection only after a long period of research and development. The benefits of such specialized research can be obtained when it is possible to use a standardized bearing of the proper size and type. Ball bearings are used in almost every kind of machine and device with rotating parts. However, such bearings cannot be used indiscriminately without a careful study of the loads and operating conditions. In addition, the bearing must be provided with adequate mounting, lubrication, and sealing.

Pillow blocks, incorporating ball or roller bearings, are available from stock, and are very convenient as the mounting and sealing of the bearing has already been done.

1. Construction and Types of Ball Bearings

A ball bearing usually consists of four parts: an inner ring, an outer ring, the balls, and the cage or separator. To increase the contact area and permit larger loads to be carried, the balls run in curvilinear grooves in the rings. The radius of the groove is slightly larger than the radius of the ball and a very slight amount of radial play must be provided. The bearing is thus permitted to adjust itself to small amounts of angular misalignment in the assembled shaft and mounting. The separator keeps the balls evenly spaced and prevents them from touching each other on the sides where their relative velocities are the greatest.

Ball bearings are made in a wide variety of types and sizes. Single-row radial bearings are made in four series, extra light, light, medium, and heavy, for each bore, as illustrated in Fig. 9-1(a), (b), and (c). The heavy series of bearings is designated by 400. Most manufacturers use a numbering system so devised that if the last two digits are multiplied by 5, the result will be the bore in millimeters. The digit in the third place from the right indicates the series number. Thus, bearing 307 signifies a medium-series bearing of 35-mm bore. Additional digits, which may be present in the catalog number of a bearing, refer to manufacturer's details. Some makers list deep groove bearings and bearings with two rows of balls.

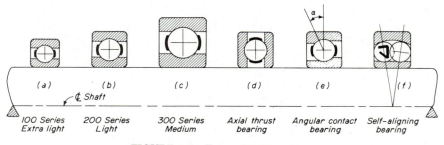

FIGURE 9-1   Types of ball bearings.

The radial bearing is able to carry a considerable amount of axial thrust. However, when the load is directed entirely along the axis, the thrust type of bearing should be used. The angular contact bearing will take care of both radial and axial loads. The self-aligning ball bearing will take care of large amounts of angular misalignment. An increase in radial capacity may be secured by using rings with deep grooves, or by employing a double-row radial bearing.

Radial bearings are divided into two general classes, depending on the method of assembly. These are the Conrad, or nonfilling-notch type, and the maximum, or filling-notch type. In the Conrad bearing, the balls are placed between the rings as shown in Fig. 9-2(a). Then they are evenly spaced and the separator is riveted in place. In the maximum-type bearing, the balls are inserted through a filling notch ground into each ring, as shown in Fig. 9-2(b). Because more balls can be placed in such bearings, their load capacity is greater than that of the Conrad type. However, the presence of the notches limits the load-carrying ability of these bearings in the axial direction.

High-carbon chromium steel 52100 is used for balls and rings. It is heat

(a) *Conrad or non-filling notch type*   (b) *Maximum or filling notch type*

FIGURE 9-2 Methods of assembly for ball bearings.

treated to high strength and hardness, and the surfaces are smoothly ground and polished. The dimensional tolerances are very small; the balls must be very uniform in size. The stresses are extremely high because of the small contact areas, and the yield point of the material may be exceeded at certain points. Because of the high values of the fluctuating stresses, antifriction bearings are not designed for unlimited life, but for some finite period of service determined by the fatigue strength of the materials. A specified speed and number of hours of expected service must therefore accompany the given load values for these bearings.

2. Selection of Ball Bearings

Ball bearings were formerly rated on the basis of the compressive stress in the most heavily loaded ball. Except for static loads, experience has shown that the actual cause of failure is fatigue. Fatigue characteristics are thus used for load rating and are dependent to a large extent on experimental results.

The life of a ball bearing is the life in hours at some known speed, or the number of revolutions, that the bearing will attain before the first evidence of fatigue appears on any of the moving elements. Experience has shown that the life of an individual ball bearing cannot be precisely predicted. Fatigue characteristics are thus used for load ratings. Suppose a number of apparently identical bearings are tested under identical conditions. The life at which 10 percent of them have failed and 90 percent are good is called the *rating life*, or L_{10} life, for this bearing.[1] Since high-grade bearings have consistent quality,

[1] See Reference 1, end of chapter.

the *median life*, or life at which 50 percent have failed and 50 percent are good, is generally not more than five times the rating life.

When the rating life is taken at one million revolutions, the load C for this life is called the *basic rating load*. Values for C in Table 9-2 thus represent the load that the bearing can carry for one million revolutions with 90 percent probability that it will be attained or exceeded.

Most ball bearings may have an intended life that is many times one million revolutions. In fact, one million revolutions are attained in 8 hr at a speed of 2,084 rpm.

The basic load rating C for a rating life of one million revolutions for radial and angular contact ball bearings, except filling slot bearings, with balls not larger than 1 in. diam is[2] given by the equation

$$C = f_c(i \cos \alpha)^{0.7} Z^{2/3} D^{1.8} \tag{1}$$

where $f_c =$ a constant from Table 9-1, as determined by the value of $(D \cos \alpha)/d_m$

$i =$ number of rows of balls in the bearing

$\alpha =$ nominal angle of contact (angle between line of action of ball load and plane perpendicular to bearing axis)

$Z =$ number of balls per row

$D =$ ball diameter

$d_m =$ pitch diameter of ball races

Table 9-2 shows dimensions for typical ball bearings, and the basic load rating C.

TABLE 9-1   Constants for Single-Row Radial Contact Groove Ball Bearings

$\dfrac{D \cos \alpha}{d_m}$	f_c	$\dfrac{D \cos \alpha}{d_m}$	f_c	$\dfrac{F_a}{iZD^2}$	X	Y
0.05	3,550	0.22	4,530	25	0.56	2.30
0.06	3,730	0.24	4,480	50		1.99
0.07	3,880	0.26	4,420	100		1.71
0.08	4,020	0.28	4,340	150		1.55
0.09	4,130	0.30	4,250	200		1.45
0.10	4,220	0.32	4,160	300		1.31
0.12	4,370	0.34	4,050	500		1.15
0.14	4,470	0.36	3,930	750		1.04
0.16	4,530	0.38	3,800	1,000		1.00
0.18	4,550	0.40	3,660			
0.20	4,550					

f_c is also valid for single- and double-row angular contact groove ball bearings.
X and Y are also valid for double-row radial contact groove ball bearings.

[2]For balls larger than 1 in. diam, the exponent for D is 1.4.

TABLE 9-2 **Dimensions and Basic Load Ratings for Conrad-Type Single-Row Radial Ball Bearings**

Bearing No.	Bore		Outside Diameter		Width		Balls		Capacity, lb	
	mm	in.	mm	in.	mm	in.	No., Z	Diam., D	Dynamic, C	Static, P_{st}
102	15	0.5906	32	1.2598	9	0.3543	9	3/16	965	550
202			35	1.3780	11	0.4331	7	1/4	1,340	760
302			42	1.6535	13	0.5118	8	17/64	1,660	930
103	17	0.6693	35	1.3780	10	0.3937	10	3/16	1,040	640
203			40	1.5748	12	0.4724	7	5/16	1,960	1,040
303			47	1.8504	14	0.5512	6	3/8	2,400	1,240
104	20	0.7874	42	1.6535	12	0.4724	9	1/4	1,620	980
204			47	1.8504	14	0.5512	8	5/16	2,210	1,280
304			52	2.0472	15	0.5906	7	3/8	2,760	1,530
105	25	0.9843	47	1.8504	12	0.4724	10	1/4	1,740	1,140
205			52	2.0472	15	0.5906	9	5/16	2,420	1,520
305			62	2.4409	17	0.6693	8	13/32	3,550	2,160
106	30	1.1811	55	2.1654	13	0.5118	11	9/32	2,290	1,590
206			62	2.4409	16	0.6299	9	3/8	3,360	2,190
306			72	2.8346	19	0.7480	8	1/2	5,120	3,200
107	35	1.3780	62	2.4409	14	0.5512	11	5/16	2,760	2,010
207			72	2.8346	17	0.6693	9	7/16	4,440	2,980
307			80	3.1496	21	0.8268	8	17/32	5,750	3,710
108	40	1.5748	68	2.6772	15	0.5906	13	5/16	3,060	2,450
208			80	3.1496	18	0.7087	9	1/2	5,640	3,870
308			90	3.5433	23	0.9055	8	5/8	7,670	5,050
109	45	1.7717	75	2.9528	16	0.6299	13	11/32	3,630	2,970
209			85	3.3465	19	0.7480	9	1/2	5,660	3,980
309			100	3.9370	25	0.9843	8	11/16	9,120	6,150
110	50	1.9685	80	3.1496	16	0.6299	14	11/32	3,770	3,260
210			90	3.5433	20	0.7874	10	1/2	6,070	4,540
310			110	4.3307	27	1.0630	8	3/4	10,680	7,350
111	55	2.1654	90	3.5433	18	0.7087	13	13/32	4,890	3,950
211			100	3.9370	21	0.8268	10	9/16	7,500	5,710
311			120	4.7244	29	1.1417	8	13/16	12,350	8,660
112	60	2.3622	95	3.7402	18	0.7087	14	13/32	5,090	4,560
212			110	4.3307	22	0.8661	10	5/8	9,070	6,890
312			130	5.1181	31	1.2205	8	7/8	14,130	10,100
113	65	2.5591	100	3.9370	18	0.7087	15	13/32	5,280	4,950
213			120	4.7244	23	0.9055	10	11/16	10,770	8,460
313			140	5.5118	33	1.2992	8	15/16	16,010	11,600
114	70	2.7559	110	4.3307	20	0.7874	14	15/32	6,580	6,080
214			125	4.9213	24	0.9449	10	11/16	10,760	8,740
314			150	5.9055	35	1.3780	8	1	18,000	13,260

Example 1. Find the value of C for a 207 radial bearing.

Solution. By Table 9-2:
$$d_m = \tfrac{1}{2}(2.8346 + 1.3780) = 2.1063 \text{ in.}$$

$$\frac{D \cos \alpha}{d_m} = \frac{0.4375}{2.1063} = 0.208$$

By Table 9-1:
$$f_c = 4,550$$

By Table 9-2:
$$D = \tfrac{7}{16} = 0.4375 \text{ in.}$$

$$\log D = 9.64098 - 10$$

$$1.8 \log D = 9.35376 - 10$$

$$D^{1.8} = 0.2258$$

$$Z = 9, \qquad Z^{2/3} = \sqrt[3]{9^2} = 4.327$$

By Eq. (1):
$$C = 4,550 \times 4.327 \times 0.2258 = 4,440 \text{ lb,}$$
load for 1 million revolutions with 90 percent probability that it will be attained or exceeded

Design equations for antifriction bearings are based largely on testing with units of inches and pounds. When data for a design are expressed in SI units it is best to first transform them into U.S. customary units. Therefore, in this chapter N will be used to represent number of revolutions. Newtons, when present, will be spelled out in full.

If two groups of identical bearings are run with different loads P_1 and P_2, their rating lives N_1 and N_2 revolutions are found to be inversely proportional to the cubes of their loads.[3]

$$\frac{N_1}{N_2} = \frac{P_2^3}{P_1^3} \tag{2}$$

This equation can be generalized as follows.

$$10^6 C^3 = N_1 P_1^3 = N_2 P_2^3 = N_3 P_3^3 = \cdots \qquad \text{a constant} \tag{3}$$

The terms in this equation are constants and are all equal to one another. They are also equal to $10^6 C^3$ as previously defined. Life N in revolutions in all equations can be replaced if desired by

$$N = 60nL \tag{4}$$

where n = speed, rpm
  L = life, hr

Example 2. For the 207 bearing of Example 1, find the radial load P for a rating life of 500 hr, at 1,500 rpm.

[3]See p. 75 of Reference 8, end of chapter.

Solution. By Eqs. (3) and (4): $P_1^3 = \dfrac{10^6 C^3}{N_1} = \dfrac{10^6 C^3}{60nL}$

Or:
$$= \frac{10^6 \times 4{,}440^3}{60 \times 1{,}500 \times 500} = 1{,}945{,}000{,}000$$

From which: $P_1 = 1{,}250 \text{ lb}$

3. Effect of Axial Load

When an axial component of load is present in addition to the radial, the *equivalent radial load P* is the larger of the values given by the two following equations.[4]

$$P = V_1 F_r \tag{5}$$

$$P = X V_1 F_r + Y F_a \tag{6}$$

where F_r = radial component of load

F_a = axial component of load

X = radial factor from Table 9-1

Y = axial or thrust factor from Table 9-1 as determined from value of F_a/iZD^2

V_1 = race rotation factor, equal to unity for inner ring rotation, and 1.2 for outer ring rotation

Equations (5) and (6) apply to radial and angular contact bearings, but not to filling-slot bearings.

A service factor C_1 can be inserted into Eqs. (5) and (6) to care for any shock and impact conditions to which the bearing may be subjected.

$$P = C_1 V_1 F_r \tag{7}$$

$$P = C_1 (X V_1 F_r + Y F_a) \tag{8}$$

Values to be used for C_1 depend on the judgment and experience of the designer, but Table 9-3 may serve as a guide.

TABLE 9-3 Shock and Impact Factors

Type of Load	C_1
Constant or steady	1.0
Light shocks	1.5
Moderate shocks	2.0
Heavy shocks	3.0 and up

[4]See Reference 1 and p. 95 of Reference 8, end of chapter.

The impact load on a bearing should not exceed the static capacity as given by Table 9-2 or the race may be damaged by Brinelling from the balls. This load may be exceeded somewhat if the bearing is rotating and the duration of the load is sufficient for the bearing to make one or more complete revolutions while the load is acting.

Example 3. Suppose the bearing in Example 1 carries a combined load of 400 lb radially and 300 lb axially at 1,200 rpm. The outer ring rotates, and the bearing is subjected to moderate shock. Find the rating life of this bearing in hours.

Solution. $\qquad \dfrac{F_a}{iZD^2} = \dfrac{300}{9 \times 0.4375^2} = 174$

By Table 9-1: $\qquad Y = 1.50$

By Table 9-3: $\qquad C_1 = 2$

By Eq. (8): $\qquad P = 2(0.56 \times 1.2 \times 400 + 1.5 \times 300)$

$\qquad\qquad\qquad\qquad = 1,440$ lb equivalent radial load

By Eqs. (3) and (4): $\quad N = \dfrac{10^6 C^3}{P^3} = 60nL$

Rating life: $\qquad L = \dfrac{10^6 C^3}{60nP^3} = \dfrac{10^6 \times 4,440^3}{60 \times 1,200 \times 1,440^3} = 410$ hr

$\qquad\qquad\qquad\qquad$ life with 90 percent probability that it will be attained or exceeded

Example 4. What change in the loading of a ball bearing will cause the expected life to be doubled?

Solution. Let N_1 and P_1 be the original life and load for the bearing. Let N_2 and P_2 be the new life and load.

Then: $\qquad N_2 = 2N_1$

By Eq. (3): $\qquad P_2^3 = \dfrac{N_1 P_1^3}{N_2} = \dfrac{N_1 P_1^3}{2N_1} = 0.5 P_1^3$

Or: $\qquad P_2 = 0.794 P_1$

Hence a reduction of the load to 79 percent of its original value will cause a doubling of the expected life of a ball bearing.

4. Design for Variable Loading

Ball bearings frequently operate under conditions of variable load and speed. Design calculations should take into account all portions of the work cycle and should not be based solely on the most severe operating conditions. The work cycle should be divided into a number of portions in each of which the speed and load can be considered as constant.

Suppose $P_1, P_2, \ldots$ are the loads on the bearing for successive intervals of the work cycle. Let N_1 be the life of the bearing, revolutions, if operated exclusively at the constant load P_1. Let there be N_1' applications of load P_1. Then N_1'/N_1 represents the proportion of the life consumed in this portion of the cycle.

Let N_2 be the life of the bearing, revolutions, if operated exclusively at load P_2. Let there be N_2' applications of load P_2. Then N_2'/N_2 represents the proportion of the life consumed by load P_2.

A corresponding statement can be made for each portion of the work cycle. The sum of these proportions represents the total life of the bearing or unity. Then

$$\frac{N_1'}{N_1} + \frac{N_2'}{N_2} + \frac{N_3'}{N_3} + \cdots = 1 \tag{9}$$

Let N_c be the life of the bearing under the combined loading. Let $N_1' = \alpha_1 N_c$ where α_1 represents the proportion of the total life, consumed under load P_1. In a similar way $N_2' = \alpha_2 N_c'$, $N_3' = \alpha_3 N_c$, and so on. Substitution in Eq. (7) yields[5]

$$\frac{\alpha_1}{N_1} + \frac{\alpha_2}{N_2} + \frac{\alpha_3}{N_3} + \cdots = \frac{1}{N_c} \tag{10}$$

Letter N in the SI refers to force, newtons, but letter N has other long-established meanings in engineering. As used in Miner's equations, Eq. (9) and Eq. (10) derived above, it denotes number of stress cycles. The reader must carefully keep this in mind and wherever letter N appears be sure he is giving it the correct meaning.

By Eq. (3): $$N_1 = \frac{10^6 C^3}{P_1^3}, \qquad N_2 = \frac{10^6 C^3}{P_2^3}$$

and so on. Substitution should be made into Eq. (10).

$$\frac{1}{N_c} = \frac{\alpha_1 P_1^3}{10^6 C^3} + \frac{\alpha_2 P_2^3}{10^6 C^3} + \cdots \tag{10}$$

or

$$\frac{10^6 C^3}{N_c} = \alpha_1 P_1^3 + \alpha_2 P_2^3 + \cdots \tag{11}$$

It is obvious that $\alpha_1 + \alpha_2 + \cdots$ must equal unity. The application of this equation will be demonstrated by the following examples.

Example 5. A ball bearing is to operate on the following work cycle.

> Radial load of 1,400 lb at 200 rpm for 25% of the time
> Radial load of 2,000 lb at 500 rpm for 20% of the time
> Radial load of   800 lb at 400 rpm for 55% of the time

[5]See p. 89 of Reference 8, end of chapter.

The inner ring rotates; loads are steady. Find the minimum value of the basic rating load C for a suitable bearing for this application if the required life is 7 years at 4 hr per day.

Solution. The equation will be applied on the basis of a work cycle of 1 min. The following table should be constructed.

	Assumed Interval, min	rpm	In Assumed Interval, rev.
$P_1 = 1,400$ lb	0.25	200	50
$P_2 = 2,000$ lb	0.20	500	100
$P_3 = 800$ lb	0.55	400	220
	1.00		370 rpm

Then
$$\alpha_1 = \frac{50}{370}, \qquad \alpha_2 = \frac{100}{370}, \qquad \alpha_3 = \frac{220}{370}$$

A working year is assumed to consist of 250 days.

No. of revolutions, $N_c = 7 \times 250 \times 4 \times 60 \times 370 = 155,400,000$

By Eq. (11): $\dfrac{10^6 C^3}{N_c} = \dfrac{50}{370} \times 1,400^3 + \dfrac{100}{370} \times 2,000^3 + \dfrac{220}{370} \times 800^3$

$$= 370,800,000 + 2,162,200,000 + 304,400,000$$

$$\frac{10^6 C^3}{N_c} = 2,837,400,000$$

$$\frac{C^3}{N_c} = 2,837.4$$

$$C^3 = 2,837.4 N_c = 2,837.4 \times 155,400,000 = 440,930,000,000,000$$

$$C = 7,610 \text{ lb}$$

Hence a bearing such as No. 308 should be satisfactory, keeping in mind there is but a 90 percent probability that the required life will be attained or exceeded.

Example 6. A 306 radial ball bearing with inner ring rotation has a 10-sec work cycle as follows.

For 2 sec	For 8 sec
$F_r = 800$ lb	$F_r = 600$ lb
$F_a = 400$ lb	$F_a = 0$
$n = 900$ rpm	$n = 1,200$ rpm
Light shock	Steady load

Find the rating life of this bearing in hours and in years of 250 working days of 2 hr each.

Solution. Equation (11) will be applied to the work cycle of 10 sec. For 2 sec:

By Eq. (7): $\qquad P = C_1 V_1 F_r = 1.5 \times 1 \times 800 = 1,200$ lb

Data from Table 9-2: $\qquad \dfrac{F_a}{iZD^2} = \dfrac{400}{8 \times 0.5^2} = 200$

By Eq. (8): $\qquad P = C_1(XV_1 F_r + YF_a)$

By Table 9-1: $\qquad Y = 1.45$

In Eq. (8): $\qquad P = 1.5(0.56 \times 1 \times 800 + 1.45 \times 400)$

$$= 1,540 \text{ lb, equivalent radial load (will be used)}$$

$$\text{No. of revolutions} = 2 \times \frac{900}{60} = 30, \quad \text{in 2 sec}$$

For 8 sec: $\qquad P = F_r = 600$ lb

$$\text{No. of revolutions} = 8 \times \frac{1,200}{60} = 160, \quad \text{in 8 sec}$$

For 10 sec:

$$\text{No. of revolutions} = 30 + 160 = 190$$

Then: $\qquad \alpha_1 = \dfrac{30}{190}, \qquad \alpha_2 = \dfrac{160}{190}$

By Eq. (11): $\qquad \dfrac{10^6 C^3}{N_c} = \alpha_1 P_1^3 + \alpha_2 P_2^3$

$$\frac{10^6 \times 5,120^3}{N_c} = \frac{3}{19} \times 1,540^3 + \frac{16}{19} \times 600^3$$

$$= 576,700,000 + 181,900,000 = 758,600,000$$

$$\frac{5,120^3}{N_c} = 758.6$$

$$N_c = \frac{5,120^3}{758.6} = 176,950,000 \text{ rev}$$

$$n = \frac{190}{10} \times 60 = 1,140 \text{ rpm}$$

By Eq. (4): $\qquad L_{10} = \dfrac{N_c}{60n} = \dfrac{176,950,000}{60 \times 1,140} = 2,580 \text{ hr}$

$$= \frac{2,580}{2 \times 250} = 5.16 \text{ yr life with 90 percent probability}$$

$$\text{that it will be attained or exceeded}$$

5. Static Capacity

Unless there is relative motion between the rings of a ball bearing, the depressions of the balls into the races will gradually enlarge, and permanent indentations will remain. The static capacity is ordinairly defined as the maximum allowable static load that does not impair the running characteristics of the bearing to make it unusable.

This permanent deformation under the balls is known as Brinnelling and takes place at moderate to high loads. The magnitude of the permissible load is found by methods given in the standard.[6] Calculations for the bearings of Table 9-2 have been made and are shown in the column headed P_{st}.

When very smooth and quiet operation is required, the loading should be no more than about one-half the static capacity.

6. Design for Different Confidence Levels

When a bearing is installed there is no way of knowing whether it is one of the 90 percent that are good or one of the 10 percent that will not attain the rating life. In other words, one can have but 90 percent confidence that the bearing will achieve or exceed its rating life, usually designated L_{10}.

In some cases a greater degree of reliability is required. The expected life will of course be reduced as the reliability is made higher. Let an adjusting factor a_1 be taken such that life L_n is equal to $a_1 L_{10}$. Factors for a_1 for different values of the reliability are given in Table 9-4. Life L_{10} is the rating life.

TABLE 9-4   Constants for Designing at Different Confidence Levels

Reliability, %	L_n	Life Adjustment Factor, a_1
90	L_{10}	1.00
95	L_5	0.62
96	L_4	0.53
97	L_3	0.44
98	L_2	0.33
99	L_1	0.21

7. Friction and Lubrication of Ball Bearings

Rolling friction prevails to a large extent in ball and roller bearings. Ball bearings, in general, have slightly less frictional resistance than high-grade partial journal bearings operating with flooded lubrication. Ball bearings

[6]See Reference 1, end of chapter.

have definitely less friction than journal bearings operating with scanty oil supply.

Average values for the coefficient of friction[7] are as follows:

Self-aligning ball bearings	0.0010
Cylindrical roller bearings with flange guided short rollers	0.0011
Thrust ball bearings	0.0013
Single-row deep-groove ball bearings	0.0015
Tapered and spherical roller bearings with flange-guided rollers	0.0018
For cageless needle bearings, in a similar manner, is obtained	0.0045

The amount and viscosity of the lubricating oil used determine to a large extent the friction of a ball bearing. Tests have shown that drip feed, which supplies a drop of oil every 2–4 hr, gives much lower friction values than a more copious oil supply. In fact, for flooded lubrication, the power lost in churning the oil may be greater than the friction of the bearing alone. A very light or thin oil will give lower friction than one whose viscosity is high.

A light coating of oil or grease is all that is required to maintain an oil film between balls and races. When oil lubrication is used, more or less elaborate seals are needed to retain the lubricant. Lubrication by oil mist has proved successful for very high-speed applications.

Grease is essentially a soap impregnated with lubricating oil. For rolling bearing service, the oil has viscosity properties similar to SAE 20 or SAE 30.

Grease tends to remain in the bearing and protect the surfaces. Therefore it does not require such elaborate retainers. At low temperature, the balls cut a channel through the grease, but enough oil usually sweats off to provide lubricant. Prelubricated sealed bearings can frequently be mounted at lower cost because sealing parts and grease fittings are eliminated. The bearings are filled at the factory with the proper quantity of grease. They have been known to run for years without servicing. For temperatures above 250°F, special high-temperature greases must be used.

Reliability of the oil supply is of the utmost importance. Lack of lubricant can cause local heating, expansion, and loss of radial play. The load on the balls may be increased to the extent that spalling and early bearing failure occur.

Back and forth rotation of the shaft through small angles can cause early failure of bearings unless the load is very light. Lubrication is difficult because the oil or grease may not be replenished back of a ball or roller before the motion is reversed.

[7]See p. 41 of Reference 8, end of chapter. See also Styri, H., "Friction Torque in Ball and Roller Bearings," *Mech. Eng.*, **62**, 1940, p. 886.

The lubricant in a ball bearing serves not only to reduce the friction, but to prevent foreign matter, which would injure the surfaces, from entering the bearing. Every effort must be made to protect the highly polished surfaces from grit, water, acids, or anything that will cause scratches or corrosion. Corrosion fatigue is particularly injurious in causing early fatigue failure of the bearing. Fatigue failures are caused by the bearing surface breaking down; small particles of the metal come out and leave pits or spalls, as illustrated in Fig. 9-3. This breakdown is preceded by minute surface cracks, which are developed by repeated stress applications until they become sufficiently large to form zones of local weakness.

FIGURE 9-3   Failure of bearing race by pitting or spalling. (*Courtesy of New Departure, Division of General Motors.*)

8. Bearing Materials and Surface Finish

As previously mentioned, high-carbon chromium steel 52100 is widely used as a ball bearing material. This is a through hardening steel, and a commonly accepted minimum hardness for bearing components is 58 Rockwell C. This material is not suitable for temperatures over 350°F. For higher temperatures, steels especially developed for high-temperature service should be used.

In contrast to ball bearings, roller bearings are usually made of case-hardened steels. The carburized case or exterior should have a hardness of 58–63 R_c. The core is softer with a hardness of 25–40 R_c. Certain plain-carbon and alloy steels have been found suitable for roller bearing service. The maximum temperature is limited to about 350°F.

The separator, cage, or retainer for conventional bearings is usually a stamping of low-carbon steel. For higher speeds or precision service, the separator is machined from a suitable copper alloy, such as bronze. Cages are also made of a solid lubricant material for use where conventional types of lubrication cannot be used.

The life of a rolling element bearing depends to a large extent on the smoothness of the contacting surfaces—the balls, rollers, and races. Typical surface roughness dimensions for production bearings are as follows.[8]

---

[8] For a discussion of root mean square (rms), average, see Chapter 13.

These are in terms of microinches or millionths of an inch, usually written μ in.

Balls	2–3 μin. rms
Ball races	6–10 μin. rms
Rollers	8–12 μin. rms
Roller races	10–20 μin. rms

Surface finish values vary considerably from manufacturer to manufacturer. For critical applications it may be necessary to obtain the actual value of the surface finish for the bearing under consideration.

9. Mounting of Ball Bearings

For a rotating shaft, relative rotation between shaft and bearing is usually prevented by mounting the inner ring with a press fit and securing it with a nut threaded on the shaft. Excessive interference of metal must be avoided in press fits, or the stretching of the inner ring may decrease the small but requisite internal looseness of the bearing.

Although the outer ring, when the shaft rotates, is mounted more loosely than the inner ring, rotational creep between the ring and housing should be prevented. When two bearings are mounted on the same shaft, the outer ring of one of them should be permitted to shift axially to care for any differential expansion between shaft and housing. Several examples of typical mounting details with oil retainers are shown in Fig. 9-4. The catalogs of the various manufacturers contain useful illustrations of this kind, as well as other practical information.

Shafts or spindles in machine tools and precision equipment that must rotate without play or clearance in either the radial or axial directions can be mounted on preloaded ball bearings. The preloading, which removes all play from the bearing, can be secured in a number of different ways. For example, suppose the outer rings of the bearings at A in Fig. 9-5 project a small but controlled amount beyond the inner rings. When the inner rings are brought into contact at B by means of the locknut, the balls will be displaced in the rings an amount sufficient to remove all looseness from the bearing. Close attention must be paid to dimensions and tolerances to secure just enough projection of the ring to remove the play, but not so much as to induce excessive pressure or binding of the balls. The bearing at the other end of the shaft must be arranged for free axial movement of the outer ring. The bearings in Fig. 9-5 can be separated if desired with one bearing at each end of the shaft. Although this arrangement will remove the looseness from both ends of the shaft, serious stresses may be induced by a temperature difference between shaft and housing. Preloaded, double-row radial bearings are made by some manufacturers.

(a)

(b)

(c)

(d)

(e)

(f)

(g)

(h)

FIGURE 9-4   Details of typical anti-friction bearing mountings.

FIGURE 9-5   Method for obtaining preloading in ball bearings.

10. Permissible Misalignment

Misalignment of ball and roller bearings can have a serious effect on their life and load-carrying capacity. Misalignment can occur from shaft deflections, from loading, or from errors in machining the supports for the rings. Sometimes the support for a bearing will deflect elastically and the design should be arranged so that the housing deflection offsets the shaft deflection rather than opposing it.

Antifriction bearings are manufactured with various amounts of internal clearance. If the misalignment does no more than remove the internal clearance of a ball bearing, in general no shortening of the life will occur. Pressing a ring on a shaft may remove a portion of the original clearance. Permissible misalignment is also influenced by such factors as types of loading, race curvature and depth, diameter and number of balls, method of lubrication, retainer geometry, and minor dimensional details of the various makes.

Roller bearings are more sensitive to the effects of misalignment so that any slope of the shaft has a bad effect on the life and load capacity. A slight crown on the rolls has proved to be beneficial in overcoming the effects of misalignment. In needle bearings the rollers depend on each other for guidance so that some resultant clearance is necessary for satisfactory operation.

Recommended misalignment limits are given in Table 9-5. In this table some makers recommend the lower ranges, while others permit the higher values. In case of doubt, the safest plan is to obtain the assistance of the maker of the particular bearing that is under consideration.[9]

TABLE 9-5   Permissible Angular Misalignment between Bearing and Shaft

Single-row radial ball bearing:	
Slow speed and loose fit	0.0020–0.0040 in./in.
High speed	0.0010 in./in.
Cylindrical roller bearing	0.0003–0.0015 in./in.
Needle bearing	0.0003–0.0010 in./in.
Tapered roller bearing	0.0005 in./in.

11. Unground Ball Bearings

The foregoing discussion has referred to ball bearings of the highest quality of materials and workmanship. Other bearings of lower quality can be purchased for installations requiring less accuracy or where cost is the controlling

[9]See Smith, R. J., "Rules for Application of Needle Bearings," *Machine Design* **30**, Jan. 9, 1958, p. 116. See also Korff, W. H., "Twelve General Considerations for Needle Bearing Application," *Prod. Eng.*, **15**, 1944, p. 389.

factor. The rings are made on automatic screw machines and are hardened
but not ground.[10]

Different types of construction are in use. The bearing of Fig. 9-6(a)
has the outer ring split by a plane perpendicular to the axis. The bearing is
assembled by spinning the edges of the bushing, which is slipped over the
outer rings. The bearing of Fig. 9-6(b) has a split inner ring, and is made by
staking the bore of the inner ring as shown. Various additional features,
such as pulleys, gears, castor wheels, and so on, can be incorporated as an
integral part of the outer ring. Sketch (c) shows a sheave-idler in which the
outer ring is formed by the stampings comprising the sheave. Unground ball
bearings are frequently cheaper than an equivalent plain bushing.

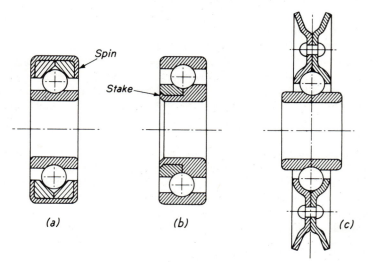

FIGURE 9-6   Types of unground ball bearings.

12. Relative Advantages of Rolling Element Bearings

Some of the advantages of rolling element are as follows.

(1) Starting friction is low—a desirable feature for intermittent service
or for starting at low temperatures.

(2) Loads can be inclined at any angle in the transverse plane.

(3) Thrust components can be carried.

(4) Maintenance costs are low.

[10]See Agnoff, C., "Multipurpose Bearings," *Machine Design*, **39**, Aug. 31, 1967, p. 98.
See also Recknagel, F. W., "Construction and Characteristics of Low-Cost Ball Bearings,"
Prod. Eng., **21**, Jan. 1950, p. 106.

(5) Bearings are easily replaced when worn out.

(6) Less axial space is required than for journal bearings. Shafts are thus shorter, and may even be smaller in diameter.

Some of the advantages of fluid film bearings are as follows.

(1) First cost is usually lower.

(2) Less radial space is required than for ball bearings.

(3) They are better suited to overload and shock conditions.

(4) Operation is quieter than with ball bearings, especially after wear has taken place.

(5) There is less difficulty with fatigue.

(6) They are less easily injured by foreign matter.

13. Roller Bearings

When shock and impact loads are present, or when a large bearing is needed, cylindrical and tapered roller bearings are usually used. A roller bearing in general consists of the same four elements as a ball bearing: the two rings, the cage, and the rollers. Some typical examples of roller bearings are shown in Fig. 9-7.

| Cylindrical roller bearing | Flexible roller bearing | Needle bearing | Tapered roller bearing | Spherical roller bearing | Roller thrust bearing |

FIGURE 9-7 Types of roller bearings.

In the cylindrical roller bearing, the flanges on the rings serve to guide the rollers in the proper direction. When the flanges are omitted from one of the rings, as shown in Fig. 9-7, the rings can then be displaced axially with respect to each other, and no thrust component can be carried.

In addition to the radial load, the tapered roller bearing can carry a large axial component whose magnitude depends on the angularity of the rollers. The radial load will also produce a thrust component. The outer ring is separable from the remainder of the bearing. In this type of bearing, it is possible to make adjustment for the radial clearance. Two bearings are usually mounted opposed to each other, and the clearance is controlled by adjusting one bearing against the other. Double-row tapered roller bearings are also available.

Roller bearings in general can be applied only where the angular misalignment caused by shaft deflection is very slight. This deficiency is not present in the spherical roller bearing. It has excellent load capacity and can carry a thrust component in either direction.

In the flexible roller bearing, the rollers are wound from strips of spring steel, and afterwards are hardened and ground to size. If desired, the rollers can bear directly on the shaft without an inner ring, particularly if the shaft surface has been locally hardened. This bearing has been successfully applied under conditions of dirty environment.

The needle bearing has rollers that are very long as compared to their diameters. Cages are frequently not used, and the inner ring may or may not be present. The outer ring may consist of hardened thin-walled metal as shown in Fig. 9-7; the housing in which the bearing is mounted must have sufficient thickness to give adequate support. The friction of needle bearings is several times as great as for ordinary cylindrical roller bearings. Because of the tendency of the unguided rollers to skew, needle bearings are particularly adapted to oscillating loads, as in wrist pins, rocker arms, and universal joints. For continuous rotation, needle bearings are usually suitable where the loading is intermittent and variable so that the needles will be frequently unloaded and thus tend to return to their proper locations. When the application involves angular misalignment of the shaft, two short bearings end to end usually are better than one bearing with long rollers. The needle bearing is low-priced and requires very little radial space.

FIGURE 9-8 Self-aligning spherical roller bearing for radial and thrust loads. (*Courtesy McGill Mfg. Co.*)

Spherical roller bearings, Figs. 9-7 and 9-8, can be used when the shaft has angular misalignment.

Thrust bearings can be constructed by the use of straight or tapered rollers.

Roller bearings are selected by a process similar to that used for ball bearings. They must be chosen, however, in accordance with the recommendations given in the catalog of the manufacturer of the particular type of bearing under consideration.

14. Contact Stress between Cylinders

When two bodies having surfaces of different curvatures are pressed together, an area of contact develops, because of course there is no such thing as line or point contact. This area is usually very small and the resulting stress can attain a very high value. In addition to rolling contact bearings, the problem is of importance in the design of cams and gear teeth.

The elastic analysis is very complex, but Fig. 9-9 gives some results for

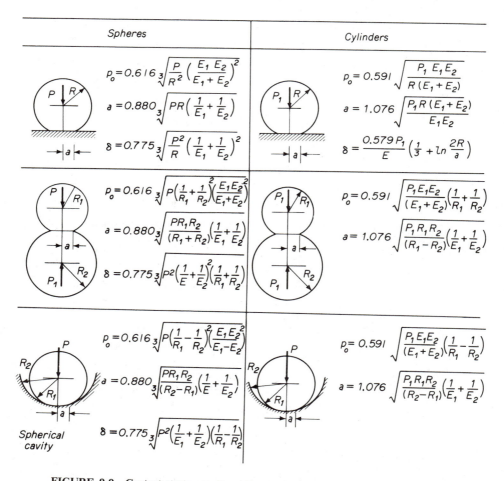

FIGURE 9-9 Contact stress equations. P_0, maximum compressive stress; a, half width of contact zone; P, total load on spheres; P_1, load per axial inch on cylinders; $\mu = 0.3$; δ, normal approach toward each other for points remote from contact zone; $\ln x = \log_e x = 2.3026 \log_{10} x$.

both spheres and cylinders. For spheres, a is the radius of the contact circle, but for rectangles, a represents the half-width of the contact rectangle. The total force pressing two spheres together is P, while P_1 represents the load per axial inch for cylinders. The maximum compressive stress in the zone of contact is represented by p_0. Poisson's ratio in these equations was taken as 0.3. Normal approach toward each other for points remote from the contact zone is given by δ. As usually written, $\ln x = \log_e x = 2.3026 \log_{10} x$.

Example 7. A $\frac{1}{2}$-in.-diam steel roller runs on the inside of a steel ring of inside diameter 3.2 in. If the load on the roller is 1,000 lb/axial inch, find the value of the maximum compressive stress p_0 and the half-width a of the contact zone.

Solution. Use the equations in the lower right-hand corner of Fig. 9-9.

$$E_1 = E_2 = E$$

$$p_0 = 0.591 \sqrt{\frac{P_1 E}{2}\left(\frac{1}{R_1} - \frac{1}{R_2}\right)} = 0.591 \sqrt{\frac{P_1 E}{2}\left(\frac{R_2 - R_1}{R_1 R_2}\right)}$$

$$= 0.591 \sqrt{\frac{1,000 \times 30,000,000(1.6 - 0.25)}{2 \times 0.25 \times 1.6}}$$

$$= 0.591 \times 225,000 = 133,000 \text{ psi}$$

$$a = 1.076 \sqrt{\frac{P_1 R_1 R_2}{R_2 - R_1} \times \frac{2}{E}} = 1.076 \sqrt{\frac{1,000 \times 0.25 \times 1.6 \times 2}{(1.6 - 0.25)30,000,000}}$$

$$= 1.076 \times 0.004\ 44 = 0.0049 \text{ in.}$$

Example 7A. A 12-mm-diam steel roller runs on the inside of a steel ring of inside diameter 80 mm. If the load on the roller is 175 newtons per axial millimeter, find the value of the maximum compressive stress p_0 and the half-width a of the contact zone.

Solution. Use the equations in the lower right-hand corner or Fig. 9-9.

$$E_1 = E_2 = E$$

$$p_0 = 0.591 \sqrt{\frac{175 \times 206,900(40 - 6)}{2 \times 6 \times 40}}$$

$$= 0.591 \times 1,602 = 947 \text{ MPa}$$

$$a = 1.076 \sqrt{\frac{175 \times 6 \times 40 \times 2}{(40 - 6)206,900}}$$

$$= 1.076 \times 0.1093 = 0.118 \text{ mm}$$

15. Elasto-hydrodynamic Lubrication

Elasto-hydrodynamic lubrication deals with the lubrication of highly loaded contacts, where the elastic deformation of the parts must be taken into account as well as the increase in viscosity of the lubricant due to the high pressure.[11] This small elastic flattening of the parts, together with the increase in viscosity, provides a film, although very thin, that is much thicker than would prevail with completely rigid parts. The theory is complex and beyond the scope of this book.

REFERENCES

1. *Load Ratings and Fatigue Life for Ball Bearings*, Arlington, Va.: The Anti-Friction Bearing Manufacturers Association, Inc., 1979.

2. ALLAN, R. K., *Rolling Bearings*, London: Sir Isaac Pitman & Sons Ltd., 1945.

3. *Bearings Reference Issue, Machine Design*, **42**, June 18, 1970.

4. *Boundary Lubrication—An Appraisal of World Literature*, New York: American Society of Mechanical Engineers, 1969.

5. DOWSON, D., and G. R. HIGGINSON, *Elasto-hydrodynamics Lubrication*, 2d ed., New York: Pergamon Press, Inc., 1977.

6. HARRIS, T. A., *Rolling Bearing Analysis*, New York: John Wiley & Sons, Inc., 1966.

7. MORTON, H. T., *Anti-friction Bearings*, 2nd ed., Ann Arbor, Mich.: H. T. Morton, 1965.

8. PALMGREN, A., *Ball and Roller Bearings*, 3rd ed., Philadelphia: SKF Industries, Inc., 1959.

9. ROTHBART, H. A., Ed., *Mechanical Design and Systems Handbook*, New York: McGraw-Hill Book Company, 1964, Section 13.

10. SHAW, M. C., and E. F. MACKS, *Analysis and Lubrication of Bearings*, New York: McGraw-Hill Book Company, 1949, Chapter 10.

PROBLEMS

In the following problems, it is assumed that the bearing has but one row of balls. A working year will be taken as 250 days.

[11]See Zaretsky, E. V., and W. J. Anderson, "EHD Lubrication," *Machine Design*, **40**, Nov. 7, 1968, p. 167. See also Burton, R. A., "An Analytical Investigation of Visco-elastic Effects in the Lubrication of Rolling Contact," *Trans. ASME*, **3**, 1960, p. 1; and Cheng, H. S., "A Numerical Solution of the Elastohydrodynamic Film Thickness in an Elliptical Contact," *Trans. ASME*, **92**, Series F., *J. Lubr. Technol.*, Jan. 1970, p. 155.

1. If a 309 bearing is installed at the left reaction in Fig. 9-10, find the rating life at 8 hr per day. Speed is 450 rpm and the bearing is subjected to light shock. Inner rings are pressed on the shaft; outer rings have free fits.   *Ans.* $L_{10} = 10.05$ yr.

FIGURE 9-10   Problem 1.

2. Use Fig. 9-10 but the rise is 5 in. vertically for 12 in. horizontally. A 308 bearing is installed at the left end. The shaft rotates 1,800 rpm for 60% of the time and 750 rpm for the remainder of the time. Loads are steady. Make a diagram and show the reaction at each end. Compute the rating life at 8 hr per day of the left bearing.   *Ans.* $L_{10} = 3.96$ yr.

3. A cross section through an industrial car, and the loading for one axle, are shown schematically in Fig. 9-11. The inner rings of the bearings are fixed on the shaft; the outer rings have free fits. Wheels are press fitted on the axle.

Calculate and show reactions at the rails. Now separate the system at the bearings and show all forces acting at the points of separation for each portion. Consider the forces from bearings and wheels to be concentrated at their respective centers. Draw and dimension the bending moment diagram for the axle.

Based on the loading conditions of the left bearing, find the rating life at 4 hr per day of a 313 bearing at a speed of 300 rpm. Assume the bearing is subjected to moderate shock.   *Ans.* $L_{10} = 10.20$ yr.

FIGURE 9-11   Problem 3.

4. Work Problem 3, except the downward load is 2,400 lb and the sidewise load is 450 lb.   *Ans.* $L_{10} = 11.39$ yr.

5. In Fig. 3-14, let the loads be steady and the rpm equal to 900. Find the rating life at 4 hr per day for a 314 bearing installed at the left reaction.

Ans. $L_{10} = 7.40$ yr.

6. A 205 bearing carries a 1,000-lb radial and a 250-lb axial load for one-sixth of the work cycle, and a 500-lb radial for the remainder of the cycle. Speed is uniform at 180 rpm. The inner ring rotates and loads are steady. Find the rating life at 4 hr per day for this bearing. *Ans.* $L_{10} = 4.84$ yr.

7. A 210 bearing has a work cycle with 1,000 rpm for one-third of the time, 2,000 rpm for one-third of the time, and 4,000 rpm for one-third of the time. The inner ring rotates. Assume light-shock conditions. The radial load is 400 lb and the axial load is 150 lb. Find the rating life at 5 hr per day for this bearing.

Ans. $L_{10} = 2.78$ yr.

8. A 206 bearing carries a radial load of 500 lb at 500 rpm for one-half the time, and 150 lb radial at 3,600 rpm for the remaining half of the time. The inner ring rotates, and the loads are steady. Find the rating life at 8 hr per day.

Ans. $L_{10} = 8.46$ yr.

9. A 206 bearing is subjected to the following work cycle.

> Radial load of 500 lb at 150 rpm for 30% of the time
> Radial load of 750 lb at 600 rpm for 10% of the time
> Radial load of 250 lb at 300 rpm for 60% of the time

The inner ring rotates; loads are steady. What is the rating life at 8 hr per day of this bearing? *Ans.* $L_{10} = 9.36$ yr.

10. What change in the loading of a ball bearing will cause the expected life to be halved?

11. A 104 bearing operates under the following schedule of loads and speeds. The inner ring rotates; loads are steady. Find the rating life at 2 hr per day for this bearing.

> Radial load of 370 lb at 2,000 rpm for 5% of the time
> Radial load of 255 lb at 3,300 rpm for 15% of the time
> Radial load of 125 lb at 1,750 rpm for 35% of the time
> Radial load of 100 lb at 2,200 rpm for 45% of the time

Ans. $L_{10} = 9.17$ yr.

12. A 307 bearing operates under the following schedule of loads and speeds. The inner ring rotates; loads are steady. Find the rating life at 8 hr per day for this bearing.

> Radial load of 450 lb at 1,000 rpm for 30% of the time
> Radial load of 1,000 lb at 300 rpm for 15% of the time
> Radial load of 2,000 lb at 100 rpm for 55% of the time

Ans. $L_{10} = 3.09$ yr.

13. The work cycle for a 304 ball bearing is as follows.

> 500-lb radial and 125-lb axial at 300 rpm for one-half the cycle
> 600-lb radial at 500 rpm for one-quarter of the cycle
> 350-lb radial at 1,000 rpm for one-quarter of the cycle

Loads are steady; the outer ring rotates. Find the rating life at 2 hr per day of this bearing.   *Ans.* $L_{10} = 7.18$ yr.

14. Find the rating life at 2 hr per day of a 307 bearing that is subjected to a steady radial load of 615 lb, and an axial load of 350 lb. The inner ring rotates. The speed is 3,600 rpm.   *Ans.* $L_{10} = 13.53$ yr.

15. An electric motor for a drill press has 104 bearings mounted as in Fig. 9-10 except that the shaft is vertical. Bearings are 10 in. center to center. The center of the V-belt pulley is located 2.5 in. from the lower motor bearing. During two-thirds of the work cycle the steady belt forces equal 128 lb at 900 rpm. During the remainder of the cycle the steady belt forces equal 88 lb at 1770 rpm. The motor armature weighs 67 lb. Find the rating life at 4 hr per day of the most heavily loaded bearing.
                                       Ans. $L_{10} = 9.08$ yr.

16. A ball bearing is subjected to a radial load of 2,000 lb for 40 percent of the time and 1,000 lb for the remainder of the load cycle. Speed is 1,760 rpm. Expected life to be 10 yr at 6 hr per day. Find the minimum value of the basic rating load C that the bearing must have. Loads are steady.   *Ans.* $C = 18,190$ lb.

17. A ball bearing is subjected to a radial load of 2,000 lb at 880 rpm for 40 percent of the time and to a radial load of 1,000 lb at 1,760 rpm for the remainder of the time. Expected life to be 10 yr at 6 hr per day. Find the minimum value of the basic rating load C that the bearing must have. Loads are steady.
                                       Ans. $C = 15,160$ lb.

18. A ball bearing carries a load of 1,000 lb at 1,760 rpm for 0.4 of the time and a load of 2,000 lb at 880 rpm for the remainder of the time. Life is to be 10 yr at 6 hr per day. Find the minimum value of the basic rating load C that the bearing must have. Loads are steady.   *Ans.* $C = 16,130$ lb.

19. A ball bearing carries a load of 1,000 lb at 880 rpm for 0.4 of the time and a load of 2,000 lb at 1,760 rpm for the remainder of the time. Life is to be 10 yr at 6 hr per day. Find the minimum value of the basic rating load C that the bearing must have. Loads are steady.   *Ans.* $C = 19,930$ lb.

20. A ball bearing carries a load of 2,000 lb at 1,760 rpm for 0.4 of the time and a load of 1,000 lb at 880 rpm for the remainder of the time. Life is to be 10 yr at 6 hr per day. Find the minimum value of the basic rating load C that the bearing must have. Loads are steady.   *Ans.* $C = 10,820$ lb.

21. A ball bearing carries a load of 1,000 lb at 880 rpm for 0.3 of the time and a load of 500 lb at 440 rpm for the remainder of the time. Life is to be 6 yr at 8 hr per day. Find the minimum value of the basic rating load C that the bearing must have. Loads are steady.   *Ans.* $C = 6,020$ lb.

22. A ball bearing carries a load of 1,000 lb at 440 rpm for 0.3 of the time and a load of 500 lb at 880 rpm for the remainder of the time. Life to be 6 yr at 8 hr per day. Find the minimum value of the basic rating load C that the bearing must have. Loads are steady.   *Ans.* $C = 5,320$ lb.

23. A ball bearing carries a load of 500 lb at 880 rpm for 0.3 of the time and a load of 1,000 lb at 440 rpm for the remainder of the time. Life to be 6 yr at 8 hr per day. Find the minimum value of the basic rating load C that the bearing must have. Loads are steady.   *Ans.* $C = 6,260$ lb.

24. A ball bearing carries a load of 500 lb at 440 rpm for 0.3 of the time and a load of 1,000 at 880 rpm for the remainder of the time. Life to be 6 yr at 8 hr per day. Find the minimum value of the basic rating load C that the bearing must have. Loads are steady.

Ans. $C = 7,690$ lb.

25. A 205 bearing is rotating 870 rpm. Find the life in hours and years at 8 hr per day for radial loads of 400, 450, 500, 550, and 600 pounds. Plot the results and note how quickly the expected life is decreased for even nominal increases in the load.

26. Same as Problem 25 except for a 106 bearing at 1,740 rpm.

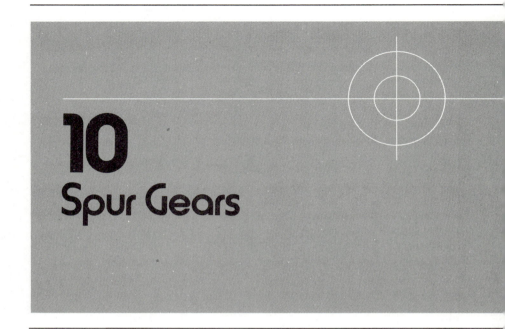

10
Spur Gears

The designer is frequently confronted with the problem of transferring power from one shaft to another while maintaining a definite ratio between the velocities of rotation of the shafts. Another requirement might be the transmission of a specified angular motion from one shaft to the other. Various types of gearing have been developed for this purpose, which will operate quietly and with very low friction losses. Smooth and vibrationless action is secured by giving the proper geometric shape to the outline of the teeth. The proportions of the gear tooth, as well as the sizes of the teeth, have been standardized. This procedure has simplified design calculations and has reduced the required number of cutting tools to a minimum. The proper material must be selected to obtain satisfactory strength, fatigue, and wear properties. Ease of manufacture and ease of inspection are necessary if production costs are to be kept at their lowest level. All these problems must be taken into account by the designer.

a,	addendum	f,	clearance
b,	face width	F_b,	bending capacity of tooth
BHN,	Brinell Hardness Number	F_d,	dynamic force
c,	center distance	F_p,	horsepower force
d,	pitch diameter	F_w,	wear capacity of tooth
e,	error; shift in position of generating rack	g,	acceleration due to gravity, 386 in./sec^2 or 9.8066 m/sec^2
E,	modulus of elasticity	hp,	horsepower, 0.7457 kW

in.,	inch, 25.4 mm		P_d,	diametral pitch
I,	mass moment of inertia		Pa,	pascal, 1 newton/m²
J,	joule, newton-meter, Nm		r,	pitch radius
k,	kilo (prefix), 1,000		rpm,	revolutions per minute
k,	spring constant of tooth		R,	radius of curvature of involute
kg,	kilogram, 2.2046 lb (mass)			
kW,	kilowatt, 1.341 horsepower		t,	time; tooth thickness
K,	constant in wear equation		T,	torque
lb,	pound;		V,	velocity
	1 lb (force) = 4.4482 newtons		W,	watt, 1 joule/sec or 1 Nm/sec
	1 lb (mass) = 0.4536 kg		y,	form factor for circular pitch
m,	meter, 39.370 in.		Y,	form factor for diametral pitch
m,	milli (prefix), one thousandth		z,	length of contact for meshing teeth
m_0,	module, diameter (mm)/ number of teeth			
mm,	millimeter		γ,	(gamma) weight per unit volume
M,	mega (prefix), 1 million			
MPa,	megapascal, 1 newton/mm²		σ,	(sigma) bending stress
n,	revolutions per minute, rpm		σ_{ec},	surface endurance stress
N,	number of teeth in gear		φ,	(phi) pressure angle
p,	circular pitch		ω,	(omega) angular velocity, radians per second
p_b,	base pitch			

1. Introduction

In the friction cylinders of Fig. 10-1, the tangential velocities of the surfaces are equal if it is assumed that no slip occurs at the point of contact O. Hence,

$$r_1 \omega_1 = r_2 \omega_2 \tag{1}$$

where r_1 and r_2 refer to the radii, and ω_1 and ω_2 are the angular velocities in radians per second. These cylinders can be transformed into spur gears by placing teeth on them that run parallel to the axes of the cylinders. The circles in Fig. 10-1 are then called the pitch circles; their diameters are called the pitch diameters of the gears. The teeth are arranged to extend both outside and inside the pitch circles. The names of some of the more important parts of spur gear teeth are shown in Fig. 10-2. When operating together, the teeth of one gear extend to the working-depth circle of the other. Clearance f is required to prevent the end of the tooth of one gear from riding on the bottomland of the mating gear.

For positive transmission of motion, the teeth need not be of any particular shape. However, for quiet and vibrationless operation, the velocities of the pitch circles of the two gears must be the same at all times. This statement refers especially to the short interval of time during which two particular teeth are in contact. If the pitch circle of the driver is moving with constant

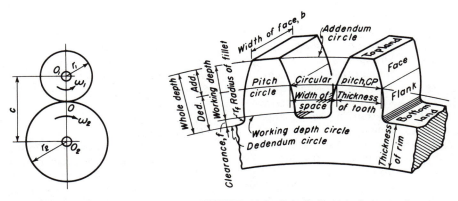

FIGURE 10-1   Friction cylinders.

FIGURE 10-2   Principal parts of gear teeth.

velocity, the shape of the teeth must be such that the velocity of the pitch circle of the driven gear is neither increased nor decreased at any instant while the two teeth are touching. When this condition is satisfied, the gears are said to fulfill the fundamental law of toothed gearing.

2. *Velocity Law of Toothed Gearing*

Portions of two gears having centers at O_1 and O_2 are shown in Fig. 10-3. The gears have angular velocities of ω_1 and ω_2, respectively. The teeth are in contact at point K_1, but no particular shape has as yet been specified for the outline of the teeth. Lines NN and TT represent respectively the normal and tangent drawn to the tooth surfaces at K_1. Normal NN intersects the line of centers at point O. The vector K_1M_1 represents the velocity of K_1 considered as a point of gear 1. The vector is perpendicular to radius O_1K_1. Similarly, K_1M_2 represents the velocity of K_1 when K_1 is considered as a point on gear 2. Since the teeth remain in contact, the projection K_1N_1 of the velocity vectors on the common normal must be the same for both gears.

From the figure, $K_1M_1 = \omega_1 \times O_1K_1$, and $K_1M_2 = \omega_2 \times O_2K_1$, so that

$$\frac{\omega_2}{\omega_1} = \frac{O_1K_1}{K_1M_1} \times \frac{K_1M_2}{O_2K_1} \tag{a}$$

From sides of similar triangles, the following relationships are secured from the figure.

$$\frac{O_1K_1}{K_1M_1} = \frac{O_1A}{K_1N_1} \quad \text{and} \quad \frac{K_1M_2}{O_2K_1} = \frac{K_1N_1}{O_2B} \tag{b}$$

Substitution in Eq. (a) gives: $\dfrac{\omega_2}{\omega_1} = \dfrac{O_1A}{O_2B}$ $\qquad$ (c)

Also, from similar triangles: $\dfrac{O_1A}{O_2B} = \dfrac{O_1O}{O_2O}$ $\qquad$ (d)

Substitution in Eq. (c) gives: $\omega_1 \times O_1O = \omega_2 \times O_2O$ $\qquad$ (2)

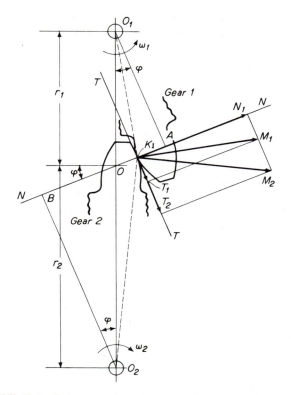

FIGURE 10-3 Component of velocity normal to tooth surface at point of contact is same for both gears.

The ratio ω_1/ω_2 from Eq. (2) must remain constant at all times if there is to be no change in the velocity ratio of the two gears. Equations (1) and (2) should be solved simultaneously, making use of the relationship

$$r_1 + r_2 = O_1O + O_2O$$

The result is:

$$O_1O = r_1$$

$$O_2O = r_2 \qquad\qquad (e)$$

Point O is therefore a fixed point through which the pitch circles are drawn. Hence, in order to fulfill the fundamental law of toothed gearing, the sides of the teeth must be so shaped that the normal drawn through the point of contact will at all times pass through the pitch point O.

The difference $K_1T_2 - K_1T_1$ between the tangential vectors is equal to the velocity of sliding of one tooth on the other. The velocity of sliding is not constant, but varies with the location of the contact point K_1. When K_1 coincides with O, the sliding velocity is zero, and pure rolling contact exists between the teeth for this point.

It can be shown that gear teeth composed of involutes or cycloids fulfill the fundamental law. For involute gearing, the normal *NN* not only passes through point *O* at all times, but maintains a constant inclination φ with respect to the common tangent to the pitch circles. For cycloidal gearing, point *O* remains fixed, but angle φ varies as the location of the point of contact K_1 changes.

3. Kinematics of Involute Gear Teeth

Assume that the required velocity ratio for a pair of gears is secured by pitch circles of radii O_1O and O_2O, respectively, in Fig. 10-4(a). Draw the line *AB* at an angle φ to the common tangent to the circles. In practice, φ is usually made either $14\frac{1}{2}°$ or 20°. Circles are drawn tangent to *AB* and are called the

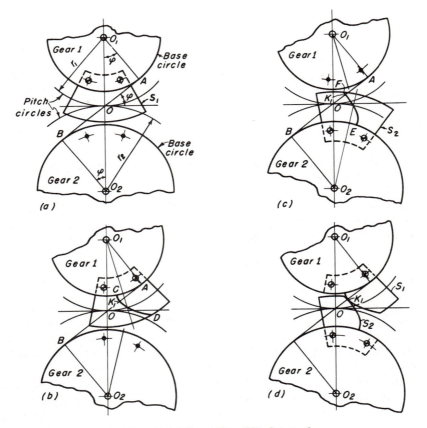

FIGURE 10-4   Generation of involute teeth.

base circles of the gears. Let cylinders be made equal to the base circles, and let a string be passed around the base circle of gear 1, then from A to B, and then around the base circle of gear 2 somewhat in the manner of one strand of a crossed belt.

In Fig. 10-4(a), let a smooth plate S_1 be attached to gear 1, and let a scribing point be fastened to the cord at location O. Keeping the string taut, give the cylinders a small rotation in both directions, thus drawing the curve CD shown in Fig. 10-4(b). For all locations of the gear, the distance K_1A is equal to the arc AC. The curve CD is therefore an involute and is at all times normal to the string. A similar procedure should be followed on gear 2 after first removing S_1 and attaching surface S_2 to gear 2. A corresponding curve EF will be obtained as shown in Fig. 10-4(c).

If certain portions of S_1 and S_2 are now removed, both plates may be attached to the gear simultaneously, as in Fig. 10-4(d). The tooth outlines are in contact at K_1, and one gear may be turned by a pressure supplied by the other at this point. The string is therefore not required for producing motion and may be removed. The common normal to the tooth outlines coincided with the string and therefore will always cross the line of centers at the fixed point O. The rule of gearing is accordingly fulfilled. The resulting velocity ratio for the gears will be the same as the velocity ratio of the given pitch circles.

In the similar triangles O_1AO and O_2BO,

$$\frac{O_2B}{O_1A} = \frac{O_2O}{O_1O} \tag{3}$$

Substitution in Eq. (2) gives: $\quad \omega_1 \times O_1A = \omega_2 \times O_2B \tag{4}$

Thus the linear velocities of the base circles of the two gears are also equal to each other.

In involute gearing, at least one pair of teeth must always be in contact with each other along the line AB. This line is accordingly called the line of action, or the pressure line, and the angle φ is usually referred to as the pressure angle.

A large-scale drawing of gear teeth is frequently helpful or necessary in studying the form or kinematic action of a proposed gear design. In laying out the involute on the drawing board, the pressure line is first drawn in accordance with the desired pressure angle φ. As shown in Fig. 10-5, the base circle is drawn tangent to the pressure line.

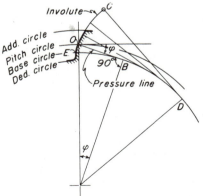

FIGURE 10-5 Drawing board layout for the involute.

Tangents are drawn to the base circle to represent various positions of the string as it unwinds.

The starting point *E* for the involute is found by making arc *EB* equal to tangent *OB* by taking short steps along these curves with a divider. Other points on the involute are found similarly. For example, arc *ED* must be equal to tangent *CD*. It should be noted that the involute curve becomes a straight line for a rack or a gear of infinite radius.

It is obvious from Fig. 10-4 that the force between a pair of teeth when transmitting power lies along *AB*. This line, which is tangent to both base circles, is therefore called the *pressure line*.

4. Cycloidal Gear Teeth

Gear teeth, when formed of portions of epicycloids and hypocycloids, also fulfill the law of tooth gearing. These curves are traced by a point on the circumference of the generating circle as it rolls without slipping along the inside and outside of the pitch circle of the gear as shown in Fig. 10-6.

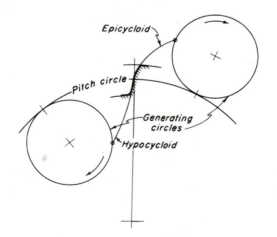

FIGURE 10-6   Generation of epicycloid and hypocycloid.

Cycloidal gearing was formerly in extensive use for cast-iron gears with cast teeth. Cycloidal gearing has been superseded by the involute tooth form, principally because the cutting tools required for manufacturing the latter are simpler in form. Variations in center distance have no effect on involute teeth, whereas for cycloidal gearing, tooth action is not correct unless the center distance is accurately maintained. The chief interest at the present time in cycloidal gearing lies in the fact that the outline for the $14\frac{1}{2}°$ composite system is partially composed of cycloids. The rotors of the Root positive blower are also cycloidal in shape.

5. *Pitches of Gear Teeth*

The use of the letter N to represent the number of teeth in a gear is so firmly established in practice that its use will be continued in this chapter. When the force newton must be designated it will be spelled out in full.

For gear calculations, the following different pitches are in use.

(a) *Circular pitch* is defined as the distance from a point on the pitch circle of one tooth to the corresponding point on the adjacent tooth measured along the pitch circle. If circular pitch is represented by p, pitch diameter by d, and the number of teeth in the gear by N, then

$$p = \frac{\pi d}{N} \tag{5}$$

(b) *Diametral pitch* is defined as the number of teeth in the gear per inch of pitch diameter. If diametral pitch is represented by P_d, then

$$P_d = \frac{N}{d} = \frac{N}{2r} \tag{6}$$

From Eq. (6), the following equation can be easily derived for center distance c.

$$c = r_1 + r_2 = \frac{N_1 + N_2}{2P_d} \tag{7}$$

The smaller of two meshing gears is usually called the *pinion*. Combination of Eqs. (5) and (6) gives the useful relationship

$N = \frac{\pi d}{p} = P_d d$

$$pP_d = \pi \tag{8}$$

$P = P_d p$

Figure 10-7 shows a number of gear teeth of different diametral pitches in actual size.

(c) *Base pitch* is defined as the distance from a point on one tooth to the corresponding point on the adjacent tooth measured around the base circle. As shown in Fig. 10-8, the base pitch is also the distance from tooth to tooth measured on a tangent to the base circle. Since the radius of the base circle is $r \cos \varphi$, where r is the pitch radius, the base pitch is equal to the circumference

FIGURE 10-7 **Actual sizes of gear teeth of various diametral pitches. Equivalent module is equal to 25.4/P_d.**

FIGURE 10-8 Base pitch.

of the base circle divided by the number of teeth in the gear. Thus,

$$P_b = \frac{2\pi r \cos \varphi}{N} = p \cos \varphi \qquad (9)$$

$$= \frac{\pi \cos \varphi}{P_d}$$

For $\varphi = 14\frac{1}{2}°$:

$$P_b = \frac{3.0415}{P_d} \qquad (10)$$

For $\varphi = 20°$:

$$p_b = \frac{2.9521}{P_d} \qquad (11)$$

Example 1. What is the thickness of a $4P_d$ tooth measured along the pitch circle?

Solution. By Eq. (8): $p = \dfrac{\pi}{P_d} = \dfrac{\pi}{4} = 0.7854$ in.

The thickness of the tooth along the pitch circle for no backlash can be taken as one-half the value of the p if clearance or backlash is neglected. Hence

$$\text{tooth thickness} = \tfrac{1}{2} \times 0.7854 = 0.3927 \text{ in.}$$

Example 2. A 20-tooth $5P_d$ gear meshes with a 63-tooth gear. Find the value of the standard center distance.

Solution. By Eq. (6):

Pitch radius: $\qquad r_1 = \dfrac{N_1}{2P_d} = \dfrac{20}{2 \times 5} = 2.0 \text{ in.}$

Pitch radius: $\qquad r_2 = \dfrac{N_2}{2P_d} = \dfrac{63}{2 \times 5} = 6.3 \text{ in.}$

Center distance: $\qquad c = r_1 + r_2 = 2.0 + 6.3 = 8.3 \text{ in.}$

Example 3. Two $8P_d$ gears are to be mounted on a center distance of 18 in. The speed ratio is to be 7:9. Find the number of teeth in each gear.

Solution. By the given conditions:

$$\frac{N_1}{N_2} = \frac{7}{9} \qquad \text{or} \qquad N_1 = \frac{7}{9} N_2$$

In Eq. (7): $\quad 2P_d c = N_1 + N_2 = \dfrac{7}{9} N_2 + N_2 = \dfrac{16}{9} N_2$

$$N_2 = \frac{9}{16} \times 2P_d c = \frac{9}{16} \times 2 \times 8 \times 16 = 144 \text{ teeth}$$

$$N_1 = \frac{7}{9} \times 144 = 112 \text{ teeth}$$

The following pitches are considered standard.

P_d 1–P_d 2 varying by $\tfrac{1}{4}$ diamental pitch
P_d 2–P_d 3 varying by $\tfrac{1}{2}$ diametral pitch
P_d 3–P_d 12 varying by 1 diametral pitch
P_d 12–P_d 32 varying by 2 diametral pitch

6. Gear Pitches in SI Units

In SI units, the size of the teeth is specified by the module m_0, where the module is taken as the pitch diameter in millimeters divided by the number of teeth, or

$$m_0 = \frac{d}{N} \qquad \text{or} \qquad d = m_0 N \tag{6a}$$

This equation can be easily remembered by noting that it is the inverse of Eq. (6) for P_d. An equation in U.S. customary units can then be changed into SI merely by replacing P_d by $1/m_0$.

In SI units the pitch diameter d and radius r must be in millimeters. N is the number of teeth. Module m_0 has the dimensions of millimeters. The numerical value of the diametral pitch P_d is equal to 25.4 divided by the module m_0 in millimeters, or

$$P_d = \frac{25.4}{m_0} \tag{6b}$$

The center distance c for a pair of meshing gears, from Eq. (6a), is

$$c = r_1 + r_2 = \frac{m_0}{2}(N_1 + N_2) \tag{7a}$$

As before, the circular pitch p is equal to the pitch circle circumference πd divided by the number of teeth N.

$$p = \frac{\pi d}{N} = \pi m_0 \tag{8a}$$

The last form of this equation is formed by the substitution of Eq. (6a).

Example 1A. What is the thickness of a module 6 tooth measured along the pitch circle?

Solution. By Eq. (8a), $p = \pi m_0 = 6\pi = 18.85$ mm

For no backlash, the thickness of tooth on the pitch circle is equal to one-half the circular pitch.

Example 2A. A 20-tooth module 5 gear meshes with a 63-tooth gear. Find the value of the standard center distance.

Solution. By Eq. (6a):

Pitch radius: $r_1 = \frac{1}{2}m_0 N_1 = \frac{1}{2} \times 5 \times 20 = 50$ mm

$r_2 = \frac{1}{2}m_0 N_2 = \frac{1}{2} \times 5 \times 63 = 157.5$ mm

Center distance: $c = r_1 + r_2 = 207.5$ mm

Example 3A. Two module 3 gears are to be mounted on a center distance of 384 mm. The speed ratio is 7:9. Find the number of teeth in each gear.

Solution. By the given conditions:

$$\frac{N_1}{N_2} = \frac{7}{9} \quad \text{or} \quad N_1 = \frac{7}{9}N_2$$

By Eq. (7a): $c = \frac{m_0}{2}\left(\frac{7}{9}N_2 + N_2\right) = \frac{3}{2} \times \frac{16}{9}N_2 = 384$

$$\frac{8}{3}N_2 = 384 \qquad N_2 = 144 \text{ teeth}$$

$$N_1 = \frac{7}{9} \times 144 = 112 \text{ teeth}$$

The following metric hobs are generally available for generating SI gears: Modules 1, 1.5, 2, 2.5, 3, 3.5, 4, 4.5, 5, 6, 8, and 10.

7. Standard Systems of Gearing

The American Gear Manufacturers Association has recommended that the basic pressure angle φ be either 20° or 25° with full-depth addendums equal to $1/P_d$. Standards are retained for the $14\frac{1}{2}$° full-depth involute system, the $14\frac{1}{2}$° composite system, and the 20° involute stub-tooth system. These latter systems are required mainly for the replacement of gears already in service.

As the number of teeth in a gear is increased, the involute tooth outline becomes straighter until at the limit, when the radius is infinite, the gear becomes a rack with straight-sided teeth. Such racks have been standardized by the AGMA as follows.

(a) 20° *Full-Depth Involute.* These gears have a 20° pressure angle and the basic rack illustrated in Fig. 10-9. This is a widely used system of gearing. The rack for φ equal to 25° is similar except for the change in the angle.

FIGURE 10-9 Basic rack for 20° full-depth involute system. 25° system is similar except for change in angle.

FIGURE 10-10 Basic rack for $14\frac{1}{2}$° full-depth involute system.

(b) $14\frac{1}{2}$° *Full-Depth Involute.* This is an involute tooth form whose basic rack has straight-sided teeth, as shown in Fig. 10-10. This system of gearing is very satisfactory, provided there is a considerable number of teeth in the gears. As will be explained, when the number of teeth is small, these gears, if made by one of the generating processes, are subject to undercutting, which may reduce the duration of contact between the teeth. They cannot be operated interchangeably with gears of the $14\frac{1}{2}$° composite system. The tooth outlines are different, and true gear action cannot take place.

(c) $14\frac{1}{2}$° *Composite System.* In this system, the tooth curve is an involute for a short distance each side of the pitch line, but is a cycloid for the inner and outer portions of the outline. Such gears are usually cut with the formed milling cutter shown in Fig. 10-13. The proportions for the basic rack for the system are given in Fig. 10-11. The least number of teeth for satisfactory action is twelve. This system is sometimes erroneously called the "standard $14\frac{1}{2}°$ involute system." However, as mentioned above, only a portion of the tooth is of involute form.

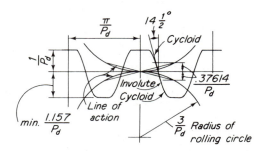

FIGURE 10-11   Basic rack for $14\frac{1}{2}°$ composite system.

(d) *20° Stub-Tooth Involute.* Gears in this system operate on a pressure angle of 20°, and have shorter addenda and dedenda than the full-depth systems. The basic rack is shown in Fig. 10-12. Although undercutting has been lessened, the short addendum reduces the duration of contact. Vibration may occur, especially in gears of few teeth, because of insufficient overlap in the gear action, as is explained in Section 14.

FIGURE 10-12   Basic rack for 20° stub-tooth involute system.

(e) *Fellows Gear Shaper System.* This system uses a pressure angle of 20° and two diametral pitches, such as $\frac{4}{5}$, $\frac{6}{8}$, and so forth. The numerator indicates the pitch, which determined the thickness of the tooth and the pitch diameter. The denominator is used for determining the addendum in the usual manner. A gear having shortened or stub teeth is thus produced. Figures 10-9 to 10-12, inclusive, illustrate the basic racks for the designated systems of gearing. The actual cutting tools, however, must have the tip of the teeth extended in order to cut the clearance f on the mating gear. This clearance is usually made equal to $0.157/P_d$ or $0.25/P_d$.

FIGURE 10-13   Formed circular cutter for gear teeth. (*Courtesy Illinois Tool Works.*)

For full-depth metric gears the addendum in millimeters is equal to the module m_0.

8. Methods of Manufacture

Gear teeth are formed by the following milling or generating processes.

(a) *Milling Cutter.* Spur gears may be made from a blank by removing the material between the teeth on a milling machine that uses the formed cutter shown in Fig. 10-13.

Gears of the $14\frac{1}{2}°$ composite system are usually made by this method. Since the geometric curves forming the sides of the teeth vary with the number of teeth in the gear, eight cutters are required for each pitch if gears of all sizes are to be cut. The eight cutters and the range of each are as follows.

No. 1:   135–rack
No. 2:   55–134 teeth
No. 3:   35-54 teeth
No. 4:   26–34 teeth
No. 5:   21–25 teeth
No. 6:   17–20 teeth
No. 7:   14–16 teeth
No. 8:   12 and 13 teeth

Cutters are theoretically correct only for the lowest number of teeth in each range. If, when gears are being cut near the higher end of the range, a more accurate tooth form is desired, cutters in half-numbers suitable for gears with such numbers of teeth are also available.

Gears with large-size teeth are frequently made by formed cutters because of the lack of suitable generating machines and tools.

Example 4. Two spur gears have a 3:1 ratio. The center distance is to be 11.600 in. Determine whether it is possible to use standard gears of $P_d = 4$ for this train.

Solution.
$$N_2 = 3N_1$$

In Eq. (7):
$$N_1 + N_2 = N_1 + 3N_1 = 2P_d c$$
$$4N_1 = 2 \times 11.6P_d$$
$$N_1 = 5.8P_d$$

When $P_d = 4$ is substituted in this equation, the right side will not be an integer. Gears with $P_d = 4$ therefore cannot be used for this train.

(b) *Rack Generation.* Gear tooth forms may be produced by another method, known as generating. Since a rack may be considered a gear of infinite radius, a tool of this shape may be constructed of hardened steel with cutting edges around the boundaries of the teeth. The tool is given a reciprocating motion parallel to the gear axis. At the same time, the gear blank is slowly rotated, and the rack is given a lateral motion equal to the pitch-line velocity of the gear. The material between the gear teeth is cut away and involute teeth are generated, as shown in Fig. 10-14. Only one tool for each pitch will be required to cut gears of any number of teeth.

(c) *Hobbing.* The hobbing process generates teeth from a straight-sided tool as shown in Fig. 10-15. The hob may be considered a cylinder around which a thread of the same cross section as a rack tooth has been helically wound. The resulting worm is gashed, the edges are relieved, and are then hardened and ground. The hob is located to give the proper depth of cut and is then rotated. Kinematically, the action of the hob on the blank is equivalent to that of a rack cutter. The lead of the helix of the hob as it rotates simulates the lateral motion of the rack. As the cutting progresses, the hob is fed axially along the blank until the teeth extend across the entire width of face. The hobbing process accounts for the major portion of gears made in quantity production.

(d) *Fellows Gear Shaper Method.* This is a generating process using a cutter that resembles a hardened gear with properly relieved edges. Cutter and blank are mounted on parallel axes and are slowly rotated; the cutter is given an additional reciprocating motion on its axis. Teeth are generated in the blank as shown in Fig. 10-16. At the beginning, the cutter is fed radially into the blank a distance equal to the depth of the tooth. The Fellows method must be used for cutting internal gears. The method is also suitable for shoulder gears where space is restricted at one end of the teeth, as in the

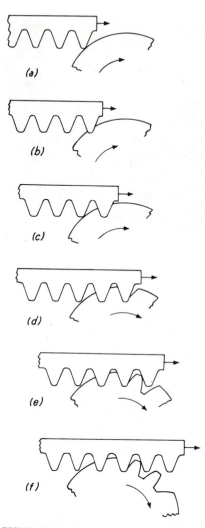

FIGURE 10-14 Rack generation of involute teeth.

cluster gears of automobile transmissions. Fellows cutters are made to cut gears of all systems.

Hobbed and shaped teeth are usually provided with a small amount of tip relief. The cutting tools are modified to cut the tooth slightly narrower in the outer portion. The oncoming pair of teeth pick up the load more gradually, and errors in tooth form and spacing have less effect in producing noise and vibration.

FIGURE 10-15   **Hob with partially cut gear blank.**

FIGURE 10-16   **Fellows gear shaper method of forming gear teeth.**

9. Backlash

When making gearing calculations, it is standard practice to assume that the thickness of the tooth measured around the pitch circle is exactly one-half the circular pitch. Because of unavoidable inaccuracies, it is necessary to cut the teeth slightly thinner to provide some clearance so that the gears will not bind but will roll together smoothly. This clearance is called *backlash*. It is illustrated in Fig. 10-17 and is defined as the amount by which a tooth space exceeds the thickness of the engaging tooth. The measurement is to be made on the pitch circle. If the measurement is made normal to the tooth profile, as by a feeler gage, the reading can be converted to a length taken along the pitch circle.

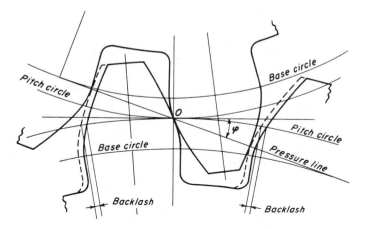

FIGURE 10-17   Backlash in gears.

TABLE 10-1   Recommended Backlash After Assembly

Center Distance	Diametral Pitch, P_d		
	2, 2½, 3	*3½, 4, 5*	*6, 7, 8, 9*
Over 5 to 10			0.010–0.020
Over 10 to 20		0.020–0.030	0.015–0.025
Over 20 to 30	0.030–0.040	0.025–0.030	0.020–0.030
Over 30 to 40	0.035–0.045	0.030–0.040	0.025–0.035
Over 40 to 50	0.040–0.055	0.035–0.050	0.030–0.040
Over 50 to 80	0.045–0.065	0.040–0.060	

Backlash is usually obtained by setting the cutting tools deeper into the blank to produce a thinner tooth and wider space. Tooth thickness is usually reduced for each gear, although it is possible to obtain the backlash by making the teeth thinner on but one of the gears. Backlash can also be increased or decreased by giving a small variation to the center distance at which the gears are mounted. Recommended backlash values are given in Table 10-1. For gear trains in precision equipment and instruments it is frequently necessary to keep the backlash at a very low value. For involute teeth, backlash and variation in center distance have no effect on theoretically correct gear tooth action.

10. Gear Finishing Methods

For high speeds and heavy loads, a finishing operation, subsequent to cutting, may be required to bring the tooth outline to a sufficient degree of accuracy and surface finish. The following methods are in use.

(a) *Gear Shaving*. The gear is run with a specially made cutter resembling a gear or rack. The cutter tooth has an axial component of motion on the tooth surface, which removes fine hair-like chips. Shaving is done prior to hardening, and sufficient stock must be allowed for this operation. Shaving is a widely used finishing method for gears made in large quantities.

(b) *Honing*. Here the hardened gear is run with an abrasive impregnated plastic helical gear-shaped tool, which will make minor tooth-form corrections and improve the smoothness of the surface.

(c) *Lapping*. The gear is run with a gear-shaped lapping tool in an abrasive-containing medium. Sometimes two mating gears are similarly run. An additional relative motion in the axial direction is required for spur and helical teeth. Excessive lapping can be detrimental to the involute shape.

(d) *Grinding*. Grinding can be used to give the final form to the teeth after heat treatment. Errors resulting from distortion in hardening can be corrected. When cutting, sufficient stock must be allowed on the tooth surface. Some fine-pitch gears are ground from solid blanks.

(e) *Burnishing*. This is a plastic smearing process resulting from rubbing pressure, which flattens and spreads minute surface irregularities. A special hardened burnishing tool is used. The process will not correct errors due to improper cutting.

11. Gear Forces and Their Effects

Design methods are required for handling the forces that occur between the teeth when a pair of gears is transmitting power. The simplified method presented in this book is founded on the principles of elementary mechanics. It should be noted, however, that far more extensive and detailed methods are available in the standards of the American Gear Manufacturers Association.

There are two forces that must be considered in the design of gears. One is the power transmitting force, F_p, and the other is a dynamic effect F_d. The latter is caused by inaccuracies in the tooth form or spacing. These will be treated separately in the following pages.

These forces have two principal effects on the teeth. One is the tendency to cause breakage in bending so that the capacity F_B of the tooth to resist bending moment must be checked. The other effect is to cause deterioration of the tooth surface. Contact stresses on the curved surfaces are very high and the teeth must be sufficiently hard not to fail in such service. The capacity of the teeth to resist such loading is generally given the name of wear capacity F_w. The design is then based on F_B or F_w, whichever is the smaller.

12. *Load on Gear Tooth from Power Transmitted*

It is obvious from Fig. 10-4 that the force between a pair of teeth always lies along AB. This line, which is tangent to both base circles, is therefore called the *pressure line*. The power produced by this force is $F_p V_b$ where F_p is the power force, pounds, and V_b is the velocity of the base circle, feet per minute. The horsepower is

$$\text{hp} = \frac{F_p V_b}{33,000} \tag{12}$$

The diameter of the base circle is equal to $d \cos \phi$, so that its velocity V_b is

$$V_b = \frac{\pi n d \cos \phi}{12} \text{ fpm} \tag{13}$$

where d is the pitch diameter of the gear in inches and n is its rotative speed revolutions per minute.

Velocity V of the pitch circle is

$$V = \frac{\pi d n}{12} \text{ fpm} \tag{14}$$

so that
$$V_b = V \cos \phi \tag{15}$$

Force F_p in Fig. 10-18 can be divided into two components, one, F_t, tangential to the pitch circle and the other directed toward the center. The latter transmits no power. In Fig. 10-18

$$F_p = \frac{F_t}{\cos \phi} \tag{16}$$

Let values for F_p and V_b from Eqs. (16) and (15) be substituted into Eq. (12).

$$\text{hp} = \frac{F_t V}{33,000} \tag{17}$$

FIGURE 10-18 **Force transmitted from one gear to another.**

The corresponding equation in SI units for kilowatts is

$$kW = \frac{F_p V_b}{1,000} = \frac{F_t V}{1,000} \tag{18}$$

where pressure line force F_p and tangential force F_t are in newtons and base circle velocity V_b and pitch circle velocity V are in meters per second.

Base circle velocity V_b is given by

$$V_b = \frac{\pi dn \cos \phi}{60 \times 1,000} \frac{m}{sec} \tag{19}$$

Pitch circle velocity V is given by

$$V_b = \frac{\pi dn}{60 \times 1,000} \frac{m}{sec} \tag{19a}$$

where diameter d is in millimeters and n is the revolutions per minute.

The relationship between horsepower and kilowatts can be easily found.

$$1\frac{ft}{min} = \frac{12 \text{ in.}}{min} = \frac{12 \times 25.4 \text{ mm}}{60 \text{ sec}} = 5.08 \frac{mm}{sec} = 0.00508 \frac{m}{sec} \tag{20}$$

$$1\frac{ft\ lb}{min} = 0.00508 \frac{m}{sec} \times 4.4482 \text{ newtons}$$

$$= 0.022597 \frac{newton\ m}{sec} \text{ or watt} \tag{21}$$

$$1 \text{ hp} = 33,000 \frac{ft\ lb}{min} = 33,000 \times 0.022597$$

$$= 745.7 \text{ watts or } 0.7457 \text{ kW} \tag{22}$$

$$1 \text{ kW} = \frac{1}{0.7457} = 1.341 \text{ hp} \tag{23}$$

Also: $$1\frac{lb\ sec^2}{in.} = \frac{4.4482 \text{ newton sec}^2}{25.4 \text{ mm}} = 0.17513 \frac{newton\ sec^2}{mm} \tag{24}$$

Example 5. A draft horse walks at a steady pace of 3 miles per hour. What steady force must he exert if the power output is exactly 1 horsepower?

Solution.   1 mile $= 5,280$ feet

$$V = \frac{3 \times 5,280}{60} = 264 \text{ ft/min}$$

By Eq. (17): $$F_t = \frac{33,000 \text{ hp}}{V} = \frac{33,000 \times 1}{264} = 125 \text{ lb}$$

Example 5A. Transform the speed and force of Example 5 into SI units and calculate the power in kilowatts.

Solution. $1 \text{ mi} = \dfrac{5,280 \times 12 \times 25.4}{1,000} = 1,609.344 \text{ m}$

$$V = 3 \frac{\text{mi}}{\text{hr}} = \frac{3 \times 1,609.344}{60 \times 60} = 1.34112 \frac{\text{m}}{\text{sec}}$$

$$F = 125 \text{ lb} = 125 \times 4.4482 = 556.025 \text{ newtons}$$

$$\text{kW} = \frac{F_t V}{1,000} = \frac{556.025 \times 1.34112}{1,000} = 0.7457$$

It is noted by Eq. (22) that this value for kW is equal to 1 hp.

The official abbreviation for newton is N. However N is so universally used for the number of teeth in a gear that, in this chapter, the word newton will be spelled out in full.

13. *Dynamic Load between Gear Teeth*

The determination of the dynamic load between gear teeth presents a difficult problem for the designer. He is working at the design stage and trying to predict what this force will be in the finished gears. The magnitude of the errors is unknown. In addition, the force will depend on the equivalent mass for the pair, the speed of rotation and the spring constant for the contacting teeth.

Many equations have been proposed for the dynamic load F_d that give widely different answers. One equation that is derived in mechanics on a reasonable set of assumptions is

$$F_d = \frac{2e}{t}\sqrt{km_e} \qquad (25)$$

where e = combined error, in both gears
t = time for the passage of a tooth through the contact zone
k = spring constant for the pair of teeth
m_e = equivalent mass for the two gears

Errors of tooth form and spacing in the cutting of gear teeth are due to various causes and are difficult to control. One common source of error arises from improper sharpening of the cutting tool or hob.[1] A typical example with error e is shown in Fig. 10-19.

Permissible manufacturing tolerances for gear teeth can be found in the publication *Gear Classification Manual* 390-02, 1964, of the American Gear Manufacturers Association. Based on the permissible tolerances, Table 10-2 has been devised based both on errors of tooth profile and spacing.

[1]See Reference 6, end of chapter.

FIGURE 10-19 Typical error in tooth outline.

TABLE 10-2 Total Tooth-to-Tooth Spacing and Profile Errors, e, Inches

AGMA Quality Number	Pitch	Pitch Diameter					
		1-1/2 in. or 38 mm	3 in. or 75 mm	6 in. or 150 mm	12 in. or 300 mm	25 in. or 625 mm	50 in. or 1,250 mm
8	$P_d = 2$ or $m_0 = 12$			0.0048	0.0050	0.0052	0.0055
	$P_d = 4$ or $m_0 = 6$		0.0032	0.0034	0.0036	0.0037	0.0040
	$P_d = 8$ or $m_0 = 3$	0.0022	0.0023	0.0024	0.0025	0.0028	0.0029
	$P_d = 16–20$, or $m_0 = 1.5–1.25$	0.0017	0.0018	0.0019	0.0020	0.0021	0.0023
9	$P_d = 2$ or $m_0 = 12$			0.0033	0.0034	0.0037	0.0039
	$P_d = 4$ or $m_0 = 6$		0.0022	0.0023	0.0025	0.0027	0.0027
	$P_d = 8$ or $m_0 = 3$	0.0016	0.0017	0.0018	0.0019	0.0019	0.0020
	$P_d = 16–20$ or $m_0 = 1.5–1.25$	0.0013	0.0013	0.0014	0.0014	0.0015	0.0016
10	$P_d = 2$ or $m_0 = 12$			0.0024	0.0025	0.0025	0.0027
	$P_d = 4$ or $m_0 = 6$		0.0016	0.0017	0.0018	0.0019	0.0024
	$P_d = 8$ or $m_0 = 3$	0.0012	0.0012	0.0012	0.0013	0.0014	0.0015
	$P_d = 16–20$ or $m_0 = 1.5–1.25$	0.0009	0.0010	0.0010	0.0010	0.0011	0.0011
11	$P_d = 4$ or $m_0 = 6$		0.0012	0.0012	0.0013	0.0014	0.0015
	$P_d = 8$ or $m_0 = 3$	0.0008	0.0009	0.0009	0.0009	0.0010	0.0011
	$P_d = 16–20$ or $m_0 = 1.5–1.25$	0.0006	0.0007	0.0007	0.0007	0.0008	0.0008
12	$P_d = 4$ or $m_0 = 6$		0.0009	0.0009	0.0009	0.0010	0.0010
	$P_d = 8$ or $m_0 = 3$	0.0006	0.0006	0.0007	0.0007	0.0007	0.0008
	$P_d = 16–20$ or $m_0 = 1.5–1.25$	0.0005	0.0005	0.0005	0.0005	0.0006	0.0006

In the table, Quality No. 8 is the lowest and No. 12 is the highest. A gear hobbing machine in good condition with a properly shaped hob can be expected to give a quality of about No. 8. This quality is widely used in the general field of power transmission. Higher-quality gears may require special finishing operations, and will be more expensive. Automotive gears, for example, have qualities of No. 10 or 11. The designer aims to use the lowest-quality number that the intended service will permit.

Time t in Eq. (25) for the application of the error will be taken as the time required for the passage of a single tooth through the zone of contact. In other words, it is merely the time t for one revolution of the gear divided by the number of teeth. The time, seconds, for one revolution is $60/n_1$ where subscript 1 refers to the pinion or smaller gear. The time t for the passage of a single tooth then is

$$t = \frac{60}{n_1 N_1} \text{ sec} \tag{26}$$

When the value for t is substituted into Eq. (25) the result is

$$F_d = \frac{en_1 N_1}{30} \sqrt{km_e} \tag{25a}$$

It was shown in Example 10, Chapter 1, that if the spring constant for a cantilever is calculated, then any other geometrically similar cantilever, but of the same breadth b, will have an equal value for the spring constant. The same is true for gear teeth since if the only change is in the pitch, the teeth are geometrically similar. It can be shown[2] that the average value of the spring constant for a pair of meshing teeth can be represented by

$$k = \frac{b}{9} \left(\frac{E_1 E_2}{E_1 + E_2} \right) \tag{27}$$

where E_1 and E_2 are the moduli of elasticity for the materials of the two gears.

When both gears are steel, $E_1 = E_2 = 30,000,000$ psi, the value of k for a meshing pair becomes

$$k = 1,667,000b \text{ lb/in.} \tag{28}$$

When one gear is steel, $E_1 = 30,000,000$ psi and the other gear is cast iron or bronze, $E_2 = 16,000,000$ psi, then

$$k = 1,160,000b \text{ lb/in.} \tag{29}$$

Here b is the width of face of the gears.

The above equations, when expressed in SI units, become the following.

Both gears steel: $$k = 11,500b \text{ newtons/mm} \tag{28a}$$

[2]See p. 31 of Reference 5 and p. 432 of Reference 1, end of chapter.

One gear steel and other gear cast iron or bronze:

$$k = 8,000b \text{ newtons/mm} \qquad (29a)$$

Width b in the above equations is in millimeters.

Gears are sometimes made in the form of a rim or ring in which the teeth are cut with the inertia of the spokes or web small enough to neglect. Two such blanks are shown in Fig. 10-20(a) where the connecting teeth have been replaced by a spring of equal spring rate k.

(a) Gear blanks with teeth replaced
by spring of constant k

(b) Gear blanks replaced by   (c) Masses of (b) replaced by
equivalent masses on pitch   equivalent masses m_e
circles

FIGURE 10-20   Model used for calculating dynamic load F_d.

The net moment of inertia of a ring about its center is

$$I = \frac{\pi b \gamma}{2g}(r_0^4 - r_i^4) \qquad (30)$$

where b is the thickness in the axial direction, γ is the weight of the material per unit volume, g is the gravitational constant, and r_o and r_i are the outside and inside radii respectively of the rim.

Instead of the rotating cylinders it is usually more convenient to replace them by masses m_1' and m_2' moving on their respective pitch circles. The moment of inertia of the rim is thus set equal to concentrated mass m' at radius r from the center. Mass m' is of sufficient size to make its moment of inertia $m'r_o^2$ equal to that of the rim, Eq. (30). Then

$$m' = \frac{\pi b \gamma}{2gr_o^2}(r_o^4 - r_i^4) \tag{31}$$

If the blank should be a solid cylinder, r_i is equal to zero, and then

$$m' = \frac{\pi b \gamma r_o^2}{2g} \tag{31a}$$

It is shown in books in mechanics that the two-mass system of Fig. 10-20(b) has similar dynamic properties to that of sketch (c) where the new mass m_e is given by

$$\frac{1}{m_e} = \frac{1}{m_1'} + \frac{1}{m_2'} \tag{32}$$

or

$$m_e = \frac{m_1' m_2'}{m_1' + m_2'} \tag{32a}$$

The dimension for mass can be found by Newton's equation, force equals mass times acceleration. In US customary units

$$m = \frac{\text{force (lb)}}{\text{accel (in./sec}^2)} \quad \text{or} \quad \frac{\text{lb sec}^2}{\text{in.}}, \quad \text{also} \quad \frac{\text{newton sec}^2}{\text{mm}}$$

Application of the foregoing equations will now be made by an example.

Example 6. A 24-tooth steel pinion, solid blank, operates with a 96-tooth cast iron gear made as a rim with internal diameter of 21 in., 20° pressure angle, $P_d = 4$, $b = 3$ in. Find the value of the equivalent mass m_e. Neglect inertia of the spokes.

Solution. $\quad d_1 = \dfrac{N_1}{P_d} = \dfrac{24}{4} = 6$ in., $\quad d_2 = \dfrac{N_2}{P_d} = \dfrac{96}{4} = 24$ in.

By Eq. (31a): $\quad m_1' = \dfrac{\pi b \gamma r_o^2}{2g} = \dfrac{\pi 3 \times 0.283 \times 3^2}{2 \times 386} 0.0311 \dfrac{\text{lb sec}^2}{\text{in.}}$

By Eq. (31): $\quad m_2' = \dfrac{\pi b \gamma}{2gr_o^2}(r_o^4 - r_i^4) = \dfrac{\pi 3 \times 0.256}{2 \times 386 \times 12^2}(12^4 - 10.5^4)$

$$= 0.00002170(20{,}736 - 12{,}155)$$

$$= 0.1863 \frac{\text{lb sec}^2}{\text{in.}}$$

By Eq. (32a): $\quad m_e = \dfrac{m_1' m_2'}{m_1' + m_2'} = \dfrac{0.0311 \times 0.1862}{0.0311 + 0.1862} = 0.0266 \dfrac{\text{lb sec}^2}{\text{in.}}$

In the simplest case both gears will be solid blanks of the same material, $\gamma_1 = \gamma_2 = \gamma$. By Eq. (31a) the equivalent moment of inertia for the gear of radius r_1 is

$$m_1' = \frac{\pi b \gamma r_1^2}{2g}$$

Let gear 2 have a radius r_2 equal to Cr_1. Equation (31a) for gear 2 becomes

$$m_2' = \frac{\pi b \gamma r_2^2}{2g} = \frac{\pi b \gamma C^2 r_1^2}{2g}$$

When this and the equations for m_1' are substituted into Eq. (32a) the result is

$$m_e = \frac{C^2}{1 + C^2} \frac{\pi b \gamma r_1^2}{2g} \tag{33}$$

Example 7. Two $20°$ full-depth steel gears with solid blanks operate on a center distance of 9 in. and a 2-to-1 speed reduction. $b = 3$ in. Speed of pinion is 860 rpm. ASME Quality No. 8. $P_d = 6$. Find the value of the dynamic load F_d.

Solution. $d_1 = 6$ in., $d_2 = 12$ in., $N_1 = 6 \times 6 = 36$

$$C = \frac{d_2}{d_1} = 2$$

By Table 10-2: $e = 0.029 + 0.031 = 0.006$ in.

By Eq. (26): $t = \dfrac{60}{n_1 N_1} = \dfrac{60}{860 \times 36} = 0.001938$ sec

By Eq. (28): $k = 1,667,000b = 1,667,000 \times 3 = 5,000,000$ lb/in.

By Eq. (33): $m_e = \dfrac{C^2}{1 + C^2} \dfrac{\pi b \gamma r_1^2}{2g} = \dfrac{4}{5} \dfrac{\pi 3 \times 0.283 \times 3^2}{2 \times 386} = 0.02488 \dfrac{\text{lb sec}^2}{\text{in.}}$

By Eq. (25): $F_d = \dfrac{2e}{t} \sqrt{km_e} = \dfrac{2 \times 0.006}{0.001938} \sqrt{5,000,000 \times 0.02488}$

$$= 6.192 \sqrt{124,400} = 2,180 \text{ lb, actual load along pressure line}$$

14. Variation in Force While Teeth Are in Contact

The operation of one gear being driven by forces arising from the mating gear is illustrated in Fig. 10-21. Consider the force acting on tooth X of gear 2, which is being driven by gear 1. Tooth X first comes in contact with a tooth of gear 1 at K_2 in sketch (a). However, at this time, providing the gears have good accuracy, gear 2 also has contact with gear 1 at K_1, and the total load is divided between the two contacts K_2 and K_1.

After some rotation the contact at K_1 is broken and K_2 carries the total load as shown in sketch (b).

This single contact carries the total load until after further rotation a new contact is about to be made at K_3 in sketch (c).

The load is now carried by contacts K_2 and K_3 until after more rotation contact K_2 is about to be broken as shown in sketch (d).

A similar sequence is repeated for each tooth as it comes into the contact zone. At first a tooth carries about half the load, then it carries the entire load, and then carries about half the load until it passes out of contact. Under some conditions distance from K_2 in sketch (a) to K_2 in sketch (d) may be so great that three or more pairs of teeth may simultaneously be in contact.

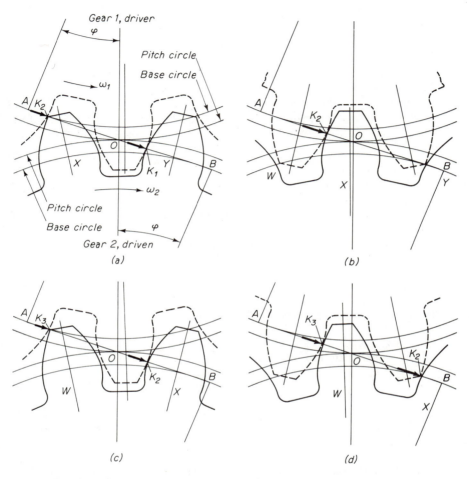

FIGURE 10-21 Transmitted forces on tooth *X* while it is passing through the contact zone.

15. Bending Capacity of Gear Tooth

It is necessary to know the force F_b perpendicular to the centerline and applied at the tip that the tooth can carry. Bending stresses in gears are calculated by the bending equation $6M/bh^2$ but the application is only approximate at best. The equation was derived for a long thin uniform beam loaded by moments at the ends—conditions quite different than those prevailing in a gear tooth. However, the application is in universal use and is given in the following manner.

The tooth is considered a cantilever beam of width *b* and length *l* from the tip to the weakest section. The capacity load F_b with lever arm l then gives the stress

$$\sigma = \frac{6M}{bh^2} = \frac{F_b}{b} \times \frac{6l}{h^2} \tag{34}$$

Length l extends from the tip to the weakest section where the thickness of the tooth is h. In the axial direction the width is b.

The factor $h^2/6l$ is a purely geometrical property of the size and shape of the tooth and may be written as a function of the circular pitch p. Therefore, let

$$py = \frac{h^2}{6l} \qquad \text{or} \qquad y = \frac{h^2}{6lp} \tag{35}$$

The term y is a pure number and is called the form or Lewis factor. It depends on the number of teeth in the gear and the system of gearing used. Dimensions h and l must be for the cross section, which makes $h^2/6l$ a minimum.

Substitution of Eq. (35) into Eq. (34) gives

$$F_b = \sigma byp \tag{36}$$

Equation (36) gives the capacity of the tooth in beam action. Values of y for gears of different numbers of teeth are given in Table 10-3. Sometimes the form factor is expressed as Y, which includes the factor π. Thus,

$$Y = \pi y \tag{37}$$

Since $p = \pi/P_d$, Eq. (36) may be written

$$F_b = \sigma b \frac{Y}{P_d} \tag{38}$$

Equation (40), it should be noted, gives merely the capacity of the tooth to resist a bending load and makes no reference to the actual load the teeth may be carrying.

TABLE 10-3   Form or Lewis Factor, y, for Spur Gears with Load at Tip of Tooth

No. of Teeth	14-1/2° Full Depth	20° Full Depth	20° Stub	No. of Teeth	14-1/2° Full Depth	20° Full Depth	20° Stub	No. of Teeth	14-1/2° Full Depth	20° Full Depth	20° Stub
10	0.056	0.064	0.083	19	0.088	0.100	0.123	43	0.108	0.126	0.147
11	0.061	0.072	0.092	20	0.090	0.102	0.125	50	0.110	0.130	0.151
12	0.067	0.078	0.099	21	0.092	0.104	0.127	60	0.113	0.134	0.154
13	0.071	0.083	0.103	23	0.094	0.106	0.130	75	0.115	0.138	0.158
14	0.075	0.088	0.108	25	0.097	0.108	0.133	100	0.117	0.142	0.161
15	0.078	0.092	0.111	27	0.099	0.111	0.136	150	0.119	0.146	0.165
16	0.081	0.094	0.115	30	0.101	0.114	0.139	300	0.122	0.150	0.170
17	0.084	0.096	0.117	34	0.104	0.118	0.142				
18	0.086	0.098	0.120	38	0.106	0.122	0.145	rack	0.124	0.154	0.175

16. Form or Lewis Factors for Spur Teeth

The form or Lewis factor is named for the American engineer who first made application of the bending equation to gear teeth.[3] Although gear materials and methods of manufacture have changed greatly since the values in Table 10-3 were published in 1893, they have remained in use to the present day with but minor changes.

Values of l and h, in Eq. (35), are somewhat uncertain since they are influenced to a large extent by the size of the fillet radius r_f. For teeth made by milling cutters, the radius may sometimes be as small as $0.05p$. For radii of this size, the actual values of y are smaller than those shown in the table, and the teeth are then not as strong as Eq. (36), using tabular values, would indicate. In fact, fillets of rather generous size are required to give the y values of Table 10-3.

For teeth that increase in thickness all the way to the base, the length of the moment arm l may be found by inscribing a parabola within the tooth outline. It should be tangent to the fillets on either side, and the vertex should be at the point where the line of action crosses the center line. The bending stresses are then computed for the cross section passing through the points of tangency.

Since the tensile strength of materials used for gears increases with the hardness,[4] working stress values for bending can be found in Table 10-4.

TABLE 10-4 Allowable Working Stresses in Bending, psi, and Values of K for Gear Materials*

| Material | Hardness | | Bending, σ, psi | | Compression, σ_{ec}, psi | K, $14\frac{1}{2}°$, psi | K, $20°$, psi |
	Average BHN	Average R_c	Spur, Helical	Bevel			
Cast iron	180		5,000	3,000	50,000	56	76
Cast iron	230		7,000	4,000	60,000	80	110
Steel	180		20,000	10,000	60,000	43	59
Steel	230		22,000	11,000	70,000	58	80
Steel	325	35	32,000	15,000	100,000	119	163
Flame or induction hardened		50	35,000	15,000	160,000	305	417
Carburized or case hardened		60	55,000	30,000	200,000	477	651

*Values for K are calculated for both gears of the same material. For combinations of steel and cast iron or bronze, see Problem 12.

[3]See Lewis, W., "Investigation of the Strength of Gear Teeth," *Proc. Engr's Club Phila.*, **10**, 1893, p. 16.

[4]See pp. 130 and 134 of Reference 4, end of chapter.

TABLE 10-4A Allowable Working Stresses, and Values of K for Gear Materials, SI Units*

Material	Hardness		Bending, σ, MPa		Com-pression, σ_{ec}, MPa	K, $14\frac{1}{2}°$, MPa	K, $20°$, MPa
	Average BHN	Average R_c	Spur, Helical	Bevel			
Cast iron	180		34.5	20.7	344.8	0.386	0.524
Cast iron	230		48.3	27.6	413.8	0.552	0.759
Steel	180		137.9	69.0	413.8	0.297	0.407
Steel	230		151.7	75.9	482.8	0.400	0.552
Steel	325	35	220.7	103.4	689.7	0.821	1.124
Flame or induction hardened		50	241.4	103.4	1,103.4	2.103	2.876
Carburized or case hardened		60	379.3	206.9	1,379.3	3.290	4.490

*Values for K apply when both gears are of same material. For combinations of steel and cast iron or bronze, see Problem 12. Stress values are those of Table 10-4 divided by 145, being the conversion factor from psi to MPa.

Example 8. Find the bending capacity F_b for a 36-tooth 20° full depth tooth if the permissible bending stress is 30,000 psi.

Solution. By Table 10-3, for $N = 36$, $y = 0.120$:

By Eq. (8): $\quad P = \dfrac{\pi}{P_d} = \dfrac{\pi}{6}$

By Eq. (36): $\quad F_b = \sigma b y p = 30{,}000 \times 3 \times 0.120 \times (\pi/6) = 5{,}655$ lb bending capacity of one tooth, perpendicular to center line

This value F_b represents the capacity at the end of the tooth and perpendicular thereto that the tooth can carry at the given value σ for the bending stress. Force F_b is illustrated in Fig. 10-22(a). The corresponding force along the pressure line is $F_b/\cos \alpha$, where $\cos \alpha$, to a good degree of approximation is

$$\cos \alpha = \frac{r \cos \phi}{r + \text{Add.}} \tag{39}$$

For full-depth teeth the addendum, Add., in US customary units is equal to $1/P_d$. In SI units the addendum is equal to the module M_0 in millimeters.

The pressure line capacity for gear teeth in the position of Fig. 10-22(a) is divided into two parts with each contact taking half the load. The total bending capacity on the pressure line for both contacts must be equal to or greater than the impressed loads. As indicated by Fig. 10-22 the following equation results.

$$\frac{2F_b}{\cos \alpha} \geq F_p + F_d \tag{40}$$

Example 9. For a pair of gears like Example 8, find the capacity in bending along the pressure line. Gear is 20° full depth, $N_1 = 36$, $P_d = 6$.

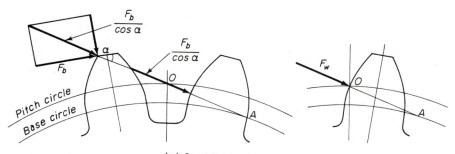

(a) Capacity loads on gear teeth

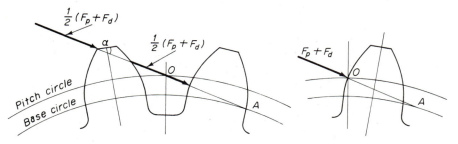

(b) Actual pressure line forces

FIGURE 10-22 Capacity and actual loading of gear teeth.

Solution. $d = \dfrac{N}{P_d} = \dfrac{36}{6} = 6$ in.

By Eq. (39): $\cos \alpha = \dfrac{r \cos \phi}{r + \text{Add.}} = \dfrac{3 \times 0.93969}{3 + \frac{1}{6}} = 0.8902$

By Eq. (40): $\dfrac{2F_b}{\cos \alpha} = \dfrac{2 \times 5{,}655}{0.8902} = 12{,}705$ lb.

capacity in bending along pressure line two pairs in contact.

A stress concentration occurs at the fillet of the tooth but since the load is carried by two contacts the stress concentration can in general be neglected.

17. Tooth Capacity in Contact Stress or Wear. Surface Durability

The gear teeth must also be sufficiently strong to carry the compressive stress between the teeth. Calculations are made by considering the teeth as two parallel cylinders in contact.[5] The radii of the cylinders are taken as the radii of curvature for the involutes when the teeth are making contact at the pitch

---

[5]See Buckingham, E., *Dynamic Loads on Gear Teeth,* New York: American Society of Mechanical Engineers, 1931, p. 41.

point O, Fig. 10-23. Approximating the teeth as cylinders is satisfactory, for although the compressive stresses are very high, they decrease rapidly at locations removed from the area of contact.

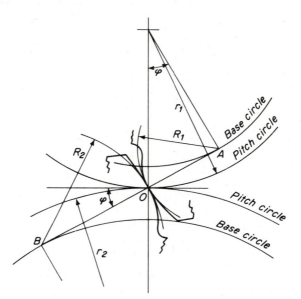

FIGURE 10-23 Radii of curvature R_1 and R_2 for tooth surfaces at pitch point O.

In Fig. 9-9 the maximum compressive stress p_0 in the contact zone is shown as

$$p_0 = 0.591 \sqrt{\frac{P_1 E_1 E_2}{E_1 + E_2}\left(\frac{1}{R_1} + \frac{1}{R_2}\right)} \tag{a}$$

where P_1 is the load per axial inch pressing the cylinders together, E_1 and E_2 are the moduli of elasticity for the materials composing the gears, and R_1 and R_2 are the radii of curvature OA and OB, respectively, for the two gears.

When contact occurs at the pitch point, the radii of curvature for the involutes are

$$R_1 = \frac{d_1}{2} \sin \varphi \tag{b}$$

$$R_2 = \frac{d_2}{2} \sin \varphi = \frac{N_2 d_1}{2N_1} \sin \varphi \tag{c}$$

The last form of Eq. (c) arises from the relationship $d_1/N_1 = d_2/N_2$.

Let p_o in Eq. (a) be called σ_{ec}, the surface endurance limit in compression for the gear materials. Substitution of Eqs. (b) and (c) gives

$$\sigma_{ec}^2 = 0.35 \frac{P_1 E_1 E_2}{(E_1 + E_2)} \cdot \frac{2}{d_1 \sin \varphi}\left(1 + \frac{N_1}{N_2}\right)$$

Both sides of this equation are multiplied by the face width b, and the total load $P_1 b$ is given the name *wear load* F_w. The equation, when solved for F_w, gives

$$F_w = \frac{\sigma_{ec}^2 b d_1 \sin \varphi}{1.40}\left(\frac{1}{E_1} + \frac{1}{E_2}\right)\left(\frac{2N_2}{N_1 + N_2}\right) \tag{42}$$

$$= d_1 b Q K \tag{43}$$

where

$$Q = \frac{2N_2}{N_1 + N_2} \tag{44}$$

and

$$K = \frac{\sigma_{ec}^2 \sin \varphi}{1.4}\left(\frac{1}{E_1} + \frac{1}{E_2}\right) \tag{45}$$

Values for σ_{ec} and K for materials of different hardnesses are shown in Tables 10-4 and 10-4A for both U.S. customary and SI units. The three values for steel are plotted in Figs. 10-24 and 10-24A. Wear capacity F_w of the tooth acts normal to the surface and is taken at the pitch point O, Fig. 10-22(b).

	K	
BHN	*US*	*SI*
180	*59*	*0.407*
228	*80*	*0.552*
326	*163*	*1.124*

FIGURE 10-24 **Values of constant K as determined by Brinell hardness number for 20° full-depth steel spur gears.**

BHN	US σ	SI
180	20,000	138
228	22,000	152
326	32,000	221

FIGURE 10-24A   Values of permissible bending stress σ in tension as determined by Brinell hardness number for 20° full-depth steel spur gears.

Example 10. Two 20° full-depth spur gears have 36 and 72 teeth. Gears are of steel hardened to 310 BHN. $P_d = 6$, width of face, $b = 3$ in. Find the capacity F_w of the tooth to withstand wear load.

Solution. By Eq. (6): $d_1 = \dfrac{N_1}{P_d} = \dfrac{36}{6} = 6$

By Eq. (44): $Q = \dfrac{2N_2}{N_1 + N_2} = \dfrac{144}{108} = \dfrac{4}{3}$

By Fig. 10-24, for BHN $= 310$, $K = 145$

By Eq. (43): $F_w = d_1 bQK = 6 \times 3 \times \dfrac{4}{3} \times 145 = 3,480$ lb. capacity in wear

When the design is based on wear capacity then F_w must be equal to or greater than the power load F_p plus the dynamic load F_d.

$$F_w \geq F_p + F_d \tag{46}$$

Example 11. Find the horsepower that can be transmitted by the gears of Examples 7 to 10. Speed $n_1 = 860$ rpm.

Solution. By Example 9: $\dfrac{2F_b}{\cos \alpha} = 12{,}705$ lb capacity of gears in bending,

two pairs in contact

By Example 10: $F_w = 3{,}480$ lb capacity in wear

The gears are weaker in wear. Then

By Eq. (46) and Example 7: $F_p = F_w - F_d = 3{,}480 - 2{,}180 = 1{,}300$ lb

By Eq. (13): $V_b = \dfrac{\pi d_1 n_1 \cos \phi}{12} = \dfrac{\pi 6 \times 860 \times 0.93969}{12} = 1{,}269.4$ fpm

By Eq. (12): $\text{hp} = \dfrac{F_p V_b}{33{,}000} = \dfrac{1{,}300 \times 1{,}269.4}{33{,}000} = 50$

Another type of problem occurs when the transmitted power is known and it is required to find the necessary hardness of the material.

Example 11A. Two 20° full-depth steel gears with solid blanks operate on a center distance of 216 mm. Speed ratio is 2 to 1. $b = 75$ mm. Kilowatt power transmitted is 35. Speed of pinion 860 rpm. $m_0 = 4$. Find the BHN for the tooth surface in wear and check gears to see if bending capacity is sufficient.

Solution. $d_1 = 144$ mm, $d_2 = 288$ mm

By Eq. (6a): $N_1 = \dfrac{d_1}{m_0} = \dfrac{144}{4} = 36$, $N_2 = \dfrac{288}{4} = 72$ teeth

By Eq. (19): $V_b = \dfrac{\pi \, dn \cos \phi}{60 \times 1{,}000} = \dfrac{\pi 144 \times 860 \times 0.93969}{60 \times 1{,}000} = 6.093 \; \dfrac{\text{m}}{\text{sec}}$

By Eq. (18): $F_p = \dfrac{1{,}000 \, \text{kW}}{V_b} = \dfrac{1{,}000 \times 35}{6.093} = 5{,}744$ newtons, actual load

By Table 10-2: $e = (0.0029 + 0.0031)25.4 = 0.152$ mm

By Eq. (33): $m_e = \dfrac{C^2}{1 + C^2} \dfrac{\pi b \gamma r_1^2}{2g} = \dfrac{4}{5} \dfrac{\pi 75 \times 0.0000768 \times 72^2}{2 \times 9{,}806.6}$

$= 0.003826 \; \dfrac{\text{newton sec}^2}{\text{mm}}$

By Eq. (25a): $F_d = \dfrac{en_1 N_1}{30} \sqrt{km_e} = \dfrac{0.152 \times 860 \times 36}{30} \sqrt{11{,}500 \times 75 \times 0.003826}$

$= 157.28 \sqrt{3{,}300} = 157.28 \times 57.45 = 9{,}036$ newtons,

actual load

By Fig. 10-22(b), the tooth surface must be hard enough to carry in wear the sum of loads F_p and F_d.

By Eq. (46): $F_w = F_p + F_d = 5{,}744 + 9{,}036 = 14{,}780$ newtons

This is the required capacity along the pressure line.

By Eq. (44): $Q = \dfrac{2N_2}{N_1 + N_2} = \dfrac{2 \times 72}{36 + 72} = \dfrac{4}{3}$

By Eq. (43): $K = \dfrac{F_w}{d_1 b Q} = \dfrac{14{,}780}{144 \times 75 \times \frac{4}{3}} = 1.026$

By Fig. 10-24:   BHN $= 310$

A check must now be made to see if the tooth is sufficiently strong in bending For BHN $= 310$, Fig. 10-24A indicates that a bending stress of 207 MPa is permissible. For $N_1 = 36$ teeth, Table 10-3 shows $y = 0.120$.

By Eq. (5): $p = \dfrac{\pi d_1}{N_1} = \dfrac{\pi 144}{36} = 12.57$ mm

By Eq. (36): $F_b = \sigma b y p = 207 \times 75 \times 0.120 \times 12.566 = 23.410$ newtons

This is the bending capacity of the tooth. It is located at the tip and is perpendicular to the tooth center line.

By Eq. (39): $\cos \alpha = \dfrac{r \cos \phi}{r + \text{Add.}} = \dfrac{72 \times 0.93969}{72 + 4} = 0.8902$

The resulting load along the pressure line would be

$\dfrac{2F_b}{\cos \alpha} = \dfrac{2 \times 23{,}410}{0.8902} = 52{,}594$ newton, along pressure line; capacity

in bending, two pairs in contact

The teeth are thus amply strong in bending to carry the impressed load of 14,780 newton.

The foregoing discussion was concerned with the force situation between the teeth in a pair of gears. The load forces F_p and F_d lie along the pressure line as shown in Fig. 10-22. When only one pair of teeth is in contact, force $F_p + F_d$ must be sustained by the wear capacity F_w as provided by Eq. (43).

When two pairs of teeth are in contact, each contact is assumed to take one-half the load, or $\frac{1}{2}(F_p + F_d)$. Each of these must be sustained by the bending capacity of a single tooth, $F_b/\cos \alpha$ as applied along the pressure line.

The total capacity of the gear to resist bending, $2F_b/\cos \alpha$, is used in comparing with the wear capacity F_w in determining whether the gear is weaker in bending or is weaker in wear.

18. Operation of Gear Trains at Maximum
Horsepower Capacity

Assume in this discussion that failure will take place by wear or surface fatigue, Fig. 10-22. The wear capacity F_w then is equal to $F_p + F_d$.

By Eq. (12): $\text{hp} = \dfrac{F_p V_b}{33{,}000} = \dfrac{(F_w - F_d)V_b}{33{,}000}$

Let Eqs. (25a) and (13) be substituted for F_d and V_b. Let the resulting equation be differentiated with respect to n_1 and the result be set equal to zero. The resulting equation reduces to

$$F_w = 2F_d \qquad (47)$$

When $F_w = F_p + F_d$ is substituted in this equation the result is

$$F_p = F_d \qquad (48)$$

A pair of gears will be transmitting their maximum power whenever the power force F_p is equal to the dynamic force F_d.

Example 12. Consider the two $6P_d$ gears of BHN = 310 of Examples 7 to 11. Let the gears be operated at speeds between 400 and 1,000 rpm. Calculate the hp output for each speed and plot the curve for hp vs. rpm.

Find the speed at which the maximum horsepower is transmitted and the horsepower for this speed.

Solution. From previous examples:

$$F_w = 3{,}480 \text{ lb}$$
$$e = 0.0060 \text{ in.}$$
$$d_1 = 6 \text{ in.}, \qquad N_1 = 36$$
$$K = 5{,}000{,}000 \text{ lb/in.}$$
$$m_e = 0.02488 \text{ lb sec}^2/\text{in.}$$

By Eq. (25a):
$$F_d = \frac{en_1 N_1}{30}\sqrt{km_e} = \frac{0.0060n_1 36}{30}\sqrt{5{,}000{,}000 \times 0.02488}$$

$$= 0.0072n_1\, 352.7 = 2.539n_1 \qquad (a)$$

By Eq. (46):
$$F_w = F_p + F_d = 3{,}480 \text{ lb} \qquad (b)$$
$$F_p = 3{,}480 - F_d$$

By Eq. (13):
$$V_b = \frac{\pi n_1 d_1 \cos\phi}{12} = \frac{\pi 6 n_1\, 0.93969}{12} = 1.476n_1 \qquad (c)$$

Values for F_d, F_p, V_b, and hp are calculated for various speeds and entered into Table 10-5. The curve for speed vs. horsepower is shown in Fig. 10-25.

The maximum power occurs when $F_p = F_d$. Then

By Eq. (b):
$$F_d = \frac{1}{2} \times 3{,}480 = 1{,}740 \text{ lb}$$

By Eq. (a):
$$2.539n_1 = F_d = 1{,}740$$

$$n_1 = \frac{1{,}740}{2.539} = 685.3 \text{ rpm for maximum power}$$

By Eq. (c):
$$V_b = 1.476n_1 = 1.476 \times 685.3 = 1{,}011.5 \text{ fpm}$$

By Eq. (12):
$$\text{hp} = \frac{F_p V_b}{33{,}000} = \frac{1740 \times 1{,}011.5}{33{,}000} = 53.3 \text{ maximum}$$

FIGURE 10-25 Variation of horsepower capacity with speed for gears of Example 12.

The power transmitted at different speeds has been determined in the above example. The speed for maximum power and the power at this speed has also been determined. It is seen that the given speed of 860 rpm is a little higher than the speed at which the maximum power occurs. It should be noted at all speeds that the power transmitting force F_p plus the dynamic force F_d is equal to the wear capacity F_w of 3,480 lb.

TABLE 10-5   Calculations for Example 12

n_1, rpm	$F_d =$ $2.535n_1$	$F_p =$ $3,480 - F_d$	$V_b =$ $1.476n_1$	$hp =$ $\dfrac{F_p V_b}{33,000}$
400	1,014	2,466	590	44.1
500	1,268	2,212	738	49.5
600	1,521	1,959	886	52.6
700	1,774	1,706	1,033	53.4
800	2,028	1,452	1,181	52.0
900	2,282	1,198	1,328	48.2
1,000	2,535	945	1,476	42.3

19. Determination of Bearing Loads

The tooth loads and bearing reactions for the various shafts of a gear train can be easily found by making suitable sketches. A typical case is shown in Fig. 10-26, the computations for which are carried out in the following example.

FIGURE 10-26   Shaft and gears of Example 13.

Example 13. Find the tooth loads and bearing reactions for the shaft shown in Fig. 10-26(a). All gears are 20° pressure angle with diametral pitch equal to 3. Make top and front views showing the loading for the shaft. Assume frictional losses to be negligible. $AB = 6$ in.; $BC = 20$ in.; $CD = 5$ in.

Solution. For gear 1: $d = \dfrac{N}{P_d} = \dfrac{24}{3} = 8$ in.

For gears 1 and 2: $V = \dfrac{\pi dn}{12} = \dfrac{\pi 8 \times 300}{12} = 628.3$ fpm at pitch circle

(17): $F_t = \dfrac{33{,}000 \text{ hp}}{V} = \dfrac{33{,}000 \times 75}{628.3} = 3{,}939$ lb,

tangential to pitch circle radial force on gear 2

$F_r = F_t \tan \varphi = 3{,}939 \times 0.36397 = 1{,}434$ lb

For gear 3: $V = \dfrac{48}{30} \times 628.3 = 1{,}005.3$ fpm at pitch circle

$F_t = \dfrac{33{,}000 \times 75}{1{,}005.3} = 2{,}462$ lb,

tangential to pitch circle radial force on gear 3

$F_r = F_t \tan \varphi = 2{,}462 \times 0.36397 = 896$ lb

The tooth loads described above are shown in Fig. 10-26(a). The bearing reactions are found by simple statics. The top and front views for the shaft loads are shown in sketches (b) and (c), respectively. Bending moments and stresses for the shaft can be found in the usual manner.

20. Number of Pairs of Teeth in Contact

During operation the teeth of the two gears remain in contact with each other during the interval from the points where the two addendum circles cross the line of action. This would be the interval $A_1 B_1$ in Fig. 10-27. This interval

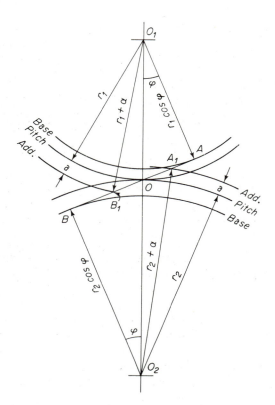

FIGURE 10-27   **Layout for determining interval of contact A_1B_1.**

must have a length equal to at least one base pitch and preferably, for smooth and continuous operation, should be not less than 1.4 base pitches. Its length can be easily found from the sides of triangles in the figure.

$$A_1B_1 = A_1O + OB_1$$

$$A_1O = A_1B - OB$$

$$= \sqrt{(r_2 + a)^2 - r_2 \cos \phi} - r_2 \sin \phi \tag{49}$$

where the value of the addendum is represented by a.

$$OB_1 = AB_1 - AO$$

$$= \sqrt{(r_1 + a)^2 - r_1 \cos \phi} - r_1 \sin \phi \tag{50}$$

A_1B_1 or the sum of A_1O and OB_1 is divided by the base pitch p_b or $p \cos \phi$ to obtain the number of tooth intervals in contact.

Example 14. Two 20° full-depth gears have a diametral pitch of 6 and operate on a center distance of 7.5 in. The speed ratio is 1.5:1. Find the number of base pitches in the length of contact.

Solution. $c = r_1 + r_2 = r_1 + 1.5r_1 = 2.5r_1 = 7.5, r_1 = 3$ in., $r_2 = 4.5$ in.

In Eq. (49):
$$OA_1 = \sqrt{\left(4.5 + \frac{1}{6}\right)^2 - 4.5^2 \times 0.93969^2}$$

$$- 4.5 \times 0.34202 = 1.9740 - 1.5391 = 0.4349 \text{ in.}$$

In Eq. (50):
$$OB_1 = \sqrt{\left(3 + \frac{1}{6}\right)^2 - 3^2 \times 0.93969^2} - 3 \times 0.34202$$

$$= 1.4424 - 1.0261 = 0.4163 \text{ in.}$$

By Eq. (11):
$$p_b = \frac{2.9521}{P_d} = \frac{2.9521}{6} = 0.4920 \text{ in.}$$

$$\text{number} = \frac{0.4349 + 0.4163}{0.4920} = 1.73 \text{ tooth intervals in length}$$

of contact

If the number of teeth in the gears is decreased, the radii become proportionally smaller and the addenda proportionally larger. Thus, point A_1 in Fig. 10-27 will eventually coincide with point A. Should gears be made with still fewer teeth, interference of metal will occur, unless the flanks are undercut, as will be explained later. The equations above for length of contact are applicable only to gears that are free from undercutting.

Points A and B in Fig. 10-27 are called the interference points for the gears. It is possible to have gear action all the way along the pressure line to the interference point. However, when the gears are highly stressed, it is good practice to have contact begin and end a considerable distance short of the interference points. Figure 10-5 shows how the radius of curvature of the involute curve becomes progressively smaller as the interference point, or the beginning of the involute, is approached. Contact stresses caused by the tooth load increase rapidly as the radius of curvature of the body is decreased. The stresses may become so high that pitting and surface failure will occur if the involute is too heavily loaded in the region of small radius of curvature.

The designer usually specifies teeth of the smallest size that will carry the load in order that the overlap of gear action will be as large as possible.

The stresses in spur gear teeth are very sensitive to the accuracy of mounting. If the two shafts are not parallel, the loading does not occur across the entire width of face, but is concentrated at one edge of the tooth, which can produce a large increase in the stresses. Sometimes the teeth are crowned, which means they are made slightly smaller toward the ends to ensure that the load is applied over the central portion of the width.

To assist the tooth in picking up its load gradually when coming into contact at K_2, Fig. 10-21(a), the thickness of the tooth is decreased slightly from the theoretical value near the tip. This is known as tip relief.[6]

21. Undercutting in Gear Teeth

A gear having few teeth, if made by one of the generating processes, has undercut flanks[7] as illustrated in Fig. 10-28. For such gears, the end of the cutting tool extends inside the interference point A or point of tangency of base circle and pressure line, and removes an excessive amount of metal. Undercutting not only weakens the tooth, but removes a small portion of the involute adjacent to the base circle; this loss of involute may cause a serious reduction in the length of contact. Gear action can take place only on the involute, which ends at the base circle. The portion of the mating tooth that projects inside the interference point is therefore not needed for the operation of the gears.

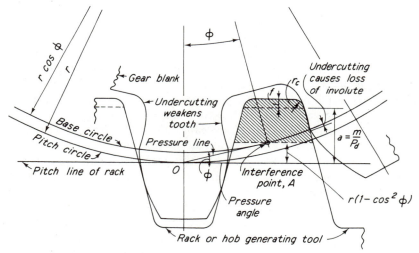

FIGURE 10-28   Undercutting in generated gear resulting from insufficient number of teeth.

Undercutting can be prevented by using more and smaller teeth in the gear. No undercutting can occur if the addendum does not extend inside the interference point A. In Fig. 10-28 the height from the interference point to the center of the gear is $r \cos^2 \varphi$. The distance from pitch line of the rack to the interference point, $r(1 - \cos^2 \varphi)$ or $r \sin^2 \varphi$, should be set equal to the

[6]See Walker, H., "Gear Tooth Deflection and Profile Modification," *The Engineer*, **166**, 1938, pp. 409 and 434; **170**, 1940, p. 102; **178**, 1944, pp. 484 and 502.

[7]See the author's article "Undercutting of Hobbed Spur Gear Teeth," *Machine Design*, **28**, Apr. 19, 1956, p. 123.

addendum a or m/P_d of the cutting tool. Pitch radius r is equal to $N/2P_d$ so that addendum a is equal to

$$a = \frac{m}{P_d} = \frac{N}{2P_d} \sin^2 \varphi \qquad \text{or} \qquad N = \frac{2m}{\sin^2 \varphi} \tag{51}$$

The minimum number of teeth that a gear must have if generated teeth are not to be undercut are as follows.

For $14\frac{1}{2}°$ full-depth gears: $\qquad m = 1; \qquad N = 32$

For $20°$ full-depth gears: $\qquad m = 1; \qquad N = 17$

For $20°$ stub teeth: $\qquad m = 0.8; \qquad N = 14$

Generated gears having fewer teeth than indicated above should be used only when the resulting loss of involute will not reduce the length of contact below a satisfactory value. The superiority of $20°$ gears, when the number of teeth is small, should be noted.

Should the gear in Fig. 10-28 be made by some method that would not undercut the flanks, there would be interference of metal, and the gear would probably neither mesh nor roll with another gear. Undercutting and interference can be avoided in gears of few teeth by using a system with a larger pressure angle, or by cutting the teeth on the long and short addendum system described in the following section. Cycloidal gear teeth neither undercut nor interfere. This advantage is retained by the $14\frac{1}{2}°$ composite system.

Involutes are of course produced by all the standard generating systems. However, in order to determine what the teeth actually look like, the method of production and the particular tool that will be used must be taken into account. Such information can be secured by making a large-scale layout of the teeth of both gears. It is especially needed when the number of teeth is small, and it is necessary to determine the extent of the undercutting. A magnified cutter should be made from a piece of tracing paper and rolled over the circle representing the blank; successive positions should be traced until the tooth outline is generated. The teeth of both gears can then be meshed together, and definite information can be obtained regarding undercutting and loss of involute, radius of fillet, length of contact, overlap, length of addendum, and depth of cut.

22. *Long and Short Addendum Gearing*

A very successful method of avoiding undercutting is to use the so-called "long and short addendum gearing." The addendum on the pinion or smaller gear is made longer than standard, and the addendum of the larger gear is made shorter by an identical amount. The gears are generated by the same rack or hob as for gears with standard addenda. The pressure angle during operation is unchanged.

FIGURE 10-29 Shift in position of generating rack to cause addendum to pass through interference point of gear. $P_d = 1$.

The outside diameter of the pinion is increased, and since the hob penetrates less deeply into the blank, undercutting is reduced or eliminated. The increase of radius for the blank at which undercutting can be just avoided is equal to the value of e, as shown in Fig. 10-29. Here the end of the rack tooth, not counting clearance f, passes through the interference point. From the figure,

$$e = \frac{1}{P_d} - r(1 - \cos^2 \varphi)$$

$$= \frac{1}{P_d} - r \sin^2 \varphi \qquad (52)$$

Example 15. Find the shift in the position of the rack necessary to eliminate undercutting in a $14\frac{1}{2}°$ full-depth gear of 18 teeth. $P_d = 1$.

Solution. $r = \frac{N}{2P_d} = \frac{18}{2} = 9$ in.

In Eq. (52): $e = \frac{1}{1} - 9(1 - 0.96815^2) = 1 - 0.5642 = 0.4358$ in.

Although the value of e can be selected at the discretion of the designer, it is frequently rounded off to a value somewhat larger than that given by Eq. (52).

The teeth in Fig. 10-29 are thicker than those of a standard gear, and the spaces are thinner. Tooth thickness can be easily determined from Fig. 10-30, which shows an oversize blank with teeth cut by a rack located a distance e from the pitch circle of the gear. The pitch circle rolls without slipping along pitch line TT' of the rack. Segment OA equals arc OE, and the tooth thickness t on the pitch circle for the chosen value of e is

$$t_1 = 2e \tan \varphi + \tfrac{1}{2}p \qquad (53)$$

Figure 10-30 illustrates the cutting of an oversize pinion. When the undersize gear blank is cut, the rack is located a distance e inside the pitch circle, and the tooth thickness is reduced to

$$t_2 = -2e \tan \varphi + \tfrac{1}{2}p \qquad (54)$$

Example 16. Find the thickness of the tooth on the pitch circle for $14\frac{1}{2}°$ gears, if e is taken as 0.5 in. for $P_d = 1$.

Solution. $p = \frac{\pi}{P_d} = 3.1416$ in.

In Eq. (53), for pinion: $t_1 = 2 \times 0.5 \times 0.25862 + 0.5 \times 3.1416$

$$= 0.2586 + 1.5708 = 1.8294 \text{ in.}$$

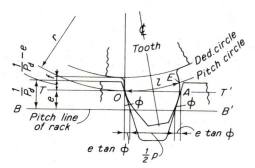

FIGURE 10-30   Change in tooth thickness resulting from nonstandard setting of generating rack.

In Eq. (54), for gear:
$$t_2 = -2 \times 0.5 \times 0.25862 + 0.5 \times 3.1416$$
$$= -0.2586 + 1.5708 = 1.3122 \text{ in.}$$

The teeth for this example, for $N_1 = 18$ and $N_2 = 62$, are illustrated in Fig. 10-31. It can be seen that the pinion tooth, which normally would be badly undercut, is not.

FIGURE 10-31   Long and short addendum gears.

Because the larger gear is undersize and the hob has been shifted inwardly, undercutting can be avoided only by having a sufficiently large number of teeth. The larger gear should have at least as many teeth in excess of the minimum number as the smaller gear has fewer than the minimum.

Because of the change in tooth thickness, the y-factors of Table 10-3 cannot be used with long and short addendum gears.

When both mating gears have few teeth, undercutting in both can be prevented by the use of positive values for e. Assembly, when both gears have thicker teeth, can be made only at an increased center distance.[8] The calculations become involved and are beyond the scope of this book.

23. Speed Ratios of Gear Trains

The speed ratio of a gear train can be easily found if the number of teeth in each gear is known. Thus in Fig. 10-32 the ratio between n_4, the rpm of the shaft for gear 4, and n_1, the rpm of the shaft for gear 1, is given by the following equation.

$$\frac{n_4}{n_1} = \frac{N_1}{N_2} \cdot \frac{N_3}{N_4} \tag{55}$$

FIGURE 10-32 Gear train.

Additional ratios can be inserted into Eq. (55) if the train consists of a larger number of gears.

The inverse problem, that of finding the number of teeth each gear should have if the ratio of the train is to equal an arbitrarily chosen value, is much more difficult. Such problems arise whenever it is necessary to give a specified relative motion to two portions of a mechanism.

A typical example is shown in Fig. 10-33, where a thread of lead l_1 is being cut by an engine lathe whose lead screw has a lead of l_2. Let gear 1 make one complete revolution. Gear 4 makes N_1N_3/N_2N_4 revolutions, and carriage A moves a distance $N_1N_3l_2/N_2N_4$ in. The distance must be equal to

[8]See Chapter 10 of Reference 8, end of chapter.

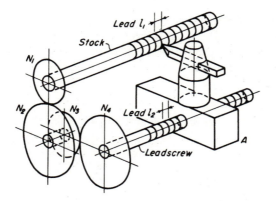

FIGURE 10-33 Thread being cut by lathe.

l_1. Hence

$$\frac{N_1 N_3}{N_2 N_4} = \frac{l_1}{l_2} \tag{56}$$

The problem is thus reduced to that of finding the numbers of teeth in the various gears.

In general, a given ratio cannot be exactly produced by a gear train because the individual members must necessarily contain integral numbers of teeth. It is possible, however, to approximate the desired ratio with a degree of accuracy that is sufficient for most applications. Although trial and error is involved, a systematic procedure helps reduce the labor to a minimum. A number of different methods can be used, of which the following is very effective.[9]

Let the given ratio between the first and last shafts be represented by G, and let u/w be a common fraction, u and w integers, whose value is close to that of G. Ratio G can be exactly represented by

$$G = \frac{uh - j}{wh} \tag{57}$$

where j is an integer, and h has the value obtained by solving the equation. Thus

$$h = \frac{j}{u - wG} \tag{58}$$

In Eq. (57), it is necessary that both numerator and denominator be integers factorable into terms suitable for use as the numbers of teeth in the gears of the train. However, the value of h, as given by Eq. (58), in general

[9]See the author's article "A Practical Method for Designing Gear Trains," *Prod. Eng.*, **24**, Feb. 1953, p. 211. See also Chapter 11 of Reference 8, end of chapter.

is not an integer. Then the numerator and denominator in Eq. (57) will not be integers either. However if an integer h', whose value is close to that of h, is used, an approximate value G' of the ratio is obtained that may be sufficiently close to the exact value. The equation for G' is

$$G' = \frac{uh' - j}{wh'} \qquad (59)$$

The use of the above equation is illustrated by the following example.

Example 17. Suppose the shaft on the left in Fig. 10-32 is to make 2.54 revolutions for each revolution of the shaft on the right. Find suitable numbers of teeth for the gears.

Solution. $$G = \frac{1}{2.54} = 0.3937008$$

By trial it is found that fraction $\frac{24}{61}$ has a value close to G. Hence

$$\frac{u}{w} = \frac{24}{61}$$

Let: $j = 1$

In Eq. (58): $h = \dfrac{1}{24 - 61 \times 0.3937008} = -63.499$

Suppose: $h' = -63$

In Eq. (59): $G' = \dfrac{-24 \times 63 - 1}{-61 \times 63} = \dfrac{1,513}{3,843} = \dfrac{17 \times 89}{61 \times 63} = 0.3937028$

error $= 0.3937028 - 0.3937008 = 0.0000020$

If the solution above is sufficiently accurate, the numbers of teeth in the gears can be made $N_1 = 17$, $N_2 = 61$, $N_3 = 89$, and $N_4 = 63$.

Another solution can be obtained by taking

$$h' = -64$$

In Eq. (59): $G' = \dfrac{-24 \times 64 - 1}{-61 \times 64} = \dfrac{1,537}{3,904} = \dfrac{29 \times 53}{61 \times 64} = 0.3936988$

Another solution can be obtained as follows.

Let: $j = 3$

$$h = \frac{3}{24 - 61 \times 0.3937008} = -190.498$$

Let: $h' = -190$

In Eq. (59): $G' = \dfrac{-24 \times 190 - 3}{-61 \times 190} = \dfrac{4,563}{11,590} = \dfrac{13 \times 13 \times 27}{10 \times 19 \times 61}$

$$= 0.3937015$$

error $= 0.0000008$

The accuracy has been improved, but a train of six gears is required.

If the numerator in Eq. (59) is not factorable, a different value of h' should be used. A factor table, such as that in *Machinery's Handbook*,[10] is very helpful. Because of the unavailability of large gears, it is frequently specified in change gear work that the largest gear should not exceed 100 or 120 teeth.

Suitable values for the fraction u/w can be found on the slide rule. Thus if the end of the C-scale is set at the value for G on the D-scale, numerous combinations for u/w can be found. Some additional values for the example above are $\frac{13}{33}$, $\frac{35}{89}$, $\frac{37}{94}$, $\frac{41}{104}$, $\frac{87}{221}$, $\frac{102}{259}$, $\frac{124}{315}$, and $\frac{137}{348}$.

Suitable values for u/w can also be found on a standard calculating machine. The given value for G should be placed on the keyboard, and successive values added until the sum is very close to an integer. Thus

$$wG \approx u \qquad (60)$$

In the example above 61 times the value of G is a number very close to 24. Hence u/w can be taken as $\frac{24}{61}$.

24. Internal or Annular Gears

A gear can also be made as a ring (see Fig. 10-34) with teeth on the inside that are somewhat like the spaces between the teeth of an external spur gear. The assembly of an internal and external gear is very compact, with a reduced distance between the shaft centers.

FIGURE 10-34   Internal or annual gear.

Internal gears are cut on the Fellows gear shaper. To prevent the corners on the inner ends of the teeth from being cut away, the gear must have considerably more teeth than the cutter. To obtain proper tooth action without undue modification of tooth shape, the smallest permissible difference

[10]New York: Industrial Press, Inc., 1964. See also Reference 10, end of chapter.

between the number of teeth in the pinion and internal gear must be as follows:

8 teeth for 20° stub-tooth form
10 teeth for 20° full-depth teeth
12 teeth for 14½° full-depth teeth

If the spokes or web of the internal gear are integral with the rim, a groove for cutter relief must be provided at the inner end of the teeth. This groove must be at least $\frac{3}{32}$ in. in width.

The teeth of internal gears are stronger than those of corresponding spur gears. Operation is smooth and quiet because a greater number of teeth are in contact.

25. Dimensioning of Gears

The drawing for a gear should show the outside diameter and thickness of the blank with suitable tolerances. The number of teeth, diametral pitch, and pressure angle should be given. Backlash should be taken into account, and the tooth thickness, as measured around the theoretical pitch circle, should be shown. Tooth thickness should be expressed with a tolerance, but the pitch circle diameter should be given as a flat dimension. The addendum and whole depth should also be shown. It is usually helpful to include the center distance, number of teeth in the mating gear, and backlash after assembly. It is good practice, however, to avoid giving superfluous interdependent theoretical dimensions.

26. Materials for Gears

Gears are made from a wide variety of materials, such as gray and alloy cast iron; cast, wrought, and forged steel; brass; bronze; and impregnated fabric. Cast iron has good wearing properties but is weak in bending, which necessitates the use of relatively large teeth. Unhardened low-carbon and free-cutting steels can be used for gears but are suitable only for applications requiring moderate strength and impact resistance.

Higher strength and surface hardness can be obtained only by some form of heat treatment. The principal production processes for heat treatment are as follows.

(a) *Through Hardening.* With a sufficiently high carbon content, 0.35 to 0.50%, the gear, after cutting, is quenched and drawn at suitable temperatures to obtain the desired physical properties. Plain carbon steels[10] such as 1040 and 1050 and free cutting steel 1137 do not harden deeply and are subject to distortion in quenching. Alloy steels 4140 and 4340 will through-harden and retain their shape better when heat-treated. The greater strength and hardness of surface and core for heat-treated materials is accompanied by a loss of ductility.

(b) *Carburizing or Case Hardening.* In this process the parts are packed in a compound rich in carbon and held for a long time at red heat in a furnace. Carbon is absorbed by the surface layers, which become very hard after quenching and tempering. The core is also strengthened, but the gear requires careful handling to prevent distortion. Alloy steels of low carbon content such as 4620 and 4320 are widely used for carburizing.

(c) *Nitriding.* For extremely hard surfaces and wear resistance the nitriding process can be used. After heat treatment the gear is held for an extended period at a temperature of 850 to 1000°F in the presence of ammonia gas. Complex nitrides form on the surfaces. The process is relatively expensive. Steels 4140, 4340, and Nitralloy N are widely used, but certain others can also be employed. Nitrided gears can be operated at higher temperatures than carburized gears.

Both carburizing and nitriding leave the surface in a state of compressive stress which is beneficial under the fatigue stresses that occur in gears.

(d) *Induction and Flame Hardening.* In this process the gear is quickly heated by an induction coil followed by quenching in oil. Localized tooth by tooth heating by coil or oxyacetylene flame is also used. Quenching is done by a water spray or an air blast. The steel must have sufficient carbon content to harden on quenching. SAE 4140 is the cheapest, but 4340 will give greater hardness.

Distortion can be eliminated if the teeth are first rough cut in a blank that has been suitably prepared for machining by normalizing or annealing. The gear is then heat-treated to the upper limit for machining, about 300–350 Brinell, and the teeth are brought to final size by taking a finish cut. No subsequent heat treatment need be done. A free-machining steel of suitable physical properties is sometimes employed with the foregoing process, and a reduction in the machining costs can thereby be effected. It is noted that the power capacity of a pair of gears increases rapidly with increase of surface hardness.

27. Lubrication and Mounting of Gears

Gears operate under a diversity of conditions, and the methods of lubrication will vary accordingly. For unenclosed or exposed gearing, the lubricant is applied by an oilcan, a drip oiler, or a brush. Frequent applications of small amounts of lubricant are preferable to large volumes at longer intervals.

If the gears are exposed to water or acids, a sticky lubricant that will adhere to the metal must be used.

When gears run in an enclosed casing, the larger gears may dip into a bath of oil, which will be carried to the wearing surfaces. Sometimes enclosed gearing is lubricated by spraying a jet of oil on the working surfaces as they revolve toward each other. When the contact pressure is very high, extreme-pressure EP lubricants must be used to prevent rupture of the oil film and the resulting metal-to-metal contact of the parts. Lubricants of the EP type contain additives that increase the load-carrying properties as well as prevent the squeezing out of the lubricant.

Foreign matter in the lubricant can cause a rapid increase in the rate of wear of the teeth.

The mounting of gears is very important. Care must be exercised that shafts are parallel or the entire load will be carried at the ends of the teeth instead of across the entire width of face. Excessive wear and danger of failure result. The same effect occurs if the teeth are not cut parallel to the axis of rotation or may have other inaccuracies. Elastic deformation of the teeth, blanks, shafts, and bearing supports can also be the cause of misalignment. Unbalanced rotating masses in a geared system can also cause unexpected loading which the teeth are called upon to carry. Noise, vibration, and shortened life can result from many causes, which may be difficult to locate and analyze.

In general, spur gears operate at high efficiencies. Good-quality commercial gears properly mounted and lubricated should not consume in friction more than 1 or 2% of the power transmitted.

28. Gear Tooth Failures

At the present time, gear design is really a combination of art and science. Gear failures can best be avoided if practical field experience is available to assist in the design. Some gears run better and last longer than would be expected by the design equations, while others fail even when operated well within the limits imposed by the calculations.[11]

It is very important that a correct analysis be made as to why a pair of gears failed since merely making the set larger may not be a cure for the original cause of the failure. The three most common types of gear tooth failure are tooth breakage, surface pitting, and scoring.

(a) *Tooth breakage* may be caused by an unexpectedly heavy load being imposed on the teeth. A more common type of failure is due to bending

[11]See Almen, J. O., in *Mechanical Wear*, J. T. Burwell, Ed., Metals Park, Ohio: American Society for Metals, 1950, Chapter 12. See also pp. 52, 141, and 291 of Reference 4, end of chapter.

fatigue, which results from the large number of repetitions of load imposed on the tooth as the gear rotates. A small value for the radius of the fillet may accentuate the bending fatigue effects.

(b) *Pitting* is a fatigue phenomenon caused by stresses exceeding the endurance limit of the surface material. After a sufficient number of repetitions of the loading cycle, bits of metal on the surface will fatigue and drop out. The process sometimes continues at an increasing rate since the remaining unpitted areas are less able to carry the load. Lubrication difficulties may contribute to pitting failures.

(c) *Scoring* can occur under heavy loads and inadequate lubrication. The oil film breaks down and metal-to-metal contact occurs. High temperatures result and the high spots of the two surfaces weld together. The welds are immediately broken, but the surfaces of the teeth undergo rapid wear.[12] Gearboxes should be broken in by preliminary operation at lower loads and smaller speeds until the tooth surfaces are highly polished. Misalignment of the shafts may shift the entire load to one edge of the tooth with resulting excessive stress and the probability of scoring. Fine-pitch teeth are largely immune to scoring. The diametral pitch should be at least 10 or greater. For coarse pitches, scoring is sometimes a greater hazard than breakage or pitting. At the present time there is no simple and reliable formula by means of which the designer can predict the possibility of scoring.

29. Planetary Gear Trains

Planetary or epicyclic gear trains provide a compact arrangement suitable for speed reducers. In Fig. 10-35, the right-hand shaft is integral with arm 0. Gears 1 and 3 are keyed to a short length of shaft, which revolves in a bearing

(c) Consider Gears 1 and 3 and connecting shaft as free body

FIGURE 10-35 Planetary gear train.

in arm 0. Gear 1 meshes with the fixed gear 2, and gear 3 meshes with gear 4, which in turn is keyed to the left-hand shaft.

Let N_1, N_2, N_3, and N_4 be the number of teeth in gears 1, 2, 3, and 4, respectively. As shown in the figure, arm 0, which was originally vertical, has been given an angular displacement α, which causes gear 1 to traverse arc AB on gear 2. Arc BC turned by gear 1 is equal to arc AB. Since angles are inversely proportional to the radii, or to the number of teeth, gears 1 and 3 are turned through angle $\alpha N_2/N_1$.

While the foregoing was taking place, gears 3 and 4 were rotating on each other through the equal arcs DE and EF. Gear 4 is turned in the reverse direction through angle $\alpha N_2 N_3/N_1 N_4$.

The net effect of these two operations is to move the point of gear 4, which was originally vertical at G, over to location F. Gear 4 has thus been rotated clockwise through angle

$$\left[1 - \frac{N_2 N_3}{N_1 N_4}\right]\alpha$$

This latter value, when divided by α, gives the ratio of the rotations of shafts 4 and 0, respectively.

$$\frac{n_4}{n_0} = 1 - \frac{N_2 N_3}{N_1 N_4} \tag{61}$$

The equation above for the speed ratio can be derived in another way by giving consideration to the various torques. In Fig. 10-35(c) consider gears 1 and 3 and the connecting shaft as the free body. Let F_0 be the force exerted on the free body from arm 0, F_{12} be the force from gear 2 upon gear 1, and F_{34} be the force from gear 4 upon gear 3. If arm 0 is the driver, the input torque T_0 is equal to $F_0 r_0$. The output torque T_4 is equal to $F_{34} r_4$.

The moments of the forces about the center of the planet gears must be in equilibrium. Let r_1 be the radius of gear 1 and r_3 the radius of gear 3.

$$F_{12} r_1 = F_{34} r_3 \tag{a}$$

The moments about the main shaft centers must also be in equilibrium.

$$F_0 r_0 + F_{12} r_2 = F_{34} r_4 \tag{b}$$

or

$$F_{12} r_2 = F_{34} r_4 - F_{12} r_2 \tag{c}$$

The value of F_{12} from Eq. (a) should now be substituted.

$$F_0 r_0 = F_{34} r_4 - \frac{F_{34} r_3 r_2}{r_1} = F_{34} r_4 \left[1 - \frac{r_2 r_3}{r_1 r_4}\right]$$

or

$$T_0 = T_4 \left[1 - \frac{N_2 N_3}{N_1 N_4}\right]$$

If friction is neglected, the input and output energies for the gear train are equal. Hence

$$T_4 n_4 = T_0 n_0$$

or

$$\frac{n_4}{n_0} = \frac{T_0}{T_4} = 1 - \frac{N_2 N_3}{N_1 N_4} \qquad (62)$$

The reaction or housing torque T_2 is equal to $F_{12} r_2$. Its value can be found by substitution for F_{34} from Eq. (a) into Eq. (c).

$$F_0 r_0 = \frac{F_{12} r_1 r_4}{r_3} - F_{12} r_2 = F_{12} r_2 \left[\frac{r_1 r_4}{r_2 r_3} - 1 \right]$$

$$T_0 = T_2 \left[1 - \frac{N_2 N_3}{N_1 N_4} \right] \frac{N_1 N_4}{N_2 N_3}$$

or

$$T_2 = \frac{N_2 N_3 / N_1 N_4}{1 - N_2 N_3 / N_1 N_4} T_0 \qquad (63)$$

The foregoing methods can be applied to all types of epicyclic systems, including those containing bevel gears. Either shaft 0 or 4 can be used as the driver.

Instead of a single arm 0, an actual gear train is usually constructed with either two or three equally spaced arms, each of which carries a pair of planet gears.

The epicyclic train can be arranged so that the fixed gear in Fig. 10-35 is rotatable and can serve as an additional input gear. A wider variety of output ratios can thus be obtained.[12]

REFERENCES

1. BUCKINGHAM, E., *Analytical Mechanics of Gears*, New York: McGraw-Hill Book Company, 1949.

2. BUCKINGHAM, E., *Spur Gears*, New York: McGraw-Hill Book Company, 1928.

3. DUDLEY, D. W., Ed., *Gear Handbook*, New York: McGraw-Hill Book Company, 1962.

4. DUDLEY, D. W., *Practical Gear Design*, New York: McGraw-Hill Book Company, 1954.

5. *Proc. Int. Conf. Gearing*, London: Institution of Mechanical Engineers, 1958.

6. "Right and Wrong of Hob Sharpening," Chicago: Illinois Tool Works, 1942.

7. ROTHBART, H. A., Ed., *Mechanical Design and Systems Handbook*, New York: McGraw-Hill Book Company, 1964, Section 32.

8. SPOTTS, M. F., *Mechanical Design Analysis*, Englewood Cliffs, N.J.: Prentice-Hall, Inc., 1964.

[12]See Chapter 11 of Reference 8, end of chapter.

9. *Standards of the American Gear Manufacturers Association*, Washington, D.C.

10. STUTZ, C.C., *Formulas in Gearing*, Providence, R.I.: Brown and Sharpe Mfg. Co., 1938.

11. TUPLIN, W. A., "Dynamic Load on Gear Teeth," *Machine Design*, **25**, Oct. 1953, p. 203.

PROBLEMS

Assume that all gear blanks are solid circular cylinders unless specified otherwise.

1. From a piece of vellum, make a rack cutter similar to that in Fig. 10-36. $\varphi = 14\frac{1}{2}°$; $P_d = 1$. Use standard clearances at top and bottom of the tooth. Use a radius of the fillet $1\frac{1}{3}$ times the clearance. The fillet is tangent to both the sloping cutting edge and the clearance line. To preserve the strength of the tool, only a small portion of the paper along the cutting edges should be removed, as shown. The numbered lines at right angles to the pitch line are spaced $\frac{1}{8}p$ apart. Also, draw Fig. 10-37 to simulate the blank of a 12-tooth gear. The numbered radial lines are also spaced $\frac{1}{8}p$ apart around the pitch circle.

FIGURE 10-36   Problem 1.

Tool and blank are synchronized by placing the pitch line of the rack on the pitch circle of the gear, and having the correspondingly numbered lines coincide. A sharp drafting pencil should be drawn along the cutting edges for each position of the rack; an involute tooth is thus generated in the manner illustrated by Fig. 10-14.

Note the large amount of undercutting when a gear of few teeth is made by this process. Draw the base circle on the gear and estimate the percentage of involute lying between the base and pitch circles that has been lost by undercutting. If greater accuracy is desired, this problem can be worked out at twice the scale mentioned above.

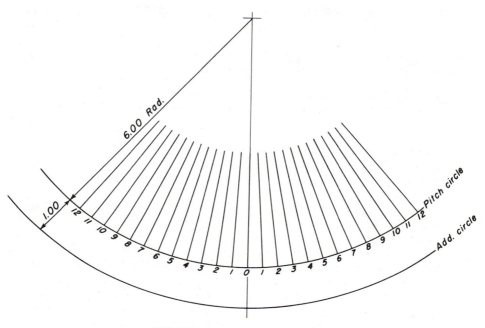

FIGURE 10-37 Problem 1.

2. For the gears of Problem 1, assume that the undercutting, as shown in Fig. 10-28, extends one-third of the way from the base to the pitch circles. Make a full-size layout and compute the number of tooth intervals in the remaining length of contact if two of these gears are meshed together. Note the seriousness of the loss of even a small portion of the involute curve.

Ans. No. of tooth intervals = 0.42.

3. How many teeth must there be in a $14\frac{1}{2}°$ gear of full-depth involute teeth if the base and working-depth circles coincide? *Ans.* $N = 62.8$.

4. Make a full-size layout of a 20° full-depth involute tooth. $N = 12$; $P_d = 1$. Make the flank of the tooth radial inside the base circle. Let the radius of the fillet be $0.05p$ on one side of the tooth and $0.15p$ on the other side. Assume that the load is located on the pressure line at the end of the tooth. Length l should be measured from point A to the cross section of minimum depth h.

Measure the tooth and compute y by Eq. (35). Do the same for several nearby cross sections until the one is found that gives the smallest value for y. Two sets of measurements and computations are needed, one for the tooth of the small fillet and the other for the large. Note how the size of the fillet as actually produced on the tooth will cause changes in the value of the y-factor. Can you obtain the tabulated value of 0.078 for y? If $r_f = 0.05p$, will the tooth be safe when carrying the load indicated by Eq. (36)?

5. Two 20° full-depth gears are made of soft steel, BHN = 200. AGMA quality No. 8. Pinion turns 172.5 rpm. $N_1 = 48$; $N_2 = 96$; $P_d = 4$, and $b = 3$ in.

Find the horsepower the gears are transmitting, and the number of base pitches in the interval of contact.   *Ans.* hp = 23.2; no. of intervals = 1.80.

6. Two 20° full-depth steel gears are heat treated to BHN = 350. AGMA Quality No. 8. Pinion turns 860 rpm. $N_1 = 30$; $N_2 = 120$; $m_0 = 5$, and $b = 60$ mm. Find the power the gears are transmitting.   *Ans.* kW = 58.7.

7. Two 20° full-depth steel gears are heat treated to BHN = 350. AGMA Quality No. 8. Pinion turns 860 rpm. $N_1 = 24$; $N_2 = 96$; $P_d = 4$, and $b = 2$ in. Find the horsepower the gears are transmitting.   *Ans.* hp = 77.2.

8. Two 20° full-depth gears are made of induction-hardened steel. AGMA Quality No. 8. Pinion turns 215 rpm $N_1 = 36$; $N_2 = 72$; $m_0 = 8$, and $b = 60$ mm. Find the power the gears are transmitting.   *Ans.* kW = 124.6.

9. Two 20° stub-tooth steel gears are heat treated to BHN = 350. AGMA Quality No. 8. Pinion turns 860 rpm. $N_1 = 30$; $N_2 = 90$; $P_d = 5$, and $b = 2$ in. Find the horsepower the gears are carrying, and the number of base pitches in the interval of contact.   *Ans.* hp = 62.2; no. of intervals = 1.43.

10. Two 20° gears are carrying 35 kW at 860 rpm for the pinion. Speed ratio is 2:1 with a center distance of 225 mm. AGMA Quality No. 9. $m_0 = 5$. Both gears hardened to BHN = 350. Calculate the necessary value for face width b.
                    Ans. $b = 35.4$ mm.

11. Two 20° full-depth gears are made of induction-hardened steel. Center distance is 7.5 in. Speed ratio is 4:1. Pinion turns 3,450 rpm. AGMA Quality No. 8. If the gears are carrying 120 hp, find suitable values for P_d and b.

12. Calculate the value of K of the wear equation for $14\frac{1}{2}°$ and 20° for the combination of steel pinion and cast iron or bronze gear. $E_1 = 30,000,000$ psi and $E_2 = 16,000,000$ psi respectively.

(a) For BHN 160–200 cast iron and steel
(b) For BHN 210–245 cast iron and steel

Ans.   For 160–200 cast iron and steel:

$$K = 42.8 \text{ psi} = 0.295 \text{ MPa for } 14\tfrac{1}{2}°$$

$$K = 58.5 \text{ psi} = 0.403 \text{ MPa for } 20°$$

For 210–245 cast iron and steel:

$$K = 61.7 \text{ psi} = 0.426 \text{ MPa for } 14\tfrac{1}{2}°$$

$$K = 84.3 \text{ psi} = 0.581 \text{ MPa for } 20°$$

13. A 20° steel pinion meshes with a cast-iron gear, BHN = 220. Pinion turns 1,140 rpm. AGMA Quality No. 9. $N_1 = 20$; $m_0 = 6$, and $b = 50$ mm. For K, see Problem 12.

(a) Find the power if $N_2 = 40$.   *Ans.* kW = 16.2.
(b) Find the power if $N_2 = 120$.   *Ans.* kW = 22.4.

14. A 20° full-depth steel pinion meshes with a cast-iron gear, BHN = 220. Center distance is 8 in. Speed ratio is 3:1. Pinion turns 430 rpm. AGMA Quality No. 8. $P_d = 4$ and $b = 1.75$ in. Find the horsepower capacity of the set, and the number of base pitches in the interval of contact.
                    Ans. hp = 9.4; no. in contact = 1.62.

15. Two 20° full-depth gears have AGMA Quality No. 8. Pinion turns 430 rpm.

$N_1 = 24$; $N_2 = 48$; $P_d = 4$, and $b = 2$ in. Find the horsepower transmitted for the following combination of materials.

(a) Steel pinion and cast-iron gear, BHN $= 220$.　　　　*Ans.* hp $= 17.9$.
(b) Both gears steel with BHN $= 200$.　　　　　　　　　*Ans.* hp $= 7.7$.
(c) Both gears steel with BHN $= 350$.　　　　　　　　　*Ans.* hp $= 41.8$.
(d) Both gears induction-hardened steel.　　　　　　　*Ans.* hp $= 108.8$.
(e) Both gears carburized steel.　　　　　　　　　　　*Ans.* hp $= 177.6$.

16. Find the horsepower for the gears of Problem 7 if (a) the quality is made No. 9, and (b) if the quality is made No. 10.

Ans. (a) hp $= 92.2$; (b) hp $= 102.3$.

17. Find the minimum value of the pressure angle for a system of gearing that will generate a 12-tooth full-depth pinion without undercutting.

Ans. $\varphi = 24°6'$.

18. A 72-tooth $14\frac{1}{2}°$ pinion meshes with a 144-tooth gear on a standard center distance of 18 in. The smaller gear has full-depth teeth, and the larger gear has standard stub teeth. Find the number of tooth intervals in the length of contact. $P_d = 6$.　　　　　　　　　　　　　　　*Ans.* No. of intervals $= 2.10$.

19. In Fig. 10-27(a), distance OA_1 is equal to three-fourths of OA. Gear 1 has 20 teeth. If gear 2 has full-depth addenda, find the (theoretical) value of r_2. Make computations for $P_d = 1$, and $\varphi = 20°$.　　　　　*Ans.* $r_2 = 22.74$ in.

20. Two $20°$ spur gears have 17 teeth each. What must be the value of the addendum if the length of contact extends two-thirds of the way from the pitch point to the interference point? $P_d = 1$. Find the number of teeth in contact.

Ans. Add. $= 0.842$ in.; no. $= 1.31$.

21. The length of contact on a $20°$ full-depth gear of 18 teeth extends two-thirds of the way from the pitch point to the interference point. Find the length of the addendum of the mating gear if it has 72 teeth. $P_d = 1$. Find the number of teeth in contact.　　　　　　　　　　　　*Ans.* Add. $= 0.752$; no. $= 1.46$.

22. In Fig. 10-38, consider shaft B together with gears 2 and 3 as the free body. Determine the tooth forces for these gears based on the horsepower. Draw a top

FIGURE 10-38 Problem 22.

view of shaft *B* and show all horizontal loads and reactions. Draw an elevation of shaft *B* and show all vertical loads and reactions. $P_d = 6$; $\varphi = 20°$.

23. Work Problem 22 but with view *AA* as given by Fig. 10-39.

FIGURE 10-39   Problem 23.

FIGURE 10-40   Problems 24 and 25.

24. In Fig. 10-40, gear 1 has 15 teeth of $P_d = 5$, and turns 2,150 rpm. Gear 4 has 43 teeth of $P_d = 4$, and turns 450 rpm. The distance between shafts *A* and *C* is 13 in. Find the number of teeth in gear 2 and 3.   *Ans.* $N_2 = 35$; $N_3 = 21$.

25. In Fig. 10-40, gear 1 has 24 teeth and gear 2 has 70 teeth of $P_d = 4$. Gear 3 has 20 teeth and gear 4 has 50 teeth of $P_d = 3$.

(a) Find the value of the center distance between shafts *A* and *C*.

(b) Find the rpm of gear 4 if gear 1 turns 1,200 rpm.

(c) Find the torque at shaft *C* if the input horsepower at shaft *A* is 50. The efficiency of each pair of gears is 98 percent.

Ans.   (a) 23.417 in.; (b) $n_4 = 164.6$ rpm;
(c) $T = 18,390$ in. lb.

26. In Fig. 10-41, $N_1 = 32$, $N_2 = 75$, and $N_3 = 24$. Find the value of the *p* for gear 3 if the rack is to be moved 0.5 ft for each revolution of gear 1.

Ans. $p = 0.586$ in.

FIGURE 10-41   Problem 26.

FIGURE 10-42   Problem 27.

27. In Fig. 10-42, gear 1 is located on the motor shaft and turns 1,200 rpm. Find the speed at which the load is being raised. *Ans.* 52.36 fpm.

28. Determine some additional solutions for a gear train of ratio 2.54: 1. *Ans.* (29 × 53)/(61 × 64); (11 × 67)/(36 × 52); (13 × 19 × 71)/(16 × 32 × 87), etc.

29. Determine the suitable numbers of teeth for the gears of a train of ratio 251: 93. *Ans.* (41 × 61)/(75 × 90); (29 × 67)/(57 × 92); (15 × 53 × 79)/(43 × 54 × 73), etc.

30. Make a front view of the reducer shown in Fig. 10-43 and show the relative displacements of all gears for a given rotation α of arm 0. Mark the values of all angles used in deriving the equation for the speed ratio. *Ans.* $\dfrac{n_1}{n_0} = 1 + \dfrac{N_3}{N_1}$.

FIGURE 10-43   Problem 30.

31. Make a front view of the epicyclic gear train shown in Fig. 10-44. By a process similar to that illustrated in Section 29, derive the equation for the speed ratio of shafts 4 and 0. A negative result means that gear 4 is rotating oppositely to arm 0. *Ans.* $\dfrac{\text{rpm}_4}{\text{rpm}_0} = -\dfrac{N_2 N_3}{N_1 N_4} + 1$.

FIGURE 10-44   Problem 31.

32. Work Problem 31 with the epicyclic system shown in Fig. 10-45.

$$Ans. \quad \frac{\text{rpm}_4}{\text{rpm}_0} = 1 + \frac{N_2 N_3}{N_1 N_4}.$$

FIGURE 10-45   Problem 32.

FIGURE 10-46   Problem 33.

33. Work Problem 31 for Fig. 10-46.

$$Ans. \quad \frac{\text{rpm}_4}{\text{rpm}_0} = 1 + \frac{N_2 N_3}{N_1 N_4}.$$

34. Work Problem 31 for Fig. 10-47.

$$Ans. \quad \frac{\text{rpm}_4}{\text{rpm}_0} = - \frac{N_2 N_3}{N_1 N_4} + 1.$$

FIGURE 10-47   Problem 34.

FIGURE 10-48   Problem 35.

35. Work Problem 31 for Fig. 10-48.

$$Ans. \quad \frac{\text{rpm}_4}{\text{rpm}_0} = 1 - \frac{N_2 N_3}{N_1 N_4}.$$

36. Two 20° gears are made of induction-hardened steel. Center distance is 216 mm. Two-to-one reduction ratio. Pinion turns 1,725 rpm. AGMA Quality No. 8. If the gears are carrying 175 kW, find suitable values for m_0 and b.

37. A 20° 6-in. solid steel pinion meshes with an 18-in. cast iron gear in the form of a ring. Inside diameter 14 in. Width of face 2.5 in. Find the equivalent mass m_e.

Ans. $m_e = 0.0217$ lb sec^2/in.

38. A 20° solid steel pinion 48 teeth, $P_d = 4$, meshes with a 96-tooth cast iron gear in the form of a ring. Inside diameter 20 in. Width of face 3 in. Find the equivalent mass m_e.

Ans. $m_e = 0.0811$ lb sec^2/in.

39. A 20° solid steel pinion, 42 teeth, $P_d = 5$, meshes with a 90-tooth cast iron gear in the form of a ring. Inside diameter 14 in. Width of face 2 in. Find the equivalent mass m_e.

Ans. $m_e = 0.0295$ lb sec^2/in.

40. A 20° solid steel pinion, 36 teeth, $m_0 = 4$, meshes with a 114-tooth cast iron gear in the form of a ring. Inside diameter 380 mm. Width of face 75 mm. Find the equivalent mass m_e.

Ans. $m_e = 0.00394$ newton sec^2/mm.

41. Show that when a pair of gears are operating at their power capacity, the speed n_1 of the pinion must be such as to make F_d equal to one-half of F_w.

42. When the gear consists of a rim and spokes, at what value for r_i will m' be one-half as great as if the blank was solid?

Ans. $r_i = 0.841\, r_0$.

43. Two meshing spur gears have 20° full depth teeth with solid blanks. $P_d = 4$, $N_1 = 24$, $N_2 = 48$, $b = 3$ in. Speed $n_1 = 860$ rpm. Horsepower $= 75$, Quality No. 8. Find the required BHN for the gears.

Ans. BHN $= 315$.

44. Two meshing spur gears have 20° full depth teeth with solid blanks. $m_0 = 6$, $N_1 = 24$, $N_2 = 48$, $b = 75$ mm. Speed $n_1 = 860$ rpm. Kilowatt power $= 50$, Quality No. 8. Find the required BHN for the gears.

Ans. BHN 320.

45. Two meshing spur gears have 20° full depth teeth. $P_d = 4$, $N_1 = 20$ teeth, and $N_2 = 80$ teeth. Pinion is solid steel. Gear is of cast iron with rim and spokes with inside diameter 17 in. $b = 2$ in. Quality No. 9, $n_1 = 860$ rpm. Find the value of the dynamic load.

Ans. $F_d = 490$ lb.

46. Two meshing spur gears have 20° full depth teeth. $m_0 = 6$, $N_1 = 20$ teeth, $N_2 = 80$ teeth. Pinion is solid steel. Gear is of cast iron with rim and spokes with inside diameter 380 mm. $b = 50$ mm. Quality No. 9, $n_1 = 860$ rpm. Find the value of the dynamic load.

Ans. $F_d = 2460$ newtons.

47. Two meshing spur gears have 20° full depth teeth. Gears are steel with solid blanks. $N_1 = 25$, $N_2 = 75$, $P_d = 5$, $b = 2.5$ in., Quality No. 10, $n_1 = 1,760$ rpm. Horsepower $= 40$. Find the required value for the BHN. Check the teeth for bending.

Ans. BHN $= 260$.

48. Two meshing spur gears have 20° full depth teeth. Gears are steel with solid blanks. $N_1 = 25$, $N_2 = 75$, $m_0 = 5$, $b = 60$ mm, Quality No. 10, $n_1 = 1,760$ rpm. Kilowatt power is 30. Find the required value for the BHN. Check the teeth for bending.

Ans. BHN $= 260$.

49. Two meshing spur gears have 20° full depth teeth. Gears are of steel with solid blanks. $N_1 = 25$, $N_2 = 75$. $P_d = 5$, $b = 2.5$, Quality No. 10, $n_1 = 1,750$ rpm. Horsepower $= 75$. Find the required value for the BHN and check the teeth for bending.

Ans. BHN $= 290$.

50. Two meshing spur gears have 20° full depth teeth. Gears are of steel with solid blanks. $N_1 = 25$, $N_2 = 75$, $b = 60$ mm, Quality No. 10, $n_1 = 1,760$ rpm. Kilowatt power transmitted is 58. Find the required value for the BHN, and check the teeth for bending.

Ans. BHN $= 300$.

51. Two 20° full-depth steel spur gears, solid blanks, of 36 teeth each are operating at their maximum power capacity, $P_d = 6$, $b = 3$, Quality No. 8. Find the speed and horsepower if the BHN is 200. Assume that teeth are sufficiently large to be safe in bending.   *Ans.* rpm = 306; hp = 8.1.

52. Same as Problem 51 except BHN = 225.   *Ans.* rpm = 362; hp = 11.4.

53. Same as Problem 51 except BHN = 250.   *Ans.* rpm = 436; hp = 16.5.

54. Same as Problem 51 except BHN = 275.   *Ans.* rpm = 520; hp = 23.4.

55. Same as Problem 51 except BHN = 300.   *Ans.* rpm = 631; hp = 34.5.

56. Same as Problem 51 except BHN = 325.   *Ans.* rpm = 733; hp = 46.6.

57. Two 20° full-depth steel spur gears, solid blanks, are operating at their maximum power capacity. $N_1 = 36$, $N_2 = 36$, $b = 3$ in., BHN = 325. Total error, $e = 0.0059$ in. Find value for the rpm and the horsepower carried. Teeth are sufficiently large to be safe in bending.   *Ans.* rpm = 734; hp = 50.6.

58. Same as Problem 57 except $N_2 = 42$.   *Ans.* rpm = 737; hp = 54.7.

59. Same as Problem 57 except $N_2 = 48$.   *Ans.* rpm = 741; hp = 58.4.

60. Same as Problem 57 except $N_2 = 54$.   *Ans.* rpm = 749; hp = 62.0.

61. Same as Problem 57 except $N_2 = 60$.   *Ans.* rpm = 759; hp = 65.4.

62. Same as Problem 57 except $N_2 = 66$.   *Ans.* rpm = 765; hp = 68.3.

63. Same as Problem 57 except $N_2 = 72$.   *Ans.* rpm = 773; hp = 70.8.

64. Same as Problem 57 except $N_2 = 78$.   *Ans.* rpm = 782; hp = 73.8.

65. Same as Problem 57 except $N_2 = 84$.   *Ans.* rpm = 791; hp = 76.4.

66. Same as Problem 57 except $N_2 = 90$.   *Ans.* rpm = 799; hp = 78.7.

67. Same as Problem 57 except $N_2 = 96$.   *Ans.* rpm = 807; hp = 81.0.

68. Two 20° full-depth spur gears are operating at their maximum power capacity. $N_1 = 24$, $N_2 = 24$, $b = 1$ in. Gears have solid blanks and are carburized. Total error, $e = 0.0047$ in. Find the value for the rpm and the horsepower carried. Teeth are sufficiently large to be safe in bending.   *Ans.* rpm = 5,588; hp = 122.

69. Same as Problem 68 except $N_2 = 30$.   *Ans.* rpm = 5,623; hp = 136.

70. Same as Problem 68 except $N_2 = 36$.   *Ans.* rpm = 5,700; hp = 149.

71. Same as Problem 68 except $N_2 = 42$.   *Ans.* rpm = 5,795; hp = 161.

72. Same as Problem 68 except $N_2 = 48$.   *Ans.* rpm = 5,891; hp = 172.

73. Same as Problem 68 except $N_2 = 54$.   *Ans.* rpm = 5,988; hp = 181.

74. Same as Problem 68 except $N_2 = 60$.   *Ans.* rpm = 6,078; hp = 190.

75. Same as Problem 68 except $N_2 = 66$.   *Ans.* rpm = 6,167; hp = 197.

76. Same as Problem 68 except $N_2 = 72$.   *Ans.* rpm = 6,250; hp = 205.

77. Same as Problem 68 except $N_2 = 78$.   *Ans.* rpm = 6,323; hp = 211.

78. Same as Problem 68 except $N_2 = 84$.   *Ans.* rpm = 6,393; hp = 217.

11

Bevel, Worm, and Helical Gears

Bevel, worm, and helical gears are advanced forms of gearing capable of meeting special requirements of geometry or strength that cannot be obtained from spur gears. Bevel gears, with straight or spiral teeth cut on cones, can be used to connect intersecting shafts. A worm gear, consisting of a screw meshing with a gear, can be used to obtain a large speed reduction. Helical gears have teeth that lie in helical paths on the cylinders instead of teeth that are parallel to the shaft axis. The geometry of these different types of gearing is considerably more involved than for spur gears, and the problems of production and inspection are also more complicated.

b,	face width of tooth	K,	constant in wear equation
b',	thickness of bevel gear parallel to axis	lb,	pound; 1 lb (force) = 4.4482 newtons 1 lb (mass) = 0.4536 kg
c,	center distance		
d,	pitch diameter	L,	lead of worm or helical gear
d',	pitch diameter, formative gear	n,	speed of rotation, revolutions per minute
e,	error		
F_b,	bending capacity	N,	number of teeth in gear
F_d,	dynamic tooth load	N',	number of teeth in formative gear
F_t,	transmitted or horsepower load		
		p,	circular pitch in plane of rotation
F_w,	capacity in wear	p_n,	circular pitch normal to teeth
hp,	horsepower	p_o,	circular pitch at outside radius of bevel gear
in.,	inch, 25.4 mm		

P_d,	diametral pitch	β,	(beta) ratio N_1/N_2
P_{dn},	diametral pitch normal to teeth	η,	(eta) efficiency
Q,	factor in wear equation, helical gears	λ,	(lambda) helix angle of worm (between teeth and plane of rotation)
Q',	factor in wear equation, bevel gears	μ,	(mu) coefficient of friction
r,	pitch radius	σ,	(sigma) bending stress
r',	pitch radius, formative gear	φ,	(phi) pressure angle in plane of rotation
r_o,	outside radius of pitch cone of bevel gear	φ_n,	(phi) pressure angle normal to teeth
T,	torque		
V,	pitch-line velocity, feet per minute	ψ,	(psi) helix angle of helical gear (between teeth and axis)
V_s,	velocity of sliding between teeth	ω,	(omega) speed of rotation, radians per second
y,	Lewis or form factor		
α,	(alpha) pitch cone angle of bevel gear		

For SI symbols, see the notation list of Chapter 10. As mentioned in Chapter 10, number of teeth is designated N. The force newton will be spelled out in full.

1. Straight-Tooth Bevel Gears

When intersecting shafts must be connected by gearing, the pitch cylinders are replaced by pitch cones tangent to each other along an element, with their apexes at the intersection of the shafts. Teeth are placed on the cones, forming bevel gears, as illustrated in Fig. 11-1. The names of some of the principal parts of a bevel gear are given in Fig. 11-2. The back cone, as shown in the latter figure, has elements that are perpendicular to those of the pitch cone. The outlines of the teeth, as generated on the back cone, are shown in view A-A. Because of difficulties in securing uniform bearing along the tooth, the width of face b is generally not made longer than one-third of the pitch cone radius. Undercutting occurs in bevel gears having few teeth.

FIGURE 11-1   Straight-tooth bevel.

The size of bevel gear teeth is usually specified by a diametral pitch P_d or circular pitch p_o, which refers to the developed back cone of the gear as

FIGURE 11-2   **Straight-tooth bevel gear.**

shown by view A-A in Fig. 11-2. Thus,

$$r_{o1} = \frac{N_1}{2P_d} \quad \text{and} \quad r_{o2} = \frac{N_2}{2P_d} \tag{1}$$

Here N_1 and N_2 represent the numbers of teeth in the gears.

2. Beam Strength of Bevel Gears

The beam strength of bevel gear teeth is computed by the Lewis equation. It is assumed that the bevel gear tooth is equivalent to a spur tooth whose cross section is the same as the cross section of the bevel tooth at the midpoint of the face b.

Let r be the pitch radius at the midpoint of the face. The circular pitch p at the midpoint of the face, then, is

$$p = \frac{2\pi r}{N} \tag{2}$$

For the pinion or smaller gear, the radius r_1 to the midpoint of the face is found by

$$r_1 = r_{o1} - \tfrac{1}{2}b \sin \alpha_1 \tag{3}$$

where α_1 is the pitch cone angle for gear 1. The corresponding equation for the larger gear is

$$r_2 = r_{o2} - \tfrac{1}{2}b \sin \alpha_2 = r_{o2} - \tfrac{1}{2}b \cos \alpha_1 \tag{4}$$

where α_2 is the pitch cone angle for gear 2.

Although the number of teeth in a bevel gear is N_1, view *A-A* in Fig. 11-2 indicates that the shape of the teeth will be the same as those of a spur gear of radius r'_{o1}. The number of teeth in a fictitious gear of this radius is called the formative or virtual number of teeth N'_1. Radius r'_{o1} is equal to

$$r'_{o1} = \frac{r_{o1}}{\cos \alpha_1} \tag{5}$$

The circular pitch p_o at the outside radius has the value

$$p_o = \frac{2\pi r'_{o1}}{N'_1} = \frac{2\pi r_{o1}}{N_1} = \frac{2\pi r'_{o1} \cos \alpha_1}{N_1}$$

The last form on the right results from the substitution of Eq. (5). This equation yields

$$N'_1 = \frac{N_1}{\cos \alpha_1} \tag{6}$$

The *y*-factor, Table 10-3, for use in the Lewis equation, is selected for a gear of N'_1 teeth. The Lewis equation, for the capacity of the tooth to carry a bending load, is

$$F_b = \sigma b y p \tag{7}$$

3. Limit Load for Wear. Dynamic Load

The limit load for wear is determined by an equation similar to that used for spur gears but is based on the virtual number of teeth. However, because of the difficulty in securing bearing along the entire face width b, only about three-quarters of b is considered as effective.[1] Then

$$F_w = 0.75 d'_1 b Q' K \tag{8}$$

Here d'_1 is the diameter of the formative gear or $2r'_1$, and Q' is obtained by the use of the formative number of teeth. Hence by Eq. (6),

$$Q' = \frac{2N'_2}{N'_1 + N'_2} = \frac{2N_2/\sin \alpha_1}{(N_1/\cos \alpha_1) + (N_2/\sin \alpha_1)} = \frac{2N_2}{N_1 \tan \alpha_1 + N_2} \tag{9}$$

Factor K is obtained from Table 10-4 for spur gears.

The dynamic load can be approximated by the same equation used for spur gears if the thickness b' of the blanks in the axial direction is specified.

Example 1. Two $20°$ full-depth steel bevel gears with solid blanks have $N_1 = 40$ and $N_2 = 60$ teeth. $P_d = 4$ at outside diameter. $b = 2.5$ in. b' for blanks equals

[1]See p. 540 of Reference 1, Chapter 10.

1.5 in. BHN = 300. AGMA Quality No. 8. Find the speed of operation and the horsepower transmitted if gears are operating at their maximum capacity.

Solution.
$$r_{o1} = \frac{N_1}{2P_d} = \frac{40}{2 \times 4} = 5 \text{ in.}, \qquad r_{o2} = \frac{N_2}{2P_d} = \frac{60}{2 \times 40}$$

$$= 7.5 \text{ in. at outside diameter}$$

$$\tan \alpha_1 = \frac{5}{7.5} = 0.66667, \qquad \alpha_1 = 33.690°$$

By Eq. (3):
$$r_1 = r_{o1} - \tfrac{1}{2}b \sin \alpha_1 = 5 - \tfrac{1}{2} \times 2.5 \times 0.55470$$

$$= 4.3062 \text{ in. at midpoint of face; see Fig. 11-3}$$

$$d_1 = 2r_1 = 8.6124 \text{ in.}$$

FIGURE 11-3 Pitch cones for gears of Example 1.

By Eq. (9):
$$Q' = \frac{2N_2}{N_1 \tan \alpha_1 + N_2} = \frac{2 \times 60}{40 \times 0.66667 + 60} = 1.3846$$

$$r'_1 = \frac{r_1}{\cos \alpha_1} = \frac{4.3062}{0.83205} = 5.1754 \text{ in.}$$

By Fig. 10-24 for BHN = 300, $K = 136$,

By Eq. (8):
$$F_w = 0.75 d'_1 b Q' K = 0.75 \times 2 \times 5.175 \times 2.5 \times 1.3846 \times 136$$

$$= 3,654 \text{ lb}$$

For maximum power:
$$F_d = F_p = \tfrac{1}{2}F_w = 1,827 \text{ lb}$$

$$C = \frac{N_2}{N_1} = \frac{60}{40} = 1.5$$

By Eq. (33), Chapter 10:
$$m_e = \frac{C^2}{1 + C^2} \frac{\pi r_1^2 b' \gamma}{2g} = \frac{2.25}{3.25} \frac{\pi 4.3062^2 \times 1.5 \times 0.283}{2 \times 386}$$

$$= 0.02218 \frac{\text{lb sec}^2}{\text{in.}}$$

By Table 10-2: $\qquad\qquad\qquad e = 0.0033 + 0.0034 = 0.0067$ in.

By Eq. (28), Chapter 10: $\quad k = 1,667,000 \times 2.5 = 4,167,000$ lb/in.

By Eq. (25a), Chapter 10: $\quad n_1 = \dfrac{30F_d}{eN_1\sqrt{km_e}} = \dfrac{30 \times 1,827}{0.0067 \times 40\sqrt{4,167,000 \times 0.02218}}$

$$= \frac{204,515}{\sqrt{92,417}} = \frac{204,515}{304} = 673 \text{ rpm}$$

By Eq. (13), Chapter 10: $\quad V_b = \dfrac{\pi d_1 n_1 \cos \varphi}{12} = \dfrac{\pi 8.6124 \times 673 \times 0.93969}{12}$

$$= 1,426 \text{ fpm at midpoint of face}$$

By Eq. (12), Chapter 10: $\quad \text{hp} = \dfrac{F_p V_b}{33,000} = \dfrac{1,827 \times 1,426}{33,000} = 79$

Example 2. Suppose the bevel gears of Example 1 must transmit 100 horsepower when operating at their maximum capacity. Find the BHN to which they must be hardened and the speed at which they must be operated.

Solution.

By Eq. (12), Chapter 10: $\quad \text{hp} = \dfrac{F_p V_b}{33,000}, \qquad V_b = \dfrac{33,000 \text{ hp}}{F_p}$

By Eq. (13), Chapter 10: $\quad n_1 = \dfrac{12V_b}{\pi d_1 \cos \varphi} = \dfrac{12}{\pi d_1 \cos \varphi} \times \dfrac{33,000 \text{ hp}}{F_p}$ $\qquad$ (a)

By Eq. (25a), Chapter 10: $\quad n_1 = \dfrac{30F_d}{eN_1\sqrt{km_e}}$ $\qquad\qquad\qquad\qquad$ (b)

Equate (a) and (b): $\quad \dfrac{30F_d}{eN_1\sqrt{km_e}} = \dfrac{12 \times 33,000 \text{ hp}}{F_d \pi d_1 \cos \varphi}$

At maximum horsepower: $\quad F_d = F_p$

Then: $\qquad\qquad\qquad F_d^2 = \dfrac{12 \times 33,000 \text{ hp } eN_1\sqrt{km_e}}{30\pi d_1 \cos \varphi}$

$$= \frac{12 \times 33,000 \times 100 \times 0.0067 \times 40 \times 304}{30\pi 8.6124 \times 0.93969}$$

$$= 4,230,000$$

$$F_d = 2,057 \text{ lb}$$

At maximum power: $\qquad F_w = 2F_d = 4,114 \text{ lb}$

By Eq. (8): $\qquad\qquad\qquad K = \dfrac{F_w}{0.75d_1' bQ'} = \dfrac{4,114}{0.75 \times 2 \times 5.175 \times 2.5 \times 1.3846}$

$$= 153$$

By Fig. 10-24: $\qquad\qquad \text{BHN} = 318$

By Eq. (12), Chapter 10: $V_b = \dfrac{33,000 \text{ hp}}{F_p} = \dfrac{33,000 \times 100}{2,057} = 1,604 \text{ fpm}$

By Eq. (13), Chapter 10: $n_1 = \dfrac{12 V_b}{\pi d_1 \cos \varphi} = \dfrac{12 \times 1,604}{\pi 8.6124 \times 0.93969} = 757 \text{ rpm}$

A somewhat higher speed is thus necessary to develop the higher horsepower.

For short tooth lengths and teeth of $5P_d$ and larger, the teeth can be formed directly in the forging dies. The gears are then carburized and placed in service without the need of any subsequent machining. A small amount of crowning of the teeth is beneficial in helping secure good tooth contact.

4. Bearing Loads for Bevel Gears

Let the force between the teeth of bevel gears be considered as concentrated and lying along the pressure line. It will therefore be inclined at the pressure angle ϕ to the plane tangent to the pitch cone at the midpoint of the face. This force, marked F_n, is shown in Fig. 11-4. The component tangential to the cone is marked F_t. Its value is marked on the figure. The plane of F_n and F_t is rotated about F_t through the cone angle α measured from the perpendicular to the shaft. This causes F_n to have additional components parallel and perpendicular to the shaft. The equations for computing these are marked in the figure. These forces can then be used for computing the bearing reaction.

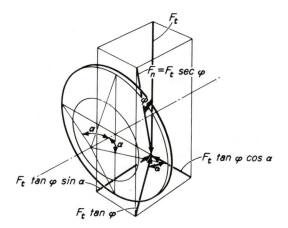

FIGURE 11-4 Forces at midpoint of bevel gear tooth.

Example 3. Let there be a force F_n of 1,586 lb applied at the midpoint of the face to gear 2 of the Examples 1 and 2. Let the base of the pitch cone for this gear be located at the midpoint of a 24-in. simply supported shaft. Make a free body sketch for the gear and shaft and show all forces acting. Let the thrust and torque be resisted at the right end of the shaft which is illustrated by Fig. 11-5.

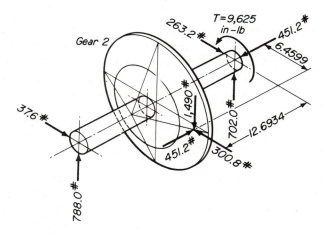

FIGURE 11-5   Tooth loads and shaft reactions on gear for Example 3.

Solution. By Example 1, $\alpha_2 = 90 - 33.690 = 56.310°$.

$$F_t = F_n \cos \varphi = 1,586 \times 0.93969 = 1,490 \text{ lb}$$
$$F_t \tan \varphi \sin \alpha_2 = 1,490 \times 0.36397 \times 0.83205 = 451.2 \text{ lb}$$
$$F_t \tan \varphi \cos \alpha_2 = 1,490 \times 0.36397 \times 0.55470 = 300.8 \text{ lb}$$

By Eq. (4): $r_2 = r_{o2} - \frac{1}{2}b \sin \alpha_2 = 7.5 - \frac{1}{2} \times 2.5 \times 0.83205$
$$= 7.5 - 1.0401 = 6.4599 \text{ in.}$$

By Fig. 11-5, distance from right bearing to centerline of face is $12 + 0.6934 = 12.6934$ in.

Torque: $1,490r_2 = 1,490 \times 6.4599 = 9,625$ in. lb

The above forces and dimensions are now used to calculate the bearing reactions by simple statics.

The forces above are shown on the gear in Fig. 11-5. The reactions at the bearings are found by simple statics.

The efficiency of bevel gearing is very high, providing the gears are properly mounted and adjusted. Not more than 1 or 2% of the power should be lost by accurately manufactured gears that are properly mounted.

5. Spiral Bevel Gears

Bevel gears can also be made with teeth lying in spiral paths on the pitch cones. More teeth are then in contact simultaneously, and smoother and quieter action can be obtained from such gears. Accuracy of adjustment between a pair of bevel gears can be more readily secured with spiral teeth than with straight teeth.

The Gleason Works has developed a system of bevel gearing, both straight-tooth and spiral, which is in wide use. The addenda are arranged on the long and short plan to avoid undercutting and to give greater strength to pinions having few teeth. The addendum of the pinion is made longer than the usual value, and the addendum of the gear is shortened a corresponding amount. A variety of pressure angles are used in order to obtain the best operating conditions for each velocity ratio.

Hypoid gears have spiral teeth and shafts that do not intersect. It is thus possible to connect continuous nonintersecting shafts by such gears, as shown in Fig. 11-6. Hypoid gears have been used in automobiles so that the drive shaft can be placed beneath the level of the floor.

FIGURE 11-6 Hypoid gears on nonintersecting shafts. (*Courtesy Gleason Works.*)

In the rear axles of automobiles, the bevel gear stresses are very low under normal conditions of service. Such gears are therefore designed on the assumption that operation in low gear at full engine torque for a relatively short period, perhaps 0.1 % of the life of the car, would normally be sufficient to cause fatigue failure. The requirements are somewhat higher for trucks and busses. Gear steels for automotive service are selected mainly on the basis of machining characteristics, cost, and resistance to stress concentration and warping in heat treatment.[2] A comparison between good and bad practice[3] in the mounting of bevel gears is shown in Fig. 11-7.

[2]See Almen, J. A., and A. L. Boegehold, "Rear Axle Gears," *Proc. ASTM*, **35**, Part 2, 1935, p. 99.

[3]See Candee, A. H., "Industrial Applications of Spiral Bevel Gears and Hypoid Gears," *Trans. ASME*, **60**, 1938, p. 549. Other types of gears for right-angle drives are available. See, for example, Nelson, W. D., "Spiroid Gearing," *Machine Design*, **33**, 1961, Feb. 16, p. 136; Mar. 2, p. 93; and Mar. 16, p. 165.

BAD DESIGN

Face width too great, more than one third of cone distance.

Metal at small end of pinion between teeth and bore too thin for proper strength. Cutter would interfere seriously with arbor in cutting operation.

Webbed section of steel ring gear adds to cost of material and machining.

Use of rivets to hold gear to hub introduces danger of runout.

A setscrew is inadequate to hold gear in correct axial position, and tends to "cock" gear on shaft.

Pinion held on shaft only by fit of bore.

An overhung pinion cannot be held in line by one double-row bearing.

No means of adjustment for definitely placing gears in correct mounting position.

GOOD DESIGN

Face width reasonable, less than one third of cone distance.

Sufficient metal at small end of pinion to provide strength and avoid interference of cutter with arbor.

Section of ring gear simpler and more direct in design. Gear is supported directly on back and is centered on large bore.

Screws to hold gear to hub are preferable.

Gear is positively held in position on shaft.

Pinion locked in position by washer and screw.

Pinion rigidly supported by addition of inboard bearing.

Adjusting washers provided, to be ground to thickness required to obtain correct position of pinion and gear.

Pinion and bearings can be assembled as a complete unit.

FIGURE 11-7 Design and arrangement of spiral bevel gears. Comparison of good and bad practice.

6. Worm Gears

Worm gears, as illustrated by Fig. 11-8, have spiral teeth and shafts at right angles. Worms can be made with single, double, triple, or more threads. The teeth of the wheel or gear envelop the worm threads and give line contact between the mating parts. The wheel is hobbed, and the worm is made by grinding or by milling with a disk cutter. Care must be exercised that the teeth of worm and gear are properly shaped to give conjugate surfaces. The geometry of worm gearing is very complicated, and reference should be made to the literature for complete information.

FIGURE 11-8 Worm and wheel. (*Courtesy D. O. James Gear Mfg. Co.*)

It is possible to secure a large speed reduction or a high increase of torque by means of worm gears. The velocity ratio does not depend upon diameters, but upon the numbers of teeth. The pressure angle should not be less than 20° for single- and double-thread worms, and 25° for triple- and multithread worms.

Case-hardened alloy steel is recommended for the worm. The worm wheel should be of bronze of approved composition. The wheel should usually contain not less than 29 teeth. The wheel should be provided with axial adjustment to obtain correct contact with the worm.

7. Geometric Relationships of Worm Gears

Figure 11-9 shows a worm of helix angle λ. Let p_1 and p_2 be the circular pitches in the planes of rotation for worm and gear, respectively. Let p_n be the circular pitch normal to the direction of the teeth. Then

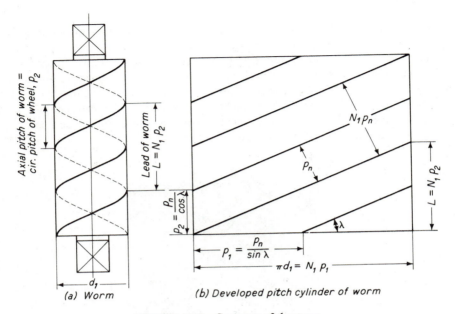

(a) Worm   (b) Developed pitch cylinder of worm

FIGURE 11-9   Geometry of the worm.

$$p_n = p_1 \sin \lambda = p_2 \cos \lambda \tag{10}$$

Equation (10) should be divided through by π, and π/p_n replaced by P_{dn}, the normal diametral pitch, to give

$$P_{dn} = \frac{\pi}{p_1 \sin \lambda} = \frac{\pi}{p_2 \cos \lambda}$$

Let N_1 be the number of starts or threads for the worm, and let N_2 be the number of teeth in the gear. Let d_1 and d_2 be the pitch diameters of worm and gear, respectively. Now $\pi d_1 = N_1 p_1$. The value of p_1 from Eq. (10) can be substituted to give

$$d_1 = \frac{N_1 p_1}{\pi} = \frac{N_1 p_n}{\pi \sin \lambda} = \frac{N_1}{P_{dn} \sin \lambda} \tag{11}$$

Similarly, $\pi d_2 = N_2 p_2$. Substitution for p_2 from Eq. (10) gives

$$d_2 = \frac{N_2 p_2}{\pi} = \frac{N_2 p_n}{\pi \cos \lambda} = \frac{N_2}{P_{dn} \cos \lambda} \tag{12}$$

Equations (11) and (12) can be combined to give

$$\tan \lambda = \frac{N_1 d_2}{N_2 d_1} \tag{13}$$

The values given above for d_1 and d_2 can be substituted into the equation for the center distance c to give

$$c = \frac{1}{2}(d_1 + d_2) = \frac{1}{2P_{dn}}\left(\frac{N_1}{\sin \lambda} + \frac{N_2}{\cos \lambda}\right) \tag{14}$$

or

$$\frac{2P_{dn}c}{N_2} = \frac{\beta}{\sin \lambda} + \frac{1}{\cos \lambda} \tag{15}$$

where

$$\beta = \frac{N_1}{N_2} \tag{16}$$

Equation (15) has been plotted in Fig. 11-10 and is helpful in the solution of problems.

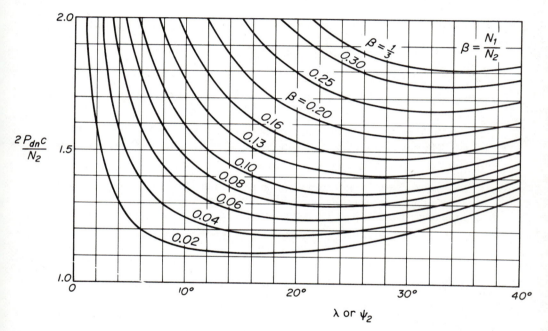

FIGURE 11-10   Design chart for worm and helical gears for 90° shafts.

Let L be the lead of the worm, or the distance the pitch circle of the wheel is advanced for each revolution of the worm. Then

$$L = N_1 p_2 \tag{17}$$

The pitch-line velocities V_1 and V_2 for the worm and wheel are

for worm:

$$V_1 = \frac{\pi d_1 n_1}{12} = \frac{n_1 N_1 p_1}{12} \text{ fpm} \tag{18}$$

for wheel:

$$V_2 = \frac{\pi d_2 n_2}{12} = \frac{n_1 N_1 p_2}{12} \text{ fpm} \tag{19}$$

where n_1 and n_2 are the speeds, rpm, for the worm and wheel, respectively. The last form of Eq. (19) is obtained by observing in Fig. 11-9 that the pitch-line velocity $\pi d_2 n_2$ of the wheel is equal to the speed $n_1 L$ or $n_1 N_1 p_2$ along the worm axis.

From Fig. 11-9 and Eq. (18) and (19),

$$\tan \lambda = \frac{L}{\pi d_1} = \frac{p_2}{p_1} = \frac{V_2}{V_1} \tag{20}$$

Example 4. A triple-thread worm has a pitch diameter of 4.7856 in. The hob for cutting the worm wheel has a diametral pitch P_{dn} equal to 2. Find the pitch diameter d_2 if the reduction is 12 : 1.

Solution. $N_1 = 3;$ $N_2 = 3 \times 12 = 36$

By Eq. (11): $\sin \lambda = \dfrac{N_1}{P_{dn} d_1} = \dfrac{3}{2 \times 4.7856} = 0.31344$

$\lambda = 18°16'$

By Eq. (12): $d_2 = \dfrac{N_2}{P_{dn} \cos \lambda} = \dfrac{36}{2 \times 0.94961} = 18.9552$ in.

Let the module in the direction perpendicular to the tooth be represented by m_{on}. The foregoing equations will be valid for SI units if P_{dn} is replaced by $1/m_{on}$.

Example 4A. A triple-threaded worm has a pitch diameter of 115 mm. The hob for cutting the worm wheel has a pitch perpendicular to the teeth of 12 modules. Find the pitch diameter d_2 if the reduction is 12: 1.

Solution. By Eq. (11): $\sin \lambda = \dfrac{N_1 m_{on}}{d_1} = \dfrac{3 \times 12}{115} = 0.313\ 04$

$\lambda = 18°14.6'$

By Eq. (12): $d_2 = \dfrac{N_2 m_{on}}{\cos \lambda} = \dfrac{36 \times 12}{0.94974} = 454.86$ mm

Example 5. A double-thread worm has a lead L of 2.18 in. The gear has 30 teeth and is cut with a hob of $P_{dn} = 3$. Find pitch diameters of the worm and gear, and center distance of the shafts.

Solution. $p_2 = \dfrac{L}{N_1} = \dfrac{2.18}{2} = 1.09$ in.

$\cos \lambda = \dfrac{p_n}{p_2} = \dfrac{1.0472}{1.09} = 0.96073$

$\lambda = 16°6.6'$

$d_1 = \dfrac{N_1}{P_{dn} \sin \lambda} = \dfrac{2}{3 \times 0.27747} = 2.4026$ in.

$d_2 = \dfrac{N_2}{P_{dn} \cos \lambda} = \dfrac{30}{3 \times 0.96073} = 10.4087$ in.

Center distance: $\qquad c = \frac{1}{2}(d_1 + d_2) = \frac{1}{2}(2.4026 + 10.4087)$

$$= 6.4056 \text{ in.}$$

Example 5A. A double-threaded worm has a lead L of 60 mm. The gear has 30 teeth and is cut with a hob of module $m_o = 8.5$ perpendicular to the teeth. Find the pitch diameters of the worm and gear, and the center distance of the shafts.

Solution. By Eq. (17): $\qquad p_2 = \dfrac{L}{N_1} = \dfrac{60}{2} = 30 \text{ mm}$

$$p_n = \pi m_{on} = 8.5\pi = 26.70 \text{ mm}$$

By Eq. (10): $\qquad\qquad \cos \lambda = \dfrac{p_n}{p_2} = \dfrac{26.70}{30} = 0.890\ 12$

$$\lambda = 27°6.7'$$

By Eq. (11): $\qquad\qquad d_1 = \dfrac{N_1 m_{on}}{\sin \lambda} = \dfrac{2 \times 8.5}{0.45572} = 37.30 \text{ mm}$

By Eq. (12): $\qquad\qquad d_2 = \dfrac{N_2 m_{on}}{\cos \lambda} = \dfrac{30 \times 8.5}{0.89012} = 286.48 \text{ mm}$

By Eq. (14): $\qquad\qquad c_1 = \frac{1}{2}(d_1 + d_2) = \frac{1}{2}(37.30 + 286.48)$

$$= 161.89 \text{ mm}$$

Example 6. A worm gear drive has a center distance of 8 in., and a P_{dn} equal to 4. The worm is quadruple and the wheel has 48 teeth. Find the pitch diameters of the worm and wheel and the value of the helix angle.

Solution. This type of problem must be solved by trial and error using Eq. (15). Approximate first trial values can be obtained from Fig. 11-10 using,

By Eq. (15): $\qquad \dfrac{2P_{dn}c}{N_2} = \dfrac{2 \times 4 \times 8}{48} = 1.3333$

By Eq. (16): $\qquad\qquad \beta = \dfrac{N_1}{N_2} = \dfrac{4}{48} = 0.0833$

The final values, after trial and error, are

$$\lambda = 16°45.7'; \qquad d_1 = 3.4676 \text{ in.}; \qquad d_2 = 12.5324 \text{ in.}$$

In general such problems have two solutions. The other one is

$$\lambda = 31°37.6'; \qquad d_1 = 1.9070 \text{ in.}; \qquad d_2 = 14.0930 \text{ in.}$$

8. Horsepower Capacity of Worm Gears

The recommended value of the face width b_2 of the worm wheel can be determined by the following equation.

$$b_2 = \frac{1}{3}c^{0.875} \tag{21}$$

The curve of Fig. 11-11 can be used if desired.

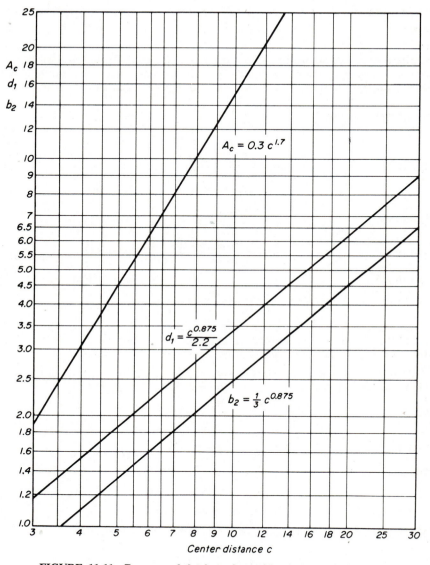

FIGURE 11-11   Recommended values of quantities used in design of worm gears.

The velocity of sliding V_s between the teeth is indicated by Fig. 11-12. Since the worm thread and gear tooth remain in contact, the normal component V_n of the worm and gear velocities V_1 and V_2 is the same for both members. Components OA and OB then represent the velocities along the

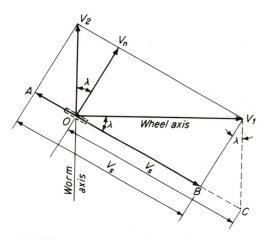

FIGURE 11-12   Relationship between V_1 and V_s.

tooth, and their sum is V_s, the total velocity of sliding between the teeth. When component OA is transplanted to BC, length OC can also be considered V_s since it has the same magnitude as AB. Velocity of sliding V_s is then given by

$$V_s = \frac{V_1}{\cos \lambda} \tag{22}$$

The horsepower capacity of a worm gear reduction can be found by the following equation.[4]

$$\text{hp} = \frac{n_1 c b_2 d_2 \beta (1 - \beta)^2 180}{330(180 + V_s^{0.85})} \tag{23}$$

where n_1 = speed of worm, rpm

$\beta = \dfrac{N_1}{N_2}$,   speed reduction, Eq. (16)

Values of $V_s^{0.85}$ can be had by calculation or from Fig. 11-13.

[4]From American Gear Manufacturers Association Standard 340.01, Sept. 1947, *Design and Rating of Worm Gears for Power Transmission.* A more recent Standard 440.03, Sept. 1959, will permit somewhat greater loadings, but the process of calculation is too involved and lengthy to be presented here.

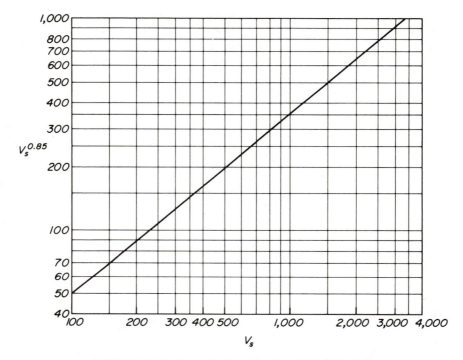

FIGURE 11-13   Curve for determination of $V_s^{0.85}$ from V_s.

Example 7. A 10:1 worm gear reduction has a quadruple-thread worm and a center distance of 8.9333 in. Worm turns 600 rpm. $P_{dn} = 3$. Find the horsepower.

Solution. $\qquad\qquad N_2 = 4 \times 10 = 40$

By Eq. (15): $\qquad \dfrac{2P_{dn}c}{N_2} = \dfrac{2 \times 3 \times 8.9333}{40} = 1.3400$

By Eq. (16): $\qquad \beta = \tfrac{4}{40} = 0.1$

A trial value for λ should now be determined from Fig. 11-10. This is brought to the final form by trial and error, which gives

$$\lambda = 24°47.0'$$

By Eq. (11): $d_1 = \dfrac{N_1}{P_{dn} \sin \lambda} = \dfrac{4}{3 \times 0.41919} = 3.1807 \text{ in.}$

By Eq. (12): $d_2 = \dfrac{N_2}{P_{dn} \cos \lambda} = \dfrac{40}{3 \times 0.90790} = 14.6859 \text{ in.}$

By Eq. (21) or Fig. 11-11:

$$b_2 = 2.265 \text{ in.}$$

By Eq. (18): $\quad V_1 = \dfrac{\pi d_1 n_1}{12} = \dfrac{\pi 3.1807 \times 600}{12} = 500 \text{ fpm}$

By Eq. (22): $\quad V_s = \dfrac{V_1}{\cos \lambda} = \dfrac{500}{0.90790} = 550 \text{ fpm}$

By calculation or Fig. 11-13:

$$V_s^{0.85} = 213.6$$

By Eq. (23): $\quad \text{hp} = \dfrac{600 \times 8.9333 \times 2.265 \times 14.6859 \times 0.1 \times 0.9^2 \times 180}{330(180 + 213.6)}$

$$= 20.0$$

The result given by Eq. (23) is usually called the mechanical horsepower of the reduction.

The worm diameter at the base of the threads must be large enough to provide sufficient rigidity. When threads are integral with the shaft, it is recommended that the pitch diameter d_1 be not less than the value given by

$$d_1 = \frac{c^{0.875}}{2.2} \tag{24}$$

This equation should be used as a check and not for making design calculations.

9. Thermal Capacity of Worm Gear Reductions

The horsepower capacity of a worm gear reduction in continuous operation is usually limited by the heat-dissipating capacity of the case or container. The cooling rate for rectangular or box-type casings can be estimated by the curves[5] in Fig. 11-14. The cooling rate is greater at high velocities of the worm shaft, which causes a better circulation of the oil within the box. The cooling rate can be increased by a fan that directs a stream of air on the sides of the box.

Cooling rate C_1 is expressed in Btu/(hr ft² deg F). Values from the curve can be divided by 60 to obtain the heat lost per minute. The total heat lost depends on the area of the housing A_c in square feet, and the temperature difference ΔT between the housing surface and the surrounding air. The energy lost in Btu can be multiplied by 778 to give the equivalent energy in foot-pounds. The lost energy, in terms of horsepower, is then

$$\text{hp}_c = \frac{778 C_1 A_c \, \Delta T}{60 \times 33{,}000} \tag{25}$$

[5]See Walker, H., "Thermal Rating of Worm Gear Boxes," *Proc. Inst. Mech. Eng.*, **151**, 1944, p. 326.

FIGURE 11-14   Heat dissipating capacity of worm gear box. *A,* **without fan;** *B,* **with fan on worm shaft.**

The housing should have a liberal clearance to avoid churning the oil and to provide for the dissipation of the heat. The cooling area can be estimated by the curve in Fig. 11-11 or by the following equation.

$$A_c = 0.3c^{1.7} \tag{26}$$

The lost hp_c is equal to the difference between the input and output horsepower. The output horsepower hp_o is equal to the input horsepower hp_i multiplied by the efficiency η. Hence

$$hp_c = hp_i - hp_o = hp_i - hp_i \times \eta = hp_i(1 - \eta)$$

or
$$hp_i = \frac{hp_c}{1 - \eta} \tag{27}$$

The oil temperature should not exceed 180°F.

The efficiency equation for a worm gear reduction is the same as that used for a screw and nut. In the notation of this chapter, Eq. (17) of Chapter 5 becomes

$$\eta = \frac{\cos \varphi_n - \mu \tan \lambda}{\cos \varphi_n + \mu \cot \lambda} \tag{28}$$

Here φ_n is the pressure angle normal to the teeth, and μ is the coefficient of friction. This equation assumes that the axial thrust of the worm is taken by a rolling element thrust bearing. Values for μ, which depends on the velocity of sliding V_s, are given in Table 11-1.

TABLE 11-1   Values of Coefficient of Friction, μ, for Different Values of Sliding, V_s*

V_s, fpm	μ	V_s, fpm	μ	V_s, fpm	μ	V_s, fpm	μ
0	0.150	90	0.0560	500	0.0295	2,000	0.0160
5	0.099	100	0.0540	600	0.0274	2,200	0.0154
10	0.090	120	0.0519	700	0.0255	2,400	0.0149
20	0.080	140	0.0498	800	0.0240	2,600	0.0146
30	0.073	160	0.0477	900	0.0227	2,800	0.0143
40	0.0691	180	0.0456	1,000	0.0217	3,000	0.0140
50	0.0654	200	0.0435	1,200	0.0200	4,000	0.0131
60	0.0620	250	0.0400	1,400	0.0186	5,000	0.0126
70	0.0600	300	0.0365	1,600	0.0175	6,000	0.0122
80	0.0580	400	0.0327	1,800	0.0167		

*From *American Gear Manufacturers Association Standard 440.04*, Oct. 1971.

Example 8. Find the input and output horsepower for the worm reduction in Example 7 for continuous operation based on the cooling of the housing in still air. The housing temperature rise is to be 90 deg F. The pressure angle of the normal section is $\varphi_n = 20°$.

Solution. By Fig. 11-11 or Eq. (26):

$$A_c = 12.4 \text{ ft}^2$$

By Fig. 11-14: $C_1 = 2.43 \text{ Btu/(hr ft}^2 \text{ deg F)}$

By Eq. (25): $\text{hp}_c = \dfrac{778 C_1 A_c \, \Delta T}{60 \times 33,000} = \dfrac{778 \times 2.43 \times 12.4 \times 90}{60 \times 33,000} = 1.07$

By Eq. (22): $V_s = \dfrac{V_1}{\cos \lambda} = \dfrac{499.6}{0.90790} = 550 \text{ fpm}$

By Table 11-1: $\mu = 0.0286$

By Eq. (28): $\eta = \dfrac{0.93969 - 0.0286 \times 0.46171}{0.93969 + 0.0286 \times 2.1659} = 0.925$

By Eq. (27): $\text{hp}_i = \dfrac{\text{hp}_c}{1 - \eta} = \dfrac{1.067}{1 - 0.925} = 14.21 \text{ input hp}$

$$\text{hp}_o = 14.21 - 1.07 = 13.14 \text{ output hp}$$

The output horsepower based on thermal capacity for this reduction is thus considerably less than the mechanical horsepower of Example 7.

A considerably greater output can be obtained in intermittent service if short periods of operation are followed by sufficiently long cooling periods. Artificial cooling also permits operation at larger loads.

The mounting of the worm and wheel must provide for the thrust loads that occur during operation. The axial position of the worm is not critical, but the wheel must be correctly located axially to secure proper contact between the teeth.

Rubbing speeds between the teeth for case-hardened worms and chill cast phosphor bronze gears cut with good accuracy are limited to about 6,000 fpm. Lower speeds should be used if the materials and accuracy are of lower quality. On the other hand higher speeds can be permitted for superior materials and greater accuracy. Reduction in loading should be made for shock and impact conditions or for operation at elevated temperatures.

The helix angle should be limited to about 6° per thread. For example, if λ is 30°, then the worm should have at least five threads. For lower values of λ, the worm is self-locking; that is, a torque applied to the wheel axis will not rotate the worm.

When designing a worm gear reduction, it is a good plan, before deciding on the diameter of the worm, to ascertain whether a hob of the proper size is available for cutting the wheel. The purchase of a special tool may be avoided by changing the diameter of the worm so that the wheel can be cut by a hob that is already on hand. Gear-cutting firms sometimes catalog their worm gear hobs to enable their customers to take advantage of available tools.

10. Helical Gears

If a number of spur gears formed from thin plates are assembled with a small angular displacement between the teeth, the stepped gear of Fig. 11-15 results. A helical gear can be considered the limiting case of a stepped gear as the thickness of the plates and the angular displacements are made progres-

FIGURE 11-15   Stepped gears. Helical gears result if the plates are made progressively thinner.

sively smaller. A typical example of a pair of helical gears is given by Fig. 11-16. Because of the helix angle there will be end thrust on the shafts of helical gears, which must be provided for in the design of the bearings. End thrust can be eliminated by cutting a right-hand spiral over one-half the face, and using a left-hand spiral for the other half; herringbone gears are thus

FIGURE 11-16 **Types of spiral gears: left, helical gears; right, herring-bone gears.** (*From* Lubrication, *published by The Texas Co.*)

formed, as illustrated in Fig. 11-16. A central groove must usually be provided around the gear for clearance for hob or cutter. Herringbone gears made on the Sykes type of generating machine do not require the central clearance groove.

The helix angle ψ, illustrated in Fig. 11-19, usually varies from 15° to 30° for helical gears, and from 23° to 30°, or even 45°, for herringbone gears. Helical gears can be cut by the same hobs used for making spur gears. When the gear shaper is used, however, a special cutter is required. The relationship between directions of rotation and thrust for helical gears is shown in Fig. 11-17.

Gears with helical teeth possess certain inherent advantages. More teeth are in contact simultaneously, and the load is transferred gradually and uniformly as successive teeth come into engagement. Helical gears thus operate more smoothly and carry larger loads at higher speeds than spur gears. The line of contact extends diagonally across the face of mating teeth. Since more teeth are in contact, undercutting causes little trouble in helical gearing.

Helical gears can also be used for transmitting power between nonparallel shaft, as shown in Fig. 11-18. When used in this way, the teeth have only point contact, which does not shift axially along the teeth during operation. Such gears are usually used only for the transmission of relatively small loads.

Right hand teeth

Left hand teeth

FIGURE 11-17   Directions of rotation and thrust for helical gears.

FIGURE 11-18   Helical gears on nonparallel shafts. (*Courtesy Socony-Vacuum Oil Co.*)

11. *Pitch Diameter of Helical Gear*

Let the pitch diameter of a helical gear be called d, and let the helix angle ψ be measured between an element of the tooth at the pitch cylinder and the center line of the shaft, as shown in Fig. 11-19. Let the symbol p represent the

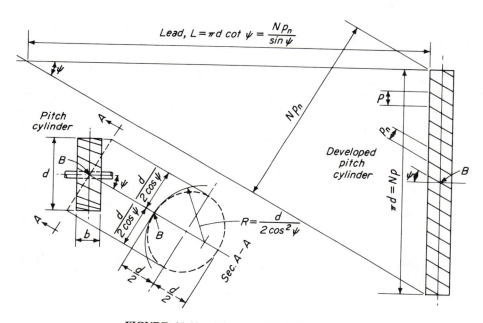

FIGURE 11-19   Geometry of helical gear.

circular pitch or distance from tooth to tooth measured on the pitch cylinder in the plane of rotation. The normal circular pitch p_n is measured normal to the tooth elements. It is shown in Fig. 11-19 that

$$p_n = p \cos \psi \qquad \text{and} \qquad \pi d = Np \tag{29}$$

The pitch diameter d of a helical gear is then equal to

$$d = \frac{Np}{\pi} = \frac{Np_n}{\pi \cos \psi} \tag{30}$$

Let P_{dn} be the diametral pitch normal to the tooth surface. Then, as for spur gears, $p_n P_{dn} = \pi$. Substitution for p_n in Eq. (30) gives

$$d = \frac{N}{P_{dn} \cos \psi} \tag{30a}$$

where N is the number of teeth in the gear, and π/p_n is replaced by P_{dn}. The pitch diameter of a helical gear thus depends upon the helix angle ψ as well as upon the normal circular pitch and the number of teeth.

When a helical gear is cut with a hob, the circular pitch of the tool is equal to p_n of Fig. 11-19. However, when the gear shaper is used, the circular pitch of the cutter and p of Fig. 11-19 coincide. Angle ψ is usually made large enough to give an overlapping of tooth action; that is, $b \tan \psi$ is greater than p, where b is the width of the face.

12. Formative Number of Teeth

A plane normal to the element of the tooth at point B in Fig. 11-19 intersects the pitch cylinder (extended) in the ellipse shown by section *A-A*. The shape of the tooth at B would be that generated on a cylindrical surface having the same radius of curvature as the ellipse at B. From analytic geometry, the radius of curvature R at the end of a semiminor axis of an ellipse is known to be

$$R = \frac{d}{2\cos^2\psi} \tag{a}$$

The formative number of teeth N' is defined as the number of teeth in a gear of radius R.

$$N' = \frac{2\pi R}{p_n} = \frac{\pi d}{p_n \cos^2\psi} \tag{b}$$

Substitution of the value for $\pi d/p_n$ from Eq. (30) gives the following equation for N'.

$$N' = \frac{N}{\cos^3\psi} \tag{31}$$

When the beam strength of helical teeth is computed, the form factor y for N' teeth should be used.

13. Center Distance of Mating Gears

Figure 11-20 shows a pair of mating helical gears having pitch diameters d_1 and d_2 and helix angles ψ_1 and ψ_2, respectively. From Eq. (30), these diameters can be written as

$$d_1 = \frac{N_1 p}{\pi} = \frac{N_1 p_n}{\pi \cos \psi_1} = \frac{N_1}{P_{dn} \cos \psi_1} \tag{32}$$

$$d_2 = \frac{N_2 p}{\pi} = \frac{N_2 p_n}{\pi \cos \psi_2} = \frac{N_2}{P_{dn} \cos \psi_2} \tag{33}$$

When meshed together, the center distance c for these gears is

$$c = \frac{d_1 + d_2}{2} = \frac{1}{2P_{dn}}\left[\frac{N_1}{\cos\psi_1} + \frac{N_2}{\cos\psi_2}\right] \tag{a}$$

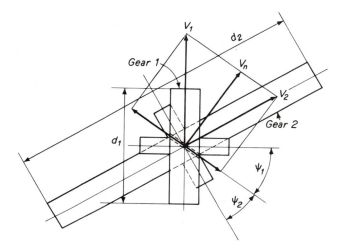

FIGURE 11-20 Components of velocity for mating helical gears.

Let the pitch-line velocities of the gears be V_1 and V_2, respectively. As illustrated by Fig. 11-20, the common velocity V_n normal to the tooth surfaces of both gears is

$$V_n = V_1 \cos \psi_1 = V_2 \cos \psi_2$$

or

$$V_2 = \frac{\cos \psi_1}{\cos \psi_2} V_1 \qquad (b)$$

The angular velocities of the two gears are

$$\omega_1 = \frac{2V_1}{d_1}$$

and

$$\omega_2 = \frac{2V_2}{d_2} = \frac{2 \cos \psi_1}{d_2 \cos \psi_2} V_1 \qquad (c)$$

The velocity ratio ω_1/ω_2 becomes

$$\frac{\omega_1}{\omega_2} = \frac{d_2 \cos \psi_2}{d_1 \cos \psi_1} \qquad (d)$$

Substitution from Eqs. (32) and (33) gives

$$\frac{\omega_1}{\omega_2} = \frac{N_2}{N_1} \qquad (34)$$

The velocity ratio of the gears thus depends on the number of teeth, and is independent of the helix angles. Substitution of Eq. (16) for velocity ratio $\beta = N_1/N_2$ into Eq. (a) gives

$$\frac{\beta}{\cos \psi_1} + \frac{1}{\cos \psi_2} = \frac{2P_{dn}c}{N_2} \qquad (35)$$

Sometimes the center distance c, the angle between the shafts $\psi_1 + \psi_2$, and the velocity ratio β are fixed. Suitable values for N_2 and P_{dn} are chosen for the conditions of the problem, and the values of ψ_1 and ψ_2 are found by trial and error in Eq. (35).

Example 9. Two helical gears with 90° shafts have a center distance of 12 in. The gears have 21 and 77 teeth. The normal diametral pitch is 6. Find the required values for the helix angles. Find the corresponding pitch diameters of the two gears.

Solution. By Eq. (15): $\dfrac{2P_{dn}c}{N_2} = \dfrac{2 \times 6 \times 12}{77} = 1.87013$

By Eq. (16): $\beta = \dfrac{N_1}{N_2} = \dfrac{21}{77} = 0.27273$

Trial values are selected from Fig. 11-10 that are refined by trial and error to give

$$\psi_2 = 19°44'$$

$$\psi_1 = 70°16'$$

By Eq. (32): $d_1 = \dfrac{N_1}{P_{dn} \cos \psi_1} = \dfrac{21}{6 \times 0.33764} = 10.3660 \text{ in.}$

14. Tooth Loads of Helical Gears

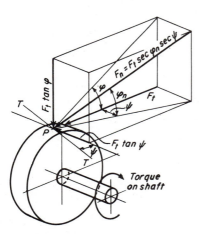

FIGURE 11-21 Forces acting on tooth of helical gear when axes are parallel.

Figure 11-21 shows the force F_n, normal to the tooth surface, for a helical tooth. It is inclined at angle φ_n to the plane tangent to the pitch cylinder and lies in a plane located at helix angle ψ to the plane of rotation. Normal force F_n has the component $F_n \cos \varphi_n$ in the tangent plane and this component has the further component F_t in the plane of rotation. Then

$$F_t = F_n \cos \varphi_n \cos \psi \qquad (36)$$

Force F_n has a component in the plane of rotation that is inclined at angle φ to the tangent plane. The relationship between φ and φ_n is given

$$\tan \varphi_n = \tan \varphi \cos \psi \qquad (37)$$

This is found by the same method used for the derivation of Eq. (13), Chapter 5.

Helical gears on parallel shafts operate at high values for the efficiency.

15. Dynamic Load and Wear Load for Helical Gears on Parallel Axes

The geometry of helical gears is involved but approximate equations will be presented for finding the loads and capacities.

The equation for dynamic load, F_d, for spur gears will be used for helical gears using Table 10-2 for error e. Force F_d is assumed inclined at angle φ_n to the tangent plane.

In the equation for wear capacity F_w, the virtual diameter, must be used to take care of the larger radius of curvature of the tooth outline. As shown by Fig. 11-19, this is equal to $d/\cos^2 \psi$, so that the equation for F_w becomes

$$F_w = \frac{d_1 bQK}{\cos^2 \psi} \tag{38}$$

where d_1 is the pitch diameter of the smaller gear in the plane of rotation. The previous equation for Q remains unchanged as no change results from the substitution of Eq. (31) for the virtual number of teeth. Then

$$Q = \frac{2N_2}{N_1 + N_2} \tag{39}$$

Example 10. Two helical gears have parallel shafts and normal diametral pitch $P_{dn} = 4$. Blanks are solid and normal pressure angle φ_n is 20°. Face width $b = 2.5$ in. Numbers of teeth are 32 and 44. Center distance c is 10 in. Both gears are steel BHN = 300. AFMA Quality No. 8. Pinion turns 860 rpm.

Find the value of the helix angle ψ_1 and the horsepower the gears are carrying.

Solution. $d_1 = \frac{32}{76} \times 20 = 8.4211$ in., $d_2 = \frac{44}{76} \times 20$

$$= 11.5789 \text{ in.}$$

By Eq. (30a): $\cos \psi = \dfrac{N_1}{P_{dn} d_1} = \dfrac{32}{4 \times 8.4211} = 0.95000$

$\psi_1 = 18.195°$

By Fig. 10-24, for: BHN = 300, $K = 136$

By Eq. (39): $Q = \dfrac{2N_2}{N_1 + N_2} = \dfrac{88}{76} = 1.158$

By Eq. (38): $F_w = \dfrac{d_1 bQK}{\cos^2 \psi} = \dfrac{8.4211 \times 2.5 \times 1.158 \times 136}{0.95000^2}$

$$= 3{,}674 \text{ lb capacity in wear, normal to the tooth}$$
surface

By Eq. (28), Chapter 10: $k = 1{,}667{,}000 \times 2.5 = 4{,}167{,}000$ lb/in.

$$C = \frac{N_2}{N_1} = \frac{44}{32} = 1.375 \quad C^2 = 1.8906$$

By Eq. (33), Chapter 10:

$$m_e = \frac{C^2}{1+C^2} \frac{\pi b \gamma r_1^2}{2g}$$

$$= \frac{1.8906}{2.8906} \times \frac{\pi 2.5 \times 0.283 \times 4.2106^2}{2 \times 386}$$

$$= 0.03339 \text{ lb sec}^2/\text{in.}$$

By Table 10-2:

$$e = 0.0035 + 0.0036 = 0.0071 \text{ in.}$$

By Eq. (25a), Chapter 10:

$$F_d = \frac{e n_1 N_1 \sqrt{k m_e}}{30}$$

$$= \frac{0.0071 \times 860 \times 32\sqrt{4,167,000 \times 0.03339}}{30}$$

$$= 6.513\sqrt{139,125} = 6.513 \times 373.0 = 2,430 \text{ lb}$$

By Fig. 10-22(b):

$$F_p = F_w - F_d = 3,674 - 2,430 = 1,244 \text{ lb}$$

By Eq. (36):

$$F_t = F_p \cos \varphi_n \cos \psi$$

$$= 1,244 \times 0.93969 \times 0.9500 = 1,111 \text{ lb}$$

By Eq. (14), Chapter 10:

$$V = \frac{\pi d n}{12} = \frac{\pi 8.4211 \times 860}{12} = 1,896 \text{ fpm}$$

By Eq. (17), Chapter 10:

$$\text{hp} = \frac{F_t V}{33,000} = \frac{1,111 \times 1,896}{33,000} = 63.8$$

A rough check will indicate that the teeth are amply strong in bending.

16. *Load Capacity of Helical Gears with Crossed Axes*

When helical gears with crossed axes are first put into service, only point contact between the teeth occurs. Contact usually increases to a short length of line after some wear has taken place. The load capacity is quite small when the gears are new, but the loading can be increased appreciably after careful wearing in.[6] The teeth should be of a size that causes two or more pairs always to be in contact. Nominal variations in center distance or axial locations of the gears have no effect on the action, and such gears are usually considered to be easy to mount.

REFERENCES

Standard reference works on gear theory are given at the end of Chapter 10.

[6]See p. 136 of Reference 4, Chapter 10, and p. 536 of Reference 1, Chapter 10.

PROBLEMS

1. Two 20° full-depth solid flame-hardened steel bevel gears have a diametral pitch P_d of 3 at the outside radius. Numbers of teeth are 24 and 60. Gears have AGMA Quality No. 8. Width of face b is 3 in., but blank thickness b' parallel to the shaft is 1.0 in. Pinion turns 570 rpm. Find the horsepower capacity of the pair.

Ans. hp = 273.

2. Two 20° full-depth steel bevel gears have a diametral pitch P_d of 4 at the outside radius, and a BHN equal to 325. Numbers of teeth are 24 and 48. Gears have AGMA Quality No. 8. Width of face b is 2 in., but thickness of blank b' parallel to the shaft is 1.0 in. Pinion turns 1,140 rpm.

(a) Find the horsepower the gears are carrying.

(b) Let the base of the pitch cone for the pinion be mounted at the midpoint of a simply supported shaft 16 in. long. Thrust and torque are resisted as in Fig. 11-5. Make a free-body diagram for F_t acting at the midpoint of the tooth. Show numerical values for all forces and moments. *Ans.* hp = 58.3.

3. Two 20° full-depth steel bevel gears have a module 6 at the outside radius and a BHN = 325. Numbers of teeth are 30 and 45. Gears have AGMA Quality No. 8. Width of face b is 62 mm, but thickness of blank b' parallel to the shaft is 35 mm. If the gears are carrying 45 kW, find the speed at which the pinion is turning.

Ans. n_1 = 604 or 1,519 rpm.

4. A triple-thread worm has a lead angle of 17° and a pitch diameter of 2.2802 in. Find the center distance when the worm is mated with a wheel of 48 teeth.

Ans. c = 6.7171 in.

5. A double-thread worm has a pitch diameter of 3 in. The wheel has 20 teeth and a pitch diameter of 5 in. Find the value of the helix angle.

Ans. λ = 9°27.7′.

6. A worm gear set gives a reduction of 18 : 1, and has a helix angle of 12°. The center distance is 4.6424 in. and the normal diametral pitch P_{dn} is 5. Find suitable values for N_1 and N_2.

7. A worm gear set has a helix angle of 19° and a center distance of 236.18 mm; m_{on} = 6.5. Find suitable values for N_1 and N_2. Speed ratio 20 : 1.

8. A double-thread worm meshes with a 25 tooth wheel. Center distance is 175 mm; m_{on} = 10. Find the value of d_1 and d_2.

9. A double-thread worm meshes with a 50-tooth wheel. Center distance is 8 in. P_{dn} = 4. Worm turns 1,750 rpm. Find values for d_1, d_2, and the mechanical horsepower. *Ans.* hp = 10.5.

10. A quadruple-thread worm meshes with a 40-tooth wheel. Helix angle is 23°. P_{dn} = 3. φ_n = 20°. Worm turns 1,050 rpm. Find the center distance, the mechanical horsepower, and the horsepower based on cooling for a temperature rise of 90 deg F. *Ans.* c = 8.9486 in.; hp_{mech} = 25.2; hp_o = 21.7.

11. A triple-thread worm meshes with a 60-tooth wheel. P_{dn} = 4; φ_n = 20°.

Pitch diameter of worm is 3.1838 in. Worm turns 1,150 rpm. Find the mechanical horsepower and the horsepower based on cooling for a temperature rise of 90 deg. F.   *Ans.* $hp_{mech} = 18.0$; $hp_o = 15.4$.

12. A triple-thread worm meshes with a 36-tooth wheel. $P_{dn} = 2$; $c = 12$ in.; $\varphi_n = 20°$. Pinion turns 860 rpm. Find values for d_1, d_2, the mechanical horsepower, and the horsepower based on cooling for a rise in temperature of 90 deg. F.
Ans. $d_1 = 5.2013$ in.; $hp_{mech} = 36.1$; $hp_o = 26.9$.

13. Two helical gears have parallel shafts and are cut with a hob of p_n equal to 0.5236 in. Speed ratio is 2 : 1 and the center distance is equal to 9 in. Find the required value of the helix angle if the smaller gear has 35 teeth.
Ans. $\psi_1 = 13°32'$.

14. Two helical gears have shafts at 90° and helix angles of 45°. Speed ratio is 3 : 1; $m_{on} = 6$. Find the center distance if the smaller gear has 20 teeth.
Ans. $c = 339.41$ mm.

15. Work Problem 14 but with the helix angle ψ_1 for the smaller gear equal to $37\frac{1}{2}°$, and $P_{dn} = 4$.   *Ans.* $c = 15.4713$ in.

16. Work Problem 14 but with the helix angle ψ_1 for the smaller gear equal to $52\frac{1}{2}°$.   *Ans.* $c = 325.45$ mm.

17. Two meshing helical gears have parallel shafts and a center distance of 9 in. Number of teeth are 35 and 70 solid blanks. $\varphi_n = 20°$; $P_{dn} = 6$; $b = 2$ in. Both gears are steel hardened to 320 BHN. AGMA Quality No. 8. Pinion turns 860 rpm. Find the horsepower capacity.   *Ans.*   hp = 45.2.

18. Two helical gears have shafts at 90° and a center distance of 364 mm. The speed ratio is 3 : 1. The smaller gear has 20 teeth; $m_{on} = 6.5$. Find suitable values for the helix angles and the corresponding pitch diameters of the gears.
Ans. $\psi_1 = 63°38'$, $\psi_2 = 26°22'$;
or $\psi_2 = 46°13.8'$, $\psi_2 = 43°46.2'$.

19. Two helical gears have shafts at a 60° angle. Numbers of teeth are 35 and 105. Center distance is 8 in.; P_{dn} equals $\pi/10$. Find suitable values for angles ψ_1 and ψ_2 and the pitch diameters of the gears.

20. A pair of 20° helical gears have parallel shafts and a center distance of 260 mm; $m_{on} = 6.5$. Numbers of teeth are 32 and 44. Make a perspective sketch and show the tooth forces at the center of the tooth when the gear is carrying a transmitted force of 9,000 newtons.

21. A helical gear has 30 teeth and a pitch diameter of 264 mm. Gear has $m_{on} = 8$ and $\varphi_n = 20°$. The force normal to the tooth surface is 5,000 newtons. Find the kW transmitted at 570 rpm.   *Ans.*   kW = 33.6.

22. A worm gear reducer has an input horsepower of 15. The worm shaft turns at 1,000 rpm. What is the efficiency of the reducer based on the cooling capacity for a 90 deg F rise in temperature for continuous operation? The center distance is 12 in.
Ans.   Eff = 0.85.

23. A worm gear reducer has a center distance of 8 in. and a helix angle of

$8°33.6'$. Worm turns 1,750 rpm and has a diameter of 3.3592 in. $\varphi_n = 20°$. Find the horsepower based on cooling for a temperature rise of 90 deg F.

Ans. $hp_o = 12.3$.

24. A single-thread worm meshes with a 50-tooth wheel. $P_{dn} = 4$; $\varphi_n = 20°$; $c = 8$ in.; $n_1 = 1,725$ rpm. Find d_1, d_2, hp_{mech}, and the horsepower based on a 90 deg F rise in temperature. *Ans.* $hp_{mech} = 5.32$; $hp_o = 2.19$.

25. A worm gear reduction has a center distance of 16 in. Worm turns 720 rpm. Worm has eight starts and wheel has 40 teeth. $P_{dn} = 2$; $\varphi_n = 25°$. Find d_1, d_2, hp_{mech}, and the horsepower based on a 90 deg F rise in temperature.

Ans. $hp_{mech} = 111.9$; $hp_o = 76.0$.

26. A worm gear reduction has a 10-in. center distance. Worm has six starts and wheel has 40 teeth. $P_{dn} = 3$; $\varphi_n = 25°$. Worm turns 1,160 rpm. Find d_1, d_2, hp_{mech} and the horsepower based on a temperature rise of 90 deg F.

Ans. $hp_{mech} = 46.6$; $hp_o = 39.1$.

27. A triple-thread worm meshes with a 24-tooth wheel on a center distance of 6 in. $P_{dn} = 3$; $\varphi_n = 20°$. Worm turns 415 rpm. Find d_1, d_2, hp_{mech}, and the horsepower based on a temperature rise of 90 deg F.

Ans. $hp_{mech} = 4.96$; $hp_o = 3.57$.

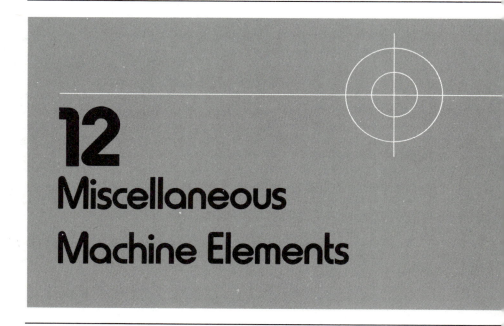

12
Miscellaneous
Machine Elements

Design methods are given in this chapter for a miscellaneous group of machine elements that require only brief treatment.

1. Stresses in a Thick Cylinder

The thick-walled cylinder shown in Fig. 12-1 is subjected to uniform pressures p_i and p_o on the internal and external lateral surfaces, respectively. The top and bottom surfaces are assumed to be free from load. Since the body and the loading are symmetrical about the axis, shear stresses in the tangential and radial directions are not present, and only normal stresses σ_t and σ_r act on the element, as illustrated in Fig. 12-1(a).

When the wall of the cylinder is thin, the tangential or hoop stress σ_t can be assumed to be uniform throughout the wall thickness, and its value can be found by the elementary equations of Chapter 7. This assumption cannot be made for a thick-walled cylinder, and the equations of the following derivation should be used for finding the stresses.[1]

[1]See Marin, J., "Designing Shrink-Fit Assemblies," *Machine Design*, **14**, 1942, June, p. 68; July, p. 72; and Aug., p. 78. See also Horger, O. J., and C. W. Nelson, "Design of Press- and Shrink-Fitted Assemblies," *Trans. ASME*, **59**, 1937, p. A-183; **60**, 1938, p. A-32; Peterson, R. E., and A. M. Wahl, "Fatigue of Shafts at Fitted Members with a Related Photoelastic Analysis," *Trans. ASME*, **57**, 1935, pp. A-1 and A-69; **58**, 1936, p. A-74; Horger, O. J., and J. L. Maulbetsch, "Increasing the Fatigue Strength of Press-Fitted Axle Assemblies by Surface Rolling," *Trans. ASME*, **58**, 1936, p. A-91; **59**, 1937, p. A-37. See also Chapter 7 of Reference 6, end of chapter.

Consider the stresses acting on the semicircular element of Fig. 12-1(b). The thickness in the direction perpendicular to the paper has been taken equal to unity. The vertical component of the inward radial stresses across the diameter of the element is equal to $2\sigma_r r$, and for the outward component of the stresses, $2(\sigma_r + d\sigma_r)(r + dr)$. The equilibrium equation for this element is then

$$2\sigma_r r + 2\sigma_t dr$$
$$= 2(\sigma_r + d\sigma_r)(r + dr) \qquad \text{(a)}$$

The right-hand side should be expanded and terms of higher order should be neglected. The equation can then be rewritten as

$$r\frac{d\sigma_r}{dr} + \sigma_r = \sigma_t \qquad \text{(1)}$$

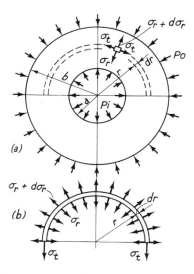

(a)

(b)

FIGURE 12-1   **Thick-walled cylinder subjected to internal and external pressures.**

The strain or unit deformation ϵ_o in the direction perpendicular to the paper can be found from the stresses σ_t and σ_r as follows:

$$\epsilon_o = -\frac{\mu\sigma_t}{E} - \frac{\mu\sigma_r}{E} \qquad \text{(b)}$$

where μ is Poisson's ratio and E is the modulus of elasticity. This equation may be rearranged as

$$\sigma_t + \sigma_r = -\frac{\epsilon_o E}{\mu} \qquad \text{(c)}$$

The right-hand side is a constant. For convenience it will hereafter be called $2C_1$. Substitution into Eq. (1) gives

$$r\frac{d\sigma_r}{dr} + 2\sigma_r = 2C_1 \qquad \text{(d)}$$

Or:
$$r^2\frac{d\sigma_r}{dr} + 2r\sigma_r = 2rC_1 \qquad \text{(e)}$$

Or:
$$\frac{d}{dr}(r^2\sigma_r) = 2rC_1$$

Integration gives:
$$r^2\sigma_r = C_1 r^2 + C_2 \qquad \text{(f)}$$

where C_2 is the constant of integration. Equation (f) may be written as follows:

$$\sigma_r = C_1 + \frac{C_2}{r^2} \qquad \text{(g)}$$

This value for σ_r should be substituted into Eq. (c), to give

$$\sigma_t = C_1 - \frac{C_2}{r^2} \tag{h}$$

At the inner boundary, $r = a$, the radial stress σ_r becomes equal to $-p_i$, and Eq. (g) is written

$$-p_i = C_1 + \frac{C_2}{a^2} \tag{i}$$

At the outer boundary, $r = b$, radial stress σ_r is equal to $-p_o$, and Eq. (g) becomes

$$-p_o = C_1 + \frac{C_2}{b^2} \tag{j}$$

Constants C_1 and C_2 can now be found by solving Eqs. (i) and (j) simultaneously:

$$C_1 = \frac{a^2 p_i - b^2 p_o}{b^2 - a^2} \tag{k}$$

$$C_2 = -\frac{a^2 b^2 (p_i - p_o)}{b^2 - a^2} \tag{l}$$

Substitution into Eqs. (g) and (h) gives[2] the values of the stresses σ_r and σ_t.

$$\sigma_r = \frac{a^2 p_i - b^2 p_o}{b^2 - a^2} - \frac{a^2 b^2 (p_i - p_o)}{r^2 (b^2 - a^2)} \tag{2}$$

$$\sigma_t = \frac{a^2 p_i - b^2 p_o}{b^2 - a^2} + \frac{a^2 b^2 (p_i - p_o)}{r^2 (b^2 - a^2)} \tag{3}$$

For many applications, the outer pressure p_o is equal to zero. Equations (2) and (3) then reduce to the following forms,

$$\sigma_r = \frac{a^2 p}{b^2 - a^2} \left(1 - \frac{b^2}{r^2} \right) \tag{4}$$

$$\sigma_t = \frac{a^2 p}{b^2 - a^2} \left(1 + \frac{b^2}{r^2} \right) \tag{5}$$

where p_i is now called p. The stresses are a maximum at the inner edge, where $r = a$, and the value of σ_r is $-p$. The tangential stress for this point is seen to be

$$\sigma_t = p \left[\frac{1 + (a/b)^2}{1 - (a/b)^2} \right] \tag{6}$$

At the inner boundary, $r = a$, the tangential elongation ϵ_t is equal to

$$\epsilon_t = \frac{1}{E_h} (\sigma_t - \mu \sigma_r) \tag{m}$$

---

[2] These are known as the Lamé equations. See p. 239 of Reference 7, end of chapter.

where E_h is the modulus of elasticity for the material. The total increase in length of the inner boundary is equal to $2\pi a\epsilon_t$, giving an increase u_h in the radius of the hole equal to $2\pi a\epsilon_t/2\pi$, or $a\epsilon_t$. Hence

$$u_h = \frac{a}{E_h}(\sigma_t - \mu\sigma_r) = \frac{ap}{E_h}\left[\frac{1 + (a/b)^2}{1 - (a/b)^2} + \mu\right] \tag{7}$$

Radial displacement u_h, is, of course, outward.

2. Shrink and Press Fit Stresses

In a shrink or press fit, pressure p is caused by the interference of metal between the shaft and the hub. Because of this pressure, the radius of the hole is increased, and the radius of the shaft is decreased. The decrease u_s in the radius of the shaft is equal to

$$u_s = a\epsilon_t = \frac{a}{E_s}(\sigma_t - \mu\sigma_r) \tag{a}$$

where E_s is the modulus of elasticity for the material composing the shaft. For a solid shaft, $a = p_i = 0$ in Fig. 12-1. Equations (2) and (3) give $\sigma_r = \sigma_t = -p_o$ or $-p$, and the radial deformation becomes

$$u_s = -\frac{ap}{E_s}(1 - \mu) \tag{8}$$

This displacement is inward, as indicated by the minus sign. The sum of the absolute values for u_h and u_s is the radial interference of metal for the shrink or press fit as shown by Fig. 12-2.

When hub and shaft are composed of the same material, the modulus of elasticity for both parts is the same and can be represented by E. The double sum of the absolute values of u_h and u_s is the diametral interference of metal Δ. By Eqs. (7) and (8), $\Delta = 2(|u_h| + |u_s|)$.

FIGURE 12-2 Interference of metal in press fit.

$$\Delta = \frac{4ap}{E}\left[\frac{1}{1 - (a/b)^2}\right] \tag{9}$$

If Eq. (9) is solved for p, the value of the radial stress corresponding to a given value for the diametral interference is obtained.

$$\sigma_r = -p = -\frac{E\Delta}{4a}\left[1 - \frac{a^2}{b^2}\right] \tag{10}$$

When this value for p is substituted into Eq. (6), the equation for σ_t for the hub is obtained.

$$\sigma_t = \frac{E\Delta}{4a}\left[1 + \frac{a^2}{b^2}\right] \tag{11}$$

Example 1. An 8 in.-diam steel shaft is to have a press fit in a 20-in.-diam cast iron disk. The maximum tangential stress in the disk is to be 5,000 psi. The modulus of elasticity for steel is 30,000,000 psi, and for cast iron, 15,000,000 psi. Poisson's ratio is equal to 0.3.
   (a) Find the required diametral interference of metal.
   (b) Plot values of σ_t along a radius of the disk.
   (c) If the disk is 10 in. thick in the axial direction, find the force required to press the parts together if the coefficient of friction is equal to 0.12. Also find the torque which the joint could carry because of the shrink fit pressure.

   Solution. (a) $a = 4$ in.; $b = 10$ in.; $b/a = 2.5$; $a/b = 0.4$.

From Eq. (6): $p = \dfrac{\sigma_t\,[1 - (a/b)^2]}{1 + (a/b)^2} = \dfrac{5,000(1 - 0.16)}{1 + 0.16} = 3,620$ psi

By Eq. (7): $u_h = \dfrac{4 \times 3,620}{15,000,000}\left[\dfrac{1.16}{0.84} + 0.3\right] = 0.00162$ in.

   increase in hole radius

By Eq. (8): $u_s = -\dfrac{4 \times 3,620}{30,000,000}(1 - 0.3) = -0.00034$ in.

   decrease in shaft radius

   $\Delta = 2(0.00162 + 0.00034) = 0.0039$ in.
   diametral interference

   (b) The application of Eq. (5) for various values of r gives the curve of Fig. 12-3 for stresses σ_t.

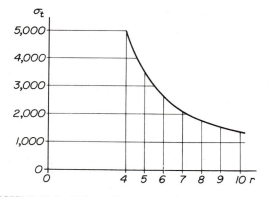

FIGURE 12-3   Values of tangential stress σ_t for Example 1.

(c) Force required for assembly,

$$F = 8\pi \times 10 \times 3{,}620 \times 0.12 = 109{,}200 \text{ lb}$$

Torque carried by press fit,

$$T = 109{,}200 \times 4 = 436{,}800 \text{ in. lb}$$

Precise measurements and a large factor of safety should be used when transmitting torque by press or shrink fits.

Example 1A. A 200-mm-diam steel shaft is to have a press fit in a 500-mm-diam cast iron disk. The maximum tangential stress in the disk is to be 35 MPa. The modulus of elasticity for steel is 206,900 MPa and half this amount for cast iron. Poisson's ratio is equal to 0.3.

(a) Find the required diametral interference of metal.

(b) If the disk is 250 mm thick in the axial direction, find the force required to press the parts together if the coefficient of friction is equal to 0.12. Also, find the torque which the joint could carry because of the shrink fit pressure.

Solution. (a) $a = 100$ mm; $b = 250$ mm; $b/a = 2.5$; $a/b = 0.4$.

From Eq. (6): $p = \dfrac{\sigma_t[1 - (a/b)^2]}{1 + (a/b)^2} = \dfrac{35(1 - 0.16)}{1 + 0.16} = 25.3$ MPa

By Eq. (7): $u_h = \dfrac{100 \times 25.3}{103{,}450}\left[\dfrac{1.16}{0.84} + 0.3\right] = 0.0412$ mm

increase in hole radius

By Eq. (8): $u_s = -\dfrac{100 \times 25.3}{206{,}900}(1 - 0.3) = -0.0086$ mm

decrease in shaft radius

$$\Delta = 2(0.0412 + 0.0086) = 0.100 \text{ mm}$$

diametral interference

(b) Force required for assembly,

$$F = 200\pi 250 \times 25.3 \times 0.12 = 477{,}700 \text{ N}$$

Torque carried by press fit,

$$T = 477{,}700 \times 100 = 47{,}770{,}000 \text{ Nmm}$$

3. Stress Concentration Caused by Press Fit

A shaft with bending stress has a stress concentration at the edge of a press fitted member as at A in Fig. 12-4(a). Some improvement in fatigue strength can be obtained by grooving the disk as shown in Fig. 12-4(b). Experiments have shown that rolling the shaft between small diameter rollers under pressure high enough to deform the surface has also resulted in improved

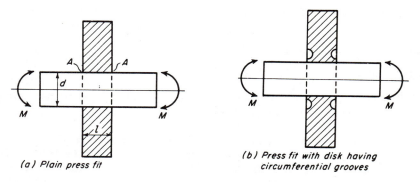

(a) Plain press fit   (b) Press fit with disk having circumferential grooves

FIGURE 12-4   Stress concentration for press fit.

fatigue strength. Stress concentration factors K for plain press fitted assemblies are given by Fig. 12-5.

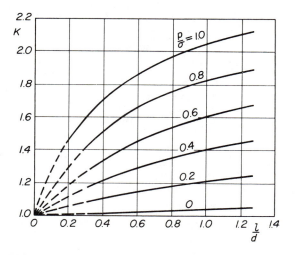

FIGURE 12-5   Stress concentration factors for press fit.

Example 2. Find the stress concentration factor for the shaft of Example 1 if the shaft carries an alternating bending moment of 350,000 in. lb.

Solution.

$$I = \frac{\pi d^4}{64} = \frac{\pi 8^4}{64} = 201.06 \text{ in.}^4$$

$$\sigma = \frac{Mc}{I} = \frac{350,000 \times 4}{201.06} = 6,963 \text{ psi}$$

$$\frac{p}{\sigma} = \frac{3,620}{6,963} = 0.52$$

$$\frac{l}{d} = \frac{10}{8} = 1.25$$

By Fig. 12-5: $K = 1.56$

Max. bending stress: $\sigma_{max} = 1.56 \times 6{,}963 = 10{,}860$ psi

Example 2A. Find the stress concentration factor for the shaft of Example 1A if the shaft carries an alternating bending moment of 40,000,000 Nmm.

Solution.
$$I = \frac{\pi d^4}{64} = \frac{\pi 200^4}{64} = 78{,}540{,}000 \text{ mm}^4$$

$$\sigma = \frac{Mc}{I} = \frac{40{,}000{,}000 \times 100}{78{,}540{,}000} = 50.9 \text{ MPa}$$

$$\frac{p}{\sigma} = \frac{25.3}{50.9} = 0.50$$

$$\frac{l}{d} = \frac{250}{200} = 1.25$$

By Fig. 12-5: $K = 1.54$

Max. bending stress: $\sigma = 1.54 \times 50.9 = 78.4$ MPa

When subjected to reversed bending, press fitted joints are subject to a mode of failure known as *fretting corrosion*.[3] Surface pitting is accompanied by an oxide debris which is caused either by a grinding action or by the alternate welding and tearing away of the high spots. The removed particles become oxidized and form an abrasive powder which continues the destructive process.

There is no completely satisfactory method for stopping fretting unless all relative motion can be prevented. When relative motion cannot be eliminated, solid lubricants such as MoS may maintain a lubricating film for a long period of time. An increase in the wear resistance of the surfaces so as to reduce surface welding may also be helpful.

4. Stresses in Disk Flywheel

Flame-cut disk flywheels are widely used because of the uniformity and high strength of steel plate.

Let Fig. 12-1(a) be a view of such a flywheel. Pressures p_o and p_i are not present. In addition to stresses σ_r and σ_t, the element is loaded by an outwardly directed centrifugal force because of the rotation at ω rad/sec. Let the material have a weight of γ lb/in.[3] Let the flywheel have unit thickness. Consider an element at radius r of length dr in the radial direction and of unit length in the tangential direction. The weight of the element is $\gamma \, dr$. It is subjected to an acceleration of $r\omega^2$. The centrifugal force on the element is $\gamma r \omega^2 \, dr/g$. This force has a total upward component on the semicircular element shown in Fig. 12-1(b) of $2\gamma r^2 \omega^2 \, dr/g$. This term should be included

[3]See Uhlig, H. H., "Mechanism of Fretting Corrosion," *Trans. ASME*, **76**, 1954, p. 401.

with the others in Eq. (a) of Section 1. When the resulting equation is reduced to lowest terms, the result is

$$\sigma_t - \sigma_r - r\frac{d\sigma_r}{dr} - \frac{\gamma r^2 \omega^2}{g} = 0 \tag{12}$$

The solution to this equation is well known[4] and will not be given here. In most cases stress σ_t controls the design. It is a maximum at the inner boundary and is equal to

$$\sigma_{t\,max} = \frac{\gamma \omega^2}{4g}[(3 + \mu)b^2 + (1 - \mu)a^2] \tag{13}$$

where μ is Poisson's ratio.

Example 3. A circular steel plate flywheel is 24 in. in outside diameter and has a 6 in. hole. The flywheel rotates at 3,000 rpm. Find the value of the maximum tangential stress. $\mu = 0.3$. $\gamma = 0.283$ lb/in.^{3}

Solution. $g = 12 \times 32.174 = 386$ in./sec^2

$$\omega = \frac{2\pi n}{60} = \frac{2\pi 3,000}{60} = 314.16 \text{ rad/sec}$$

In Eq. (13): $\sigma_{t\,max} = \dfrac{0.283 \times 314.16^2}{4 \times 386}[(3 + 0.3) \times 12^2 + (1 - 0.3) \times 3^2]$

$$= 18.09(475.2 + 6.3) = 8,710 \text{ psi}$$

Example 3A. A circular steel plate flywheel is 600 mm in outside diameter and has a 150-mm hole. The flywheel rotates 3,000 rpm. Find the value of the maximum tangential stress. $\mu = 0.3$, $\gamma = 0.000\ 0768$ N/mm^3.

Solution. $g = 9,807$ mm/sec^2

$$\omega = \frac{2\pi n}{60} = \frac{2\pi 3,000}{60} = 314.16 \text{ rad/sec}$$

By Eq. (13): $\sigma_{max} = \dfrac{0.000\ 0768 \times 314.16^2}{4 \times 9,807}[(3 + 0.3)300^2 + (1 - 0.3)75^2]$

$$= 0.000\ 1932[297,000 + 3,938] = 58.1 \text{ MPa}$$

If the plate is drilled near the inner boundary for bolting to a flange, stress concentration occurs at the bolt holes.

The polar moment of inertia J about the axis of a disk flywheel with a hole is given by

$$J = \frac{1}{2}\frac{W_o}{g}b^2 - \frac{1}{2}\frac{W_i}{g}a^2$$

where W_o represents the weight of a disk of radius b, and W_i is the weight of a disk of radius a.

[4]See p. 245 of Reference 7, end of chapter.

$$W_o = \gamma \pi l b^2 \quad \text{and} \quad W_i = \gamma \pi l a^2$$

where l represents the thickness of the plate in the axial direction. Substitution gives

$$J = \frac{\gamma \pi l}{2g}(b^4 - a^4) \tag{14}$$

The energy ΔKE delivered by a change in speed from ω_{max} to ω_{min} is given by

$$\Delta KE = \tfrac{1}{2}J(\omega_{max}^2 - \omega_{min}^2)$$

5. Flywheel with Spokes and Rim

Figure 12-6 shows a cast flywheel with spokes and rim. When the wheel rotates, a uniformly distributed force F_c acts outwardly on the rim and produces the tensile force F. The spokes exercise a restraining influence on the expansion of the rim, and cause moments M to act as shown. Equations for F and M, as well as for the tensile force F_1 in the spoke,[5] are given in Table 12-1. Here w is the weight of the rim per inch of circumference, and

FIGURE 12-6   Forces and moments on segment of fly wheel.

TABLE 12-1   Equations for Forces and Moments in Flywheel Spokes and Rim

Spokes	H	F_1, lb	F, lb	M, in. lb
4	$\dfrac{2/3}{(0.0730r^2/h^2) + 0.643 + (A/A_1)}$	$\dfrac{wr^2n^2}{35,200} \times H$	$\dfrac{wr^2n^2}{35,200}(1 - 0.500H)$	$0.1366F_1r$
6	$\dfrac{2/3}{(0.0203r^2/h^2) + 0.957 + (A/A_1)}$	$\dfrac{wr^2n^2}{35,200} \times H$	$\dfrac{wr^2n^2}{35,200}(1 - 0.866H)$	$0.0889F_1r$
8	$\dfrac{2/3}{(0.0091r^2/h^2) + 1.274 + (A/A_1)}$	$\dfrac{wr^2n^2}{35,200} \times H$	$\dfrac{wr^2n^2}{35,200}(1 - 1.207H)$	$0.0662F_1r$

[5]For derivation, see p. 98 of Reference 7, end of chapter.

symbols r, n, h, A, and A_1 have the meanings indicated in Fig. 12-6. In these equations, the depth of section h is assumed to be small as compared to radius r.

The tensile stress in the spoke is equal to F_1/A_1. The tensile stress in the rim is equal to F/A plus the effect of the bending moment $6M/bh^2$. Stress concentration effects are present at the junction between the spoke and rim. Sudden changes in the velocity of rotation, as well as belt tensions, cause additional bending stresses in spokes and rim.

Care must be exercised to assure that dimensions and speed of rotation for cast-iron flywheels have such values that the resulting tensile stresses do not exceed safe working values for this material. Tensile stress values for cast iron vary over a considerable range, depending on the composition and quality of the material.

Example 4. Find the stresses in a cast-iron flywheel with 6 spokes rotating uniformly at 600 rpm; radius $r = 36$ in. to center of rim; width of rim $b = 12$ in.; depth of rim $h = 4$ in.; and area of spokes $A_1 = 10$ in.^{2} Cast iron weighs 0.256 lb/in.^{3}

Solution. $\dfrac{r}{h} = 9$, and $\dfrac{A}{A_1} = 4.8$

By Table 12-1: $H = \dfrac{2/3}{0.0203 \times 81 + 0.957 + 4.8} = 0.0901$

$$w = 0.256 \times 12 \times 4 = 12.29 \text{ lb per in. of rim}$$

$$F_1 = \frac{12.29 \times 36^2 \times 600^2}{35,200} \times 0.0901$$

$$= 162,900 \times 0.0901 = 14,680 \text{ lb}$$

Tensile stress in spoke: $\sigma = \dfrac{14,680}{10} = 1,470 \text{ psi}$

$$F = 162,900(1 - 0.866 = 0.0901) = 150,200 \text{ lb}$$

Tensile stress in rim: $\sigma = \dfrac{150,200}{12 \times 4} = 3,130 \text{ psi}$

$$M = 0.0889 \times 14,680 \times 36 = 46,980 \text{ in. lb}$$

Bending stress in rim: $\sigma = \dfrac{6M}{bh^2} = \dfrac{6 \times 46,980}{12 \times 16} = 1,470 \text{ psi}$

Resultant tensile stress in rim:

$$\sigma = 3,130 + 1,470 = 4,600 \text{ psi}$$

Suppose the spokes in the foregoing example are not present and the rim freely expands under the influence of centrifugal force. The weight per lineal inch of rim is w, and the acceleration is $r\omega^2$. The centrifugal force F_e per inch of rim is

$$F_c = \frac{w}{g}r\omega^2 = \frac{wr}{g}\left(\frac{2\pi n}{60}\right)^2 = \frac{wrn^2}{35,200} \tag{15}$$

Force F in the rim is equal to $F_c r$.

Example 5. Find the stresses for the data of Example 4 if the spokes are not present and the flywheel consists of a freely rotating ring.

Solution. By Eq. (15): $F_c = \dfrac{12.29 \times 36 \times 600^2}{35,200} = 4,520 \text{ lb/in.}$

Force in rim: $F = F_c r = 4,520 \times 36 = 162,900 \text{ lb}$

Stress: $\sigma = \dfrac{F}{A} = \dfrac{162,900}{48} = 3,390 \text{ psi}$

It can be seen from this example that the elementary method which considers the flywheel as a rotating ring free to expand gives stresses on the unsafe side. Also, the elementary method gives no indication of the value of the stress in the spokes.

6. Flywheel Requirements

The shafts in many kinds of machinery are subjected to torque loading that varies throughout the work cycle. A smaller driving motor can be used and smoother operation obtained, if the shaft is provided with a flywheel. During the portion of the work cycle that requires a high torque, energy from the flywheel assists the driving motor, but at a reduction in speed. This energy, as well as the speed, are returned to their original values during those portions of the cycle where the torque requirements are low.

Such a shaft, and a typical loading diagram, are shown in Fig. 12-7(a) and (b). It is assumed that the shaft is very stiff, and the operating speed

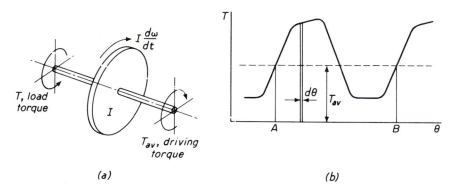

(a)　　　　　　　　　　　　　　　　　　　(b)

FIGURE 12-7 Flywheel and torque loading diagram.

sufficiently low so that any effects from torsional vibration are negligible. The shaft is in equilibrium from the driving torque T_{av}, the load torque T, and the inertia torque $Id\omega/dt$. Driving torque T_{av} is assumed to be constant. Then

$$I\frac{d\omega}{dt} = T - T_{av} \tag{a}$$

but

$$\omega\, dt = d\theta \tag{b}$$

These equations should be multiplied together:

$$I\omega\, d\omega = (T - T_{av})\, d\theta$$

The equation can be integrated between any two points A and B.

$$I\int_A^B \omega\, d\omega = \int_A^B (T - T_{av})\, d\theta \tag{c}$$

The right side represents the energy supplied to the shaft by the flywheel between points A and B. It can be represented by U_{AB}. The left side should be integrated, and the limits substituted.

$$\tfrac{1}{2}I(\omega_B^2 - \omega_A^2) = U_{AB} \tag{16}$$

On the diagram for torque loading, Fig. 12-7(b), let points A and B be selected so that the energy supplied by the flywheel is a maximum for the chosen interval along the θ-axis. Equation (16) can then be written

$$\tfrac{1}{2}I(\omega_{max}^2 - \omega_{min}^2) = U_{max} \tag{17}$$

Equation (17) represents the maximum change in kinetic energy of the system because of variation in speed. Equation (17) can be written

$$I\left(\frac{\omega_{max} + \omega_{min}}{2}\right)(\omega_{max} - \omega_{min}) = U_{max} \tag{18}$$

Let a coefficient of speed fluctuation, C_f, be defined as

$$C_f = \frac{\omega_{max} - \omega_{min}}{\omega_{av}} \tag{19}$$

where

$$\omega_{av} = \frac{\omega_{max} + \omega_{min}}{2} \tag{20}$$

When these are substituted,[6] Eq. (18) can be arranged as

$$I = \frac{U_{max}}{C_f\omega_{av}^2} \tag{21}$$

[6]See Biezeno, C. B., and R. Grammel, *Engineering Dynamics*, Vol. 4, Glasgow: Blackie & Son Ltd., 1954, p. 107. See also Timoshenko, S., and D. H. Young, *Advanced Dynamics*, New York: McGraw-Hill Book Company, 1948, p. 185; Den Hartog, J. P., *Mechanics*, New York: McGraw-Hill Book Company, 1948, p. 268; and Chapter 19 of Reference 6, end of chapter.

Equations (19) and (20) can be combined to give

$$\omega_{max} = \omega_{av}(1 + \tfrac{1}{2}C_f) \tag{22}$$

$$\omega_{min} = \omega_{av}(1 - \tfrac{1}{2}C_f) \tag{23}$$

Typical values for the coefficient of fluctuation, C_f, for different types of equipment are shown in Table 12-2.

TABLE 12-2   Coefficient of Fluctuation*

Type of Equipment	C_f
Crushing machinery	0.200
Electrical machinery	0.003
Electrical machinery, direct-driven	0.002
Engines with belt transmission	0.030
Flour-milling machinery	0.020
Gear-wheel transmission	0.020
Hammering machinery	0.200
Machine tools	0.030
Paper-making machinery	0.025
Pumping machinery	0.030–0.050
Shearing machinery	0.030–0.050
Spinning machinery	0.010–0.020
Textile machinery	0.025

*From *Kent's Mechanical Engineers' Handbook*, 12th ed., "Design and Production," pp. 7–40.

Example 6. The variation in torque loading for a work cycle of one revolution for a shaft is shown in Fig. 12-8. Input torque T_{av} is assumed to be constant. Average speed is 860 rpm. Coefficient of fluctuation C_f is 0.10. Find the diameter of the required flywheel if it is cut from a 1-in. steel plate.

FIGURE 12-8   Torque loading diagram for Example 6.

Solution. $2\pi T_{av} = \dfrac{\pi}{3}(200 + 1,600 + 400 + 900 + 900 + 200)$

$$T_{av} = 700 \text{ in. lb}$$

The areas of the torque diagram above and below the value of T_{av} are shown in the figure. The maximum value of U occurs between points A and B. It has the value

$$U_{max} = 942.48 - 314.16 + 209.44 + 209.44$$

$$= 1,047.20 \text{ in. lb}$$

$$\omega_{av} = \frac{2\pi n}{60} = \frac{2\pi 860}{60} = 90.06 \text{ rad/sec}$$

By Eq. (21): $I = \dfrac{U_{max}}{C_f \omega_{av}^2} = \dfrac{1,047.20}{0.10 \times 90.06^2}$

$$= 1.2911 \text{ in. lb sec}^2$$

By Eq. (14) for $a = 0$: $b^4 = \dfrac{2gI}{\gamma\pi l} = \dfrac{2 \times 386 \times 1.2911}{0.283\pi \times 1} = 1,121.1$

$$b^2 = 33.48, \qquad b = 5.78 \text{ in.}$$

$$\text{Diam} = 11.57 \text{ in.}$$

7. Impact of Elastic Bodies

Let a uniform stress σ_0 be suddenly applied to the end of a long bar as shown in Fig. 12-9. At the first instant, an infinitely thin layer of material in the bar is compressed.[7] This compression is transferred to the next layer, and so on. In other words a compression wave, moving with velocity c, travels along the bar. After time t, a length ct is compressed and the remainder of the bar is at rest in the unstressed condition.

FIGURE 12-9   Compression wave in long elastic bar.

[7]See Donnell, L. H., "Longitudinal Wave Transmission and Impact," *Trans. ASME*, **52**(1), 1930, p. APM-52-14-153. See also Johnson, R. C., "Impact Forces in Mechanisms," *Machine Design*, **30**, June 12, 1958, p. 138; Arnold, R. N., "Impact Stresses in Freely Supported Beams," *Proc. Inst. Mech. Eng.*, **137**, 1937, p. 217; and Chapter 18 of Reference 6, end of chapter.

The deformation δ of the compressed zone is

$$\delta = \frac{\sigma_0 ct}{E} \tag{24}$$

The velocity V of the particles in the compressed zone is

$$V = \frac{\delta}{t} = \frac{\sigma_0 c}{E} \quad \text{or} \quad c = \frac{VE}{\sigma_0} \tag{25}$$

The velocity V of the particles should be distinguished from the velocity c of the wave front.

Let the material in the bar have a weight of γ lb/in.^{3} The density ρ, mass/in.^{3}, is equal to γ/g, where g is the gravitational constant, 386 in./sec^2. The mass of the moving particles in the wave is equal to ρAct, where A is the cross-sectional area of the bar. The force on the end of the bar is equal to $A\sigma_0$. The momentum and impulse equation states that mass times velocity is equal to force times time. Hence

$$\rho ActV = A\sigma_0 t \quad \text{or} \quad \sigma_0 = c\rho V$$

Substitution of Eq. (25) gives: $\sigma_0 = V\sqrt{E\rho}$ $\tag{26}$

The force k required to compress the bar a unit distance is equal to AE/l, where l is the length of the bar; hence $E = kl/A$. The weight W_b of the bar is equal to $Al\gamma$ or $Alg\rho$; hence $\rho = W_b/Alg$. Substitution in Eq. (26) gives

$$\sigma_0 = \frac{V}{A}\sqrt{\frac{kW_b}{g}} \tag{27}$$

where F is the force due to impact. Equations (25) and (26) can be combined to give

$$c = \sqrt{\frac{E}{\rho}} \tag{28}$$

Example 7. Find the velocity of propagation for a compression wave in steel. $\gamma = 0.283$ lb/in.^{3}

Solution. In Eq. (28): $\quad c = \sqrt{\dfrac{E}{\rho}} = \sqrt{\dfrac{Eg}{\gamma}} = \sqrt{\dfrac{30{,}000{,}000 \times 386}{0.283}}$

$$= 202{,}280 \text{ in./sec} = 3.19 \text{ miles/sec}$$

Example 7A. Find the velocity of propagation for a compression wave in steel. $\gamma = 0.000\ 0768$ N/mm.^{3}

Solution. In Eq. (28): $\quad c = \sqrt{\dfrac{E}{\rho}} = \sqrt{\dfrac{Eg}{\gamma}} = \sqrt{\dfrac{206{,}900 \times 9{,}807}{0.000\ 0768}}$

$$= 5{,}140{,}000 \text{ mm/sec} = 5.14 \text{ km/sec}$$

8. Impact Stresses by Energy Method

Impact stresses can be caused by a moving weight striking the end of the bar as shown in Fig. 12-10. A compression wave travels along the bar and is reflected at the wall; it returns to weight W if it is still in contact and is again reflected; and so on. The maximum stress may thus be several times as great as that given by Eq. (26). The exact analysis of this problem is very involved. However an approximate solution, based on energy, can be obtained if simplifying assumptions are made.

FIGURE 12-10 Free end of horizontal bar struck by moving weight.

FIGURE 12-11 Impact when bar of Fig. 12-10 is replaced by weight at end of spring.

Weight W has velocity V before striking, but after impact the velocity is reduced to V_a. If it is assumed that the weight and bar remain in contact after striking, the end of the bar also has the velocity V_a. The solution to the problem by the energy method is based on the assumption that all elements in the bar upon impact are instantly given velocities which are proportional to their distances from the wall. Thus, if u is the velocity for the element at distance x from the wall, the following proportion will hold:

$$\frac{u}{x} = \frac{V_a}{l} \quad \text{or} \quad u = \frac{V_a}{l} x \tag{a}$$

The kinetic energy of the system after impact is given by

$$KE = \frac{W}{2g} V_a^2 + \int_0^l \frac{u^2}{2g} w_1 \, dx$$

where w_1 is the weight per lineal inch of the bar.

Substitution of the value of u from Eq. (a) and integration gives

$$KE = \frac{WV_a^2}{2g} \left[1 + \frac{W_b}{3W} \right] \tag{29}$$

Thus, by Eq. (29), the kinetic energy after impact would be unaffected if the bar were replaced by a weight of magnitude $W_b/3$ at the end of a weightless spring having the same elastic stiffness as the bar. Such a system is shown in Fig. 12-11(a). The situation before impact has occurred, when W has its original velocity V and when $W_b/3$ is not moving, is illustrated in Fig. 12-11(b). The total momentum before and after impact remains unchanged. Hence,

$$WV = \left[W + \frac{W_b}{3} \right] V_a \quad \text{or} \quad V_a = \frac{1}{1 + W_b/3W} V \tag{b}$$

Substitution in Eq. (29) gives the value of the kinetic energy of the system after impact.

$$KE = \frac{WV^2}{2g} \left[\frac{1}{1 + W_b/3W} \right] \tag{30}$$

Upon impact the force in the bar increases uniformly from zero to a maximum value of F. The average value is $\frac{1}{2}F$. If δ represents the maximum shortening of the bar under impact, the strain energy of the bar at the instant of maximum compression is $\frac{1}{2}F\delta$. If k represents the force required to compress the bar a unit distance, than $\delta = F/k$ and the energy is equal to $\frac{1}{2}F^2/k$. This is equal to the KE in Eq. (30). When these values are equated, the following expression results.

$$F = V \sqrt{\frac{kW}{g} \left[\frac{1}{1 + (W_b/3W)} \right]} \tag{31}$$

The stress in the bar is found by dividing both sides of this equation by the cross-sectional area A. The equation can then be rearranged as follows.

$$\sigma_e = \frac{V}{A} \sqrt{\frac{kW_b}{g}} \sqrt{\frac{W}{W_b} \left[\frac{1}{1 + (W_b/3W)} \right]}$$

or

$$\frac{\sigma_e}{\sigma_0} = \sqrt{\frac{W}{W_b} \left[\frac{1}{1 + (W_b/3W)} \right]} \tag{32}$$

where σ_0 is given by Eq. (27). Equation (32), derived from the energy assumption, is plotted in Fig. 12-12.

The exact solution to this problem is obtained by more involved methods. The results can be closely represented by the following empirical equation, which is also plotted in Fig. 12-12.

$$\frac{\sigma}{\sigma_0} = \sqrt{\frac{W}{W_b} + \frac{2}{3}} + 1 \tag{33}$$

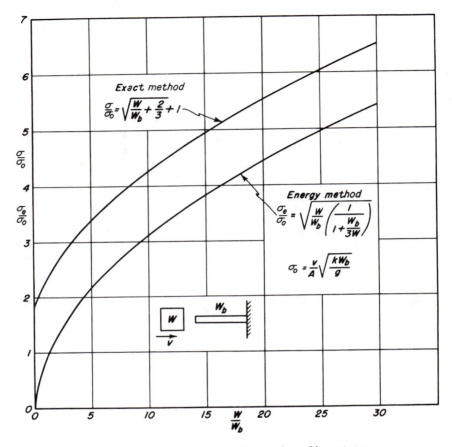

FIGURE 12-12   Exact and approximate values of impact stresses.

Example 8. A 1-in.-diam round steel bar 20 in. long is struck by a 40 lb weight moving at a velocity of 30 in./sec. Find the value of the stress due to impact by the energy method and compare with the results of the exact solution.

Solution.
$$W_b = \frac{\gamma \pi d^2 l}{4} = \frac{0.283 \pi 1^2 \times 20}{4} = 4.445 \text{ lb}$$

$$\frac{W}{W_b} = \frac{40}{4.445} = 9$$

$$k = \frac{AE}{l} = \frac{0.7854 \times 30,000,000}{20} = 1,178,000 \text{ lb/in.}$$

By Eq. (27):
$$\sigma_0 = \frac{30}{0.7854} \sqrt{\frac{1,178,000 \times 4.445}{386}} = 4,449 \text{ psi}$$

By Eq. (32), energy: $\sigma_e = 4,449\sqrt{9\left[\dfrac{1}{1 + 1/(3 \times 9)}\right]} = 4,449 \times 2.946$

$$= 13,107 \text{ psi}$$

By Eq. (33), exact: $\sigma = 4,449\left[\sqrt{9 + \dfrac{2}{3}} + 1\right] = 4,449 \times 4.109$

$$= 18,281 \text{ psi}$$

The energy method thus gave a stress much less than the exact value.

Unless the striking weight is many times as great as the struck bar, the energy method gives results that can be grossly too low. If, in Fig. 12-12, the striking weight is only twice as great as the struck bar, the energy method gives a stress only half as much as that which actually occurs. The assumption in Fig. 12-10 that all cross sections of the bar are instantly given the velocities indicated is, of course, impossible. The energy method is presented in detail here not only because it is widely used but also because it is possible to make a comparison with the exact method based on stress waves.

Example 8A. A round steel bar 25 mm in diameter, 500 mm long with fixed end, is struck by a weight of 200 N moving at a velocity of 750 mm/sec. Find the value of the stress due to impact and compare with the results of the exact solution.

Solution. $\qquad W_b = \dfrac{\gamma\pi d^2 l}{4} = \dfrac{0.000\ 0768\pi 25^2 \times 500}{4} = 18.85 \text{ N}$

$$\dfrac{W}{W_b} = \dfrac{200}{18.85} = 10.61 \qquad A = \dfrac{\pi d^2}{4} = \dfrac{\pi 25^2}{4}$$

$$= 490.9 \text{ mm}^2$$

$$k = \dfrac{AE}{l} = \dfrac{490.9 \times 206,900}{500} = 203,100 \text{ N/mm}$$

By Eq. (27): $\qquad \sigma_0 = \dfrac{V}{A}\sqrt{\dfrac{kW_b}{g}} = \dfrac{750}{490.9}\sqrt{\dfrac{203,100 \times 18.85}{9,807}}$

$$= 1.528 \times 19.76 = 30.19 \text{ MPa}$$

By Eq. (32), energy: $\sigma_e = 30.19\sqrt{10.61\left[\dfrac{1}{1 + (10.61 \times 3)}\right]}$

$$= 30.19 \times 3.207 = 96.82 \text{ MPa}$$

By Eq. (33), exact: $\qquad \sigma = \sigma_0\left[\sqrt{\dfrac{W}{W_b} + \dfrac{2}{3}} + 1\right]$

$$= 30.19\left[\sqrt{10.61 + \dfrac{2}{3}} + 1\right]$$

$$= 30.19\ [3.358 + 1] = 131.6 \text{ MPa}$$

9. Force Produced by Falling Weight

The solution will be obtained by the energy method. Let the weight W in Fig. 12-13 fall through the height h, and let W and the bolt head remain in contact after striking. Let δ be the maximum value for the stretch of the bolt, and let F be the corresponding impact force. The stiffness of the bolt is such that k lb are required for a stretch of 1 in.

After impact, and at the lowest point of travel, the elastic energy gained by the bolt is equal to $F\delta/2$, as illustrated by the force-deformation graph in Fig. 12-13. This energy comes from the kinetic energy of the weight, as given by Eq. (30), together with the potential energy $W\delta$ given up by the additional lowering of the weight. Hence,

FIGURE 12-13   Force and deformation caused by falling weight.

$$\frac{F\delta}{2} = \frac{WV^2}{2g} \times \frac{1}{1+(W_b/3W)} + W\delta \qquad (a)$$

When the substitutions for the impact deflection, $\delta = F/k$, and for the statical deflection, $\delta_{st} = W/k$, are made in the expression above, a quadratic equation results, which, upon solution for F/W, gives the following result.

$$\frac{F}{W} = 1 + \sqrt{1 + \frac{2h}{\delta_{st}} \times \frac{1}{1+(W_b/3W)}} \qquad (34)$$

The value of the impact force F can be obtained by substitution in this equation. It can be written in terms of velocity V for the falling weight, if desired, by use of the relationship $V^2 = 2gh$, where g is the gravitational constant. Equation (34) indicates that if the weight is held just above the bolt head, $h = 0$, and then released, $F = 2W$.

The value of k for a cylindrical bar is equal to AE/l, where A is the cross-sectional area, and E is the modulus of elasticity. This value is found by taking the equation for deformation, $\delta = Fl/AE$, and letting δ be equal to unity when the load F becomes equal to k. In all the foregoing, it is assumed, of course, that the elastic limit for the bolt material is not exceeded.

Example 9. In Fig. 12-13 let the steel bar be round of diameter 1 in. and length 20 in. If the weight W is 40 lb, find the stress in the bolt if height h is 1.166 in., which corresponds to a striking velocity of 30 in./sec.

Solution. Data from Table 2-1 should be used.

Area: $A = \dfrac{\pi d^2}{4} = 0.7854$ in.

$$\delta_{st} = \frac{Wl}{AE} = \frac{40 \times 20}{0.7854 \times 30,000,000} = 0.000\ 033\ 95 \text{ in.}$$

$$W_b = \gamma Al = 0.283 \times 0.7854 \times 20 = 4.445 \text{ lb}$$

By Eq. (34):
$$\frac{F}{W} = 1 + \sqrt{\frac{2 \times 1.166}{0.000\ 033\ 95} \times \frac{1}{1 + 4.445/(3 \times 40)}}$$

$$= 1 + \sqrt{1 + 66,218} = 1 + 257.3 = 258.3$$

$$F = 258.3W = 258.3 \times 40 = 10,333 \text{ lb}$$

$$\sigma_e = \frac{F}{A} = \frac{10,333}{0.7854} = 13,157 \text{ psi}$$

This example again indicates that high stresses are easily produced by impact loads. If the yield point of the material is exceeded, a portion of the energy of the falling weight is used in giving a permanent stretch to the bar. If the lower end of the bolt in the foregoing example were threaded, and if the weight were stopped in its fall by the nut, the entire impact force would pass through the net section of the threads. The stress would thereby be increased, particularly if stress concentration would be present.

The impact stress can be reduced if the bolt were made more flexible by turning the entire body above the nut to the diameter of the stress area of the threads. The static deflection is increased and the impact force decreased. Hence when the cross section is decreased to that of the area through which the impact force must pass a reduction in the stress will occur. Any further reduction of cross section will cause an increase in the stress.

When the bar is rigid, deflection δ in Fig. 12-13 is a very small number. If the last term in Eq. (a) is small enough to neglect, this equation immediately becomes Eq. (31). This is done by replacing δ on the left by F/k. The resulting stress will be very nearly the same as that obtained by Eq. (34). This is confirmed by comparing the results for σ_e as found in Examples 8 and 9. The same effect is secured in Eq. (34) when the two unity terms (to the right of the equality sign and under the radical) are neglected. Then h is replaced by $V^2/2g$ and δ_{st} by W/k. On the other hand, when the weight strikes a flexible object such as a coiled spring, then δ is not relatively small and Eq. (34) must be used.

10. Impact of Weight on Beam

Impact between a weight and a simply supported beam is illustrated in Fig. 12-14. The energy method can also be applied to this problem; it gives the following equation for the ratio between the impact force F and the weight W.

FIGURE 12-14 Impact of weight on beam.

$$\frac{F}{W} = 1 + \sqrt{1 + \frac{2h}{\delta_{st}} \times \frac{1}{1 + (17W_b/35W)}} \qquad (35)$$

where W_b is the weight of the beam, and h and δ_{st}, have the same meanings as in the preceding section. Equation (35) is known as the Cox equation. It has been in use for almost a century. Because of the previously mentioned deficiencies of the energy method, it is of somewhat limited usefulness.

For most metals, the yield point and ultimate strength are raised when the load is very quickly applied and is of short duration. Because of the many uncertainties present, designers are rarely inclined to take advantage of these raised properties when deciding upon suitable values for the working stresses.

11. Gaskets and Seals

Gaskets and seals are divided into two main classes: static and dynamic. The static gasket is used to prevent the loss of fluid in a pressure vessel. A dynamic seal is used to prevent loss of fluid in a sliding or rotating joint. Numerous types of materials are used such as organic and mineral fibers, leather, natural and synthetic rubber, cork, paper, and soft metals.

For static seals, a flat rigid surface and a thin gasket are preferable, and a minimum amount of packing surface should be exposed to the fluid. Sometimes the sealing is created by friction. Sometimes the gasket is confined to a chamber and distorted to effect a seal by jamming the packing across the leakage path. Allowance must sometimes be made for a change in volume if the gasket material swells upon contact with the fluid. For a dynamic seal it is most important that the sliding member have a very smooth hard surface.

Various shapes for a number of different gaskets and seals are shown in Fig. 12-15. The stuffing box shown in (a) has the disadvantage of high friction when the pressure of the gasket is high enough to prevent loss of fluid. The V-ring or chevron packing shown in (b) has less friction and tends to seal more tightly with increase of pressure. The molded U-cup packing is widely used for low and medium pressures. The O-ring shown in (d) has the advantage of low friction and simplicity of design. The ring should be slightly compressed after assembly. This seal is popular in aeroplane hydraulic mechanisms. The plastic back-up rings prevent extrusion of the soft rubber into the clearance space under very high pressures. The flange or hat packing shown in (f) is used in a number of different forms. Sometimes a circumferential spring is used to press the lip of the packing against the shaft.

The mechanical seal[8] of Fig. 12-16 is very effective for sealing a rotating

[8]See Alexander, T. E., "Mechanical Seals—Uses and Abuses," *Mech. Eng.*, **84**, Dec. 1962, p. 51. See also Mayer, E., "Leakage and Wear in Mechanical Seals," *Machine Design*, **32**, Mar. 3, 1960, p. 106; and Brkich, A., "Mechanical Seals," *Prod. Eng.*, **21**, Apr. 1950, p. 85.

(a) Compression packing

Gland
Packing box
Moving member
Packing

(b) V-ring packing

Adapters
Packings

(c) U-cup packing

Packing
Fiber filler or
metal pedestal ring

(d) O-ring packing

Back-up rings

(e) T-ring packing

(f) Flange or hat packing

(g) Cup packing

(h) Confined packing

FIGURE 12-15 Various types of gaskets and seals.

Sealing surface
lapped to high
degree of flatness

Carbon sealing element

FIGURE 12-16 Mechanical seal for rotating shaft.

shaft against gas or liquid under pressure. The sealing member may be either stationary or rotating and is made from refined carbon. It is spring loaded to compensate for wear. The mating surface is made of a hard material such as Stellite and must be lapped to a high degree of flatness. The edge of the carbon member should be sharp and square to prevent the entrance of foreign matter which may cause a rapid increase in the rate of wear.

The inflatable seal, Fig. 12-17, is convenient for sealing large openings with a relatively large amount of clearance.

Fluid-tight joints can also be formed by coating the surfaces with a thermosetting liquid plastic[9] which cures in the absence of air after the connection has been tightened. The material will flow into surface irregularities and machining costs can sometimes be reduced.

In piping service, rubber and nonmetallic materials, as well as lowmelting metals, are generally limited to temperatures below 250°F. Materials such as asbestos, with or without metal jacketing, may be used up to 850°F. For temperatures higher than 850°F, allmetal gaskets must be used.

FIGURE 12-17   **Inflatable seal.**

12. Design of Gaskets for Static Loads

Various kinds of gaskets are illustrated in Table A. To maintain a fluid-tight joint it is necessary that the parts be tightly bolted together. The initial or bolting-up pressure must be enough to cause local yielding of the gasket where it is in contact with the asperities of the metal flange surfaces. This minimum contact pressure, necessary to secure a tight joint even for low values of the internal pressure, is called the "yield" value y of the gasket. Values are given in Table A.

The internal fluid pressure in the pipe or vessel reduces the gasket con-

[9]See Whittemann, R. F., and M. B. Pearce, "Anerobics—A New Approach to Gaskets," *Mech. Eng.*, **91**, Aug. 1969, p. 26.

TABLE A

TABLE A: GASKET FACTORS m AND YIELD VALUES y (REFER TO TABLE B)

GASKET MATERIAL		GASKET FACTOR m	YIELD VALUE y	SKETCH & NOTES	FACING LIMITATION	USE COL.
GUM RUBBER SHEET		.50	0			
HARD RUBBER SHEET		1.00	180			
CLOTH INSERTED SOFT RUBBER		.75	50		USE ①④⑥ ONLY	
CLOTH INSERTED HARD RUBBER		1.25	400			
VEGETABLE FIBRE SHEET (HEMP OR JUTE)		1.75	1120			
RUBBERIZED WOVEN WIRE INSERTED ASBESTOS	3 – PLY	2.25	2200			
	2 – PLY	2.50	2880			
	1 – PLY	2.75	3650			
ASBESTOS COMPOSITION OR COMPRESSED ASBESTOS	1/8" THICK	2.00	1620		USE ①④⑥ ONLY	
	1/16" THICK	2.75	3650			
	1/32" THICK	3.50	6480			
SPIRAL-WOUND METAL, ASBESTOS FILLED	CARBON STEEL	2.50	2880			
	KA25 OR TYPE 316	3.00	4500			
SERRATED STEEL, ASBESTOS FILLED		2.75	3650			II
CORRUGATED METAL, ASBESTOS INSERTED OR CORRUGATED METAL JACKET, ASBESTOS FILLED	SOFT ALUMINUM	2.50	2880		USE ① ONLY	
	SOFT COPPER OR BRASS	2.75	3650			
	IRON OR SOFT STEEL	3.00	4500			
	MONEL OR 4-6% CHROME	3.25	5450			
	11-13% CHROME, KA25 OR TYPE 316	3.50	6480			
CORRUGATED METAL	SOFT ALUMINUM	2.75	3650			
	SOFT COPPER OR BRASS	3.00	4500			
	IRON OR SOFT STEEL	3.25	5450			
	MONEL OR 4-6% CHROME	3.50	6480			
	11-13% CHROME, KA25 OR TYPE 316	3.75	7600			
FLAT METAL JACKET, ASBESTOS FILLED	SOFT ALUMINUM	3.25	5450			
	SOFT COPPER OR BRASS	3.50	6480			
	IRON OR SOFT STEEL	3.75	7600			
	MONEL OR 4-6% CHROME	4.00	8820			
	11-13% CHROME, KA25 OR TYPE 316	4.25	10120			
GROOVED IRON OR SOFT STEEL CORE, METAL JACKETED	SOFT ALUMINUM	3.25	5450		USE ①④⑥ ONLY	
	SOFT COPPER OR BRASS	3.50	6480			
	IRON OR SOFT STEEL	3.75	7600			
	MONEL OR 4-6% CHROME	4.00	8820			
	11-13% CHROME, KA25 OR TYPE 316	4.25	10120			
SOLID METAL, 1/8" THICKNESS OR MORE (FOR THINNER GASKETS SEE NOTE)	SOFT ALUMINUM	4.00	8820	NOTE: FOR EACH 1/32" REDUCTION IN THICKNESS BELOW 1/8" INCREASE m BY .25, COMPUTE $y = 180 (2m-1)^2$		I
	SOFT COPPER OR BRASS	4.75	13000			
	IRON OR SOFT STEEL	5.50	18000			
	MONEL OR 4-6% CHROME	6.00	21780			
	11-13% CHROME, KA25 OR TYPE 316	6.50	25920			

tact pressure. Experience has shown that the ratio between the resultant contact pressure and the fluid pressure should not be less than a certain value if the joint is to remain tight. This ratio is called the gasket factor m. As indicated by Table A, it varies for different kinds of gaskets.

The bolting-up and fluid pressures cause a distortion in the flanges and adjacent parts of the joint. The chief effect is a rotation that tends to concentrate the entire gasket force on the outer edge of the gasket. Gasket pressures are accordingly not computed for the entire width, but only for a narrow band on the outer edge, called the effective gasket yielding width b. Equations for determining this width are found in Table B. Although these

TABLE B

TABLE B: EFFECTIVE GASKET WIDTH b

FACING SKETCH	BASIC GASKET YIELDING WIDTH b_o	
	COL. I	COL. II
(1a)	$\dfrac{n}{2}$	$\dfrac{n}{2}$
(1b)	$\dfrac{w+t}{2}\left(\dfrac{w+n}{4}\,max\right)$	$\dfrac{w+t}{2}\left(\dfrac{w+n}{4}\,max\right)$
(2)	$\dfrac{w+n}{4}$	$\dfrac{w+3n}{8}$
(3)	$\dfrac{w}{2}\,;\left(\dfrac{n}{4}\,min\right)$	$\dfrac{w+n}{4}\,;\left(\dfrac{3n}{8}\,min\right)$
(4)	$\dfrac{3n}{8}$	$\dfrac{7n}{16}$
(5)		
(6)	$\dfrac{n}{4}$	$\dfrac{3n}{8}$
(7)		
(8)	$\dfrac{w}{8}$	

EFFECTIVE GASKET YIELDING WIDTH b

$$b = b_o, \text{ when } b_o \leqq \tfrac{1}{4}"$$

$$b = \frac{\sqrt{b_o}}{2}, \text{ when } b_o > \tfrac{1}{4}"$$

LOCATION OF GASKET LOAD ATTACK

| INSIDE GASKET | WITH NUBBIN | FULL FACE GASKET |

equations are empirical, they are based on long experience in the design of fluid-tight joints.[10]

The area subjected to fluid pressure is assumed to extend to the edge of the effective gasket area.

Figure 12-18 shows the forces acting in a gasketed joint. Before the fluid pressure is acting, the force F_g in the bolt and on the gasket is

$$F_g = A_g q \qquad (36)$$

[10]For Tables A and B see Rossheim, D. B., and A. R. C. Markl, "Gasket Loading Constants," *Mech. Eng.*, **65**, 1943, p. 647; **66**, 1944, p. 72. See also Roberts, I., "Gaskets and Bolted Joints," *Trans. ASME*, **72**, 1950, p. 169; Smoley, E. M., "Sealing with Gaskets," *Machine Design*, **38**, Oct. 27, 1966, p. 172; and Linderoth, L. S., "Selecting Hydraulic Seals," *Machine Design*, **16**, Sept. 1944, p. 119.

where A_g is the effective area of the gasket, and q is pressure on the gasket caused by tightening the bolts.

After the fluid pressure p is acting, a force $A_i p$ is produced, where A_i is the area subjected to internal pressure. The force on the gasket required to prevent loss of pressure is $A_g m p$. The bolt force F_b then is

$$F_b = p(A_i + A_g m) \qquad (37)$$

Force F_g must be equal to or greater than force F_b. They can therefore be equated.

$$A_g q = p(A_i + A_g m) \qquad (38)$$

(a) Bolting up force. No fluid pressure.

(b) Forces when fluid pressure is acting.

FIGURE 12-18 Equilibrium of gasketed joint.

FIGURE 12-19 Gasketed manhole cover. Example 10.

Example 10. A cross section through a gasketed manhole joint is shown in Fig. 12-19. The internal fluid pressure is 500 psi gage. If 24 bolts are used for holding the cover in place, find the tensile stress in the net area of the threads.

Solution. From Table B:

$$b_0 = \frac{n}{2} = \frac{2}{2} = 1 \text{ in.}$$

$$b = \frac{\sqrt{b_o}}{2} = \frac{\sqrt{1}}{2} = 0.5 \text{ in., effective gasket width}$$

Effective gasket area: $A_g = 27.5\pi \times 0.5 = 43.2 \text{ in.}^2$

Area subjected to internal pressure:

$$A_i = \frac{\pi}{4} \times 27^2 = 572.6 \text{ in.}^2$$

By Table A: $m = 3.5$

In Eq. (38): $q = \dfrac{500(572.6 + 43.2 \times 3.5)}{43.2} = 8{,}380 \text{ psi}$

This value for the gasket pressure is satisfactory, since Table A indicates that a compression of 6,480 psi is sufficient to seat the gasket properly.

For $1\frac{1}{2}$ in.-12 UNF: stress area $= 1.58$ in.^{2}

By Eq. (37): $F_b = 500(572.6 + 43.2 \times 3.5) = 361,900$ lb

Tensile stress: $\sigma = \dfrac{361,900}{24 \times 1.58} = 9,540$ psi

For small diameters or low pressures, the initial gasket contact pressure, when computed as in the foregoing example, may be smaller than the y-value given in Table A. Additional bolting-up force should be applied until the contact pressure is equal to, or slightly greater than, the tabulated value. However, for large diameters or high pressures, care must be exercised to assure that the bolting-up pressure on the effective area does not exceed the true yield point value in compression for the gasket material. The gasket must retain sufficient elasticity to maintain a tight joint after the initial compression has been reduced by the fluid pressure. Published information is lacking at the present time on the maximum permissible contact pressure for the various types of gaskets. For solid metal gaskets, the yield point value in compression would be used.

Considerable experience is required to design high-pressure joints properly. The maximum internal pressure that a particular type of gasket can retain depends on a number of factors, such as flange design, type of facing, bolt location and arrangement, and relaxation of bolts at elevated temperature. Stresses and deformations at the joint arising from the thermal expansion of the connected parts must also be considered. For severe service conditions, the ring-type joint illustrated in Fig. 12-20 has proved very successful. It is easily made tight and the gasket can be used over and over again. The cross section is either oval or octagonal, and the temper should be dead soft. This type of gasket will not blow out if a leak develops in service.

FIGURE 12-20 Ring-type flange joint for high pressure service.

It is sometimes difficult, in high-pressure joints, to secure sufficient initial bolt force. Bolts with larger diameters will carry more force, but it is difficult to tighten larger bolts to the same stress that is easily secured in bolts of smaller diameter. A torque wrench must, of course, be used if a predetermined stress is to be secured in the bolt. Experience has shown that installation stresses due to torque loads for various sizes of alloy-steel studs made in accordance with the 8-pitch thread series in sizes larger than 1 in. are approximately equal to those given by the following empirical equation.

$$\sigma = \frac{45,000}{\sqrt{d}} \qquad (39)$$

where d is the nominal diameter of bolt stud.

13. Wire Rope

Wire rope is used in conveying and hoisting equipment, and also in such stationary applications as guy wires and stays.[11] A number of wires, such as 7, 19, or 37, are first twisted into a strand. A number of strands, usually 6 or 8, are then twisted about a core or center to form the rope. Cores may be of hemp or of wire. Several typical cross sections for wire rope are given in Fig. 12-21. The wires are made of various grades of steel, although wrought iron is also used. For long life the core, as well as the entire rope, must be continuously saturated with lubricant.

FIGURE 12-21 Cross sections of typical wire rope.

Flexibility in wire ropes is secured by using a large number of small-diameter wires. Ropes of few wires of relatively large size, such as the 6 × 7, are used for guy wires, but are too stiff for hoisting service unless the sheaves have very large diameters. The 6 × 19 and 6 × 37 types are the most widely used ropes for hoisting service. The radius of the groove in the sheave should be just large enough to provide clearance for the rope without pinching. The circular cross section of the rope is preserved and the stresses are more evenly distributed. The condition and proper alignment of the sheaves are also important factors affecting the useful life. When wire rope is wound on drums in two or more layers, abrasion and severe crushing stresses are

[11]See Drucker, D. C., and H. Tachau, "A New Design Criterion for Wire Rope," *Trans. ASME,* **67,** 1945, p. A-33; Discussion, **68,** 1946, p. A-75. See also Starkey, W. L., and H. A. Cress, "An Analysis of Critical Stresses and Mode of Failure of a Wire Rope," *Trans. ASME,* **81,** 1959, *J. Eng. Ind.,* p. 307.

induced, particularly where the rope must cross over the depressions made by the preceding layers.

Wire ropes are subjected to various kinds of stresses. There is the direct tension stress T/A, where T is the force in the rope and A is the cross-sectional area. Bending stresses also occur in the wires when the rope is passing over a sheave. From elementary mechanics, it is known that

$$\frac{M}{EI} = \frac{1}{r} = \frac{2}{d_s}, \qquad \sigma = \frac{Mc}{I} = \frac{Md_w}{2I} \tag{a}$$

where d_w is the diameter of the wire, d_s is the diameter of the sheave, I is the moment of inertia of a single wire, and E is the modulus of elasticity for the material. Elimination of M/I between these equations gives the following expression for the bending stress.

$$\sigma = \frac{d_w E}{d_s} \tag{40}$$

An increase in wire size is thus seen to give an increase in stress. A decrease in sheave size also gives an increase in bending stress, and care must be exercised that wire rope is not operated on excessively small sheaves.

The pressure p between rope and sheave can be determined by consideration of Fig. 12-22. This pressure is assumed to be uniformly distributed over

FIGURE 12-22   Forces in wire rope passing around sheave.

the curved surface of the groove which receives the rope and which runs circumferentially around the sheave. The value of the tensile force T in the rope is then equal to $\frac{1}{2}pd_r d_s$, where d_r is the diameter of the rope. Hence

$$p = \frac{2T}{d_r d_s} \tag{41}$$

Because of the small areas in actual contact between the wires, or between wires and groove, merely nominal values for p cause very high compressive stresses in the materials.

Failure of a wire rope occurs as a result of fatigue and wear in passing over the sheaves. Recent investigations have shown that the ratio of pressure p to the ultimate strength of the material is a significant variable for determining the fatigue life. Figure 12-23 shows the results of plotting p/σ_{ult} vs. bends to failure for various kinds of wire rope. A flexing and unflexing of the rope in passing over a sheave counts as a single bend. This figure indicates, for example, that if a 6 × 19 rope has a p/σ_{ult} value of 0.0014, a very long life can be expected so far as fatigue failure is concerned. The product $d_r d_s$ in the denominator of the variable indicates that a decrease in the diameter of the sheave permits an increase in the size of the rope; however, a practical lower limit exists for sheave size beyond which it is not safe to go because of high bending stresses and abrasion of the wires.

FIGURE 12-23 Fatigue life of wire rope as determined by experiments.

Dimensional data for ropes and information on the physical properties of the materials are given in Table 12-3. The minimum diameters for sheaves for iron rope are 50% greater than the values for steel rope. Values for the minimum factor of safety for various kinds of applications are shown in Table 12-4.

TABLE 12-3   Wire Rope Data

Metallic Cross-Sectional Area (d_r = diam rope)		Diameter of Outer Wire, d_w	Minimum Diameter d_s for Sheaves for Steel Rope	Modulus of Elasticity of Steel Rope	Breaking Stress		
Type	*in.*^{2}	*in.*	*in.*	*psi*	*Type of Material*		*psi*
6 × 7	$0.380d_r^2$	$d_r/9$	$42d_r$	14,000,000	Improved plow steel		200,000
6 × 19	$0.404d_r^2$	$d_r/16$*	$24d_r$	12,000,000	Plow steel		175,000
6 × 37	$0.404d_r^2$	$d_r/22$	$18d_r$	11,000,000	Extra-strong cast steel		160,000
8 × 19	$0.352d_r^2$	$d_r/19$	$20d_r$	10,000,000	Cast steel		140,000
					Iron		65,000

*Filler-wire type. For Warrington-type large outer wires, $d_w = \frac{1}{14}$.

TABLE 12-4   Minimum Factors of Safety

Type of Service	F_s	Type of Service			
Track cables	3.2	Elevators			
Guys	3.5				
Mine shafts		*Car Speed, fpm*	*Passenger*	*Freight*	*Dumbwaiters*
500-ft depth	8				
1,000–2,000-ft depth	6				
3,000 ft-depth and more	4	50	7.50	6.67	5.33
Miscellaneous hoisting equipment	5	150	8.20	7.32	5.98
Haulage ropes	6	300	9.17	8.20	6.88
Overhead and gantry cranes	6	500	10.25	9.14	8.00
Jib and pillar cranes	6	800	11.25	10.02	
Derricks	6	1,100	11.67	10.43	
Small electric and air hoists	7	1,500	11.87	10.61	
Hot-ladle cranes-slings	8				

Example 11. (a) What diameter of 6 × 19 wire rope of improved plow steel would be required for a 5-ton load supported on 8 lines? Compute for long life and continuous operation.

(b) What is the permissible static load for the rope of part (a)? Use a factor of safety of 6.

(c) If the rope is always loaded to the full static value, find the number of bends, using minimum size pulleys, which the rope can sustain before fatigue failure may be expected. On the basis of 100 bends per working day, what would be the expected life of the cable?

(d) What minimum size sheave would be required for the rope to be free from fatigue if operation takes place at the static load value?

Solution. (a) By Table 12-3: $\sigma_{ult} = 200,000$ psi, and $d_s = 24d$.

$$T = \frac{5 \times 2,000}{8} = 1,250 \text{ lb}$$

From Fig. 12-23: $\dfrac{2T}{\sigma_{ult}d_r d_s} = 0.0014$

Hence: $d_r d_s = 24d_r^2 = \dfrac{2T}{0.0014\sigma_{ult}}$

$$= \frac{2 \times 1,250}{0.0014 \times 200,000} = 8.93$$

Solving: $d_r = 0.61$ in. Use $\frac{5}{8}$-in.-diam rope

$d_s = 24 \times \frac{5}{8} = 15$ in. recommended

minimum diameter of sheave

(b) By Table 12-3: $A = 0.404d_r^2 = 0.404 \times 0.625^2$

$= 0.158$ in.^{2} metallic cross-sectional area

Max, static load $= \dfrac{0.158 \times 200,000}{6} = 5,200$ lb

The maximum permissible static load is over 4 times as great as the working load when fatigue is considered.

(c) $\dfrac{2T}{\sigma_{ult}d_r d_s} = \dfrac{2 \times 5,200}{200,000 \times 0.625 \times 15} = 0.00555$

From Fig. 12-23: No. of bends to failure = 65,000

Expected life $= \dfrac{65,000}{100} = 650$ working days

(d) $d_s = \dfrac{2T}{\sigma_{ult}d_r \times 0.0014} = \dfrac{2 \times 5,200}{200,000 \times 0.625 \times 0.0014} = 59.5$ in.

This result is the diameter of the sheave for operation at the permissible static load and for freedom from fatigue. It is obviously an impractical result. Therefore, a reduced load must be used, as was done in part (a), if sheaves of reasonable size are to be used, and if freedom from fatigue is to be achieved as well.

Catalogs of wire rope manufacturers contain a large amount of valuable information on the selection and maintenance of rope. Their knowledge of field conditions, collected over many years of experience, should be utilized by the purchaser.

14. Curved Beams

Curved beams in the form of hooks and brackets are frequently used as machine elements.[12] When such bodies are subjected to bending moments, the stress distribution is not linear on either side of the neutral axis, but

[12]See p. 65 of Reference 7, end of chapter. See also Wahl, A. M., "Calculation of Stress in Crane Hooks," *Trans. ASME*, **68**, 1946, p. A-239.

increases more rapidly on the inner side. For curved beams, the assumption can usually be made with sufficient accuracy that cross sections normal to the curved elements of the unloaded beam remain plane and perpendicular to the elements after the bending moments are acting. The deformations thus vary directly with the distance from the neutral axis.

Consider the elements of the curved beam of Fig. 12-24 lying between two axial planes AB and DE separated by the angle φ. Let the total angular deformation of plane AB with respect to DE be represented by $d\varphi$. A shorter element on the inner side thus deforms the same amount as a longer element, symmetrically located with respect to the neutral surface, deforms on the outer side. The inner element, however, has the higher stress, since its deformation has taken place over a shorter length element.

FIGURE 12-24   Curved beam in pure bending.

The cross sections of the beam of Fig. 12-24 are assumed to be symmetrical with respect to the v-axis, as is indicated by view A-A. Because of the higher stresses on the inner elements, the neutral surface for a curved beam no longer passes through the center of gravity of the cross section, but shifts inward a small amount, designated e in the figure. The total deformation for an element located a distance v below the neutral surface is $v\,d\varphi$. This deformation takes place in an element whose length is $(r - v)\varphi$. The elongation or unit deformation ϵ then is

$$\epsilon = \frac{v\,d\varphi}{(r - v)\varphi} \tag{a}$$

Multiplication by the modulus of elasticity E gives the stress at this point. Thus

$$\sigma = \frac{Ev\,d\varphi}{(r - v)\varphi} = \frac{E\,d\varphi}{\varphi}\left[\frac{r}{r - v} - 1\right] \tag{b}$$

This equation indicates that the distribution of stress over the cross section is hyperbolic.

From the condition of static equilibrium, the sum of the stresses over a cross section must add to zero, and the moment made by the stresses must be equal to the applied moment M. Hence

$$\int \sigma \, dA = \frac{E \, d\varphi}{\varphi} \int \frac{v \, dA}{r - v} = 0 \qquad \text{(c)}$$

$$\int \sigma v \, dA = \frac{E \, d\varphi}{\varphi} \int \frac{v^2 \, dA}{r - v} = M \qquad \text{(d)}$$

The integrations in the equations above are to extend over the entire cross section. By division of numerator by denominator, Eq. (d) becomes

$$M = -\frac{E \, d\varphi}{\varphi} \int v \, dA + \frac{Er \, d\varphi}{\varphi} \int \frac{v \, dA}{r - v} \qquad \text{(e)}$$

By Eq. (c), the second integral of Eq. (e) is equal to zero. The first integral of Eq. (e) represents the moment of area of the cross section about the neutral axis, that is, $-eA$. The minus sign is required because the center of gravity lies on the negative side of the neutral axis. Therefore

$$M = \frac{E \, d\varphi}{\varphi} eA \qquad \text{(f)}$$

Substitution in Eq. (b) yields: $\quad \sigma = \frac{M}{eA} \frac{v}{r - v} \qquad \text{(g)}$

Substitution of the limits for v gives the following equations.

$$\sigma_{max} = \frac{M h_1}{A e a} \qquad \text{when} \qquad v = h_1 \qquad (42)$$

$$\sigma_{min} = -\frac{M h_2}{A e c} \qquad \text{when} \qquad v = -h_2 \qquad (43)$$

Since these equations are sensitive to small variations in the value of e, considerable care must be exercised in determining this quantity accurately.

Additional useful equations can be obtained by letting $v = v_1 - e$ in Eq. (c).

$$\int \frac{v}{r - v} \, dA = \int \frac{v_1 - e}{r - v_1 + e} \, dA = \int \frac{v_1 - e}{R - v_1} \, dA = 0 \qquad \text{(h)}$$

The integral on the right may be written

$$\int \frac{v_1 \, dA}{R - v_1} = e \int \frac{dA}{R - v_1} \qquad \text{(i)}$$

$$= \frac{e}{R} \int \frac{R - v_1 + v_1}{R - v_1} \, dA \qquad \text{(j)}$$

$$= \frac{e}{R} \left[\int dA + \int \frac{v_1 \, dA}{R - v_1} \right] \qquad \text{(k)}$$

Let
$$\int \frac{v_1 \, dA}{R - v_1} = m_1 A \tag{44}$$

where A is the area of the cross section, and m_1 is another constant. Substitution in Eq. (k) will give the following equation for e.

$$e = \frac{m_1 R}{1 + m_1} \tag{45}$$

Substitution in Eq. (i) gives:
$$\int \frac{dA}{R - v_1} = \frac{A}{R}(1 + m_1) \tag{46}$$

The integral of Eq. (46) can be evaluated for simple types of cross sections and the value of m_1 can then be found. Eccentricity e can then be found by use of Eq. (45), and the stress by use of Eqs. (42) and (43).

15. Curved Beam of Rectangular Cross Section

FIGURE 12-25   Curved beam of rectangular cross section.

The foregoing general equations for the stresses in a curved beam are now adapted to the rectangular cross section, Fig. 12-25. Equation (46) can be written

$$A(1 + m_1) = \int_{-h/2}^{+h/2} \frac{dA}{1 - (v_1/R)} \tag{a}$$

When the numerator is divided by the denominator, the following result is obtained.

$$A(1 + m_1) = \int_{-h/2}^{+h/2} \left[1 + \frac{v_1}{R} + \frac{v_1^2}{R^2} + \frac{v_1^3}{R^3} + \frac{v_1^4}{R^4} + \cdots \right] b \, dv_1$$

This expression should now be integrated term by term and the limits substituted.

$$A(1 + m_1) = b \left[h + \frac{h^3}{12R^2} + \frac{h^5}{80R^4} \right] \tag{b}$$

or
$$m_1 = \frac{h^2}{12R^2} + \frac{h^4}{80R^4} \tag{c}$$

Let $c_1 = 2R/h$, where c_1 is called the index of curvature. When this substitution is made, Eq. (c) becomes

$$m_1 = \frac{1}{3c_1^2} + \frac{1}{5c_1^4} = \frac{1}{3c_1^2}\left[1 + \frac{0.6}{c_1^2} \right]$$

$$= \frac{1}{3c_1^2}\left[\frac{c_1^2 + 0.6}{c_1^2} \right] \approx \frac{1}{3c_1^2}\left[\frac{c_1^2}{c_1^2 - 0.6} \right] = \frac{1}{3c_1^2 - 1.8} \tag{d}$$

The eccentricity e is found by substitution into Eq. (45).

$$e = \frac{hc_1}{2(3c_1^2 - 0.8)} \tag{47}$$

Substitution for e, h_1 and a in Eq. (42) gives the maximum stress.

$$\sigma_{max} = \frac{6M}{bh^2} \times \frac{3c_1^2 - c_1 - 0.8}{3c_1(c_1 - 1)} = K_c \frac{6M}{bh^2} \tag{48}$$

where
$$K_c = \frac{3c_1^2 - c_1 - 0.8}{3c_1(c_1 - 1)} \tag{49}$$

Equation (49) gives the value of the stress concentration factor for curvature for a rectangular cross section.

Example 12. Let the width b of the rectangle be 4 in. and the depth of section h be 6 in. The radius of curvature R to the center of gravity is also equal to 6 in. Find the value of the eccentricity e, the stress concentration factor K_c, and the maximum stress.

Solution.
$$c_1 = \frac{2R}{h} = 2$$

In Eq. (47):
$$e = \frac{6 \times 2}{2(3 \times 2^2 - 0.8)} = 0.536 \text{ in.}$$

In Eq. (49):
$$K_c = \frac{3 \times 2^2 - 2 - 0.8}{3 \times 2(2 - 1)} = 1.533$$

In Eq. (48):
$$\sigma_{max} = 1.533 \frac{6M}{4 \times 6^2} = 0.0639M$$

Example 12A. Let the width b of the rectangle be 100 mm and the depth of section h be 150 mm. The value of radius R to the center of gravity is also equal to 150 mm. Find the value of the eccentricity e, the stress concentration factor K_c, and the maximum stress.

Solution.
$$c_1 = \frac{2R}{h} = \frac{300}{150} = 2$$

In Eq. (47):
$$e = \frac{hc_1}{2(3c_1^2 - 0.8)} = \frac{150 \times 2}{2(3 \times 2^2 - 0.8)} = 13.39 \text{ mm}$$

In Eq. (49):
$$K_c = \frac{3c_1^2 - c_1 - 0.8}{3c_1(c_1 - 1)} = \frac{12 - 2 - 0.8}{6 \times 1} = 1.533$$

In Eq. (48):
$$\sigma_{max} = K_c \frac{6M}{bh^2} = 1.533 \frac{6M}{100 \times 150^2} = 0.000\ 409M$$

Sometimes the cross section of a curved beam is so shaped that no analytic expression can be derived for the eccentricity e. The integration of Eq. (44) can be performed numerically if the equation is written

$$m_1 A = \sum \frac{v_1 \Delta A}{R - v_1} \tag{50}$$

Accurate results can be obtained by this method if the cross section is divided into a sufficient number of elemental strips ΔA.

Example 13. Find the value of the maximum stress for the beam of Example 12 by use of Eq. (50) and compare with the results obtained by use of Eq. (48).

FIGURE 12-26 Cross section of Example 13 divided into elementary strips.

Solution. Divide the cross section into 12 strips, as shown by Fig. 12-26. The computations of Eq. (50) have been carried out in Table 12-5. Substitution of the summation of the right-hand column in Eq. (50) gives

$$m_1 A = 2.3422$$

Solving: $m_1 = \dfrac{2.3422}{24} = 0.0976$

In Eq. (45): $e = \dfrac{0.0976 \times 6}{1 + 0.0976} = 0.534$ in.

Hence: $h_1 = 2.466$ in.

$h_2 = 3.534$ in.

In Eq. (42): $\sigma_{max} = \dfrac{2.466M}{24 \times 0.534 \times 3} = 0.0642M$

TABLE 12-5

No.	v_1	$R - v_1$	$\dfrac{v_1 \, \Delta A}{R - v_1}$
1	2.75	3.25	1.6923
2	2.25	3.75	1.2000
3	1.75	4.25	0.8235
4	1.25	4.75	0.5263
5	0.75	5.25	0.2857
6	0.25	5.75	0.0870
7	−0.25	6.25	−0.0800
8	−0.75	6.75	−0.2222
9	−1.25	7.25	−0.3448
10	−1.75	7.75	−0.4516
11	−2.25	8.25	−0.5454
12	−2.75	8.75	−0.6286

$$\sum \frac{v_1 \, \Delta A}{R - v_1} = 2.3422$$

Numerical integration thus gives accurate results, and the method can therefore be used for odd-shaped cross sections.

16. Curved Beam of Circular Cross Section

By a similar process, the equation for the maximum stress in a curved beam of round cross section can be shown to be

$$\sigma_{max} = K_c \frac{32M}{\pi d^3} \tag{51}$$

where
$$K_c = \frac{4c_1^2 - c_1 - 1}{4c_1(c_1 - 1)} \tag{52}$$

The index of curvature c_1 is equal to $2R/d$.

17. Angular Deflection of Curved Bar

Unless the index of curvature, $2R/d$ or $2R/h$, is very small, the angular change between two cross sections of a curved beam caused by the moments M is approximately the same as would take place if the beam were straight. For a straight beam, the relationship between bending moment and radius of curvature is given by the equation $1/r = M/EI$. Since the moment M is the same at all points, the radius r is also constant, and the elastic line is a circle, as shown by Fig. 12-27. Since $r\theta = 1$, substitution for r gives the value of θ as

$$\theta = \frac{MI}{EI} \tag{53}$$

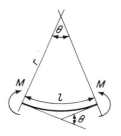

FIGURE 12-27 Angular deformation caused by moments.

18. Cams

A cam provides a convenient means for transforming rotary motion into reciprocating motion. Since the cam outline can be given a wide variety of shapes, many different types of motion can be secured. The plate cam, illustrated in Fig. 12-28, moves the follower in a direction at right angles to the cam axis. For the cylindrical cam of Fig. 12-29, the follower moves in a direction parallel to the cam axis. The cam must have the proper shape if the desired type of motion is to be imparted to the follower.

For many commonly used cams, the rise of the follower is parabolic, harmonic, or cycloidal with respect to the rotation of the cam. Equations for the displacement, velocity, and acceleration of the follower of such cams are given in Table 12-6.

Lift h of the cam takes place during a rotation β. The rise y for a rotation θ can be found by substitution in the equations of the table.

Curves showing the rise y for the three types of cams, plotted for the nondimensional coordinate θ/β, are shown in Fig. 12-30. The velocity dy/dt is shown in Fig. 12-31, and the acceleration d^2y/dt^2 is shown in Fig. 12-32.

FIGURE 12-28 Plate cam with mushroom follower.

FIGURE 12-29 Cylindrical cam.

TABLE 12-6 Equations for Displacement, Velocity, and Acceleration for the Pitch Curve of Various Types of Cams

Type of Cam	Displacement	Velocity	Acceleration
Parabolic or constant acceleration	For $\dfrac{\theta}{\beta} \leq 0.5,\ y = 2h\dfrac{\theta^2}{\beta^2}$ For $\dfrac{\theta}{\beta} \geq 0.5,\ y = h\left[1 - 2\left(1 - \dfrac{\theta}{\beta}\right)^2\right]$	$\dfrac{dy}{dt} = \dfrac{4h\omega\theta}{\beta^2}$ $\dfrac{dy}{dt} = \dfrac{4h\omega}{\beta}\left(1 - \dfrac{\theta}{\beta}\right)$	$\dfrac{d^2y}{dt^2} = \dfrac{4h\omega^2}{\beta^2}$ $\dfrac{d^2y}{dt^2} = -\dfrac{4h\omega^2}{\beta^2}$
Harmonic	$y = \dfrac{h}{2}\left(1 - \cos\dfrac{\pi\theta}{\beta}\right)$	$\dfrac{dy}{dt} = \dfrac{\pi h\omega}{2\beta}\sin\dfrac{\pi\theta}{\beta}$	$\dfrac{d^2y}{dt^2} = \dfrac{\pi^2 h\omega^2}{2\beta^2}\cos\dfrac{\pi\theta}{\beta}$
Cycloidal	$y = h\left(\dfrac{\theta}{\beta} - \dfrac{1}{2\pi}\sin\dfrac{2\pi\theta}{\beta}\right)$	$\dfrac{dy}{dt} = \dfrac{h\omega}{\beta}\left(1 - \cos\dfrac{2\pi\theta}{\beta}\right)$	$\dfrac{d^2y}{dt^2} = \dfrac{2\pi h\omega^2}{\beta^2}\sin\dfrac{2\pi\theta}{\beta}$

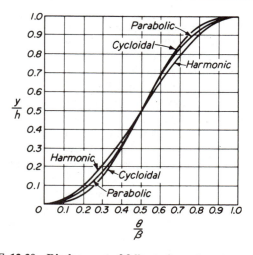

FIGURE 12-30 Displacement of follower for various types of cams.

FIGURE 12-31 Velocity of follower for various types of cams.

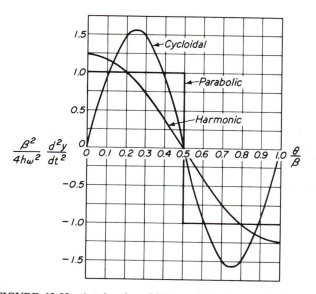

FIGURE 12-32 Acceleration of follower for various types of cams.

The inertia forces in a cam-driven system are proportional to the acceleration of the follower. The parabolic cam has the smallest maximum acceleration, but has the undesirable property of having sudden changes in value at the start, the midheight, and the end of the rise. The harmonic cam has a gradual change in the value of the acceleration except at the start and end of cam action. The cycloidal cam has a higher peak value for the acceleration than the others, but its value does not change as abruptly.

19. Plate Cam with Central Roller Follower

A plate cam with a central roller follower is shown in Fig. 12-33. The center of the follower roller has the path of the pitch curve, and the actual cam is everywhere tangent to the roller.

FIGURE 12-33   **Radial cam with translating roller follower.**

The force exerted by the cam is directed normal to the cam surface, which in general, for a roller follower, is not in the direction of the follower motion. The angle between these directions is known as the pressure angle α as illustrated in Fig. 12-33(b). The sidewise force exerted by the follower on its guides depends on the magnitude of the pressure angle. It is sometimes specified that the pressure angle should not exceed 30°. The permissible value is influenced by the speed of operation and the weight of the connected parts.

Reference to Fig. 12-33(b) indicates that the pressure angle α can be obtained by the following equation.

$$\tan \alpha = \frac{1}{r} \frac{dy}{d\theta} \qquad (54)$$

where r is the distance between the centers for cam and roller.

Example 14. A cycloidal cam with a central roller follower has a rise h of 1 in. in an angle β of 70°. Base circle radius R_b is 4 in. and the follower roller radius R_r is 0.8 in.

Calculate the values of the pressure angle α at intervals of θ/β of 0.1. Find the maximum value of the acceleration of the follower. Speed of rotation is 600 rpm.

Solution. By Table 12-6, the equation for the rise y of the cycloidal cam is

$$y = h\left(\frac{\theta}{\beta} - \frac{1}{2\pi}\sin\frac{2\pi\theta}{\beta}\right)$$

To find the pressure angle α, the derivative $dy/d\theta$ is required for use in Eq. (54). Thus

$$\frac{dy}{d\theta} = \frac{h}{\beta}\left(1 - \cos\frac{2\pi\theta}{\beta}\right)$$

$$\beta = 70° = 1.22173 \text{ rad}$$

$$\frac{h}{\beta} = 0.81851$$

The calculations can be conveniently carried out in tabular form as shown in Table 12-7.

$$\omega = \frac{2\pi n}{60} = \frac{2\pi 600}{60} = 20\pi \text{ rad/sec}$$

TABLE 12-7   Calculations for Pressure Angle α for Cycloidal Cam of Example 14

$\dfrac{\theta}{\beta}$	$\dfrac{2\pi\theta}{\beta}$	$\sin\dfrac{2\pi\theta}{\beta}$	$\cos\dfrac{2\pi\theta}{\beta}$	$\dfrac{\sin(2\pi\theta/\beta)}{2\pi}$	y	$r = R_b + y$	$\dfrac{dy}{d\theta}$	$\tan\alpha = \dfrac{1}{r}\dfrac{dy}{d\theta}$	α
0	0	0	1	0	0	4	0	0	0
0.1	36°	0.58779	0.80902	0.09355	0.00645	4.00645	0.15632	0.03902	2°14.1'
0.2	72°	0.95106	0.30902	0.15137	0.04863	4.04863	0.56558	0.13970	7°57.2'
0.3	108°	0.95106	-0.30902	0.15137	0.14863	4.14863	1.07145	0.25827	14°28.9'
0.4	144°	0.58779	-0.80902	0.09355	0.30645	4.30645	1.48070	0.34383	18°58.5'
0.5	180°	0	-1	0	0.5	4.5	1.63702	0.36378	19°59.4'
0.6	216°	-0.58779	-0.80902	-0.09355	0.69355	4.69355	1.48070	0.31548	17°30.6'
0.7	252°	-0.95106	-0.30902	-0.15137	0.85137	4.85137	1.07145	0.22086	12°27.2'
0.8	288°	-0.95106	0.30902	-0.15137	0.95137	4.95137	0.56558	0.11423	6°31.0'
0.9	324°	-0.58779	0.80902	-0.09355	0.99355	4.99355	0.15632	0.03130	1°47.6'
1.0	360°	0	1	0	1	5	0	0	0

By Fig. 12-32, the maximum value of the follower acceleration will occur for θ/β equal to 0.25. Then by Table 12-6,

$$\left(\frac{d^2 y}{dt^2}\right)_{max} = \frac{2\pi h\omega^2}{\beta^2} = \frac{2\pi \times 1 \times 400\pi^2}{1.22173^2} = 16{,}620 \text{ in./sec}^2$$

To obtain a smooth surface, the calculations for an actual cam would be made for much finer intervals—perhaps for a degree or less apart.

20. *Manufacture of Cam with Central Roller Follower*

One method for making a plate cam with a central roller follower is to mill or grind the surface on a milling machine. The radius R_g of the cutter or grinding wheel, in general, will not be the same as the radius R_r of the follower roller.

The situation will be as in Fig. 12-34 where the system of Fig. 12-33 is rotated to make the cam center O, and the cutter center B on the same vertical line. To find the distance r_g between the axes of cam and cutter, let the cosine theorem be applied to triangle OAB.

$$r_g^2 = r^2 + (R_g - R_r)^2 + 2r(R_g - R_r) \cos \alpha \qquad (55)$$

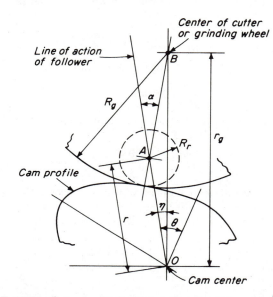

FIGURE 12-34   Geometry for locating center of cutter or grinding wheel for cam of Fig. 12-27.

When the cam is inclined at angle $\theta - \eta$ from the initial point, then the cutter should be brought into the cam until the distance between the axes at O and B is equal to r_g.

Angle η can be found by application of the sine theorem to triangle OAB. Thus

$$\sin \eta = \frac{R_g - R_r}{r_g} \sin \alpha \qquad (56)$$

Example 15. Calculate the values of r_g and η for the plate cam of Example 14 for the same θ/β values. Radius R_g of the cutter is equal to 2.5 in.

Solution. $R_g - R_r = 2.5 - 0.8 = 1.7$ in.

The calculations can be conveniently carried out in tabular form and are shown in Table 12-8.

TABLE 12-8   Calculations for Location of Cutter Center for
Cycloidal Cam of Example 15

$\dfrac{\theta}{\beta}$	$\sin \alpha$	$\cos \alpha$	r^2	$2r(R_g - R_r) \times \cos \alpha$	r_g^2	r_g	$\sin \eta$	η
0	0	1	16	13.60000	32.49000	5.70000	0	0
0.1	0.03899	0.99924	16.05164	13.61158	32.55322	5.70554	0.01162	0°40.0′
0.2	0.13836	0.99038	16.39140	13.63292	32.91432	5.73710	0.04100	2°21.0′
0.3	0.25006	0.96823	17.21113	13.65722	33.75835	5.81019	0.07316	4°11.8′
0.4	0.32515	0.94566	18.54551	13.84629	35.28180	5.93985	0.09306	5°20.4′
0.5	0.34186	0.93975	20.25	14.37818	37.51818	6.12521	0.09488	5°26.6′
0.6	0.30086	0.95367	22.02941	15.21873	40.13814	6.33547	0.08073	4°37.8′
0.7	0.21566	0.97647	23.53579	16.10654	42.53233	6.52168	0.05622	3°13.4′
0.8	0.11350	0.99354	24.51606	16.72591	44.13197	6.64319	0.02904	1°39.8′
0.9	0.03129	0.99951	24.93554	16.96975	44.79529	6.69293	0.00795	0°27.3′
1.0	0	1	25	17.00000	44.89000	6.70000	0	0

21. *Plate Cam with Flat-Faced Follower*

A plate cam with a flat-faced follower is shown in Fig. 12-35. The follower
is located at B at the rise y given by the cam equation for the pitch curve.
The actual cam outline contacts the follower at point C located a distance
q to the left of the center line. Point C is at radius r_c from center of rotation

FIGURE 12-35   Plate cam with flat-faced follower.

A, and at angle η from the center line. The upward velocity of the follower is represented by dy/dt. Then, by similar triangles,

$$\frac{dy/dt}{r_c\omega} = \frac{q}{r_c}$$

Or:

$$q = \frac{1}{\omega}\frac{dy}{dt} = \frac{dy}{d\theta} \tag{57}$$

Also:

$$r_c = (r^2 + q^2)^{1/2} \tag{58}$$

And:

$$\tan\eta = \frac{q}{r} \tag{59}$$

Example 16. A cycloidal cam with a flat-faced follower has a rise h of 1 in. in an angle β of 70°. Radius R_b of the base circle is equal to 4 in. Calculate the pitch curve and coordinates r_c and η for the cam for values of θ/β of 0, 0.2, 0.4, 0.6, 0.8, and 1.0.

Solution. By Table 12-6:

$$y = h\left(\frac{\theta}{\beta} - \frac{1}{2\pi}\sin\frac{2\pi\theta}{\beta}\right)$$

$$q = \frac{dy}{d\theta} = \frac{h}{\beta}\left(1 - \cos\frac{2\pi\theta}{\beta}\right) = \frac{1}{1.22173}\left(1 - \cos\frac{2\pi\theta}{\beta}\right)$$

The calculations are carried out in Table 12-9. The reader should plot the data for both the pitch curve and the cam outline.

TABLE 12-9   Calculations for Pitch Curve and Cam Outline of Cycloidal Cam with Flat-Faced Follower of Example 16

$\dfrac{\theta}{\beta}$	$2\pi\dfrac{\theta}{\beta}$	$\sin\dfrac{2\pi\theta}{\beta}$	$\cos\dfrac{2\pi\theta}{\beta}$	y	r^2	q	r_c	$\tan\eta = \dfrac{q}{r}$	η
0	0	0	1	0	16	0	4	0	0
0.2	72°	0.95106	0.30902	0.04863	16.39140	0.56558	4.0879	0.13970	7° 57.2′
0.4	144°	0.58779	−0.80902	0.30645	18.54551	1.48070	4.5539	0.34383	18° 58.5′
0.6	216°	−0.58779	−0.80902	0.69355	22.02941	1.48070	4.9216	0.31548	17° 30.5′
0.8	288°	−0.95106	0.30902	0.95137	24.51606	0.56558	4.9836	0.11423	6° 31.0′
1.0	360°	0	1	1	25	0	5	0	0

22. Polydyne Cam

Many other types of curves are used for cams. The rise y, for example, can be expressed as a polynomial[13] in various powers of x or θ/β. The general equation for the so-called 3-4-5 cam would then be written

[13]See Dudley, W. M., "New Methods in Cam Design," *SAE Quart. Trans.*, **2**, 1948, p. 19. See also Stoddard, D. A., "Polydyne Cam Design," *Machine Design*, **25**, Jan. 1953, p. 121; Feb., p. 146; and Mar., p. 149.

$$y = h(C_0 + C_1 x + C_2 x^2 + C_3 x^3 + C_4 x^4 + C_5 x^5) \qquad (60)$$

The various contants are evaluated from the desired boundary conditions for the follower. For example, suppose it is desired that displacement y, velocity dy/dx, and acceleration d^2y/dx^2 be zero for $x = 0$. Also let it be desired that rise y be equal to h, and dy/dx and d^2y/dx^2 be zero when $x = 1$. These initial conditions should be substituted into the above equation, and the equations obtained by differentiation, and the resulting equations solved for the C's. The result is

$$y = h(10x^3 - 15x^4 + 6x^5) \qquad (61)$$

The equation can be differentiated to obtain expressions for velocity dy/dt and acceleration d^2y/dt^2. When the results are plotted, the resulting curves are very close to those of the cycloidal cam.

23. Remarks on Cam Design

Cam design is a broad and involved subject, and only a few of the more obvious points have been touched upon in this chapter.[14] The object of the cam is to have the follower deliver a prescribed motion in a smooth and vibrationless manner. Vibration may be instigated by a sudden break in the acceleration curve. It is beneficial if the acceleration curve be continuous and have the smallest maximum values. The rise should be kept as small as possible and achieved in the longest possible time. Acceleration varies as the square of the velocity.

The accuracy with which the cam is cut will influence the vibration effects. Play or backlash should be kept to a minimum. Vibration, together with noise and wear, will occur in the dynamical system if it is operated at the speed of its natural frequency. The forces and stresses from vibration are usually superposed on those resulting from normal operation. The cycloidal form is perhaps the best cam so far as vibration is concerned.

The operation is affected by the flexibility or elastic deformation of the parts of the system. The parts of the system act as springs of various stiffnesses. The moving parts should be both as rigid and as light as possible. Faulty operation can occur from the difference between the movement at the end of the mechanical chain and the initial movement imposed by the cam. The polydyne cam permits the design of the cam shape to be such that the desired motion will occur at the end of the follower chain.

[14]For a complete and detailed treatment of the subject, see Rothbart, H. A., *Cams*, New York: John Wiley & Sons, Inc., 1956. This book contains extensive bibliographies. See also Mischke, C. R., "Optimal Offset on Translating Follower Plate Cams," *Trans. ASME*, **92**, 1970, *J. Eng. Ind.*, p. 172; and Chapter 12 of Reference 6, end of chapter.

An increase in the pressure angle usually means an increase in the forces which must be provided for. The guide for a translating follower should have adequate bearing length. Offsetting the follower may reduce the pressure angle during that portion of the cycle where the forces are the greatest. The mathematical calculations, however, are more involved. A pivoted oscillating follower will reduce the side thrust and permit a smaller cam to be used.

The spring of a compression system must be strong enough to keep the follower in contact with the cam at all times. This is important in flexible high-speed systems. The follower must not be allowed to jump or leave the cam surface. Sometimes a ramp or small precam is located at the start of the main part of the cam to remove the play and elastic deformation from the system so that the end of the chain will start to move as soon as the main part of the cam begins to act.

For light loads and slow speeds, a cam can be composed of sections of circular arcs. Such cams are relatively easy to manufacture and check dimensionally. The acceleration curve, however, has abrupt changes in value.

24. Snap Rings

Snap rings made of hardened steel have many useful applications in holding cylindrical parts in position. The external ring of Fig. 12-36(a) and the internal ring of sketch (b) have ears for expanding or contracting the ring during assembly. Special pliers are required for this operation. Sometimes the ring is merely made of hardened steel wire of rectangular or circular cross section. Sketch (c) shows how a ring can be used for holding a hub in the desired location along a shaft. In (d) the ring is used for retaining the closure at the end of a tube.

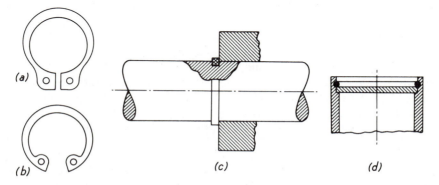

(a)

(b)   (c)   (d)

FIGURE 12-36 Snap rings and applications.

25. Flat Plates

Flat plates frequently occur in machines and structures, and the designer should be able to calculate the stresses and deflections produced by the loading. Unfortunately, the derivation of the necessary equations is very complex and beyond the scope of this book. Some results, for simple cases, will be set down here for ready reference.

(a) *Circular Plate, Uniform Load, Simply Supported Edge*

Deflection at center:
$$w = \frac{(5 + \mu)qa^4}{64(1 + \mu)D} \tag{62}$$

Moment at center:
$$M = \frac{(3 + \mu)qa^2}{16} \tag{63}$$

Maximum stress (at center):
$$\sigma = \frac{3(3 + \mu)qa^2}{8h^2} \tag{64}$$

where a = radius of circular plate

$$D = \frac{Eh^3}{12(1 - \mu^2)}, \text{ flexural rigidity of plate} \tag{65}$$

E = modulus of elasticity, psi
h = thickness of plate, in.
M = bending moment, in. lb/in.
q = uniform load, psi
μ = (mu) Poisson's ratio (usually 0.3 for metals)

(b) *Circular Plate, Uniform Load, Clamped Edge*

Deflection at center:
$$w = \frac{qa^4}{64D} \tag{66}$$

Moment at center:
$$M = \frac{(1 + \mu)qa^2}{16} \tag{67}$$

Maximum stress (at edge):
$$\sigma = \frac{3qa^2}{4h^2} \tag{68}$$

(c) *Circular Plate, Concentrated Load at Center, Simply Supported Edge*

Deflection at center:
$$w = \frac{(3 + \mu)Pa^2}{16\pi(1 + \mu)D} \tag{69}$$

Stress at center, on lower surface:

$$\sigma = \frac{P}{h^2}\left[(1 + \mu)\left(0.485 \ln \frac{a}{h} + 0.52\right) + 0.48\right] \tag{70}$$

where P = concentrated load
$\ln$ = logarithm to base e

(d) *Circular Plate, Concentrated Load at Center, Clamped Edge*

Deflection at center:
$$w = \frac{Pa^2}{16\pi D} \qquad (71)$$

Stress at center, on lower surface:

$$\sigma = \frac{P}{h^2}(1 + \mu)\left(0.485 \ln \frac{a}{h} + 0.52\right) \qquad (72)$$

Equations are available for numerous other types of loadings and edge conditions.[15]

(e) *Rectangular Plate, Uniform Load, Simply Supported Edges.* The corners of a uniformly loaded rectangular plate tend to rise if the plate is merely supported at the edges. For a simply supported edge, vertical movement must be prevented, but the edge must be free to rotate. An edge constructed as in Fig. 12-37 fulfills these conditions. The coordinate system for this plate is shown by Fig. 12-38. A good approximation for the stress and deflection can be had by the following empirical equations.

FIGURE 12-37   Simple support for plate.

FIGURE 12-38   Simply supported rectangular plate.

Maximum stress:
$$\sigma_x = \frac{0.75qa^2}{h^2[1.61(a/b)^3 + 1]} \qquad \text{at} \qquad x = \frac{a}{2}, y = 0 \quad (73)$$

Deflection at center:
$$w = \frac{0.142qa^4}{Eh^3[2.21(a/b)^3 + 1]} \qquad (74)$$

[15]See References 3, 6, 7, and 8, end of chapter. See also, Wahl, A. M., and G. Lobo, "Stresses and Deflections in Flat Circular Plates with Central Holes," *Trans. ASME,* **52**(1), 1930, *J. Appl. Mech.,* p. 29; Georgian, J. C., "Uniformly Loaded Circular Plates with a Central Hole and Both Edges Supported," *Trans. ASME,* **79**, 1957, *J. Appl. Mech.,* p. 306; and Chapter 16 of Reference 6, end of chapter.

(f) *Rectangular Plate, Uniform Load, Clamped Edges.* The coordinate system for this plate is shown by Fig. 12-39.
Maximum stress:

$$\sigma_x = \frac{qa^2}{2h^2[0.623(a/b)^6 + 1]} \qquad \text{at} \qquad x = \frac{a}{2}, y = 0 \qquad (75)$$

Deflection at center:

$$w = \frac{0.0284qa^4}{Eh^3[1.056(a/b)^5 + 1]} \qquad (76)$$

These two equations are also empirical, but give reasonably good results.

(g) *Rectangular Plate, Concentrated Load at Center, Simply Supported Edges.* Instead of a point load, the load P is assumed to be uniformly distributed over a small circle of radius e, as shown in Fig. 12-40.

FIGURE 12-39 Rectangular plate, clamped edges.

FIGURE 12-40 Rectangular plate, simply-supported edges, concentrated load at center.

Deflection at center:

$$w = \alpha \frac{Pa^2}{Eh^2} \qquad (77)$$

Maximum stress:

$$\sigma_x = \frac{1.5P}{\pi h^2}\left[(1 + \mu) \ln \frac{2a}{\pi e} + 1 - \gamma_1\right] \qquad (78)$$

Constants α and γ_1 are given below for different values of b/a.

b/a	1.0	1.1	1.2	1.4	1.6	1.8	2.0	3.0	∞
α	0.1265	0.1381	0.1478	0.1621	0.1714	0.1769	0.1803	0.1846	0.1849
γ_1	0.564		0.349	0.211	0.124	0.072	0.041		0

(h) *Rectangular Plate, Concentrated Load at Center, Clamped Edges.* The coordinate system for this plate is given by Fig. 12-39.

Deflection at center: $w = \alpha \dfrac{Pa^2}{Eh^3}$ $\qquad\qquad\qquad\qquad$ (79)

Maximum stress: $\sigma_x = \beta \dfrac{P}{h^2}$ $\qquad$ at $\qquad$ $x = \dfrac{a}{2}, y = 0$ $\qquad$ (80)

Constants α and β are given below for different values of b/a.

b/a	1.0	1.2	1.4	1.6	1.8	2.0
α	0.0611	0.0706	0.0755	0.0777	0.0786	0.0788
β	0.754	0.894	0.962	0.991	1.000	1.004

In general, the foregoing plate equations are valid only for small deflections—deflections which are only a fraction of the plate thickness. In cases where the deflection is not small compared to the thickness, but is small compared to the other dimensions, the strain of the middle surface of the plate must be considered and the theory becomes more involved. For still larger deflections, the bending effects are of less significance. The plate becomes a membrane and the load is carried largely by the tension stresses resulting from the stretching of the plate.

REFERENCES

1. FAUPEL, J. H., and F. E. FISHER, *Engineering Design*, 2d ed., New York: John Wiley & Sons, Inc., 1980.

2. MARSHEK, K. M., "On the Analyses of Sprocket Load Distribution," *Mech. Machine Theory*, **14**, 1979, pp. 135–139.

3. ROARK, R. J., and W. C. YOUNG, *Formulas for Stress and Strain*, 5th ed., New York: McGraw-Hill Book Company, 1976.

4. ROSSHEIM, D. B., and MARKE, A. R. C., "Gasket Loading Constants," *Mech. Eng.*, **65**, 1943, p. 647; **66**, 1944, p. 72.

5. ROTHBART, H. A., ed., *Mechanical Design and Systems Handbook*, New York: McGraw-Hill Book Company, 1964.

6. SPOTTS, M. F., *Mechanical Design Analysis*, Englewood Cliffs, N.J.: Prentice-Hall, Inc., 1964.

7. TIMOSHENKO, S., *Strength of Materials*, 2d ed., Part II, New York: Van Nostrand Reinhold Company, 1956.

8. TIMOSHENKO, S., and WOINOWSKY-KRIEGER, S., *Theory of Plates and Shells*, 2d ed., New York: McGraw-Hill Book Company, 1959.

PROBLEMS

1. (a) A steel disk is shrunk on a steel shaft. The radius of the disk is very large as compared to the radius of the shaft. The diametral interference is equal to 0.001 times the shaft diameter. Find the values of tangential and normal stress. $E = 30,000,000$ psi.

(b) Work part (a) but with $b = 2a$.

(c) Work part (a) but with the disk made of cast iron, $E = 15,000,000$ psi. The shaft is steel. $\mu = \frac{1}{3}$.

(d) Repeat part (c) except that $b = 2a$.

   Ans. (a) $\sigma_t = \sigma_r = 15,000$ psi; (b) $\sigma_t = 18,750$ psi, $\sigma_r = 11,250$ psi.

      (c) $\sigma_t = \sigma_r = 9,000$ psi; (d) $\sigma_t = 10,710$ psi, $\sigma_r = 6,430$ psi.

2. A steel shaft is pressed on a steel disk twice the shaft diameter, and with a thickness $1\frac{1}{2}$ times the shaft diameter. The shearing stress in the shaft caused by the torque which the pressed fit can carry is equal to 80 MPa. Find the value of the diametral interference. $E = 206,900$ MPa. The coefficient of friction is 0.12.

                  Ans. $\Delta = 0.00143a$.

3. A 2-in. shaft has a torsional shear stress of 12,000 psi. It has a shrink fit in an 8-in. steel disk 3 in. thick. Find the tangential stress and diametral interference if the entire torque of the shaft is to be transmitted through the shrink-fit friction. $E = 30,000,000$ psi. The coefficient of friction is 0.12. *Ans.* $\Delta = 0.00118$ in.

4. A disk whose diameter is 3 times that of the shaft, and whose thickness is twice the shaft diameter, has a diametral interference of 0.0005 times the shaft diameter. For the shaft, $E = 206,900$ MPa. For the disk, $E = 103,450$ MPa. Poisson's ratio $= 0.25$. If the torsional stress in the shaft is equal to 69 MPa, find the coefficient of friction required if the entire torque is to be carried by the shrink-fit friction. *Ans.* Coefficient of friction $= 0.156$.

5. A disk of Class 30 cast iron is to be shrunk on a 6-in. steel shaft. The tangential stress in the disk is not to exceed 8,000 psi. Diametral interference of the metal is 0.004 in. $\mu = 0.3$. Find the shrink-fit pressure between disk and shaft. Find the minimum permissible outside diameter of the disk. Find the factor of safety.

      Ans. $p = 2,610$ psi; $b = 4.21$ in.; $F_s = 3.44$.

6. A 250-mm-diam steel shaft is to be pressed into a steel disk 450 mm in diameter and 200 mm thick. The interference of the metal must be such that the force required to press the parts together must lie between 1.28 and 1.80 MN. If the coefficient of friction is 0.15, find the maximum and minimum values for the diametral interference of metal. *Ans.* 0.19–0.27 mm.

7. A bronze bushing 2 in. OD and $1\frac{3}{8}$ in. ID is to be pressed into a hollow cylinder of 4 in. OD. The diametral interference of the metal is 0.002 in. Find the tangential and normal stresses for the steel and bronze at the boundary between the two parts.

  Ans. $p = 4,320$ psi; bronze, $\sigma_t = -12,050$ psi; steel, $\sigma_t = 7,190$ psi.

8. A 100-mm-diam steel shaft is pressed into a steel disk 300 mm in diameter and 125 mm thick. The shaft has a shearing stress equal to 55 MPa with the torque resisted by the shrink fit. The coefficient of friction is 0.15. Find the diameter of the hole in the disk before assembly. *Ans.* 99.96 mm.

9. A 3-in.-diam steel shaft is to be press fitted into a Class 30 cast-iron hub of 7 in. OD. Find the diametral interference of the metal if the F_s against breakage of the hub from the press fit is to be 4. *Ans.* $\Delta = 0.0019$ in.

10. A disk flywheel is to be cut from 75-mm rolled steel plate. The bore is 0.2 of the outside diameter. Maximum speed is 3,000 rpm with a maximum stress from rotation equal to 83 MPa. Find the diameter of the flywheel and the amount of energy delivered for a 10 percent drop in speed.
Ans. Diam $= 718$ mm; $\Delta KE = 143,900,000$ Nmm.

11. A disk flywheel is cut from $2\frac{1}{2}$-in. steel plate. Outside diameter is 36 in., bore is 6 in. Rpm $= 3,000$. Find the value of maximum stress in the material. Find *KE* delivered for a 5 percent drop in speed.
Ans. $\sigma_t = 19,460$ psi; $\Delta KE = 121,100$ ft lb.

12. A rolled steel disk flywheel is 600 mm in outside diameter and 75 mm thick. It rotates on a 60-mm-diam shaft. Normal speed is 2,400 rpm with a 20 percent drop during the working cycle. Find the maximum stress in the material and the energy delivered per cycle. *Ans.* $\sigma_t = 36.8$ MPa; $\Delta KE = 84,955,000$ Nmm.

13. A 42-in. outside diameter cast-iron flywheel has 4 spokes and rotates 800 rpm. The rim is 4×4 in. The spoke is 2×3 in. Find the stress in rim and spokes.
Ans. Rim, $\sigma = 2,450$ psi; spoke, $\sigma = 600$ psi.

14. The load cycle for each revolution of a shaft is shown in Fig. 12-41. The shaft is driven by a belt and can be assumed to receive a constant torque from the driving motor. Find the energy delivered per cycle for a flywheel mounted on the shaft for a coefficient of fluctuation of 0.05. The speed is 600 rpm. Find the outside diameter of the flywheel if it is solid and cut from a 1-in. steel plate.
Ans. $\Delta KE = 155,500$ Nmm; diam $= 450$ mm.

FIGURE 12-41 Problem 14.

15. The load cycle for a shaft is shown in Fig. 12-42. The shaft is driven by a belt and can be assumed to receive a constant driving torque. Find the energy delivered per cycle for a flywheel mounted on the shaft for a coefficient of fluctuation of 0.10. Speed is 150 rpm. Find the outside diameter if the flywheel is solid and cut from $2\frac{1}{2}$-in. steel plate. *Ans.* $\Delta KE = 1,100$ ft lb; diam $= 41.5$ in.

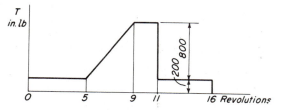

FIGURE 12-42   Problem 15.

16. Let the bolt in Fig. 12-13 be 20 mm in diameter, and let length l be equal to 400 mm. The threads at the bottom have a pitch of 2.5 mm, stopping immediately above the nut. Ignore the effect of the weight of the bolt. The bolt is made of steel.

(a) If a mass of 30 kg drops 2.5 mm to the nut, find the stress in the gross and net sections of the bolt. Take root diameter as gross diameter minus $1.3p$.

(b) Find the stress if the entire body of the bolt above the nut is reduced to the stress area of the threads.

(c) Find the stress if one-half the length of the bolt is reduced to the stress area.

(d) If bolt is made of phosphor bronze, find the stress in the net section of the thread.   *Ans.*   (a) $\sigma = 50.2$ MPa; $\sigma = 71.5$ MPa; (b) $\sigma = 60.1$ MPa; (c) $\sigma = 65.0$ MPa; (d) $\sigma = 51.0$ MPa.

17. If the bolt in Fig. 12-13 is $\frac{1}{2} \times 18$ in., and weight W is 10 lb, find the height h which will cause an impact stress of 40,000 psi in the gross area of the bolt. $E = 30,000,000$ psi; $\gamma = 0.283$ lb/in.^{3} *Ans.* $h = 9.7$ in.

18. A 100 mm $\times$ 100 mm $\times$ 1,200 mm wooden post stands on a rigid foundation. From what height must a 225-kg weight be dropped on the upper end of the post to give an impact stress of 27.5 MPa? Ignore the effect of the weight of the post. $E = 10,000$ MPa.   *Ans.* $h = 202$ mm.

19. In Fig. 12-43, the crane had been lowering the load at the uniform rate of 2 fps when the brakes on the drum were suddenly applied. Assuming that the upper

FIGURE 12-43   Problem 19.

end of the cable had been stopped instantaneously, find the additional stretch in the cable due to stopping. Let E for the twisted wire cable be 16,000,000 psi. $A = 0.5$ in.^{2} *Ans.* $y = 0.473$ in.

20. The steel beams in Fig. 12-44 are 75 mm wide. Find the maximum stress due to impact.   *Ans.* $\sigma = 439$ MPa.

FIGURE 12-44   Problem 20.

21. What must be the rate of the spring, in pounds per inch, if the force caused be the falling weight in Fig. 12-45 is equal to 1,000 lb? Rod and spring are of brass
                                                                      Ans. $k_s = 557$ lb/in.

FIGURE 12-45   Problem 21.   **FIGURE 12-46   Problem 22.**

22. In Fig. 12-46 all parts are of steel. Find the stress in the bolt and in the beam due to impact.   *Ans.*   Bolt, $\sigma = 356$ MPa; beam, $\sigma = 243$ MPa.

23. The gasket in Fig. 12-47 is of cloth-inserted hard rubber. If the bolts are only tightened sufficiently to seat the gasket, find the permissible value of the internal pressure.   *Ans.* $p = 16.2$ psi.

24. The asbestos composition gasket shown in Fig. 12-48 is $\frac{1}{8}$-in. thick. The cover is held in place by 24 bolts. If the bolts are tightened only sufficiently to seat the gasket, find the permissible value of the internal pressure. By use of Eq. (19), Chapter 5, find the force at the end of a 2-ft wrench handle required to tighten the bolts.   *Ans.* $p = 118.5$ psi; $F = 12$ lb.

FIGURE 12-47   Problem 23.   FIGURE 12-48   Problem 24.

25. The gasket shown in Fig. 12-49 is of soft copper $\frac{3}{16}$-in. thick, held in place by 24 bolts. If the bolts are only tightened sufficiently to seat the gasket, find the permissible value of the internal pressure. Find the torque required to tighten the bolts.   *Ans.* $p = 580$ psi; $T = 1,920$ in. lb.

FIGURE 12-49   Problem 25.   FIGURE 12-50   Problem 26.

26. The cable in Fig. 12-50 is 6×37 of plow steel wire. Its diameter is $1\frac{1}{4}$ in. The load is 20,000 lb and the sheave has a 24-in. pitch diameter.
   (a) Find expected life on the basis of 300 bends per week.
   (b) Find the F_s on the basis of direct stress in the cable.
   (c) Find the resultant stress when the cable is passing around the sheave.
   (d) Find the minimum pitch diameter of the sheave and drum if fatigue effects are to be avoided.
         Ans.   (a) 7.7 years; (b) $F_s = 11$; (c) $\sigma = 86,870$ psi, $d_s = 51$ in.

27. The cable in Fig. 12-50 is 6 × 19 and is 25 mm in diameter. The sheave is 600 mm in diameter. Resulting bending and direct stress in the cable when passing around the sheave is equal to 700 MPa. Find the force in the cable.

Ans. $T = 40,700$ N.

28. A 25-mm-diam 6 × 19 wire rope is made of improved plow steel wire. The resultant bending and direct stress is 690 MPa when passing around the sheaves. Expected life is 8 years at 400 bends per week and 50 weeks per year. Find the force in the rope and compute the minimum permissible sheave diameter.

Ans. $T = 32,870$ N, $d_s = 578$ mm.

29. Find the value of *m* in Fig. 12-51 that gives a resultant stress value of 10,000 psi on cross section *AB*.

Ans. $m = 11.8$ in.

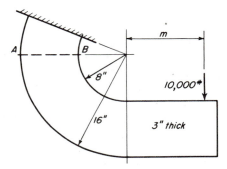

FIGURE 12-51   Problem 29.

30. Figure 12-52 shows the shackle at one end of a symmetrically arranged leaf spring. The spring supports a load of 18,000 N at its center. Find the maximum normal stress on cross section *AB*.

Ans. $\sigma = -122.0$ MPa.

FIGURE 12-52   Problem 30.

31. Find the value of length m in Fig. 12-53 that causes the tensile stress at A to be numerically equal to the compressive stress at B. *Ans.* $m = 181$ mm

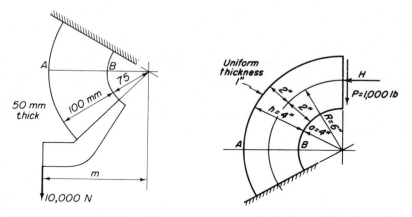

FIGURE 12-53 Problem 31.   **FIGURE 12-54 Problem 32.**

32. Find the value of force H if the resultant stresses normal to the horizontal cross section in Fig. 12-54 are to be numerically equal at points A and B.

Ans. $H = 1,465$ lb.

33. (a) Starting with Eq. (48) for the maximum bending stress in a curved beam of rectangular cross section, add the effect of the direct stress, and arrive at an expression for the resultant normal stress for point B in Fig. 12-55 in terms of load P, area of cross section A, and ratio c_1.

(b) If $\sigma = 50$ MPa at B, $P = 25,000$ N, $h = 100$ mm, and thickness equals 60 mm, find the corresponding value of a. *Ans.* (a) $\sigma = \dfrac{P}{A}\left[\dfrac{3c_1^2 - 1.8}{c_1 - 1}\right]$;

(b) $a = 88.7$ mm.

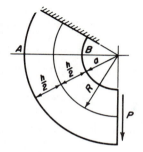

FIGURE 12-55 Problem 33.

34. What value of c_1 will make the bending stress at the inner edge of a curved beam of rectangular cross section loaded in pure bending 5% higher than that in a straight beam of the same cross section? 10% higher?

Ans. $c_1 = 13.95$; $c_1 = 7.30$.

35. A crane hook has the loading and dimensions given in Fig. 12-56.

(a) Draw cross section AB to full-size scale and divide into 18 strips $\frac{1}{4}$-in. wide running at right angles to line AB.

(b) Locate the center of gravity by use of the integral $A\bar{x} = \int x \, dA$, where A is the area of the cross section, x is the distance from the center of gravity to one end, say point B, x is the distance from point B to the center of gravity of each strip, and dA is the area of the strip. This operation should be carried out in tabular form.

(c) Find the value of m_1 by use of Eq. (50), and the value of the maximum and minimum stresses.   *Ans.* $\bar{x} = 2.06$ in.; $\sigma_{max} = 14,600$; $\sigma_{min} = -6,200$ psi.

SEC. A-B

FIGURE 12-56   Problem 35.

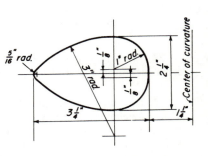

FIGURE 12-57   Problem 36.

36. Perform the operations of Problem 35 for the cross section of Fig. 12-57. Divide into 13 quarter-inch strips. Load is 5,000 lb.

Ans. $\bar{x} = 1.46$ in., $\sigma_{max} = 11,000$ psi; $\sigma_{min} = -3,750$ psi.

37. A harmonic cam has a central roller follower. $\beta = 120°$, $h = 1$ in., $R_b = 4$ in., $R_r = 0.625$ in., and $R_g = 2.5$ in. Calculate, in tabular form, values of α, r_g, and η for θ/β values of 0, 0.2, 0.4, 0.6, 0.8, and 1.0.

Ans. For $\theta/\beta = 0.4$: $\alpha = 9°19.3'$; $r_g = 6.20316$ in.; $\eta = 2°48.4'$.

38. A cycloidal cam has a central roller follower. $\beta = 120°$, $h = 1$ in., $R_b = 4$ in., $R_r = 0.625$ in., and $R_g = 2.5$ in. Calculate, in tabular form, values of α, r_g, and η for θ/β values of 0, 0.2, 0.4, 0.6, 0.8, and 1.0.

Ans. For $\theta/\beta = 0.4$: $\alpha = 11°20.5'$; $r_g = 6.1559$ in.; $\eta = 3°26.0'$.

39. A harmonic cam has a flat-faced follower. $\beta = 150°$, $h = 1$ in., $R_b = 1.5$ in. Calculate, in tabular form, values of r_c, q, and η for θ/β of 0, 0.2, 0.4, 0.6, 0.8, and 1.0.   *Ans.* For $\theta/\beta = 0.6$: $r_c = 2.2288$ in.; $q = 0.5706$ in.; $\eta = 14°50.0'$

40. A harmonic cam has a flat-faced follower. $\beta = 105°$, $h = 1$ in., $R_b = 1.5$ in. Calculate, in tabular form, values for r_c, q, and η at 15-deg intervals.

Ans. For $\theta = 60°$: $r_c = 2.2706$ in.; $q = 0.8356$ in.; $\eta = 21°35.6'$.

41. A harmonic cam has a flat-faced follower. $\beta = 90°$, $h = 0.8$ in., $R_b = 1.6$ in. Calculate, in tabular form, values of r_c, q, and η for θ/β of 0, 0.2, 0.4, 0.6, 0.8, and 1.0.

Ans. For $\theta/\beta = 0.4$: $r_c = 2.0248$ in.; $q = 0.7608$ in.; $\eta = 22°4.3'$.

42. Calculate and plot the values for y/h, $(dy/dx)/h$, and $(d^2y/dx^2)/h$ for the 3-4-5 polynomial cam. Make calculations, in tabular form, for $x = \theta/\beta$ of 0, 0.2, 0.4, 0.6, 0.8, and 1.0.

Ans. For $\theta/\beta = 0.4$: $y/h = 0.3174$; $(dy/dx)/h = 1.728$; $(d^2y/dx^2)/h = 2.88$.

43. Derive the constants in the displacement equation for the 3-4-5 polynomial cam for the initial conditions stated in Section 21.

44. A 50-mm-diam steel shaft is to be press-fitted into a 175-mm-diam × 75-mm-thick steel disk. Diametral interference of the metal is to be 0.05 mm. Find the force required to assemble them if the coefficient of friction is 0.15.

45. A 2-in.-diam steel shaft is to be press-fitted into a 6-in.-diam × 2-$\frac{1}{2}$in.-thick steel disk, with a Class 7 fit. Find the maximum required assembly force for a coefficient of friction of 0.15. Assume the tightest permissible fit.

46. A 100-mm-diam steel shaft is to be press-fitted into an 200-mm-diam cast-iron hub. F_s for hub is to be 5. Find the value of the radial and tangential stresses for the hub at the hole. For cast iron take $\sigma_{ut} = 345$ MPa and $\sigma_{uc} = 1,100$ MPa.

47. A steel shaft 3.5 in. in diameter is to be press-fitted into a Class 50 cast-iron hub 8 in. in diameter. Find the required diametral interference of the metal if the hub has a F_s of 4 from the press-fit stresses.

48. A hardened steel beam is 25 in. long with simple supports. An 80-lb weight drops 4 in. and strikes the beam at the center. Find the value of the bending stress for the following conditions. Ignore W_b in the calculations.

(a) The beam is $\frac{1}{2}$ × 2 in. set on edge.

(b) The beam is 1 × 1 in.

(c) The beam is $\frac{1}{2}$ × 2 in. laid on the flat.

49. Derive the equation for displacement for the 2-3 polynomial cam for the initial conditions: when $x = 0$: $y = 0$, and $dy/dx = 0$; and when $x = 1$: $y = h$, and $dy/dx = 0$.

13
Dimensioning and Details

The design of a machine includes many factors other than those of determining the loads and stresses and selecting the proper materials. Before construction or manufacture can begin, it is necessary to have complete assembly and detail drawings to convey all necessary information to the shop men. The designer usually does not make such drawings, but he frequently is called upon to supervise the draftsmen who do this detailing or to check the drawings before manufacture can begin. Much experience and familiarity with manufacturing processes are needed before one can become conversant with all phases of production drawings.

Drawings should be carefully checked to see that the dimensioning is done in a manner that will be most convenient and understandable to the production departments. It is obvious that a drawing should be made in such a way that it has one and only one interpretation. In particular, shop personnel should not be required to make trigonometric or other involved calculations before the production machines can be set up.

Dimensioning is an involved subject and long experience is required for its mastery.[1]

---

[1] For a complete treatment of dimensioning, see Reference 12, end of chapter.

1. Dimensioning

Tolerances must be placed on the dimensions of a drawing to limit the permissible variations in size because it is impossible to manufacture a part exactly to a given dimension. Although small tolerances give higher quality work and a better operating mechanism, the cost of manufacture increases rapidly as the tolerances are reduced, as indicated by the typical curve of Fig. 13-1. It is therefore important that the tolerances be specified at the largest values that the operating or functional considerations permit.

FIGURE 13-1 Relationship between size of tolerance and cost of production.

The development of production processes for large-volume manufacture at low cost has been largely dependent upon interchangeability of component parts. Thus the designer must determine both the proper tolerances for the individual parts, and the correct amount of clearance or interference to permit assembly with the connecting parts. The manner of placing tolerances on drawings depends somewhat on the kind of product or type of manufacturing process. If the tolerance on a dimension is not specifically stated, the drawing should contain a general note which gives the value of the tolerance for such dimensions.

2. Redundant Dimensioning

In a given direction, a point should be located by one and only one dimension. Much confusion and expense arise from violation of this rule.

For example, consider the horizontal dimension of the part shown in Fig. 13-2(a). For a part made as in sketch (b), lengths AB and AC are in accord with the drawing, but BC is not. Perhaps length BC is the important one for proper functioning, but a production man could argue that technically he had followed the drawing by making AB and AC correctly. Similarly, in sketch (c), lengths BC and AC are in accord with the drawing, but AB is not.

The difficulty can be corrected merely by omitting one of the dimensions in Fig. 13-2(a). The two dimensions that should be retained are determined by the functional requirements of the design.

(a) Detail drawing	(b) Actual part	(c) Actual part
	Measurements AB and AC are within spec. but BC is not.	Measurements BC and AC are within spec. but AB is not.

FIGURE 13-2   Example of incorrect production parts caused by redundant dimension.

3. Dimensioning of Clearance Fit—Maximum Material

The dimensions for a typical clearance fit for a hole and shaft are shown in Fig. 13-3. The tolerances shown here have been determined by the designer from consideration of the functioning of the assembly and the permissible cost of production. The bar diagram or diagram showing tolerances s and h and allowance a to a large scale for this fit is shown in Fig. 13-4.

The hole dimension of 3.000 in. and the shaft dimension of 2.998 in. represent parts at the maximum material condition, *MMC*. An assembly of

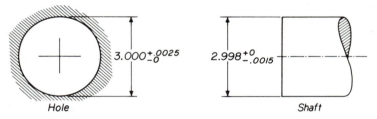

FIGURE 13-3   Maximum material tolerancing of clearance fit.

FIGURE 13-4   Exaggerated view of conditions of clearance fit of Fig. 13-3.

two parts with these dimensions will give the tightest fit, or the fit with the least possibility of assembly. In Fig. 13-3 one tolerance is zero, and the other tolerance gives parts which recede from the tightest assembly or the most dangerous condition. Clearance fits are usually dimensioned on the basis of *MMC*.

This type of dimensioning has another advantage. Should the workman aim at the principal dimensions, but through error produce an oversized hole or an undersized shaft, the parts might still be acceptable providing the dimensions do not exceed the limits specified by the drawings.

Figure 13-3 represents what would nominally be called a "3-in. shaft." It should be noted, however, that the 3-in. dimension is placed, not on the shaft, but on the dimension for the hole. When this is done, the dimensioning is said to be on the *basic hole system*. This method is advantageous because the manufacturing department can use standard-sized reamers for the holes, and then machine the shafts to fit.

On the other hand, the 3-in. dimension could just as well have been placed on the shaft, and the other dimensions adjusted to fit. When this is done, the dimensioning is known as the *basic shaft system*.

4. *Dimensioning of Interference Fit—Minimum Material*

Many joints are made as shrink or press fits, in which there is an interference of metal caused by the hole being smaller than the shaft. Here the most dangerous condition arises when the interference of metal is the least, giving a joint with the smallest holding ability. Such fits are frequently dimensioned on the basis of the least material condition, *LMC*, as shown by Fig. 13-5. Dimensions of 2.000 in. and 2.001 in. represent parts with the minimum amount of material. As indicated by Fig. 13-6, each dimension has one tolerance equal to zero with the other tolerance extending in the direction that will give tighter fit.

In minimum material dimensioning, should the workman aim at dimensions of 2.000 in. and 2.001 in., but by error produce an oversized hole or an undersized shaft, the parts will then not be in accord with the drawings.

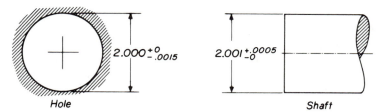

$2.000^{+0}_{-.0015}$ $2.001^{+.0005}_{-0}$

Hole   Shaft

FIGURE 13-5 **Minimum material tolerancing of interference fit.**

FIGURE 13-6 Exaggerated view of condition of interference fit of Fig. 13-5.

5. Unilateral and Bilateral Tolerances

The tolerances in Figs. 13-3 and 13-5 are known as unilateral tolerances. One tolerance is zero, and all the variations of the dimensions are given by the other tolerance. This method has the advantage that a tolerance revision can be made with the least disturbance to the other dimensions. Suppose in Fig. 13-7 that a fit was originally dimensioned with the hole and shaft tolerances h_1 and s_1 and with an allowance or minimum clearance of a_1. Suppose experience with this fit indicates that larger tolerances can be accepted. These can be easily changed on the drawings to h_2 and s_2 without disturbing other dimensions already present.

In bilateral dimensioning, a mean dimension is used which extends to the midpoint of the tolerance zone with equal plus and minus variations extending each way from this dimension, as in Fig. 13-8. It should be noted that a revision in the tolerance involves a change in at least one of the mean

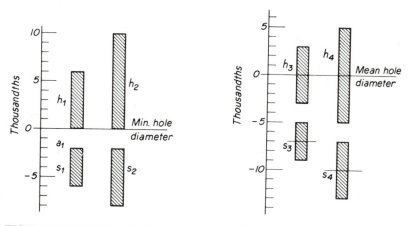

FIGURE 13-7 Unilateral tolerances.   **FIGURE 13-8** Bilateral tolerances.

dimensions. Drawings made with unilateral tolerances are usually easier to check than those made with bilateral tolerances.

In some situations, the bilateral method of tolerancing is very appropriate. The locations of hole centers, for example, when the variation from the basic dimension is equally critical in both directions, are usually so dimensioned. For practical manufacturing purposes, welded assemblies may be dimensioned with bilateral tolerances. The same applies to loosely toleranced dimensions. If the tolerances are large, it is sometimes more convenient to give the mean dimension, and the variation each way as plus and minus values.

Sometimes, instead of showing a dimension with tolerances, the two extreme limits of the dimension are given on the drawing.

6. Standardized Cylindrical Fits

Fits between cylindrical bodies and holes have been standardized into eight classes by the American Standards Association. When the class number is specified, the hole tolerance, shaft tolerance, and allowance or average interference can be determined from the diameter by means of the formulas given in Fig. 13-9. Classes 1 to 4 are clearance fits, and Classes 7 and 8 are interference fits. However, Classes 5 and 6 may be either clearance or inter-

Class of fit	1	2	3	4	5	6	7	8
Bar diagram (basic hole) system	Loose fit	Free fit	Medium fit	Snug fit	Wringing fit	Tight fit	Medium force fit	Heavy force and shrink fit
Hole tolerance, h	$.0025\sqrt[3]{d}$	$.0013\sqrt[3]{d}$	$.0008\sqrt[3]{d}$	$.0006\sqrt[3]{d}$	$.0006\sqrt[3]{d}$	$.0006\sqrt[3]{d}$	$.0006\sqrt[3]{d}$	$.0006\sqrt[3]{d}$
Shaft tolerance, s	$.0025\sqrt[3]{d}$	$.0013\sqrt[3]{d}$	$.0008\sqrt[3]{d}$	$.0004\sqrt[3]{d}$	$.0004\sqrt[3]{d}$	$.0006\sqrt[3]{d}$	$.0006\sqrt[3]{d}$	$.0006\sqrt[3]{d}$
Allowance, a	$.0025\sqrt[3]{d^2}$	$.0014\sqrt[3]{d^2}$	$.0009\sqrt[3]{d^2}$	0				
Av. interference					0	$.00025\,d$	$.0005\,d$	$.001\,d$

FIGURE 13-9 ASA B4a-1925 classification of cylindrical fits.
d = diameter.

ference fits, depending on the random assembly of the parts. They are accordingly called transitional fits.

As shown by the bar diagrams, Class 1 is the loosest fit and has the largest tolerances on the parts. It is used for fits where accuracy is not essential. The allowance and tolerances for Class 2 are also liberal. Class 3 is used for sliding fits and for the more accurate machine tool and automotive parts. Considerable precision is required in making parts for Class 4 fits. Selective assembly is usually practiced with Class 5 fits. Class 6 is suitable for drive fits for gears, pulleys, rocker arms, and so forth. It is also used for drive fits in thin sections or for extremely long fits in other sections. It is suitable for automotive, ordnance, and general machine manufacturing. Considerable pressure is required for the assembly of parts for Class 7. This fit is used for fastening the shafts to locomotive and railroad car wheels. It is also used for crank disks and for the armatures of motors and generators. It is the tightest fit that is recommended when the part with the hole is of cast iron. Class 8 is suitable for heavy force or shrink fits when the part with the hole is made of steel.

The fits shown in Fig. 13-9 have long since been superceded by larger and more flexible systems which permit a far wider variety of fits to be specified.[2] This old standard, however, does serve a useful purpose in helping form a mental picture as to how a cylindrical fit can vary from a very loose clearance fit to one where heavy pressure is required for assembly.

In order to reduce the expense of tools and gages, cylindrical sizes called for on drawings should be kept as few as possible. Perferred basic sizes are given in Table 13-1. The number of tolerances that may be used with each diameter should also be kept to a minimum to save the expense of an excessive number of gages. Recommended value for tolerances and allowances are given in Table 13-2.

TABLE 13-1 Preferred Basic Sizes, Inches

	0.0100		0.0500
	0.0125	1/16,	0.0625
1/64,	0.015625		0.0800
	0.0200	3/32,	0.09375
	0.0250		0.1000
1/32,	0.03125	1/8,	0.1250
	0.0400	5/32,	0.15625

3/16 to 3/4 by 1/16ths
7/8 to 3 by 1/8ths
3-1/4 to 4 by 1/4ths

TABLE 13-2 Recommended Tolerances and Allowances, Inches†

0.0001	0.0012	0.0100*
0.00015	0.0015	0.0120
0.0002*	0.0020*	0.0150
0.00025	0.0025	0.0200*
0.0003	0.0030	0.0250
0.0004	0.0040	0.0300*
0.0005*	0.0050*	
0.0006	0.0060	
0.0008	0.0080	
0.0010*		

†Preferred values are indicated by asterisks.

[2]See, for example, USAS B4-1-1967.

7. Eccentricity

Coaxial cylindrical parts must have a tolerance on the permissible deviations of the axes of the cylinders from each other. Inspection of such parts is frequently done by rotating the part in a V-block with an indicator resting on the other cylinder. As shown in Fig. 13-10, the full indicator movement, FIM, is double the displacement of the axes of the cylinders from each other.

(a) Detail drawing of part

(b) Maximum material part   (c) Minimum material part

FIGURE 13-10   Concentricity tolerance specified by runout. Indicator controls runout regardless of diameters. Note runout $2e_1$ is double the displacement e_1 of the axes.

8. Cumulative and Noncumulative Dimensioning

The dimensions to successive points can be placed on a drawing in two different ways. In Fig. 13-11 are shown the details of two parts that will eventually be assembled. The horizontal dimensions with tolerances can run from point to point as shown in sketch (a). This is called chain or cumulative dimensioning. In production it is possible for such parts to have the measurements of sketch (b). It is seen that these parts cannot be assembled because of the interference due to the cumulation of tolerances.

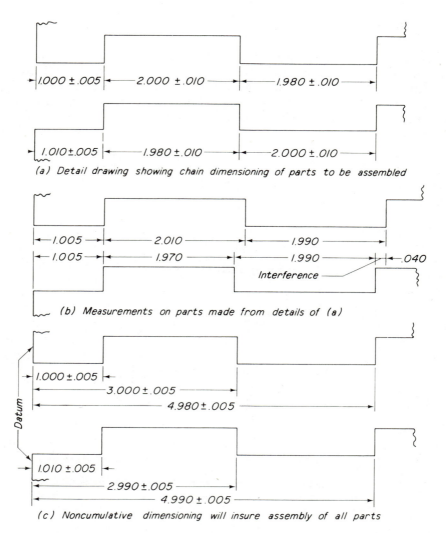

(a) Detail drawing showing chain dimensioning of parts to be assembled

(b) Measurements on parts made from details of (a)

(c) Noncumulative dimensioning will insure assembly of all parts

FIGURE 13-11   Ill effects resulting from cumulative dimensioning.

On the other hand, every dimension locating a point can start at a datum or plane of common origin as shown in Fig. 13-11(c). When such dimensioning is used, all parts made within the given tolerances will assemble and the interference problem is eliminated. Cumulative tolerancing should therefore not be used on dimensions involved in the assembly of two or more parts.

Assembly difficulties also occur when hole centers are dimensioned consecutively one to the other as shown in Fig. 13-12. The tolerances soon cumulate until it becomes impossible to insert the assembly bolt. To avoid the difficulty, the dimension locating each hole should originate at the datum.

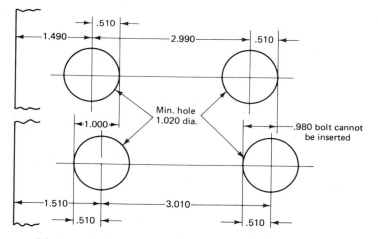

(a) Detail drawing of parts - cumulative tolerancing

(b) Two parts with minimum diameter holes and dimensions within tolerances of (a); holes at correct vertical dimension from datum B

FIGURE 13-12   Ill effects resulting from cumulative tolerances.

9. Datums

All bodies are three-dimensional, and on engineering drawings, a body can be assumed to be placed in a system of three perfectly smooth planes oriented exactly 90° to each other. Such planes are called *datums*. Theoretically, perfect planes cannot be produced. However, surface plates, angle plates, machine tables, and other equipment used in manufacturing and inspection are usually sufficiently accurate that they may be considered theoretical planes and thus used as datums.

An edge view of a datum plane appears as a line on a drawing and is marked by letters such as A, B, C, . . . placed in a box attached to the line representing the edge view of the datum. If a plane in a frontal view is a datum, it can be marked if desired by placing its letter in a box and running a leader to an oversize dot placed on the plane.

Datums should be actual physical surfaces from which measurements can be made. Abstract concepts such as center lines, center planes, or axes of symmetry should not be used as datums. Neither should a datum be taken through a row of holes. Not only can the hole diameter vary and be at an angle to the surface, but the location of the centers can shift anywhere within the tolerance zones.

The datum is the theoretical plane at which the dimensions originate. The surface of the body shown in contact with a datum is called the *datum feature*. A mental distinction must be made between the theoretically perfect datum plane and the datum feature or surface of the body shown in contact with the datum. All important dimensions should originate at datums. However, a drawing may have unimportant dimensions not concerned with functioning. Such dimensions can be arranged in any convenient way and need not start at a datum.[3]

10. Parts with Holes for Assembly with Loose Bolts

A part with two holes that is to be assembled to a similar part with loose bolts is shown in Fig. 13-13(a). Conventional rectangular dimensioning is used. Datum A is taken on the bottom of the part, datum B on the front face, and datum C along the left side of the part. The datums are assumed to be exactly 90° to each other.

Horizontal tolerances for the hole centers are ± 0.007 in. and vertical tolerance is ± 0.004 in. As shown in Fig. 13-13(a), these give a rectangular tolerance zone 0.014 by 0.008 in. However, it is not likely that chance variations for the location of the center will all neatly lie within this rectangle. It is more likely that they will fall within a circle. Thus if a hole center located at M_1 is satisfactory, then points M_2 and M_3, at the same distance from the mean, should be equally acceptable. The drawing shows that the tolerance circle has a radius of 0.008 in. or a diameter T of 0.016 in. This variation in the location of the center can be in any direction around the theoretical location. Such reasoning provides a greater area in which the center is permitted to lie without affecting assembly conditions.

Consider now a worst-fit case of assembly, as shown in Fig. 13-13(c). Both parts are in contact with datums B and C. The holes are assumed to be at the correct location from datum B, but horizontally they are at the extreme variation of 0.008 in. in opposite directions. The space remaining for the insertion of the assembly bolt is shown to be $2(0.2655 - 0.008)$, or 0.515 in., which is the maximum body diameter of a $\frac{1}{2}$-in. hexagonal head bolt.

It is noted that the tolerance $+0.005$, -0.000 on the hole diameter was

---

[3]The following abbreviations will be used when convenient: MMC, maximum material condition; LMC, least material condition.

2 HOLES, $.531^{+.005}_{-.000}$ DIA

Tolerance circle for hole center

Radius $= \sqrt{0.007^2 + 0.004^2}$

$= \sqrt{0.000\ 065}$

$= 0.008$

(a) Detail of parts: assembly by loose bolts

(b) Tolerance zones for hole centers of (a)

MMC bolt dia. $= 0.531 - 0.016$

$= 0.515$

(c) Parts with MMC holes at greatest permissible displacements in opposite directions (holes at mean location 0.700 in. from datum B)

FIGURE 13-13 Hole details and dimensions specified by rectangular dimension.

not involved in the calculations above. Should the hole be larger than the minimum, 0.531 in., assembly becomes easier. However, calculations are always made for the most difficult conditions.

11. True-Position Dimensioning

Another widely used method for dimensioning holes is called the *true-position method*. This method first establishes a theoretical network of lines for the location of the centers of the holes. These dimensions are exact and therefore not toleranced—and to so indicate, they are placed in a box as shown in Fig. 13-14. The same effect can be had without the box by marking each dimension BSC, basic, or by placing a note on the drawing saying "Untoleranced dimensions locating hole centers are basic." Untoleranced basic dimensions may extend individually from the datums or may run chain fashion from hole to hole without resulting in cumulative tolerances.

FIGURE 13-14   True-position dimensioning for the part of Fig. 13-13.

After the hole center is established, the hole diameter with its tolerance is given and beneath is placed a box indicating the permissible variation from the exact center. The small circle with vertical and horizontal cross lines means true position. This is followed by a small circle with a sloping cross line meaning diameter. Then comes the diameter of the tolerance circle, in this case 0.016 in. The modifier, uppercase M in a circle, means that the 0.016 in. tolerance circle applies only when the holes are at MMC, maximum material condition, or when they have their minimum diameter, at which time assembly is most difficult.

12. Hole Dimensioning Using Letter Symbols

Figures 13-13 and 13-14 considered the hole tolerancing situation from a numerical standpoint. If desired, such dimensions can be denoted by algebraic letters, and frequently it is advantageous to do so. The two previous figures, when written in algebraic terms, are shown in Fig. 13-15.

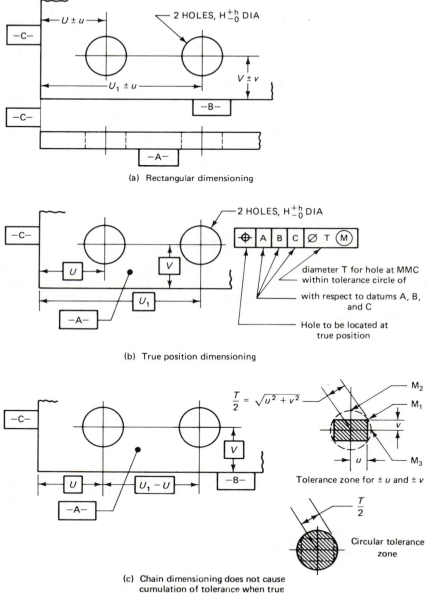

(a) Rectangular dimensioning

(b) True position dimensioning

(c) Chain dimensioning does not cause
cumulation of tolerance when true
position dimensioning is used.

**FIGURE 13-15 Details of part with holes and methods for specifying
tolerances for hole locations.**

Horizontal dimensions, for rectangular dimensioning, are denoted by $U \pm u$ and vertical dimensions by $V \pm v$. For true-position dimensioning, the mean dimension is placed in a box as shown. The MMC hole or minimum diameter is called H, and its tolerance is $+h$, -0. As in Fig. 13-15, the tolerance for the hole location is placed in a box beneath the hole diameter. As before, the true position symbol is placed first. It is usually good practice to specify the datums to which the hole dimensions refer. The presence of datum A in the box will be explained later. Datums B and C are needed, as dimensions U and V refer to these. The diameter symbol comes next for the tolerance circle followed by its diameter, which is denoted by T. Modifier M is needed if it is the designer's wish to have T apply for the MMC condition or the minimum hole diameter.

As before, the rectangle, $2u \times 2v$, is replaced by a tolerance circle of diameter T. In doing this the following equation will apply:

$$\frac{T}{2}\sqrt{u^2 + v^2} \tag{1}$$

13. Inspection of Hole Location by Fixed Gage

Hole diameters can be easily checked by a go/no go plug gage. A simple and easily made fixed gage for checking the location of the hole is shown schematically in Fig. 13-16. The fixed gage pin is located at the true position of the hole and the stops along the edges correspond to the datums. The gage will accept a MMC part with the maximum permissible deviation for the location of the holes which can occur in any angular direction about the true position.

By the tolerance box of Fig. 13-15(b), the center of a MMC hole of diameter H can be allowed to depart from the true position by the amount $T/2$. If

(a) Gage with part with hole of minimum diameter H; tolerance circle diameter of T

(b) Part with maximum diameter hole; tolerance circle diameter $T + h$

FIGURE 13-16 Inspection of part with hole by fixed gage. Gage pin diameter $H - T$ located at true position. Part of (b) acceptable when modifier M is specified.

the maximum diameter of the loose bolt is called F, Fig. 13-16(a) shows that the following equation results.[4]

$$\frac{H}{2} - \frac{T}{2} = \frac{F}{2} \quad \text{or} \quad H - T = F \tag{2}$$

Hence in use, the part must accept the gage pin of diameter $H - T$ and make contact with both datums. For this to occur, the dimensions must be in accord with Eq. (2).

Dimensions H and T appear on the drawing. Hence Eq. (2) permits the determination of the diameter F of the loose assembly bolt, which is also the diameter of the fixed pin of the gage.

The situation when the gage is inspecting a part with the maximum $H + h$ diameter hole is shown in Fig. 13-16(b). It is seen here that the tolerance circle for the hole center has a tolerance $T + h$, but this is permissible because of the presence of modifier M in the tolerance box for the hole. No difficulty will occur on assembly. It is characteristic of fixed gages to pass such technically out-of-tolerance parts. Modifier M does not prescribe the use of a fixed gage. It merely gives permission to use one if desired.

Tabular values for the unthreaded body diameters of standard hexagonal head bolts and cap screws are given in Table 13-3. It is noted that for bolts the

TABLE 13-3   Diameters of Unthreaded Body of Hexagonal Head Fastners*

Bolt		Cap Screws	
Min.	Max.	Min.	Max.
1/4 = 0.2500	0.2600	0.2450	1/4
5/16 = 0.3125	0.3240	0.3065	5/16
3/8 = 0.3750	0.3880	0.3690	3/8
7/16 = 0.4375	0.4520	0.4305	7/16
1/2 = 0.5000	0.5150	0.4930	1/2
9/16 = 0.5625		0.5545	9/16
5/8 = 0.6250	0.6420	0.6170	5/8
3/4 = 0.7500	0.7680	0.7410	3/4
7/8 = 0.8750	0.8950	0.8660	7/8
1 = 1.0000	1.0220	0.9900	1
1-1/8 = 1.1250	1.1490	1.1140	1-1/8
1-1/4 = 1.2500	1.2770	1.2390	1-1/4
1-3/8 = 1.3750	1.4040	1.3630	1-3/8
1-1/2 = 1.5000	1.5310	1.4880	1-1/2
1-3/4 = 1.7500	1.7850		
2 = 2.0000	2.0390		

*See ASA Standard B18.2-1955, New York: American Standards Association; and *Machinery's Handbook*, New York: Industrial Press, Inc., 1964.

[4]See p. 107 of Reference 1, end of chapter.

minimum body diameter and the nominal diameter coincide. For cap screws, however, the maximum body diameter and the nominal coincide.

Maximum and minimum external diameters for threaded machine screws are shown in Table 13-4. Acceptable tolerances on the diameters of drilled holes are given in Table 13-5.

TABLE 13-4   Machine Screws: Diameter of External Threads

	Max.	*Min.*		*Max.*	*Min.*
0–80	0.0595	0.0563	5–44	0.1243	0.1195
1–64	0.0724	0.0686	6–32	0.1372	0.1312
1–72	0.0724	0.0689	6–40	0.1372	0.1321
2–56	0.0854	0.0813	8–32	0.1631	0.1571
2–64	0.0854	0.0816	8–36	0.1632	0.1577
3–48	0.0983	0.0938	10–24	0.1890	0.1818
3–56	0.0983	0.0942	10–32	0.1891	0.1831
4–40	0.1112	0.1061	12–24	0.2150	0.2078
4–48	0.1113	0.1068	12–28	0.2150	0.2085
5–40	0.1242	0.1191			

TABLE 13-5   Diametral Tolerances for Drilled Holes*

1/4 to 1/2 in.	Std. to +0.004 to 0.005
33/64 to 3/4 in.	Std. to +0.005 to 0.007
25/32 to 1 in.	Std. to +0.008 to 0.010
1-1/64 to 2 in.	Std. to +0.010 to 0.012

*See p. 656 of Wilson, W. A., Ed., *Tool Engineers Handbook,* first edition, New York: McGraw-Hill Book Company Inc., 1949. Used by permission.

Example 1. Two parts such as those shown in Fig. 13-17 are to be assembled by a $\frac{1}{4}$-in. loose hexagonal bolt. The hole diameter is $\frac{9}{32}$ in. Both parts are to maintain contact with both datums. Determine the value of tolerance T that should be placed on the drawing.

Solution. $H = \frac{9}{32} = 0.2812$ in. By Table 13-3, a $\frac{1}{4}$-in. hexagonal head bolt has a maximum body diameter of 0.2600 in. By Eq. (2),

$$T = H - F = 0.2812 - 0.2600 = 0.0212 \text{ in.}$$

The tolerance circle for the hole center thus has a diameter of 0.0212 in.

Assembly conditions become easier when the hole diameter is larger than the minimum H and the bolt diameter is smaller than the maximum F. However, calculations should be made for the most difficult condition of assembly.

It should be noted that hole diameter tolerance h is not involved in the derivation or use of Eq. (2). Maximum diameter F for the loose bolt that is

FIGURE 13-17 Example 1: dimensions for parts to be assembled by a loose $\frac{1}{4}$-in. hexagonal bolt.

to be used for assembly does not appear on the detail drawings. However, the gage designer obtains the gage pin diameter F merely by subtracting tolerance circle diameter T from minimum hole diameter H.

Assembly will, of course, be possible when the left side of Eq. (2) is larger than the right side, so that the equation, if desired, can be written

$$H - T > F \tag{2a}$$

Use of this equation will provide clearance in the assembly. It is also useful in checking drawings to determine if assembly is possible.

Example 2. Two parts that are to be assembled by a $\frac{1}{2}$-in.-diameter loose bolt are dimensioned so that the hole centers coincide. The hole diameter is $\frac{17}{32}$ in. The tolerance circle for the hole location has a diameter of 0.0160 in. Determine if these dimensions are satisfactory to ensure assembly.

Solution. $H = \frac{17}{32} = 0.5312$ in. By Table 13-3, a $\frac{1}{2}$-in. hexagonal bolt has a maximum diameter of 0.5150 in. By Eq. (2a),

$$H - T > F$$
$$0.5312 - 0.0160 > 0.5150$$
$$0.5152 > 0.5150$$

Assembly will be possible.

No term for clearance at the tightest condition of assembly appears in Eqs. (2) and (2a). A term for clearance may not be necessary. The majority of the holes may have diameters larger than the minimum, with a relatively small proportion at the smallest value H. Similarly, the proportion of bolts at the maximum diameter F may be rather small. Then the probability of a hole of diameter H and a bolt of diameter F coming together for an assembly may be very remote. In other words, although Eq. (2) assumes theoretically zero clearance, in practically all cases clearance is present, and this simple equation will give valid results.

The variation in the perpendicularity of a hole with respect to the surface is sometimes of importance. The fixed gage will pass a part where the tangent of the angle for the hole is equal to the clearance between pin and hole divided by the thickness of the part. Should this angle be too great, a rearrangement of the tolerances may be required.

14. Use of Fixed Gage Confined to Modifier M

If for any reason the hole center must be arbitrarily held to the tolerance circle of diameter T, modifier S enclosed in a circle, regardless of feature size, is used in the tolerance box to replace modifier M. The corresponding dimensioning of the hole is shown in Fig. 13-18(a).

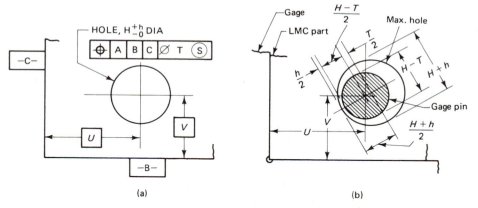

(a)   (b)

FIGURE 13-18   Conditions when tolerance for hole center is restricted to diameter T. Fixed gage cannot check hole location when modifier S is specified.

In Fig. 13-18(b) is shown a part with a maximum diameter hole, $H + h$, on the fixed gage of Fig. 13-16. The oversize tolerance circle of $T + h$ diameter is no longer permissible since modifier S restricts its diameter to T. It is seen that the ordinary fixed gage is no longer able to inspect the part for the location of the hole and other, and probably more expensive, methods must be used if the hole center is to be restricted to a tolerance circle of diameter T.

15. Dimensioning of Tapped Holes

A part with tapped holes, rectangular dimensioning, is shown in Fig. 13-19(a). The holes are specified by giving the nominal body diameter of the cap screw to be used, the number of threads per inch, the thread series of type of thread, and the class of fit required.

As before, in Fig. 13-19(d) the horizontal and vertical tolerances 0.007 in. and 0.004 in. are used to find the radius of 0.008 in. for the appropriate tolerance circle. Although variations in hole location may occur in any angular direction about the mean position, Fig. 13-19(b) shows the part when the variations have their maximum amounts horizontally but with correct locations vertically with respect to datum B.

Figure 13-19(c) on page 630 shows the part when true-position dimensioning is used. The tolerance circle diameter, 0.016 in., for the hole location is given in the box beneath the specifications for the tapped holes. Modifier M indicates that this tolerance applies at *MMC*, smallest holes, but that the hole

(d) Tolerance circle for
for hole center

$$\text{Radius} = \sqrt{.007^2 + .004}$$
$$= \sqrt{.000065}$$
$$= 0.008$$

(a) Part with tapped holes: rectangular dimensioning

(b) Part with extreme hole locations: hole centers may be anywhere within tolerance circle of diameter 0.016

FIGURE 13-19   Methods of dimensioning parts with tapped holes.

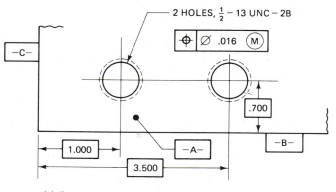

(c) Part of (a) but with true-position dimensioning

FIGURE 13-19 (cont.)

location tolerance can be larger for holes larger than the minimum as demonstrated in Fig. 13-16(b).

A comparable drawing using letter symbols rather than numerical dimensions is given in Fig. 13-20. True-position dimensioning is used.

When parts are threaded and assembly made by cap or machine screws, the dimensioning situation is somewhat different. To avoid confusion with assemblies made with loose bolts, the notation used will be as follows:

H, minimum diameter of through hole of the attached part
T, diameter of tolerance circle for position of through hole
F_1, maximum body diameter of cap or machine screw used for assembly
T_1, diameter of tolerance circle for location of tapped hole

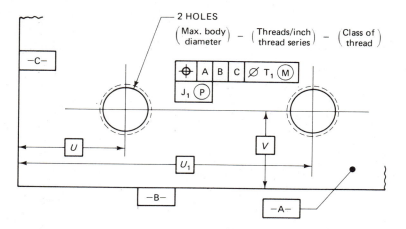

FIGURE 13-20 True-position dimensioning for part with tapped holes.

The tolerance box in Fig. 13-20 thus shows the diameter T_1 of the tolerance circle for the tapped hole. The meaning of the attached lower box will be explained later.

Pitch diameters of threaded holes can be easily checked by go/no go thread gages. A fixed gage for checking the location of the tapped holes is shown schematically in the lower part of Fig. 13-21. The gage plate contains holes at the true position of the tapped holes. These holes have the diameter $F_1 + T_1$, the maximum diameter of the fastener plus the diameter of the tolerance circle for the tapped hole.

FIGURE 13-21 Gages for checking locations of through and tapped holes. MMC parts with holes located at maximum permissible distances from true position. Gages control hole center locations within radii of $T/2$ and $T_1/2$ from true position in any angular inclination.

The gage plugs have a body diameter F_1 equal to the maximum body diameter of the cap screw to be used and threads equal to those of the tapped hole at MMC. In operation, the gage plate is placed in contact with both datums and, to be passed, the part must accept all gage plugs for the full length of threads. The center for a tapped hole can never be more than $\frac{1}{2}T_1$ in any direction from the true position.

The gage checking the part with the through holes that will be assembled to the part with the tapped holes in Fig. 13-21 is identical to the gage of Fig. 13-16 and is shown in the upper part of Fig. 13-21.

16. Assembly of Part with Through Holes to Part with Tapped Holes

The assembly of a part with through holes to a part with tapped holes with machine or cap screws is shown in Fig. 13-22. Both parts are to maintain contact with their respective datums. The through holes have minimum diameter H, and the screw has its maximum body diameter F_1. Both holes and screws have the maximum displacements in opposite directions permitted by the gages. Note that the drawing as well as Fig. 13-21 indicates that the fol-

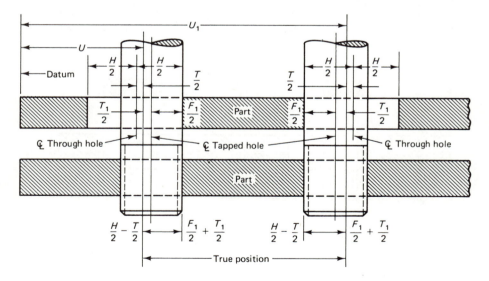

FIGURE 13-22 Worst condition of assembly. Minimum diameter through holes at greatest variation from true position. Maximum diameter screws at maximum variation from true position in opposite direction. Note that $H - T = F_1 + T_1$. Note that assembly is possible with centers of holes and screws displaced within specified limits in any direction.

lowing equation must be fulfilled to ensure assembly[5] under the worst conditions.

$$H - T = F_1 + T_1 \tag{3}$$

Note that the tolerances on the diameters of the through hole and on the fastener diameter were not involved in the derivation of Eq. (3).

Example 3. Details for a part with a through hole and a part with a tapped hole to be assembled with a $\frac{1}{4}$-in. cap screw are shown in Fig. 13-23. Determine suitable values for location tolerance T for the through hole and location tolerance T_1 for the tapped hole.

$$H - T = F_1 + T_1$$
$$.2812 - .0192 = .2500 + .0120$$

(a)　　　　　　　　　　　　　　(b)

FIGURE 13-23　Example 3. Dimensions of parts to be assembled by $\frac{1}{4}$-in. cap screw.

Solution. $H = \frac{9}{32} = 0.2812$ in. By Table 13-3, maximum diameter F_1 of a $\frac{1}{4}$-in. cap screw is 0.2500 in. By Eq. (3),

$$T + T_1 = H - F_1 = 0.2812 - 0.2500 = 0.0312 \text{ in.}$$

This value can be divided arbitrarily between T and T_1. Let

$$T_1 = 0.0120 \text{ in.}$$

Then　　　　　$T = 0.0312 - T_1 = 0.0312 - 0.0120 = 0.0192$ in.

Assembly is assisted when clearance is present. This can be had by making the left side in Eq. (3) larger than the right side. The equation is then written

$$H - T \geq F_1 + T_1 \tag{3a}$$

---

[5]See p. 107 of Reference 1, end of chapter.

In this form the equation can be used to provide some allowance to ensure assembly should the main body of the fastener be eccentric with the threaded portion. The equation is also useful in checking drawings to ascertain if assembly is possible.

Example 4. An assembly is to be made with a No. 10–24 machine screw. The minimum through hole has a specified diameter of $\frac{7}{32}$ in. with a tolerance circle diameter T equal to 0.015 in. The tapped hole has a specified tolerance circle diameter T_1 of 0.014 in. True positions for the two centers are dimensioned to coincide. Determine if assembly is always possible.

Solution. $H = \frac{7}{32} = 0.2188$ in. By Table 13-4, a No. 10–24 machine screw has a maximum diameter of 0.1890 in. By Eq. (3a),

$$H - F_1 > T + T_1$$
$$0.2188 - 0.1890 > 0.015 + 0.014$$
$$0.0298 > 0.0290$$

Assembly will be possible.

Example 5. Find the maximum-size machine screw that can be specified for the assembly of the two parts of Fig. 13-24.

FIGURE 13-24   Example 5.

Solution. $H = 0.1875$ in., $T = 0.0150$ in., $T_1 = 0.0175$ in. Equation (3a) can be written

$$F_1 < H - T - T_1$$
$$< 0.1875 - 0.0150 - 0.0175$$
$$< 0.1550 \text{ in.}$$

By Table 13-3, a No. 6–32 machine screw, with a maximum diameter of 0.1372 in., is the largest that can be used.

If Figs. 13-21 and 13-22 it is seen that no hole can invade a cylinder of diameter $H - T$ about the true position and no threaded stud can extend outside the cylinder of diameter $F_1 + T_1$. Then if the dimensions are such as to keep these two expressions equal, as required by Eq. (3), there can never be any interference on assembly.

17. Three-Plane Concept for Datums. Geometric Tolerancing

The constituents of a part such as surfaces, planes, edges, holes, slides, lugs, offsets, and so on are called its features. To completely locate a point in space, the dimensions should in general originate at three mutually perpendicular planes which form the datum system of the drawing. The datums may either pass through the body or be external to it, and may either be implied or be specifically marked. As previously mentioned, a datum is designated by a letter enclosed in a box attached to the line representing the projection of the plane. Letters I, O, and Q should not be used as datums.

A datum is a theoretically flat surface, something impossible to manufacture. However, the accuracy of surface and angle plates, machine tables, and inspection equipment is generally sufficiently good that such surfaces can be considered perfect planes.

A minimum of six contact points are necessary to provide complete restraint to a body in space. Three points will locate with respect to the first plane, two additional points will locate with respect to the second plane, and one additional point will locate with respect to the third plane. The datums should be chosen with the above requirements in mind. This is especially true for bodies of irregular shape. The planes[6] containing the three, two, and one locating points are sometimes called the primary, secondary, and tertiary datums, respectively.

In most cases selection of a datum is not arbitrary, but is made to coincide with the locating points that will be used in the production and inspection fixtures. Datums should be marked whenever there is a possibility that the

[6]See p. 33 of Reference 1, end of chapter.

tool design or production departments might choose another surface as the datum.

The three datum surfaces, each exactly 90° to the other two, are shown in Fig. 13-25(a). A six-sided rectangular body with three of its surfaces coinciding with the datums is also shown. Sketches (b), (c), and (d) show the three orthogonal projections of the body with the datum planes appearing as lines.

When only size tolerances are specified, the drawing can be interpreted as meaning that the body may be of any shape just so that it falls within the given tolerance zones. Sometimes a more precise control over the form is

FIGURE 13-25 Tolerances for flatness and parallelity of six-sided rectangular body by three-plane concept.

required so that tolerances for the flatness and angularity of the surface must be specified. Typical notes for such control for the six-sided body are shown in Fig. 13-25. Here the body is resting upon a surface plate, primary datum -A-, and is in contact with datums -B- and -C-. Vertical measurements are made from datum -A- so that any irregularities in the bottom surface of the part are disregarded.

A flatness tolerance serves to confine the roughness and undulations of a surface between two parallel planes a specified distance apart. The angularity of these planes can then be controlled by parallelity notes to the corresponding datums. The surface in contact with the primary datum, -A- in Fig. 13-25, requires no angularity tolerance since surface and primary datum are parallel. The remaining surfaces due to production variations may all be inclined at small angles to the datums and require angularity tolerances even though some of them are shown on the drawing as coinciding with the datums. Examples of the latter are the surfaces nominally in contact with datums -B- and -C- in Fig. 13-25.

It is necessary always to have a mental distinction between the theoretically perfect datum and the actual surface of the part, even though the two on the drawing may be shown as coinciding.

Care must be taken that the tolerances on the lengths are great enough to accommodate the variations arising from angularity.

The three perpendicular dimensions, running from the three datums, to locate point D in Fig. 13-25 are 4.000 ± 0.005, 3.000 ± 0.005, and 2.500 − 2.480.

18. Symbols Used in Geometric Dimensioning

Symbols are usually more convenient than notes and effect a saving in space on the drawing. A brief listing of symbols used in geometric tolerancing is shown in Fig. 13-26. The figure also gives several of the modifiers used in conjunction with appropriate dimensions.[7] The symbol is placed in a box adjacent to the dimension to which it refers. The box also gives the appropriate datums and the geometric tolerance for the feature. As before, maximum material condition is abbreviated MMC and least material condition by LMC. Use of symbols is illustrated by the following example.

A simple rectangular part with four holes is shown in Fig. 13-27. Discuss the symbols required on the drawing for the part and design a fixed gage for its inspection.

The drawing indicates that the part is to be in contact with datums *B* and *C* along the bottom and left edges. To do this the angle at the lower left

[7]See Reference 1, end of chapter.

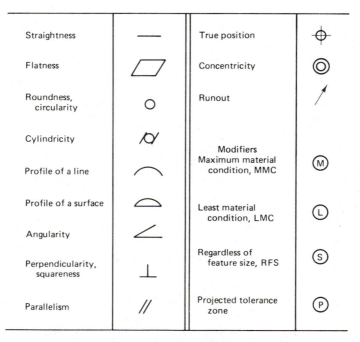

Straightness	▬	True position	⌖
Flatness	▱	Concentricity	◎
Roundness, circularity	○	Runout	↗
Cylindricity	⌭		
Profile of a line	⌒	Modifiers Maximum material condition, MMC	Ⓜ
Profile of a surface	⌓	Least material condition, LMC	Ⓛ
Angularity	∠		
Perpendicularity, squareness	⊥	Regardless of feature size, RFS	Ⓢ
Parallelism	//	Projected tolerance zone	Ⓟ

FIGURE 13-26   Geometric symbols used on drawings.

corner must necessarily be exactly 90°. Since it is not possible to obtain this in production, the lower and left edges will not coincide with the datums and the permissible variation from such contact must be shown on the drawing. This can be done by a parallelity tolerance which establishes a tolerance zone in which the actual edge must lie. The two sides may then contact the datums in a single point each or along one datum and a point on the other.

The right and top sides must also have parallelity tolerances as well as tolerances on the overall lengths. The tolerance zones for these edges must be broad enough to contain the corresponding parallelity tolerances.

It is obvious from the above that a toleranced dimension plus a parallelity tolerance are required to locate a flat surface.

This part is intended to be attached to another part by $\frac{1}{2}$-in. bolts. Table 13-3 shows that such a bolt can have a maximum diameter of 0.515 in. If the designer wishes the hole centers to lie within a tolerance circle of 0.016 in. diameter, the minimum hole diameter, H, by Eq. (2) then is

$$H = F + T = 0.515 + 0.616 = 0.531 \text{ in.} = \tfrac{17}{32} \text{ in.}$$

This is shown on the drawing. In the box beneath is given the diameter

(a) Detail drawing of part

(b) Fixed gage for inspection of part of (a)

FIGURE 13-27 Detail of flat part and gage for its inspection.

0.016 in. of the true position tolerance circle. The modifier *M*, enclosed in a circle, as previously explained, indicates that parts with larger diameter holes will be acceptable even though their centers lie outside the true position tolerance circle of 0.016 in. Hole diameters are easily controlled by a go and no-go plug gage.

An acceptable part must receive all gage pins, contact both datums, and not exceed the specified tolerances for the right and upper edges. The inspector must be provided with feeler gages to control the latter.

A datum should coincide with some physical feature of the part. For example, in the T-side of Fig. 13-28 it appears both natural and convenient to take one of the datums on the vertical center line of each member, and to run the horizontal dimensions each way from such datums. Manufacture and inspection, however, may be difficult if these imaginary planes are devoid of all physical features.

FIGURE 13-28   Tolerance zones for T-slide.

A vertical functional surface could, of course, be used as a datum, but the symmetry would then be lost. In most cases, however, it is found that once a datum is chosen and the tolerance zones established in relationship to it, it is not possible to shift the dimensioning to a new datum and preserve the same tolerance zones as originally planned.

For bodies of revolution, the datum system can consist of a cylinder about the axis, a plane through the axis, and a plane perpendicular thereto. To locate a point, the dimensions would consist of a length parallel to the axis and an angle and radius in the view perpendicular to the axis. If desired, ordinary rectangular dimension can be used in this latter view.

The control of properties such as straightness, flatness, squareness, parallelism, perpendicularity, angularity, and so on is known as geometric tolerancing. A system of symbols for their control has come into use which usually effects a saving in space on the drawing. Careful handling is required to avoid confusion and mistakes.[8]

---

[8]For further information see Reference 1, end of chapter.

19. Third-Angle Projection. First-Angle Projection

The familiar arrangement for the top, front, and right-side views of an object, as in Fig. 13-29(a), has been made by what is known as the third-angle projection.

Right side view   Front view

Top view

Front view   Right side view

Top view

(a) Arrangement for third angle projection   (b) Arrangement for first angle projection

FIGURE 13-29 Methods for locating the views in third and first angle projections.

Most metric countries use what is called first-angle projection. The individual views are exactly the same as in third-angle projection, but on a drawing the views are located differently with respect to each other as is shown in Fig. 13-29(b).

20. Dimensioning of Tapers

Tapers or conical surfaces must be carefully dimensioned to avoid ambiguities and conflicts of dimensions that might permit the drawing to be interpreted in more than one way. In general, a taper can be completely specified by means of three dimensions, covering location, diameter, and amount of taper. Several examples are illustrated in Fig. 13-30. Strictly speaking, no tolerance on the taper or angle in sketches (a) and (b) would be required in laying out either the maximum metal condition or the minimum metal condition. A compound tolerance is introduced by use of a tolerance on these taper and angle dimensions.

In sketch (c) dimensions for both large and small diameters and the length are given. The tolerance zone specified by these three dimensions is represented by Fig. 13-30(d). Since no tolerance for the taper is given in sketch (c), the interpretation can be placed on the drawing that any part falling within the tolerance zone will be acceptable. The angle for the taper could

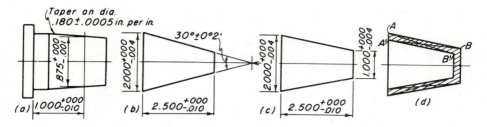

FIGURE 13-30   Various methods of dimensioning tapers.

then vary from that represented by line AB' to line $A'B$. Thus the tolerances on the diameters can cause a considerable variation in the angle of the taper. This variation must be checked carefully by the designer to make certain that it will be acceptable.

When dimensions are made to corners, as in Fig. 13-30(b) and (c), it is understood that the theoretically sharp corner is referred to. Actual measurements on a part must be made by means of rolls or by a comparator, since a theoretically sharp corner does not exist. When a fourth dimension is placed on a drawing for a taper, ambiguities immediately arise. For example, suppose an additional dimension for the small diameter were given in Fig. 13-30(b). Examination indicates that a variety of interpretations could then be placed on the drawing and much confusion arise.

21. Dimensioning of Assemblies

The assembly of a number of parts into a final product is shown schematically in Fig. 13-31. If parts 1, 2, 3, and 4 have the dimensions shown, the mean value $\bar{w}$ of the resultant dimension is $-\bar{a} + \bar{b} + \bar{c} - \bar{d}$. The maximum value of the opening occurs when 1 and 4 are at the low limit, and 2 and 3 are at the high limit. Hence

FIGURE 13-31   Tolerances for assembly.

$$\bar{w} + w = -(\bar{a} - a) + (\bar{b} + b) + (\bar{c} + c) - (\bar{d} - d) \tag{a}$$

The minimum value occurs when 1 and 4 are at the high limit and 2 and 3 are at the low limit. Hence

$$\bar{w} - w = -(\bar{a} + a) + (\bar{b} - b) + (\bar{c} - c) - (d + d) \tag{b}$$

From the two equations above, tolerance w is found to be

$$w = a + b + c + d \tag{4}$$

The equation assumes that the tolerances for an assembly are arithmetic or additive.

The simultaneous occurrence of all parts at the extreme limits assumed by Eqs. (a) and (b) is exceedingly rare. Suppose the dimensions of the different members of an assembly are in no way dependent on each other. In addition, let the lengths of the production parts be distributed normally, with the mean at the center of the tolerance zone. Also, let it be assumed that the ratio of the natural tolerances of the production process to the tolerances of the drawing be constant for all members. Then the following equation may give a more realistic value for the tolerance w of the assembly.

$$w = \sqrt{a^2 + b^2 + c^2 + d^2 + \cdots} \tag{5}$$

Example 6. In Fig. 13-31 let $a = b = c = d = 0.002$ in. Find variation w of the assembly by the additive and Gaussian rules.

Solution. By additive rule: $w = 4 \times 0.002 = 0.008$ in.

By Gaussian rule: $w = \sqrt{4 \times 0.002^2} = 0.004$ in.

It must be carefully noted that the above equation can be applied only when the lengths for every part of the assembly have the normal or Gaussian distribution.[9] Since this is rarely true in practice, assembly tolerance w will lie intermediate between the results given by Eqs. (4) and (5).

22. Tolerances in SI Units

During the transition from inches to millimeters, it may be helpful to show the dimensions in both systems on a drawing. When this is done[10] it is recommended that the millimeter dimension be placed above or to the left of the inch dimension.

[9]See Rice, W. B., "Setting Tolerances Scientifically," *Mech. Eng.*, **66**, 1944, p. 801; Burr, I. W., "Some Theoretical and Practical Aspects of Tolerances for Mating Parts," *Ind. Quality Control*, **15**, Sept. 1958, p. 18; and Acton, F. S., and E. G. Olds, "Tolerances—Additive or Pythagorean," *Ind. Quality Control*, **5**, Nov. 1948, p. 6. See also Reference 11, end of chapter.

[10]See Reference 10, end of chapter.

Although the inch is exactly 25.4 mm, the converted inch values rarely come out as exact numbers in millimeters, so rounding off is required. The recommended number of decimal places to be retained depends on the size of the tolerance and is indicated in Table 13-6. For example, by the table,

TABLE 13-6   **Practical Round Off of Decimal Places after Conversion***

Total Tolerance, in.		Millimeter Conversion Rounded to	Total Tolerance, mm		Inch Conversion Rounded to
At Least	*Less Than*		*At Least*	*Less Than*	
0.00004	0.00040	4 decimal places	0.002	0.020	5 decimal places
0.0004	0.0040	3 decimal places	0.02	0.20	4 decimal places
0.004	0.040	2 decimal places	0.2	2.0	3 decimal places
0.04	0.40	1 decimal place	2 and over		2 decimal places
0.4 and over		Whole mm			

*From SAE Standard J390, 1971.

an inch dimension of 0.0075, which lies between 0.004 and 0.040, would retain two decimal places in the equivalent value for millimeters. Thus 0.19 mm would be shown on the drawing as the converted tolerance.

23. Preferred Numbers

The factor of size or capacity must be considered in all manufactured articles or items of commerce. The term "size" may refer to linear dimensions, volume, weight, speed, power, or other features of the product. When an item is produced in a number of sizes, it is logical that the progression of sizes should be geometric rather than arithmetic. Each size is then larger than the preceding by a fixed percentage. Small sizes will then differ from each other by small amounts, and large sizes by larger amounts.

It is important from an economic standpoint that the number of sizes in a series be standardized at the smallest number that permit the product to take care of its intended range. The value of the step from one size to the next will of course depend upon the product under consideration. The system of *preferred numbers* has been devised in order that the various sizes of a series can be determined in an orderly fashion.[11]

In the range 10–100, the 5-series has five numbers, each of which is approximately 60% greater than the preceding. The 10-series gives 10 numbers approximately 25% apart, and the 20-series gives 20 numbers approximately 12% apart. These numbers are shown in Table 13-7.

[11]See "Preferred Numbers," ASA Z17.1-1936. New York: American Standards Association.

TABLE 13-7   Basic Preferred Numbers, Decimal System, 10–100

5-Series, 60 Percent Steps	10-Series, 25 Percent Steps	20-Series, 12 Percent Steps
10	10	10
		11.2
	12.5	12.5
		14
16	16	16
		18
	20	20
		22.4
25	25	25
		28
	31.5	31.5
		35.5
40	40	40
		45
	50	50
		56
63	63	63
		71
	80	80
		90

Perferred numbers below 10 are formed by dividing the numbers between 10 and 100 by 10. If still smaller numbers are desired, division is by 100. Preferred numbers above 100 are formed by multiplying the numbers between 10 and 100 by 10 or by 100. Tables for 40-series and 80-series numbers are available when the steps must be closer together. Tables using fractional numbers rather than decimal have also been constructed.

24. Surface Roughness

The properties and performance of machine parts may be affected in many ways by the degree of roughness of the various surfaces.[12] Among these effects, the following may be mentioned.

[12]See Broadston, J. A., "Standards for Surface Quality," *Prod. Eng.*, **15**, 1944, pp. 622, 704, 756, and 806; and Norton, M. R., "Development of Standards for Army Ordnance Finishes," *Mech. Eng.*, **64**, 1942, p. 703.

(1) The friction and wear between unlubricated surfaces are increased with increase of roughness.

(2) Friction and danger of seizure in lubricated surfaces are reduced for smooth surfaces.

(3) Interference fits made on rough surfaces may have a reduced area of contact and a consequent reduction in the holding ability of the joint.

(4) Fatigue strength of parts is increased as the surfaces become more smooth.

(5) Fluid flow in small passageways is reduced as the roughness of the walls is increased.

A demand has gradually grown up for a more exact method of specifying roughness than simply by marking the drawing "finish," "grind," or "lap." The roughness of a surface can be specified by calling for the height of the irregularities above and below the mean plane.

A true profile section, magnified equally in the horizontal and vertical directions, is shown in Fig. 13-32(a). The condition of the surface can be studied in more detail by exaggerating the roughness effect. Figure 13-32(b) shows the same profile when an additional magnification of 25 times has been given to the vertical direction with no change in the horizontal direction. Waviness also occurs in surfaces, as shown in Fig. 13-32(c). No exact differentiating point exists between roughness and waviness, although the pitch

(a) True profile section magnified equally in horizontal and vertical directions

(b) Exaggerated appearance of surface (a) by further magnification of 25 times in vertical direction only, no change in horizontal magnification

(c) Roughness consisting of fine irregularities superposed on larger or wavelike variations

FIGURE 13-32 Cross section of surface showing irregularities of roughness and waviness.

of the undulations must be about 0.040 in. or more before they are spoken of as waves. Specimens can be prepared for viewing under the microscope by a process called taper sectioning, in which the surface is ground at a small angle, thus accentuating the heights of the irregularities. An angle of 2°17' gives a magnification of 25. It is necessary to support the surface with a coating of nickel plating to prevent smearing of the contour during the sectioning process. A surface that has been prepared in this way[13] is illustrated in Fig. 13-33.

FIGURE 13-33   Taper section of cylindrical surface finished by grinding.

Heights above and below the mean surface are sometimes expressed in microinches or millionths of an inch, 0.000001 in., in terms of a special kind of average distance called the root-mean-square height, h_{rms}. This height is found by taking the ordinates at equidistant points on the profile, squaring, adding, dividing by the number of terms, and then taking the square root of this average. This method of calculating gives more weight to the higher peaks of the surface, since it is felt that high, narrow peaks may have considerable influence on the surface qualities, but would have negligible effect in shifting the position of the mean line. The root mean square height is about 10% larger than the arithmetic average of the ordinates.

A symbol for surface finish in wide use is the check mark with the bar on the top shown in Fig. 13-34. The sides of the mark are inclined at 60° to the surface, and the roughness height in microinches is represented as shown. The drawing should contain a note stating whether root mean square, arithmetic average, or peak-to-valley height is referred to. Sometimes the rough-

FIGURE 13-34   Symbol and marking for surface finish.

ness width and height of wave for a wavy surface are given by means of additional numbers on the symbol. The lay of the surface, or the direction

[13] See Tarasov, L. P., "Relation of Surface Roughness Readings to Actual Surface Profile," *Trans. ASME*, **67**, 1945, p. 189.

of the predominant surface pattern as formed by the machining or grinding operation, can also be placed on the symbol if desired.

Instruments are available for measuring surface roughness. A diamond tracing point is slowly drawn over the surface, and the roughness is indicated by a trace on a moving paper tape or by a meter. Some instruments give results in terms of the arithmetic average from the mean surface or in terms of rms roughness.

Sometimes roughness can be gaged with sufficient accuracy by comparing the surface with a calibrated specimen which has been prepared by the same machining operation. This comparison can be made by rubbing the fingernails across the work piece and then over the sample. Samples can be obtained for this purpose in a variety of roughnesses, such as produced by turning, milling, grinding, lapping, and so forth.

Different types of maching operations give surfaces of various degrees of roughness with a considerable range of variation for any particular operation. The approximate range of such variations[14] is shown in Fig. 13-35. Among the numbers that are recommended for specifying the average height of irregularities, the following may be mentioned: 1, 2, 4, 8, 16, 32, 63, and 125 microinches.

The inverse problem is also of interest to designers. In other words, when the rms height of a surface has been specified, how far below the peaks are the deepest valleys? It is easy to visualize a surface in terms of peak-to-valley distance, but it is difficult to determine what this range would probably be when only the arithmetic average or rms height is known. Experiments have shown that the average peak-to-valley roughness can be obtained from the rms value by mutliplying by $4\frac{1}{2}$ for cylindrical ground surfaces, by 6 or 7 for other types of fixed-abrasive finishes, and by 10 for loose-abrasive lapped surfaces. The factors quoted give average values for "predominant peak" roughness which occurs more or less uniformly over the entire surface. Another type of irregularity that occurs at intervals that are large as compared to the width of the irregularities themselves is termed "deepest maximum" roughness. The foregoing factors should be doubled to obtain deepest maximum values.

25. Detailing

When the design has been completed, detail or shop drawings must be made before actual manufacture or construction can begin. The detail draftsmen perform a very important function in industry, since the success of a project

---

[14]See *SAE Handbook*, Warrendale, Pa.: Society of Automotive Engineers, Inc.

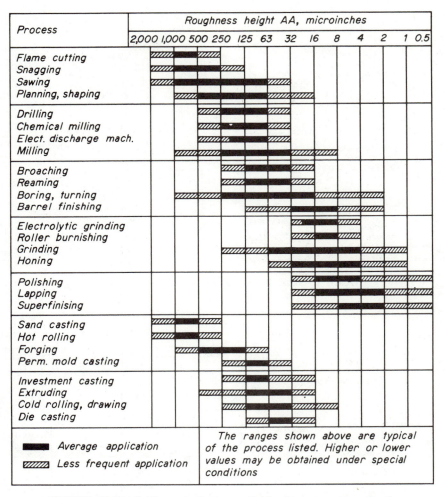

Process	Roughness height AA, microinches												
	2,000	1,000	500	250	125	63	32	16	8	4	2	1	0.5

Flame cutting
Snagging
Sawing
Planning, shaping

Drilling
Chemical milling
Elect. discharge mach.
Milling

Broaching
Reaming
Boring, turning
Barrel finishing

Electrolytic grinding
Roller burnishing
Grinding
Honing

Polishing
Lapping
Superfinising

Sand casting
Hot rolling
Forging
Perm. mold casting

Investment casting
Extruding
Cold rolling, drawing
Die casting

▬▬▬ Average application

▨▨▨ Less frequent application

The ranges shown above are typical of the process listed. Higher or lower values may be obtained under special conditions

FIGURE 13-35 Surface roughness available by common production methods. Arithmetic average of deviations from mean surface.

depends on correct details and shopwork as well as on design calculations. Skill and cleverness in arranging the mechanical details can be acquired only after considerable time has been spent in actual practice in this line of work. A few of the more obvious examples of good and bad detailing are illustrated in Figs. 13-36(a), 13-36(b), and 13-36(c) inclusive.[15]

[15]See Geiger, E., "The Rights and Wrongs of Details," *Prod. Eng.*, **12**, 1941, pp. 72, 122, and 248.

FIGURE 13-36(a) Suggested right and wrong ways of detailing.

FIGURE 13-36(b) Suggested right and wrong ways of detailing.

Wrong   *Right*   *Wrong*   *Right*

Always design an undercut for tool runout when threading up to shoulders

Clearance grooves or undercuts avoid impracticability of machining or grinding into sharp corners

Difficult to lay out hole center and start drill. Flat surface necessary

Drill may snag when breaking through

Clearance for drills and taps should be checked

FIGURE 13-36(c) Suggested right and wrong ways of detailing.

REFERENCES

1. *Dimensioning and Tolerancing for Engineering Drawings*, ANSI Y14.5M-1982, New York: American Society of Mechanical Engineers.
2. FOSTER, L. W., *Geometric Dimensioning and Tolerancing*, Reading, Mass.: Addison-Wesley Publishing company, Inc., 1970.
3. HUME, K. J., *Engineering Metrology*, 2nd ed., London: Macdonald & Co. (Publishers) Ltd., 1963.
4. KISSAM, O., *Optical Tooling*, New York: McGraw-Hill Book Company, 1962.
5. LEVY, S. J., *Applied Geometric Tolerancing*, Beverly, Mass.: TAD Products Corp., 1974.
6. *Military Standard, Dimensioning and Tolerancing*, MIL-STD-8C, Oct. 16, 1963, Washington, D.C.: U.S. Government Printing Office.
7. MILLER, L., *Engineering Dimensional Metrology*, London: Edward Arnold (Publishers) Ltd., 1962.

8. ROLT, F. H., *Gauges and Fine Measurements*, Vols. I and II, London: Macmillan Publishers Ltd., 1929.

9. ROTHBART, H. A., Ed., *Mechanical Design and Systems Handbook*, New York: McGraw-Hill Book Company, 1964, Section 19, "Fabrication Principles."

10. *Dual Dimensioning*, SAE J390, Warrendale, Pa.: Society of Automotive Engineers, Inc., 1971.

11. SPOTTS, M. F., "Allocation of Tolerances to Minimize Cost of Assembly," *Trans. ASME*, **95**, 1973, Ser. B, *J. Eng. Ind.*, p. 762.

12. SPOTTS, M. F., *Dimensioning and Tolerancing for Quantity Production*, Englewood Cliffs, N.J.: Prentice-Hall, Inc., 1983.

13. WHITEHEAD, T.N., *Design and Use of Instruments and Accurate Mechanisms*, New York: Dover Publications, Inc., 1933.

14. WILSON, F. W., Ed., *Tool Engineers Handbook*, 2d ed., New York: McGraw-Hill Book Company, 1959, Section 86, "Fits Tolerances and Allowances," and Section 87, "Inspection Equipment and Methods."

PROBLEMS

1. Draw and dimension the bar diagram for a $3\frac{3}{8}$-in. shaft, Class 3 fit. Draw and dimension a typical hole and shaft for this fit. Use unilateral tolerances and the basic hole system for maximum metal. Use dimensions to nearest 0.0001 in.

2. Draw and dimension the bar diagram for a $3\frac{3}{4}$-in. shaft, Class 2 fit. Draw and dimension a typical hole and shaft for this fit. Use unilateral tolerances and the basic shaft system for minimum metal. Use dimensions to nearest 0.0001 in.

3. Draw a typical detail for a $3\frac{7}{16}$-in. shaft and bearing with an allowance of 0.0025 in. Shaft tolerance is 0.0015 in.; bearing tolerance is 0.0020 in. Use unilateral tolerances.

(a) Dimension for basic shaft system and maximum metal.
(b) Dimension for basic shaft system and minimum metal.
(c) Dimension for basic hole system and maximum metal.
(d) Dimension for basic hole system and minimum metal.

4. Draw and dimension the bar diagram for an 8-in. shaft, Class 8 fit. Draw and dimension a typical hole and shaft for this fit. Use unilateral tolerances and the basic hole system for maximum metal. Use dimensions to nearest 0.0001 in.

5. Draw and dimension the bar diagram for a 4-in. shaft, Class 3 fit. Compute the value of the ratio of the diametral clearance to the shaft diameter for the fit at the tightest condition. Do the same for the loosest permissible fit. Draw and dimension the bar diagram for a 4-in. bearing for a minimum value for the c/r ratio of 0.0015 and for a maximum value of c/r of 0.0020. Compare the tolerance conditions of the bearing with those of a Class 3 fit.

6. A part having a hole of 1.000 in. diam is mated with a shaft of diameter 0.995 in. What class or classes of standardized cylindrical fits can the resulting assembly be designated? Answer the question for shaft diameters of 0.997, 0.9975, 0.9988, and 0.9992 in.

7. (a) Draw and dimension the bar diagram for a 6-in. shaft, Class 7 fit. Draw and dimension a typical hole and shaft for this fit. Use unilateral tolerances and basic shaft system for minimum metal. Use dimensions to nearest 0.0001 in.

(b) Suppose the shaft is of steel, and the disk is of cast iron 12 in. in diameter. What is the value of the maximum tangential stress in case the maximum size shaft is assembled with a minimum diameter hole? What is the value of the stress if the minimum diameter shaft is assembled with the maximum size hole? $\mu = 0.3$. For cast iron, $E = 15,000,000$ psi.   *Ans.*   (b) Tightest fit, $s_t = 7,370$ psi; Loosest fit, $s_t = 3,420$ psi.

8. Two parts with through holes are to be assembled with a $\frac{3}{4}$-in. loose bolt. Centers are dimensioned to coincide. The specified minimum hole diameter is $\frac{25}{32}$ in. For true position dimensioning find the tolerance T that should be shown in the tolerance box.

9. A part with a through hole is to be assembled to another with a $\frac{5}{8}$-in. cap screw. Centers are dimensioned to coincide. Minimum hole diameter is $\frac{21}{32}$ in. If the tolerance circle for the center of the through hole is 0.010 in. find the diameter of the tolerance circle for the tapped hole.

10. Find the value of e_1 in Fig. 13-37 if it is desired that the stem of the stud have a clearance of 0.001 in. along each side when two maximum material parts are assembled with the maximum permissible eccentricities facing each other. Make a drawing showing this situation and dimension completely.

FIGURE 13-37   Problem 10.

11. In the assembly of Fig. 13-38, find the maximum and minimum values of dimension d if the tolerances are additive. Find the values by the Gaussian rule.

12. In the assembly of Fig. 13-39, find the maximum displacement of the right ends with respect to each other if the tolerances are additive. Find the displacement if the tolerances are Gaussian.

FIGURE 13-38   Problem 11.   **FIGURE 13-39   Problem 12.**

13. In the assembly of Fig. 13-40, find the maximum variation of the right end of *D* with respect to the datum if the tolerances are additive. Find the variation if the tolerances are Gaussian.

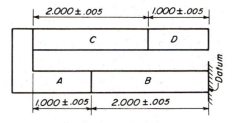

FIGURE 13-40   Problem 13.

14. The tolerance $\pm w$ on the gap $\bar{w}$ in Fig. 13-41 must be equal to ± 0.009 in.

(a) If the tolerances on the separate parts are all equal, what tolerance must be placed on each part if it is assumed that the additive rule applies?

(b) Suppose the production lengths fulfill all the necessary statistical assumptions. Find the tolerance that can now be placed on each part.

Ans.   (b) ± 0.003 in.

FIGURE 13-41   Problem 14.

15. A hole is located by the dimensions given in Fig. 13-42. Find the *x*- and *y*-coordinates of the hole.   *Ans.* $x = 0.45$ in.; $y = 0.60$ in.

FIGURE 13-42   Problem 15.

16. A hole is located by the dimensions given in Fig. 13-43. Find the x- and y-coordinates correct to the nearest 0.0001 in.

Ans. $x = 0.8996$ in.; $y = 1.4790$ in.

FIGURE 13-43   Problem 16.

17. A $\frac{3}{4}$-in. loose bolt is to be used for an assembly. Find the minimum diameter H required for the hole if its location tolerance T is 0.0132 in.

18. The parts of Fig. 13-44 are to be assembled with $\frac{5}{8}$-in.-diameter loose bolts. Find the tolerance T required for the location of the holes.

FIGURE 13-44   Problem 18.

19. Suppose that one of the parts of Fig. 13-44 is to be attached by means of $\frac{5}{8}$-in. cap screws. Find the tolerance T_1 for the location of the tapped holes if tolerance T is kept the same as T for Problem 18.

20. A loose No. 10–24 machine screw and nut are to be used in an assembly. The hole is dimensioned with tolerances ± 0.005 in. in both the vertical and horizontal directions. Find the minimum diameter H of the required hole.

21. A loose $\frac{3}{8}$-in. hexagonal bolt is to be used for assembly. The minimum hole diameter is $\frac{13}{32}$ in. The tolerance circle diameter for the hole is 0.015 in. The true positions for the hole centers coincide. Determine if assembly is possible.

22. The two parts of Fig. 13-45 are to be assembled with a No. 4–40 NC machine screw. Find the value of tolerance T_1 that should be shown on the drawing for the lower part. *Ans.* $T_1 = 0.0078$ in.

FIGURE 13-45   Problem 22.

23. Two parts are to be assembled with a $\frac{3}{8}$-in. cap screw. True-position dimensions for the through hole and tapped hole coincide. The minimum diameter H for the through hole is $\frac{13}{32}$ in. Find the values for tolerances T and T_1 that should appear on the drawing if T and T_1 are to be equal.

24. A No. 12–24 machine screw is to be used for assembly in a tapped hole. The minimum hole diameter is $\frac{1}{4}$ in. The tolerances T and T_1 are each 0.015 in. The true positions for the centers are dimensioned to coincide. Check if assembly is possible.

25. A series of parts such as those in Fig. 13-46 are to be assembled next to each other. Determine the value of tolerance T_1 to ensure assembly.

26. Two parts are to be assembled as shown in Fig. 13-47. Determine if the tolerancing for the sides of the notch and tab have been correctly done to ensure assembly. Determine the width of the tab on the gage for inspecting the groove location. Determine the width of the slot on the gage for inspecting the tab location.

27. The minimum diameter H of a hole is $\frac{21}{32}$ in. It is dimensioned 2.000 ± 0.008 in. horizontally and 1.000 ± 0.006 in. vertically. The part is to be assembled to another with a $\frac{5}{8}$-in. cap screw with T_1 equal to 0.010 in. Centers for the through

FIGURE 13-46   Problem 25.

FIGURE 13-47   Problem 26.

hole and the tapped hole are dimensioned to coincide. Determine if assembly will be possible.

28. Two parts are assembled with a loose $\frac{1}{2}$-in. hexagonal head bolt. Suppose that the design is changed and the assembly is to be made by a $\frac{1}{2}$-in. cap screw. What tolerance T_1 for the tapped hole location should appear on the revised drawing?

Ans. $T_1 = 0.015$ in.

29. A hole center is dimensioned 3.2183–3.2054 in. horizontally and 1.8271–1.8142 in. vertically. Assembly is to be made by a $\frac{3}{8}$-in. loose hexagonal head bolt. Find the minimum hole diameter H that should appear on the drawing.

Ans. $H = \frac{13}{32}$ in.

30. Two parts are to be assembled with a loose $\frac{3}{4}$-in. hexagonal head bolt. The centers are dimensioned 6.0356–6.0000 in. horizontally and 8.0267–8.0000 in. vertically. Find the minimum hole diameter H that should appear on the drawing.

Ans. $H = \frac{13}{16}$ in.

31. Two parts are to be assembled by a loose hexagonal head bolt. The hole diameters are $\frac{17}{32}$ in. Centers have the horizontal dimension 4.0097–4.000 in. The vertical dimension is 5.0130–5.0000 in. Calculate and find a suitable-size bolt for this connection.

32. Two parts are to be assembled with a $\frac{1}{4}$-in. loose hexagonal head bolt. The hole diameter is $\frac{9}{32}$ in. The horizontal dimension to center is 2.0170–2.0000 in. The vertical dimension to the center is X.XXXX–1.5000 in. Find the value of dimension X.XXXX that should appear on the drawing.

33. A part with a $\frac{5}{8}$-in. through hole is to be assembled with a $\frac{9}{16}$-in. cap screw. The centers for the hole and the cap screw are dimensioned to coincide. Tolerances on hole dimensions are ±0.012 in. horizontally and ±0.009 in. vertically. If the horizontal and vertical tolerances for the cap screw are equal, find their values.

Ans. $u_1 = v_1 = 0.0115$ in.

34. A part with a $\frac{3}{16}$-in.-diameter through hole is to be assembled with a machine screw. The tolerance circle for the hole has a diameter of 0.010 in. The tolerance circle for the screw has a diameter of 0.012 in. Determine the largest standard machine screw that can be specified for the assembly.

35. The through hole for an assembly, to be made by a $\frac{1}{4}$-in. cap screw, has a minimum diameter of $\frac{5}{16}$ in. The tolerance circle for the through hole has a diameter of 0.015 in. and for the tapped hole 0.020 in. Determine if assembly will be possible.

36. $H - T = F_1 + T_1$ or $H - F_1 = T + T_1$ or $H - F_1 = 2T$ if $T = T_1$. Note that when $H - F_1$ has values of $\frac{1}{32}, \frac{1}{16}, \frac{3}{32}, \frac{1}{8}$, and so on, that the value of T or T_1 is equal to $\frac{1}{64}, \frac{1}{32}, \frac{3}{64}, \frac{1}{16}$, and so on, respectively. Tabulate these values.

37. Determine in Fig. 13-48 whether the width of the notch and the width of the tab have been correctly dimensioned for assembly.

FIGURE 13-48   Problem 37.

38. A part is assembled with a $\frac{5}{8}$-in. cap screw. The through hole has a minimum diameter of $\frac{21}{32}$ in. The tolerance circle for both hole and stud are 0.015 in. diameter each. Check to see if assembly will be possible.

39. Let the top and bottom surface for the rectangular body in Fig. 13-49 be planes. They were intended to be parallel but after fabrication the corner heights were measured as $z_a = 2.000$ in., $z_b = 1.995$ in., $z_c = 1.987$ in., and $z_d = 1.992$ in. Find the angle that the top surface makes with the bottom surface.

Ans. 0.00126°.

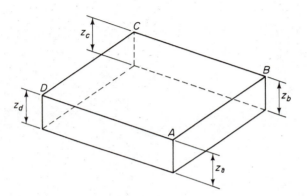

FIGURE 13-49 **Problem 39.**

40. Same as Problem 39 except $z_a = 2.000$ in., $z_b = 1.992$ in., $z_c = 1.987$ in., and $z_d = 1.995$ in. *Ans.* 0.00138°.

41. Same as Problem 39 except $z_a = 1.750$ in., $z_b = 1.730$ in., $z_c = 1.718$ in., and $z_d = 1.738$ in. *Ans.* 0.00342°.

42. Same as Problem 39 except $z_a = 1.750$ in., $z_b = 1.738$ in., $z_c = 1.718$ in., and $z_d = 1.730$ in. *Ans.* 0.00310°.

43. Same as Problem 39 except $z_a = 1.500$ in., $z_b = 1.490$ in., $z_c = 1.475$ in., and $z_d = 1.485$ in. *Ans.* 0.00243°.

44. Same as Problem 39 except $z_a = 1.500$ in., $z_b = 1.485$ in., $z_c = 1.475$ in., and $z_d = 1.490$ in. *Ans.* 0.00262°.

14
Engineering Materials

The composition and physical properties of the principal materials used in machine construction will be discussed in this chapter. It is fully as important that a machine element be made of a material that has properties suitable for the conditions of service as it is for the loads and stresses to be accurately determined. Frequently, the limitations imposed by the material are the controlling factors in a design. The designer must be acquainted with the effects that the methods of manufacture and heat treatment have on the properties of engineering materials. The manufacturing process used for fabrication of the part will also influence the type of material that can be used. Sometimes the help of a professional metallurgist is needed to ensure the best possible choice of material and heat treatment.

1. The Tension Test

The tension test is widely used for determining yield point stress, ultimate or tensile stress, breaking or rupture stress, elongation, and reduction of area for various materials. Mild- or low-carbon steel is unique among the metals in having a pronounced elongation at the yield point. The phenomenon of the upper and lower yield points, shown in Fig. 14-1(b), is sometimes explained by defining the upper yield point as the stress which will start the flow process, and the lower yield point as the lowest stress at which the extension proceeds still further. The shape of the stress-strain curve is influenced to a large extent by the type of machine on which the tests are made. The constant-strain-rate

(a) Constant strain rate type of testing machine

(c) Constant load rate type of testing machine

(b) Stress-strain diagram from constant strain rate type machine

(d) Stress-strain diagram from constant load rate type machine

FIGURE 14-1 Schematic drawings of testing machines and corresponding stress-strain diagrams for low-carbon steel.

type of machine, shown in Fig. 14-1(a), stretches the specimen at a uniform rate by means of power-driven screws. As soon as yielding starts, the poise must be reverted to the left to keep the beam in balance between the stops at the right end. The material continues to yield until it is strengthened by strain or work hardening. A higher stress then is necessary for additional stretching.

In the second type of machine, illustrated in Fig. 14-1(c), the beam is allowed to drop as rapidly as the specimen stretches from the effect of a constantly increasing load.[1] No reduction in stress can occur with this kind of loading. The stress-strain diagram which appears looks like that shown in Fig. 14-1(d). The diagram from the constant-load-rate type of machine depends on the rapidity with which the load is applied. Experiments have shown that the yield point stress σ_1 and yield point elongation ϵ_1 are increased as the speed of loading increases. Some diagrams obtained on $\frac{1}{2}$ by $\frac{3}{16}$ in. mild-steel strips are shown in Fig. 14-2.

FIGURE 14-2   Stress-strain diagrams for constant load rate tests for mild steel.

The conditions under which constant-strain-rate tests are made are somewhat analogous to drawing or rolling operations in metal forming. However, machine parts are employed in such a manner that if an overload causes the yield point of the material to be exceeded, the deformations can take place as rapidly as the stretch of the material permits.

Alloy and heat-treated steels, steels which have been cold worked, cast iron, and nonferrous metals do not exhibit the well-defined yield point of mild steel. The yield strength for such materials is generally taken as the stress corresponding to some arbitrarily selected value of permanent deformation which the material can have without undergoing appreciable structural damage. This elongation is often taken as 0.2%, as shown in Fig. 14-3, although smaller values are sometimes used.

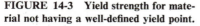

FIGURE 14-3   Yield strength for material not having a well-defined yield point.

---

[1] See Davis, E. A., "The Effect of Speed of Stretching and the Rate of Loading on the Yielding of Mild Steel," *Trans. ASME,* **60,** 1938, p. A-137; **61,** 1939, p. A-89.

Stretching of a soft steel specimen is accompanied by a decrease in diameter. After the ultimate stress is reached, a pronounced reduction of area, called necking down, occurs at some point in the stressed region. For the constant-strain-rate type of machine, the total load which the specimen can then support with the beam in balance is reduced. It is standard practice to compute the stresses for the stress–strain diagram using the original cross-sectional area. There is thus an apparent decrease of stress in the right-hand portion of the diagram. This supposed reduction does not occur if the actual values of the cross sections are determined and used for computing the stress as shown by the dashed curve in Fig. 14-1(b).

Failure under torsional loading is not accompanied by appreciable elongation or reduction of area. For ductile material the fracture occurs on a plane normal to the axis. For brittle materials the failure surface is helical and is inclined at approximately 45° to the axis.

2. Physical Constitution of Steel

Steel is an alloy of iron and carbon, or of iron, carbon, and other alloying elements. Carbon must be present to the extent of about 0.05% by weight in order for the material to be known as steel rather than commercial iron. The quantity of carbon is much higher in cast iron, about 2 to 4%. Steel, like other metals, has a crystalline structure when viewed under the microscope, with grains varying in size from about 0.001 in. to 0.010 in. in diameter.

A number of different metallurgical terms are used in describing the structure of steel. Pure iron, for example, is known as ferrite. Ferrite is soft and ductile and has a tensile strength of about 40,000 psi. Cementite is the chemical compound of iron and carbon, Fe_3C. It is nonductile and very hard. Another constituent of steel is pearlite. As shown in Fig. 14-4, pearlite grains have a laminar structure of alternate layers of ferrite and cementite. The carbon content of pearlite is 0.83% by weight. When slowly cooled from above the upper critical temperature, a steel of less than 0.83% carbon is composed of a mixture of grains of ferrite and pearlite. For carbon percentages higher than 0.83, the steel is a mixture of grains of pearlite and cementite upon slow cooling. When the percentage of carbon is exactly 0.83, the so-called "eutectoid structure" results, which is wholly composed of grains of pearlite.

When steels are heated, transformations in the crystal structure take place at certain temperatures. The magnified grains of a 0.30% carbon steel are shown schematically in Fig. 14-5. When the steel is heated to the lower critical temperature of 1330°F, a new constituent, called austenite, appears. Austenite is formed from pearlite by the carbon of the cementite going into solid solution in the ferrite. The situation is analogous to the solution of a

FIGURE 14-4   Pearlite, 0.9% carbon steel, magnification 1,000 diameters.

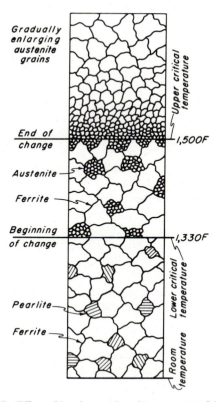

FIGURE 14-5   Effect of heating on the microstructure of 0.30% C steel.

salt in water, except that the solvent, in the case of metals, is in the solid condition. As the temperature continues to rise, the austenite dissolves the free ferrite until at the upper critical temperature the steel consists wholly of austenite.

The upper critical temperature depends on the carbon content, as indicated by a portion of the iron-carbon equilibrium diagram shown in Fig. 14-6. The upper critical temperature, for any percentage of carbon, is given by line *ACE*, and the lower critical temperature is represented by line *BCD*. Above the lower critical temperature, the pearlite, for all values of the carbon content, has become austenite. In region *ACB*, the austenite dissolves the free ferrite as the temperature rises, and in region *ECD*, the austenite dissolves

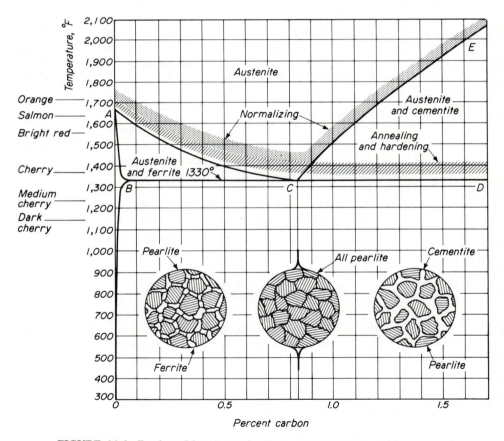

FIGURE 14-6 Portion of iron-cementite diagram showing compositions, critical temperatures, and temperature ranges for normalizing, annealing, and hardening.

the free cementite. Above the upper critical temperature, the steel is composed wholly of austenite. A reversal of these changes takes place if the steel is slowly cooled from above the upper critical temperature. Austenite rarely exists at room temperature except in certain highly alloyed steels.

The grain structure at room temperature, for different amounts of carbon, is shown by the three circles in the lower part of the figure.

3. Types of Steel Used in Machine Construction

The selection of a suitable steel for a machine part depends on a full understanding of the distribution and fluctuation of the stress. Knowledge is also required of any unusual conditions that may be present, such as stress concentration, impact, corrosion, abrasion, and high or low temperatures. The type of steel selected depends on whether or not the part is to be heat treated. The following summary gives a general classification of steels as used in machine construction:

(1) Not Heat Treated
   (a) Plain, low-carbon steel, hot rolled or cold drawn.
   (b) Free cutting steels or steels resulphurized for ease of machining.
   (c) Low-carbon sheets for forming, drawing, or spinning.
   (d) High-strength low-alloy steels, HSLA.
(2) Heat Treated
   (a) Carburized. These are low-carbon or low-alloy steels.
   (b) Oil or water quenched and then tempered. These steels contain considerable amounts of carbon or alloying elements.

4. Numbering Systems for Carbon and Alloy Steels

Many different carbon and alloy steels are used for the construction of machinery. The numbering system of the Society of Automotive Engineers and the American Iron and Steel Institute is based on the chemical composition, and provides a simple means whereby any particular steel can be specified. In general, this catalog system uses a number composed of four digits. The first two digits indicate the type or alloy classification; the last two (and in some cases, three) digits give the carbon content. Carbon has such a pronounced effect on the strength and hardness of steel that the content is thus incorporated in the catalog number. Plain carbon steel, for example, is denoted by the basic numeral 10. Thus, steel 1045 indicates a plain carbon steel containing 0.45% carbon. The basic numerals for the various types of steels are given in Table 14-1. The arrangement of Table 14-2 may serve as an aid in committing some of the series numbers to memory.

TABLE 14-1   Basic Numerals for SAE and AISI Steels

Carbon steels	1xxx	Molybdenum steels	4xxx
Plain carbon	10xx	Carbon-molybdenum	40xx
Free-cutting screw stock	11xx	Chromium-molybdenum	41xx
Manganese steels	13xx	Chromium-nickel-molybdenum	43xx
Nickel steels	2xxx	Nickel-molybdenum; 1.75% Ni	46xx
3.50% Ni	23xx	Nickel-molybdenum; 3.50% Ni	48xx
5.00% Ni	25xx	Chromium steels	5xxx
Nickel-chromium steels	3xxx	Low chromium	51xx
1.25% Ni, 0.60% Cr	31xx	Medium chromium	52xxx
1.75% Ni, 1.00% Cr	32xx	Corrosion and heat resisting	51xxx
3.50% Ni, 1.50% Cr	33xx	Chromium-vanadium steels	6xxx
Corrosion and heat resisting	30xxx	1.00% Cr	61xx
		Silicon-manganese steels	9xxx
		2.00% Si	92xx
			AISI
		Chromium-nickel-molybdenum steels	86xx
		Chromium-nickel-molybdenum steels	87xx

TABLE 14-2   Basic Numerals for Several Widely Used Steels

Nickel	23xx, 25xx
Nickel-chromium	31xx, 32xx, 33xx
Chromium- molybdenum	41xx
Chromium-nickel-molybdenum	43xx
Nickel-molybdenum	46xx, 48xx

The AISI number for a steel is similar to the SAE number, but capital letter prefixes are included to indicate the process of manufacture as follows:

      A—Basic open-hearth alloy steel
      B—Acid Bessemer carbon steel
      C—Basic open-hearth carbon steel
      D—Acid open-hearth carbon steel
      E—Electric furnace alloy steel

The chemical compositions for a selected number of steels used for machinery are given in Table 14-3. The mechanical properties for these materials are given in Tables 14-4 and 14-5. The SAE and AISI lists of steels are constantly being revised, and it is necessary to consult an up-to-date list if one is in doubt as to whether or not a particular steel is in regular production.[2]

TABLE 14-3   Chemical Composition of Selected Carbon and Alloy AISI and SAE Steels (Prefixes A, B, or C Omitted in the SAE System)

AISI	Carbon	Manganese	Sulphur
Plain Carbon Steels			
C1010	0.08–0.13	0.30–0.60	0.05 max
C1020	0.18–0.23	0.30–0.60	0.05 max
C1035	0.32–0.38	0.60–0.90	0.05 max
C1045	0.43–0.50	0.60–0.90	0.05 max
C1060	0.55–0.65	0.60–0.90	0.05 max
C1095	0.90–1.05	0.30–0.50	0.05 max
Free-Cutting Steels			
B1112	0.08–0.13	0.70–1.00	0.16–0.23
B1113	0.08–0.13	0.70–1.00	0.24–0.33
C1117	0.14–0.20	1.00–1.30	0.08–0.13
C1137	0.32–0.39	1.35–1.65	0.08–0.13
C1141	0.37–0.45	1.35–1.65	0.08–0.13
Nickel Steels			*Nickel*
A2317	0.15–0.20	0.40–0.60	3.25–3.75
High Strength Low-Alloy Steel*			
Cor-Ten	0.12 max / 0.25–0.55 Cu	0.20–0.50 / 0.30–1.25 Cr	0.25–0.75 Si / 0.65 max Ni
T₁ type A	0.12–0.21 / 0.40–0.65 Cr	0.70–1.00 / 0.15–0.25 Mo	0.20–0.35 Si / 0.03–0.08 V
T₁ type B	0.12–0.21 / 0.30–0.70 Ni / 0.03–0.08 V	0.95–1.30 / 0.40–0.65 Cr	0.20–0.35 Si / 0.20–0.30 Mo

AISI	Carbon	Manganese	Nickel	Chromium	Molybdenum
Nickel-Chromium Steels					
A3140	0.38–0.43	0.70–0.90	1.10–1.40	0.55–0.75	
Chromium-Molybdenum Steels					
A4140	0.38–0.43	0.75–1.00		0.80–1.10	0.15–0.25
Nickel-Chromium-Molybdenum Steels					
A4320	0.17–0.22	0.45–0.65	1.65–2.00	0.40–0.60	0.20–0.30
A4340	0.38–0.43	0.60–0.80	1.65–2.00	0.70–0.90	0.20–0.30
Nickel-Molybdenum Steels					
A4620	0.17–0.20	0.45–0.65	1.65–2.00		0.20–0.30
A4640	0.38–0.43	0.60–0.80	1.65–2.00		0.20–0.30
A4815	0.13–0.18	0.40–0.60	3.25–3.75		0.20–0.30
Plain Chromium Steels					
E52100	0.98–1.10	0.25–0.45		1.30–1.60	
Chromium-Vanadium Steels					
A6150	0.48–0.53	0.70–0.90		0.80–1.10	0.15 min. V
Nickel-Chromium-Molybdenum Steels					
A8620	0.18–0.23	0.70–0.90	0.40–0.70	0.40–0.60	0.15–0.25
A8640	0.38–0.43	0.75–1.00	0.40–0.70	0.40–0.60	0.15–0.52
Silicon-Manganese Steels					
A9260	0.56–0.64	0.70–1.00			1.8–2.2 Si

*Proprietary steels, USS Corp. Similar products of other makers are marketed under different trade names.

[2]"Hot-rolled" and "as-rolled" are usually considered synonymous terms. Similarly "cold-drawn," "cold-rolled," and "CRS" (if steel) are used interchangeably.

Tests have shown that the tensile strength and Brinell Hardness Number have a direct relationship. For most wrought steels, the tensile strength is found to be within 450 to 500 times the BHN. Tests show that the ductility becomes less as the hardness increases.

TABLE 14-4   Mechanical Properties of Typical Low-Carbon and Case-Hardening Steels*

Steel	Condition	Tensile Strength, psi	Yield Strength, psi	% Elong. in 2 in.	% Red. in area	BHN	Rock-well	Machin-ability, %
1010	Hot-rolled	51,000	29,000	38	70	101		40
	Cold-drawn	56,000	33,000	35	65	113		45
1020	Hot-rolled	65,000	43,000	36	59	143	B79	50
	Cold-drawn	78,000	66,000	20	55	156	B83	60
1112	Hot-rolled	67,000	40,000	27	47	140		
	Cold-drawn	82,500	71,000	15	43	170	B87	100
1113	Cold-drawn	83,500	72,000	14	40	170	B87	130
1117	Hot-rolled	71,000	45,000	33	64	143	B78	90
	Cold-drawn	86,000	75,000	22	52	170	B87	95
2317†	Hot-rolled	85,000	56,000	29	60	163		50
	Cold-drawn	95,000	75,000	25	58	197	C12	
4320†	Hot-rolled	87,000	59,000	30	60	179	C10	55
	Cold-drawn	99,000	65,000	23	54	207	C16	
4620	Hot-rolled	85,000	63,000	28	64	183	B90	58
	Cold-drawn	101,000	85,000	22	60	207	C15	64
4815†	Hot-rolled	105,000	73,000	24	58	212	C15	55
	Cold-drawn	110,000	78,000	23	55	217	C17	
8620	Hot-rolled	89,000	65,000	25	63	192	B90	60
	Annealed	102,000	85,000	22	58	212	C16	63
HSLA‡								
	Cor Ten	70,000	50,000	24				
	T₁ type A	115–135,000	100,000	18	40			
	T₁ type B	115–135,000	100,000	18	40			

*Taken mainly from *Ryerson Data Book*, Chicago, Ill.: Joseph T. Ryerson & Son, Inc. Table indicates probable median expectation of the steels listed based on a 1-in. round section. In individual tests, these steels may develop results differing considerably from those shown, particularly for cold drawn materials.

†Steels 2317, 4320, and 4815 usually not stocked by steel warehouses.

‡High Strength Low Alloy. These are proprietary steels, USS Corp. Similar products of other makers are marketed under different trade names.

TABLE 14-5   Mechanical Properties of Typical Medium-Carbon
or Direct-Hardening Steels*

Steel	Condition	Tensile Strength, psi	Yield Strength, psi	% Elong. in 2 in.	% Red. in area	BHN	Rock- well	Machin- ability, %
1035	Hot-rolled	85,000	54,000	30	53	183	B90	65
	Cold-drawn	92,000	79,000	25	50	201	B94	67
	Water Q1550°F Drawn 1000°F	103,000	72,000	23	59	201	B94	
1045	Hot-rolled	98,000	59,000	24	45	212	C16	56
	Cold-drawn	103,000	90,000	14	40	217	C18	60
1060†	Hot-rolled	95,000	59,000	25	52	197	C14	53
	Oil Q1550°F Drawn 1000°F	122,000	90,000	19	53	255	C25	
1095	Hot-rolled	142,000	83,000	18	38	293	C28	
	Water Q1450°F Drawn 800°F	200,000	138,000	12	37	388	C42	
1137	Hot-rolled	92,000	57,000	27	61	192	B92	70
	Cold-drawn	105,000	90,000	15	38	207	C15	75
	Oil Q1550°F Drawn 1000°F	112,000	88,000	21	56	255	C25	
3140†	Hot-rolled	96,000	64,000	26	56	195	C12	57
	Cold-drawn	115,000	98,000	17	45	248	C24	
4140	Hot-rolled	89,000	62,000	26	58	187	B91	57
	Cold-drawn	102,000	90,000	18	50	223	C19	66
4340	Hot-rolled	101,000	69,000	21	45	207	C15	45
	Cold-drawn	111,000	99,000	16	42	223	C19	55
	Oil Q1550°F Drawn 1000°F	182,000	162,000	15	40	363	C39	
4640†	Hot-rolled	100,000	87,000	21	50	201	C12	60
	Cold-drawn	126,000	97,000	14	39	269	C27	
52100†	Hot-rolled	109,000	80,000	25	57	235	C22	45
	Oil Q1550°F Drawn 1000°F	185,000	170,000	9	34	415	C43	
6150	Hot-rolled	103,000	70,000	27	51	217	C18	50
	Cold-drawn	118,000	94,000	20	43	255	C25	
8740	Hot-rolled	95,000	64,000	25	55	190	B92	56
	Cold-drawn	107,000	96,000	17	48	223	C19	66
9260†	Hot-rolled	142,000	92,000	18	38	302	C31	

*See notes at bottom of Table 14-4.
†Steels 1060, 3140, 4640, 52100, and 9260 usually not stocked by steel warehouses.

5. Plain Carbon Steel

When the only alloying element in steel is carbon, it is known as plain carbon steel. Carbon is a powerful alloying agent, and wide changes in strength and hardness can be secured by varying the amount of this element.

The range of desired properties can be further enhanced by heat treatment. Some of the many uses for plain carbon steels are given in Table 14-6. This

TABLE 14-6   Different Uses of Plain Carbon Steels

Carbon Range, %	Uses of Carbon Steel
0.05–0.10	Stampings, sheets, wire, rivets, welding stock, cold-drawn parts
0.10–0.20	Structural shapes, machine parts, carburized parts, screws
0.20–0.30	Gears, shafting, levers, welded tubing, carburized parts
0.30–0.40	Seamless tubing, shafts, connecting rods, crane hooks, axles; can be heat treated
0.40–0.50	Forgings, shafts, gears, studs
0.60–0.70	Drop hammer dies, set screws, locomotive tires, lock washers, hard-drawn spring wire
0.70–0.80	Plow beams, cultivator disks, anvil faces, band saws, hammers, wrenches
0.80–0.90	Plow shares, shovels, harrow blades, punches, rock drills, cold chisels, hand tools, music wire, leaf springs
0.90–1.00	Springs, knives, axes, dies, hay rake teeth, harrow blades
1.00–1.10	Drills, taps, milling cutters, knives
1.10–1.20	Drills, lathe tools
1.20–1.30	Files, reamers, knives, metal cutting tools
1.25–1.40	Razors, saws, wire-drawing dies, metal-cutting saws

type of steel is usually easier to anneal and machine than is alloy steel. Plain carbon steel is the cheapest, and the designer will normally use it except where severity of service conditions or difficulty of heat treatment necessitates the use of a more expensive alloy steel. Carbon steels are manufactured in larger quantities than any of the others.

The thickness of the standard gages of carbon steel sheets are shown in Table 14-7.

TABLE 14-7   Thickness of Standard Gages for Carbon Steel Sheets, Inches

Gage	Thickness	Gage	Thickness	Gage	Thickness	Gage	Thickness
7	0.1793	13	0.0897	19	0.0418	25	0.0209
8	0.1644	14	0.0747	20	0.0359	26	0.0179
9	0.1494	15	0.0673	21	0.0329	27	0.0164
10	0.1345	16	0.0598	22	0.0299	28	0.0149
11	0.1196	17	0.0538	23	0.0269	29	0.0135
12	0.1046	18	0.0478	24	0.0239	30	0.0120

Carbon steels can be hardened and therefore strengthened by heat treatment when the carbon content is about 0.30% or more. Heat treatment generally consists of heating the part above the upper critical temperature, as shown by the zones in Fig. 14-6, and then quenching by submerging in a cooling medium such as water. After quenching, the steel will be very strong, but it may not be suitable for use because of lack of ductility and the presence of residual stresses. Quenching stresses are removed and ductility is restored, but tensile strength is reduced, by a subsequent operation called tempering or drawing. This operation consists of reheating the part and then quenching or cooling it in air. The reduction in the tensile strength and the increase in the ductility depend on the temperature used for tempering.

During quenching, the heat can be dissipated only through the surface where the part is in contact with the cooling medium. For parts of large cross section and large amounts of heat, the cooling rate for the interior may be so low that the metallurgical changes required for hardening cannot be secured. Under such conditions, only the material near the surface will be hardened and therefore strengthened; the interior of the body will largely be unaffected. The situation is generally referred to as mass effect or the effect of section size.

Curves for the mechanical properties of plain carbon steel 1045 for various tempering temperatures and various section sizes are given in Fig. 14-7. The curves show average values only. Fluctuations are to be expected; they are due principally to the permissible variations in composition, as indicated by Table 14-3.

The principal shortcoming of plain carbon steel lies in the shallow penetration of the hardening. The cooling rate must be very rapid, and high residual stresses, distortion, lack of ductility, and even quenching cracks will occur. The use of plain carbon steel in applications requiring only a hardened surface is sometimes prevented by the foregoing disadvantages.

Wrought iron, although its method of manufacture is different, is essentially a very low carbon steel with slag inclusions. Due to rolling, the slag inclusions are elongated into long stringers and do not greatly reduce the strength in the longitudinal direction. Wrought iron is tough and welds easily. It is widely used for making iron pipe.

6. Alloy Steels

When a suitable alloy is present in addition to the carbon, the metallurgical changes which take place during quenching occur at a faster rate, the cooling effects penetrate deeper, and a larger portion of the part is strengthened. The quenching can usually be done in oil. The heat is removed less drastically, and the residual stresses are smaller than for carbon steel. Hence there is less danger of distortion or formation of hardening cracks. A carbon steel

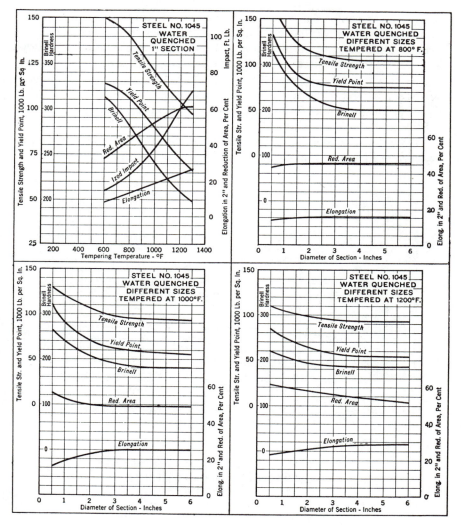

In section ½" to 2" incl., quenched from 1475/1525° F.; over 2" to 4" incl., from 1500/1550° F.; over 4", from 1525/1575° F.

FIGURE 14-7 Properties of water quenched and tempered carbon steel 1045 in different sizes. (*Courtesy International Nickel Co.*)

part that might distort or crack from water quenching can sometimes be made of an alloy steel and quenched in oil. This ability to confer depth of hardening, or hardenability, is one of the principal reasons for the use of alloys. Of the various alloying elements, manganese is used in almost all alloy steels to confer depth of hardening. Nickel, chromium, and molybdenum, singly or in various combinations, are widely used. The maximum hardness obtainable in a thin section of heat-treated steel depends upon the carbon content and is largely unaffected by the lack or presence of any alloys.

The mechanical properties for 1 in.-round test pieces of chromium-molybdenum steel 4140 at various tempering temperatures are shown in Fig. 14-8. The properties for this material in different-sized sections are given in Table 14-8.

Curves showing average mechanical properties for steels 4340 and 8742 are given in Figs. 14-9 and 14-10, respectively. The properties of steel A8640

FIGURE 14-8 **Properties of normalized and tempered chromium-molybdenum steel 4140.** (*Courtesy Climax Molybdenum Co.*)

TABLE 14-8   **Effect of Mass on the Physical Properties of**
Chromium Molybdenum Steel 4140*
(*Normalized and tempered at 1000°F*)

Size, in.	Outside Edge of Section						Center of Section					
	Tensile Strength	Yield Point	Elong. in 2 in., %	Red. Area, %	BHN	Izod, ft lb	Tensile Strength	Yield Point	Elong. in 2 in., %	Red. Area, %	BHN	Izod, ft lb
2	122,650	93,775	18.8	49.1	252	25.4	117,425	87,450	18.7	48.9	241	31.8
3	124,200	90,850	18.5	50.0	255	24.9	118,525	75,625	19.9	51.1	238	30.8
6	108,950	62,150	24.0	56.8	217	44.3	96,265	53,020	25.0	50.6	192	38.3
9	107,290	58,310	21.0	47.8	210	32.2	90,145	48,730	16.9	52.6	175	36.4
12	104,050	58,205	21.8	50.8	207	25.8	93,060	50,215	20.3	37.3	185	27.4

*Data courtesy Climax Molybdenum Co.

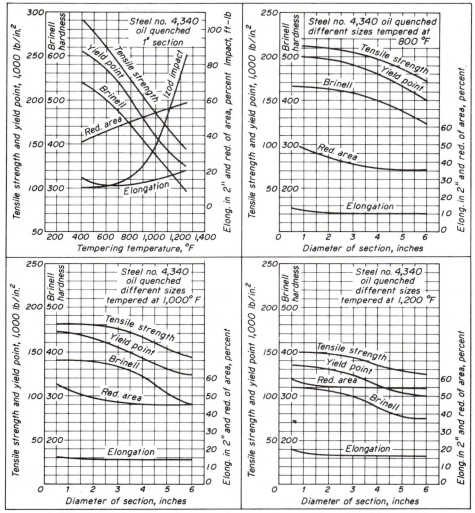

In sections 1/2" to 2" incl., quenched from 1,500 to 1,550 °F; over 2" to 4" incl., from 1,525 to 1,575 °F; over 4" from 1,550 to 1,600° F.

FIGURE 14-9 Properties of oil quenched and tempered nickel-chromium-molybdenum steel 4340 in different sizes. (*Courtesy International Nickel Co.*)

are similar to those of 8742 for small sections, but 8742 is somewhat deeper hardening in the larger sizes.

The larger steel warehouses and the various alloy producers will supply expert metallurgical advice upon request.

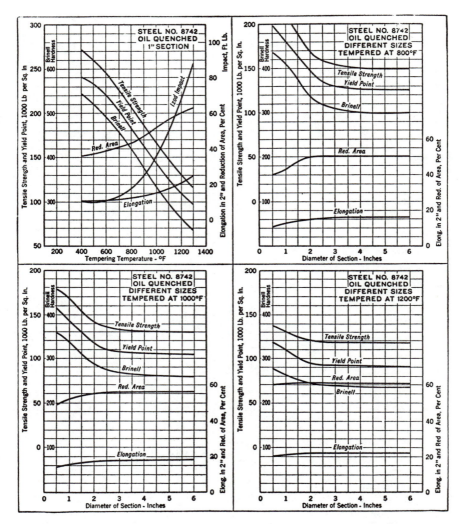

In sections ½″ to 2″ incl., quenched from 1500/1550° F . over 2″ to 4″ incl., from 1525/1575° F . over 4″, from 1550/1600° F.

FIGURE 14-10 Properties of oil quenched and tempered nickel-chromium-molybdenum steel 8742 in different sizes. (*Courtesy International Nickel Co.*)

7. High-Strength Low-Alloy Steels, HSLA

These steels have superior strength qualities, and, because of the reduced dead weight, are extensively used for mobile equipment. They exhibit good cold-forming and fabricating properties, and are readily welded with no

tendency to harden on rapid cooling. In general, preheating and stress relieving are not required. Corrosion resistance is several times as great as for
ordinary structural steel. Resistance to abrasion, fatigue, and impact is also
good. The over-all cost is somewhat more than for ordinary structural steel.

8. Cost of Steel

The cost of materials is usually of prime importance in machine construction.
In general, alloy steels are not employed if a satisfactory design can be secured
by the use of plain carbon steel. If an alloy steel must be used, the lower-
priced alloys should be given first consideration. When steel is ordered in
large quantities, it can be purchased at the mill at what is known as the mill
base price. Such prices are published in the metal trade journals for various
products such as bars, shapes, plates, and sheets. The price of an alloy steel
is determined by an "alloy extra" that is added to the base price. Table 14-9

TABLE 14-9 **Approximate Steel Prices Compared to Prices of Hot-Rolled Carbon Steel Bars**

Hot-rolled carbon steel bars, angles, structural shapes	100
Hot-rolled alloy steel bars	135–210
Cold-finished carbon steel bars	160
Carbon steel plates	100
High-strength low-alloy steel plates	125 and up

gives the relative costs of certain classes of steel based on a price of 100 for
hot-rolled plain carbon bars. Pound prices for aluminum and stainless steel
are usually about three times the price for a similar product of low-carbon
steel.

When steel is not purchased at the mill, but from a steel warehouse, the
warehouse base price for such materials applies. This price is higher than
the mill base because of extra handling charges, transportation, and ware-
houseman's profit.

Various other factors add to the price of steel. One of these is the "size
extra." The base price applies only to sections of nominal size, and very small
or very large sections cost an additional amount. A "quantity extra" is added
to the price when the amount ordered is less than a specified minimum
amount. A "cutting extra" is also added when stock lengths are not ordered.
Other extras, such as annealing, machine straightening, or truck delivery,
may further increase the pound price.

For many products where the cost of labor is relatively small, and where
the cost of the materials is important, a small increase in the mill price of one
steel is all that is needed to cause a widespread change to another type of
steel which also gives satisfactory results. When labor costs are relatively

large, as, for example, in aircraft parts, a change in the price of steel may have practically no effect on the composition used.

9. *Heat Treatment of Steel*

The microstructure that is formed when steel is quenched at a sufficiently rapid cooling rate from above the upper critical temperature is called *martensite*. Steel in the martensitic condition is nonductile and is very hard and strong. Needle-like crystals in angular arrangement appear when prepared specimens are viewed under a microscope. Martensite is unstable and is generally regarded as being a supersaturated solid solution of carbon in iron. Fully hardened steel having the martensitic structure is not only too brittle to use, but contains quenching stresses. As was previously mentioned, a second operation of heating, called *tempering*, is required to relieve the internal stress and restore a suitable amount of ductility. The products formed when martensite is tempered have the structure of very finely divided pearlite. Tempering is usually carried out at temperatures between 800°F and 1200°F. Ductility is increased and tensile strength reduced for an increase of tempering temperature. The tempering temperature must therefore be properly chosen so that the physical properties of the steel after heat treatment will be suitable for the working stress and design of the part. Steel must be heated to a temperature high enough to change the structure into austenite if hardening is to take place upon quenching. Quenching has no effect upon ferrite and pearlite.

When the cooling rate is not sufficiently rapid to form martensite, a finely divided pearlitic structure results from quenching. The ductility is greater and the strength less than for martensite. A further decrease of the cooling rate causes the thickness of the lamellae to increase until the coarse pearlitic structure of annealed or furnace-cooled steel is secured.

Prolonged heating of steel within the range 1250–1350°F causes the cementite to coalesce into spherical particles. The resulting structure is called *spheroidized cementite*. Steel in this condition is soft and ductile and is readily machined. Heat treatment after machining may be required to secure the requisite degree of strength and hardness.

Steel, as received from the mill, is frequently given a treatment called *normalizing* to secure a uniform grain structure before it is machined. Normalizing consists of heating the steel above the upper critical temperature, removing it from the furnace, and cooling it in air. Sometimes the mechanical properties developed from normalizing are very satisfactory, and no other heat treatment need be given. The results obtained from annealing and hardening are usually improved if the steel is first normalized. Normalizing is frequently applied to forgings and castings to ensure a uniform grain structure.

When the greatest softness and ductility are desired, the steel should be annealed. The heating is done slowly in the furnace and the maximum temperature, as shown by Fig. 14-6, must be maintained long enough to refine the grain structure. Cooling is done very slowly in the furnace or in an insulated container.

By a recent discovery in heat treating, it has been found possible to produce steels of the desired structure and mechanical properties by direct transformation of austenite. The method is patented and is called *austempering*. It consists of giving the heated steel a hot quench in a bath of molten lead or salt maintained at suitable temperatures. The tempering or drawing operations are eliminated when this treatment is used.

The heated parts must be protected from the oxygen of the atmosphere during the heat-treating process to prevent decarburization or loss of carbon by the surface material. If the part in service is subjected to fluctuating tension stress which exceeds the endurance limit, the weak surface layer of low-carbon material will be vulnerable to the formation of a fatigue crack. This crack will propagate inward and cause ultimate failure, even though the interior of the body has been strengthened by the heat treatment. Heat-treating furnaces can be constructed which surround the parts at the quenching temperature with an atmosphere of inert gases. Resistance to decarburization is a valuable property for a heat-treating steel. For good success in heat treatment, a part should be free from reentrant angles, sudden transitions from thin to thick sections, and thin projections attached to heavy masses.

Permanent deformation of metals is nearly always a shearing phenomenon. The atoms in the crystals are arranged in definite and repeating patterns. This uniformity of arrangement produces planes of weakness to shearing stress. The portions of the crystal, although adhering tightly together, will glide or slip on each other along these planes whenever the loading becomes sufficiently great.

The reasons for the increased strength and hardness resulting from quenching have been of great interest to investigators. The slip-interference theory is perhaps the best known. It is believed that during quenching exceedingly small cementite particles are thrown out of solution and dispersed throughout the crystal structure. These particles serve as mechanical keys which resist any motion along the slip planes and thereby give a large increase to the elastic strength of the material.

Hardenability, or depth of hardening, in steels is a question of great importance to the designer. Curves such as those shown in Figs. 14-7 to 14-10 show the average heat-treating properties for these steels. It is frequently necessary to have precise information on the hardening ability for each particular lot of incoming steel. The Jominy test has been devised for this purpose.

A 1 in.-diam specimen is made from the steel whose hardness after quenching is to be determined. The specimen is heated and then quenched by hanging it vertically over a jet of water that impinges gently on the bottom surface. Flats are ground on the sides of the specimen, Rockwell C hardness readings are taken at $\frac{1}{16}$-in. intervals from the quenched end, and the results are plotted. The resulting curve shows the manner in which the hardness decreases with distance from the quenched end. When a round bar of any size, or a machine part, is quenched in the conventional manner, the surface and interior points have cooling rates equal to those of different points of the Jominy specimen. Cooling rate curves and tables are available[3] which show the correlation that exists between distances from the quenched end of the Jominy specimen and the diameters of bars of various sizes for cooling of the surface, the center, and intermediate points. The hardness, as given by the Jominy curve of the specimen at the proper distance from the quenched end, will be the hardness of the chosen point of the quenched object. The strength of the particular lot of steel from which the specimen was taken can thus be estimated.

As a typical example, Jominy curves for steel 8642 are shown in Fig. 14-11. The effect of tempering temperatures on the hardness is illustrated

FIGURE 14-11   Typical and quench hardenability curves of AISI Steel 8642.

[3] See "Selection of Construction Steels," *Metal Progr.*, **66**, July 15, 1954, p. 1.

by the various curves. The mechanical properties for this steel as interpreted from the hardness results are given in Table 14-10.

TABLE 14-10 **Mechanical Properties Interpreted from Hardenability**
Results of Fig. 14-11 for Steel 8642
[*Quenched in oil from 1550°F and tempered (drawn as shown)*]

Size of Round	Tempering Temperature, deg F	Tensile Strength, psi	Yield Point, psi	Elong. in 2 in., %	Red. Area, %	Brinell Hardness	Rockwell Hardness, C
1 in.- round center	800	198,000	189,000	13	46	401	43
	1000	170,000	159,000	16	52	341	37
	1200	131,000	115,000	20	61	262	27
2 in.- round 1/2 radius	800	185,000	174,000	14	51	375	40
	1000	158,000	146,000	17	56	321	34
	1200	127,000	110,000	21	62	255	25
3 in.- round 1/2 radius	800	161,000	150,000	17	55	331	35
	1000	145,000	133,000	19	58	293	31
	1200	120,000	102,000	22	63	241	23
4 in.- round 1/2 radius	800	145,000	133,000	19	58	293	31
	1000	131,000	115,000	20	61	262	27
	1200	114,000	96,000	22	64	229	21

10. Residual Stresses from Heat Treatment

The increased strength and hardness resulting from the heat treatment of machine parts is usually accompanied by residual stresses. Since these may be either favorable or unfavorable to the engineering functioning of the part, the subject should be given careful consideration by the designer. Such stresses arise from plastic deformation which takes place while the material is at elevated temperature.

Consider first the purely thermal effects caused by quenching a through-heated steel from a temperature somewhat below the lower critical temperature of 1330°F. Upon immersion in the cooling medium, the surface cools at a very rapid rate. Contraction of the outer layers is largely prevented by the rigid core at the higher temperature. Tensile stresses are thus set up on the surface which are balanced by compressive stresses in the interior. With the decrease in temperature of the core, and the accompanying thermal contraction, the tension on the surface may be reduced to zero and then turned into compression. There will then be a balancing tensile stress in the interior. The

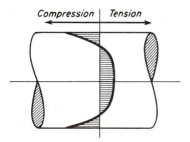

Compression | Tension

FIGURE 14-12   Distribution of residual stress along diameter of a long cylindrical body.

situation is as shown in Fig. 14-12, which represents the final stress distribution across the diameter of a long cylindrical body.

Now suppose the material is through-heated above the critical temperature into the austenitic range. Let it also be assumed that the material contains a sufficient proportion of carbon or alloying elements so that martensite will be formed upon sudden cooling. The thermal tension on the surface forms as before. However the phase change to martensite throughout the body involves an increase in volume which tends to increase the tension on the surface and the compression in the interior. The continued thermal shrinkage, however, reduces the tension on the surface or even turns it into compression.

These thermal and transformation effects occur simultaneously, but the exact manner in which they combine is highly complex and has never been explained to the extent that easily followed rules have been made available. The final quenching stresses on the surface may therefore be either tension or compression, depending on the quenching temperature, cooling rate, size and shape of the part, carbon and alloy content, and the temperature at which the transformation to martensite occurs. The stress distribution may also be more complex than that shown in Fig. 14-12. Quenching must be very carefully carried out. A large surface tensile stress may cause warping and distortion or even cracking of the part. The tendency to crack is augmented by the presence of stress raisers such as notches, grooves, holes, reentrant corners, and so on. Oil quenching is less severe than water quenching. The heat is removed less quickly, and the difference in temperature between the surface and the center is less than for water quenching. The tendency for distortion and cracking is therefore less with oil quenching.

A residual surface compressive stress is beneficial for parts with repeated or fatigue loading, and such a stress can effect a worthwhile increase in the factor of safety. It is known, however, that surface residual stresses tend to lessen or fade gradually with repeated stressing. Tempering to restore a portion of the ductility causes a loss of the surface compression. In fact, a tempering temperature of 1000°F may reduce the compression to zero or turn it into tension.

A surface with residual tension is vulnerable to the formation of fatigue cracks. Such conditions have undoubtedly been the cause of many failures. Residual tension can be removed and the surface left in compression by mechanical operations such as rolling between rollers at high pressure, shot peening, tumbling, or peen hammering.[4]

11. Carburizing and Nitriding

Sometimes a machine part requires a hard surface to resist wear, abrasion, or local deformation from impact loads, combined with a tough, strong core for strength. The case hardening or carburizing process can be used for obtaining such a structure. The parts are packed in a compound rich in carbon and held for a number of hours in the furnace at a temperature within the austenitic range. Carbon is absorbed by the surface layer. The same effect is secured by heating the steel in the presence of gases rich in carbon. The high-carbon surface material is transformed to martensite by quenching, followed by tempering at 300–350°F for relief of quenching stresses. The period of heating may extend from 3 to 24 hr. A depth of case of about $\frac{1}{16}$ in. can be secured after about 8 hr in the furnace. Freedom from warping during the extended heating period is a very desirable property for a carburizing steel. Carburizing steels in general are low in carbon and have good machinability ratings, but all machine work must be completed before the part is carburized. The core properties for a number of steels used for carburizing are given in Table 14-11.

TABLE 14-11   Core Properties of Various Carburizing Steels*

Steel	Tensile Strength	Yield Strength	% Elong. in 2 in.	% Red. Area	Izod, ft lb	Steel	Tensile Strength	Yield Strength	% Elong. in 2 in.	% Red. Area	Izod, ft lb
1020	96,000	60,000	19	41	48	4119	137,000	112,000	16	42	
1117	96,000	59,000	23	53	33	4615	110,000	75,000	23	66	67
2315	138,000	108,000	22	61	42	6115	104,000	62,000	28	58	
3115	125,000	85,000	22	54	36	A8620	135,000	108,000	17	48	

*One-inch rounds carburized 8 hr 1,650–1,700°F, cooled in box, reheated to 1,475°F. Steels 1020 and 1117 are water quenched and tempered at 350°F. Others are oil quenched and tempered at 300°F.

Nitriding produces a very hard surface layer by the absorption of nitrogen during heating that lasts for an extended period at 900–1000°F in the presence of ammonia gas. Special nitriding steels must be used, but no heat treatment subsequent to the hardening is required. Both the carburizing and nitriding processes leave the part with a residual stress of compression throughout the hardened zone. If the working load produces a fluctuating tensile stress,

[4]For additional information on residual stress see Hetényi, M., Ed., *Handbook of Experimental Stress Analysis*, New York: John Wiley & Sons, Inc., 1950, p. 495.

the presence of the residual surface compression gives a favorable stress condition for preventing the formation of a fatigue crack.

The cyaniding process forms a very hard surface layer 0.002 in.–0.010 in. thick through immersion for a short period, 15 min to 1 hr, in a bath of molten sodium cyanide, and then by quenching in water or oil. The hardness is derived from compounds of nitrogen and carbon absorbed by the surface layer.

12. Flame Hardening

Localized surface heating with a flame or induction coil, followed by rapid cooling, usually leaves the surface in tension. The cool interior of the body does not expand, and the attached heated layer must expand plastically outward. This enlarged material is unable to return to its original size when cooled, and residual tension results.

Grinding cracks can be accounted for by the process above. Brake drums, clutch plates, railroad wheels and rails, and other friction surfaces are affected in the same way. Flame-cut parts are usually left with a surface layer of tension. Contraction of weld metal deposited on relatively cold-base metal produces a residual stress of tension. Bent shafts can be straightened by heating them on the outside of the bend. If properly carried out, the residual tensile stress in the region that has been heated will cause the shaft to straighten.

Flame or induction hardening leaves the surface in a state of residual compression only when the steel has a sufficient carbon or alloy content to undergo the phase change to martensite upon quenching from the austenitic condition. The increase in volume from the martensite formation must be sufficient to overcome the thermal contraction of the heated layer.

At the edges of the heated region, the martensite gradually merges into the weaker surrounding material. When the part must be heated in sections, the junctions between the heated regions should be located at points where the loading stresses are low. For example, in hardening the teeth of a gear, the junctions should be placed at the tips of the teeth and not at the base where the maximum bending stress occurs.

13. Strain Hardening

Plastic deformation, resulting from drawing metals through a die, or rolling them between rollers, raises the values of the yield point and tensile strength of ductile materials such as mild steel, stainless steel, copper, brass, and aluminum. This increase in strength is accompanied by a loss of ductility and an increase of hardness. The effect of cold drawing on the properties of steel can be estimated by the curves in Fig. 14-13. The strengthening becomes less at distances removed from the surface, and the properties of bars of large cross section undergo less change because of the cold work. The stress-strain diagram for low-carbon steel loses its characteristic shape if the material is cold worked. The large elongation at the yield point does not take place, but a gradual bending of the curve occurs after the proportional limit has been passed.

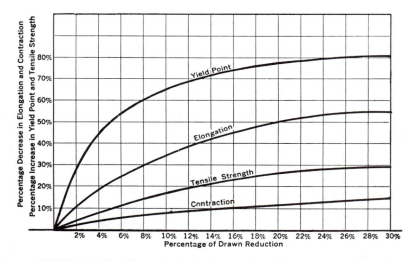

FIGURE 14-13 Effect of cold drawing on the tensile properties of steel. (*Courtesy Bethlehem Steel Co.*)

Materials that are too weak and soft for use in the annealed condition are sometimes strengthened by cold working. After such treatment, the mechanical properties may no longer be the same in all directions, and usually a definite loss of ductility results in the direction normal to the deformation. In addition, any increase of the yield point in tension produced by stretching is usually accompanied by a decrease of the yield point in compression. This is known as the Bauschinger effect.

In general, cold working leaves the part with residual stress on the surface. This is an undesirable situation, since parts made from such materials will warp if a portion of the surface is removed by machining. A tensile residual stress on the surface causes the part to be vulnerable to the formation of a fatigue crack should the stress be fluctuating.

The hardening and loss of ductility produced by plastic deformation disappears if the material is subjected to annealing temperatures. When the fabrication process involves successive steps of cold working, the material may become so brittle that rupture may occur. This danger can be avoided by restoring the ductility periodically by annealing the metal between the cold-working operations. Cold rolling and cold drawing are sometimes used for producing a smooth, bright surface and dimensional accuracy on bars and sheets.

An improvement in the strength, ductility, and impact resistance of steel can be secured by hot working at the forging temperature. Hot working gives a finer and more uniform grain structure, and improves the soundness of the material.

14. Hardness

The term hardness is used to designate certain mechanical properties of a material, such as resistance to penetration, scratching, abrasion, or cutting. Since a rigorous definition of hardness has not as yet been written, this quality must be specified by numbers which are dependent upon the method of making the test. Manufacturing operations such as cold work, quenching and tempering, and precipitation heat treating change the hardness of many engineering materials. Hardness testing is therefore one of the principal methods for determining the suitability of a material for its intended purpose. Hardness testing is a valuable inspection tool for maintaining uniformity of quality in heat-treated parts. A useful correlation exists between the hardness of a material and the tensile strength, endurance limit, and wear resistance.

The Rockwell machine is widely used in industry for determining hardness. It is rapid, nondestructive, and free from personal errors of the operator. It is a penetration method and the hardness is read directly on a dial on the machine. There are two types of Rockwell readings, R_b and R_c. The R_b reading uses a $\frac{1}{16}$-in. steel ball as the penetrator with a load of 100 kg. A 10-kg load is first applied to seat the specimen. It is used for soft steel and the harder nonferrous metals. The R_c reading uses a 120° diamond cone called a brale as the penetrator with a load of 150 kg. The initial load is first used to seat the specimen. It is used for the harder, heat-treated steels and is able to distinguish small differences in hardness. In all tests the thickness should be at least 10 times the indentation.

For very thin specimens, or for parts covered with a shallow, hardened skin, the B or C Rockwell tests penetrate too deeply, and the superficial Rockwell test must be used. In the superficial test, both the initial and major loads are considerably reduced, depending on the thickness of the specimen or of the hardened zone.

The Brinell is another well-known penetration method for determining hardness. In the standard test, a hardened 10-mm-diam steel ball is pressed into the specimen by a force of 3,000 kg. The diameter of the impression is determined by a hand microscope, and the area of the contact surface between ball and material is computed. The Brinell Hardness Number, BHN, is found by dividing the load in kilograms by the area in square millimeters. The harder the material, the smaller the indentation, and the higher the Brinell number. The tensile strength of wrought steel in pounds per square inch is usually from 450 to 500 times the BHN.

The relationship between Brinell and Rockwell numbers is shown in Table 14-12.

TABLE 14-12 Approximate Equivalent Hardness Numbers

Steel

R_c	BHN	R_c	BHN	R_b	R_c	BHN	R_b
65	739	45	421		25	253	101.5
64	722	44	409		24	247	101.0
63	705	43	400		23	243	100.0
62	688	42	390		22	237	99.0
61	670	41	381		21	231	98.5
60	654	40	371		20	226	97.8
59	634	39	362		18	219	96.7
58	615	38	353		16	212	95.5
57	595	37	344		14	203	93.9
56	577	36	336	109.0	12	194	92.3
55	560	35	327	108.5	10	187	90.7
54	543	34	319	108.0	8	179	89.5
53	525	33	311	107.5	6	171	87.1
52	512	32	301	107.0	4	165	85.5
51	496	31	294	106.0	2	158	83.5
50	481	30	286	105.5	0	152	81.7
49	469	29	279	104.5			
48	455	28	271	104.0			
47	443	27	264	103.0			
46	432	26	258	102.5			

BHN values for 10-mm tungsten carbide ball, 3000 kg load.

Another instrument for hardness testing is the Shore Scleroscope, in which a small diamond-tipped weight is allowed to fall on the specimen. The height of the rebound is used as the measure of the hardness of the material. This instrument has the desirable feature of portability, and can be used on specimens too large to be placed in a regular hardness-testing machine.

The Vickers pyramid test uses a small diamond pyramid to indent the specimen. The hardness number, as for the Brinell test, is found by dividing the load by the area of the indentation. Since the loads are light, a high magnification is required to determine the width of the impression.

Scratch hardness tests are also in use. A diamond or sapphire point is drawn over a polished surface under a definite load and the width of the scratch is measured with a microscope. The hardness number is usually considered as varying inversely with the square of the width of the scratch.

15. Machinability

The relative ease with which a given material may be machined, or cut with sharp-edged tools, is called its machinability. The ratings given in Tables 14-4 and 14-5 are based on Bessemer screw stock B 1112 as 100%. These refer to steels "when turned with a suitable cutting fluid at 180 ft per min under normal cutting conditions."

When a part requires a large amount of machining, the designer tries to specify a composition and grain structure best adapted to the cutting operations. Sometimes certain elements, principally sulfur and manganese, are added to steel to improve the cutting properties. Steels with very low carbon content usually have poor machinability. For greatest cutting ease, the carbon content should be about 0.10% for Bessemer steels and 0.20% for open-hearth steels. Alloy steels, because of their higher strength and toughness, are usually harder to machine than plain carbon steels. In general, the machining properties of a steel can be materially improved by cold drawing. Annealing and normalizing heat treatments are beneficial for medium- and high-carbon and alloy steels. For best machining, a steel should be soft and brittle. This is not possible since steel is usually soft and tough or hard and brittle. For high surface speeds, medium-size pearlite grains, with each grain surrounded by an envelope of ferrite, are desirable. For low speeds and heavy feeds and depths of cut a spheroidal microstructure is best. In general, the machinability rating cannot be correlated with the Brinell Hardness Number. Steels of best machinability have BHN of 187 to 217.

16. Grain Size

A relationship exists between the mechanical properties of steel and the size of the grains or crystals. Coarse-grained steels, in general, harden deeper and are best for rough machining operations. They have good hot-forming characteristics and offer better resistance to high-temperature creep. Fine-grained steels, in contrast, are tougher and are superior for impact resistance, especially at low temperatures. They exhibit less distortion, internal stress, and cracking from heat treatment.

Grain sizes in steel are designated by ASTM grain size numbers 1 to 8. Number 1 is the largest, and has a calculated diameter of equivalent spherical grain of 0.0113 in. Number 8 is the smallest, with a spherical grain diameter of 0.0010 in. Steels of grain sizes 1 to 5, inclusive, are generally called coarse-grained, and fine-grained steels have grain sizes 6 to 8, inclusive.

The grains of austenite that are formed when steel is heated have the smallest size at the upper critical tempe. ture, as indicated by Fig. 14-5. A fine-grained steel therefore results if the material is cooled from the fine-grained austenitic condition. If the steel is heated above the upper critical temperature, the grains of austenite increase in size and a coarse-grained steel results upon cooling. The grain size is therefore to a large extent under the control of the heat treater. Hence, if a fine-grained steel is desired in a hot-rolled or forged product, the hot working must be continued until the temperature drops almost to the upper critical point, or the grains will coarsen in cooling from the high temperature at which hot working was stopped.

Grain size in steel refers to the size of the previously existing austenite grains at the quenching temperature. Therefore, the sample must be quenched from the desired temperature and tempered lightly, and a specimen must then be prepared for examination. The martensite grains reflect the austenitic grain size. The behavior of carburizing steel is usually evaluated by means of a standardized method called the McQuaid–Ehn test.

An uncertainty sometimes exists in the purchase of steel as to whether lots from different sources, but of the same chemical composition, will have identical physical properties when given the same heat treatment. The physical properties, as mentioned in the foregoing, depend on the grain size, which in turn depends upon the grain-coarsening characteristics of the material. It is possible for one piece of steel to remain fine-grained for a considerable temperature range above the upper critical temperature, whereas another piece of steel of the same chemical composition, but from another lot, may coarsen when raised to this temperature. A knowledge of the inherent grain growth tendency is thus very important if consistent physical properties are to be expected after heat treatment. In fact, the term fine-grained is sometimes applied to a steel which does not begin to coarsen until the temperature rises a considerable amount above the upper critical. A coarse-grained steel, then, is one which coarsens rapidly as the temperature rises beyond the upper critical. Closer temperature control is required for the heat treatment of coarse-grained steel. Fine-grained steels have a wider range of safe hardening temperatures, and more predictable results can be obtained in the heat treatment process. The tendency for grain growth at elevated temperatures can be closely controlled by the addition during steel making of such inhibiting agents as minute quantities of finely divided aluminum.

17. Corrosion

Corrosion is the deterioration or destruction of a material because of reaction with its environment. All environments are corrosive to some degree: all atmospheres and most kinds of water, chemical and fuel gases, industrial and food acids. Contamination by the products of corrosion may affect the purity and market value of manufactured products. Failure of mechanical equipment due to corrosion may be hazardous to the operating personnel. Corrosion is electrochemical in nature and is frequently a very complex process.

Rusting is a term reserved for the oxidation of iron and steel. It has been estimated that the loss from the rusting of iron and steel products in the United States is $6 or $7 billion per year. An iron surface exposed to dry air forms a thin film of iron oxide. So long as the atmosphere remains dry, the reaction is seldom destructive as the film prevents the further penetration of oxygen. However, with increasing humidity, the amount of rust produced may be many times that formed in a dry atmosphere. The oxidation rate also increases with increase in temperature.

On the other hand, the oxidation rate is slow for a tightly adhering scale, such as occurs in aluminum. Aluminum oxidizes readily, but the scale adheres so tightly that it gives good protection to the parent metal.

Corrosion can take place uniformly over a surface, or may be highly localized in nature. The placing of two dissimilar metals in contact provides an electron flow between them which results in an increase in the corrosion rate. Corrosion is sometimes confined to crevices formed by holes, bolt and rivet heads, lap joints, and gasket surfaces. Here the crack is wide enough to admit the liquid, but small enough to provide a stagnant zone. Smooth surfaces will sometimes pit and the corrosion goes on at an increasingly rapid rate as the pits become deeper.

Intergranular corrosion, or the failure of grain boundaries, may occur especially if impurities are present. Selective leaching may remove one component of an alloy without affecting the others. Corrosion is sometimes hastened by relative movement between the metal and corrosive fluid which prevents the formation of a passive surface film on the metal. Impingement, when the fluid must change directions (as in an elbow), can cause failure in the metal.

Stress corrosion cracking may occur from the simultaneous presence of a tensile stress and a specific corrosive medium. The metal as a whole is largely unaffected, but a network of fine cracks spreads over its surface. This type of failure includes what is called "season cracking" of brass and "caustic embrittlement" of steel. The cracking starts at tensile stress smaller than the yield point value, but the process is not as yet fully understood. It may proceed along the grain boundaries or may occur across the crystals. Hydrogen embrittlement of steel can also be classified as a corrosion phenomenon.

A corrosive environment may reduce the fatigue strength of a metal to but a fraction of its normal value. The pits and cracks serve as stress raisers.

18. Prevention of Corrosion

There are a number of natural material combinations where corrosion can be reduced to a very low value. In general, such combinations will give the maximum amount of corrosion protection at the lowest cost. These[5] are listed in Table 14-13. Pure metals are generally more resistive to corrosion than those containing impurities or small amounts of other elements.

TABLE 14-13 Natural Corrosive-Resistive Combinations

Corrosive Element	Resisting Medium
Concentrated sulfuric acid	Steel
Diluted sulfuric acid	Lead
Nitric acid	Stainless steels
Hot hydrochloric acid	Hastelloys, Chlorinets*
Hydrofluoric acid	Monel metal
Caustic	Nickel and nickel alloys
Distilled water	Tin
Hot strong oxidizing solutions	Titanium
Nonstaining atmospheric exposure	Aluminum
Ultimate resistance of all materials	Tantalum

*Nickel-base alloys with molybdenum.

Careful attention should be given to the details of construction for the containers of corrosive materials. Low stresses should be used, and stress concentrations and uneven heat distribution should be avoided. Vessels should be welded rather than riveted or bolted. Bottoms should be rounded or sloped with round corners for good drainage. Drain-out valves and plugs should be flush with the bottom. Baffles and stiffners should provide for free drainage. Whenever possible the same material should be used throughout. Means should be provided for easy cleaning and easy replacement of failed elements. Air should be excluded from the system whenever possible.

Many processes are in use for applying a corrosion-resistant coating to the surface of steel and iron products. Coatings of zinc, tin, lead, copper, nickel, chromium, and cadmium are in wide use. Application is made by

[5]See p. 195 of Reference 5, end of chapter. This book should be consulted for information on the details of application of these materials.

hot dipping, electroplating, or by a molten metal spray, depending on the metals involved. The Parkerizing and Bonderizing processes utilize a coating of iron phosphate as the protective agent or as a base for the application of paint or enamel.

Resistance of steel to atmospheric corrosion can be increased by the addition of about 0.20% of copper. The HSLA steels described in Section 7 are still more resistant to this destructive action.

Paint, varnish, lacquer, and other organic coatings will provide limited protection against certain kinds of corrosion. Success will depend not only on the selection of the proper top coat, but on satisfactory preparation of the surface, and on a suitable prime coat. Cathodic and anodic protection can reduce corrosion substantially by means of electric currents.[6]

19. Stainless Steel

The so-called stainless steels are in wide use for resisting corrosion. There are many alloys of this kind, and the proper alloy must be selected for its intended purpose. Chromium is the principal alloying element, and occurs in percentages from 11.5 to 30. In some alloys nickel is also present in amounts up to 20%. Close control must be exercised on the carbon content of these materials. There are three general classifications of stainless steels, and Table 14-14 shows a few typical alloy numbers in each class. Because of the alloys, the cost is considerably more than for plain carbon steel.

Under corrosive environments the film that forms on the surface is very stable, and is self-healing. However some stainless steels are susceptible to localized corrosion in stagnant areas. These materials should not be used for chloride-containing media.

Considerable experience may be required for the successful fabrication of these materials. Unless a free-cutting type is specified, machining is difficult with a resulting poor surface.

(a) *Austenitic, 18% Cr–8% Ni (302, 304, 316, 347)*. About one-half of all stainless steel produced is austenitic, and of these, the greatest tonnage is number 304. The nickel is used principally to increase the ease of fabrication. When high strength is required it must be produced by cold work, but this reduces the corrosion resistance. It is possible by cold work to achieve strengths several times those shown in Table 14-14 for annealed materials. As usual, this increase in strength is accompanied by a reduction in ductility and an increase in hardness.

At temperatures 950–1450°F, these steels are susceptible to intergranular

[6]See p. 205 of Reference 5, end of chapter.

TABLE 14-14 Composition and Mechanical Properties of Selected Stainless Steels

Type No.	Austenitic (Cr 18%, Ni 8%) Hardenable by Cold Working; Nonmagnetic when Fully Annealed				Martensitic (Cr 12%) Hardenable by Heat Treatment		Ferritic (Cr 17%) Nonhardenable; Can Be Hot Worked; Magnetic	
	302	304	316	347	410	440A	430	446
Cr	17.0–19.0	18.0–20.0	16.0–18.0	17.0–19.0	11.5–13.5	16.0–18.0	14.0–18.0	23.0–27.0
Ni	8.0–10.0	8.0–12.0	10.0–14.0	9.0–12.0	0.5 max	0.5 max	0.5 max	0.5 max
C, max.	0.15	0.08	0.10	0.08	0.15	0.60–0.75	0.12	0.2
Mn, max.	2.0	2.0	2.0	2.0	1.0	1.0	1.0	1.5
Si, max.	1.0	1.0	1.0	1.0	1.0	1.0	1.0	1.0
Others			Mo 2.0–3.0	Cb stab.		Mo .75 max		N. 25 max
Yield str.*	30,000	30,000	30,000	30,000	32,000	55,000	35,000	45,000
Tensile str.	80,000	80,000	75,000	80,000	60,000	95,000	60,000	75,000
El. in 2 in. %	50	50	40	40	20	20	20	20
Red. Area	60	60	50	50	50	40	40	40
Brinell max.	180	180	200	200	200	240	200	200
Yield str.†					35–180,000	55–240,000		
Tensile str.					60–200,000	95–275,000		
El. in 2 in. %					25	20		
Brinell					120–400	200–555		
Rockwell					B70–C45	B95–C55		

*Strength values are for annealed or furnace-cooled material.
†Strength values are for heat-treated material.

corrosion unless special alloys or treatments are provided. This temperature range occurs in welding and may result in a decay zone on either side of the weld and somewhat removed from it. The spot-welding characteristics, however, are very good.

(b) *Martensitic (410, 440A).* These steels can be hardened by quenching, and then tempered to get the desired mechanical properties. For corrosion resistance the surface must be smooth with all scale and foreign matter removed. A softening treatment must usually be given before machining, although free-cutting grades are available. These steels are difficult to weld because of the strong air-hardening tendency. Welding must usually be followed by immediate annealing. Stainless steel cutlery is usually of the martensitic type. When the carbon content is high, the properties resemble those of tool steel. With lower carbon content, the steel is similar to ordinary machine steel.

(c) *Ferritic (430, 446).* The forming and drawing properties of these materials resemble those of mild steel. Number 430 is used in architectural work, automobile trim, hardware, and equipment in the chemical and food industries. It has many applications in high-temperature service, and has good scaling resistance up to 1550°F. Welding may cause grain coarsening which reduces the toughness and impact resistance. Number 446 has good heat resistance and is used in furnaces and heat-treating equipment.

20. Wear

Wear and corrosion are the two most adverse factors for reducing the life of mechanical equipment.[7] The loss of a relatively small amount of material at certain critical locations may be the only difference between a new machine and one that is worn out. High-carbon martensite, or martensite tempered below 450°F has the best wear resistance of the steel structures.

(a) *Cutting Wear.* Cutting wear results when a hard rough surface is rubbed over a softer one with or without a lubricant. The projections on the harder metal act like cutting tools to remove material from the softer one. The action is more severe at low speeds. Unless the metals have widely different hardnesses, such wear soon ceases as the parts become worn in.

(b) *Abrasive Wear.* This is the most common type of wear, as there is often some grit such as metal particles, oxides, or dust carried between the rubbing surfaces. Rapid and continuing wear may result. Particles too small to be removed by a filter can produce wear. The abrasive can be either embedded or rolling. Abrasive embedded in the softer material can wear the harder member. Cast-iron plates are used for lapping hard steel since the abrasive is embedded in the graphite flakes of the iron. Although abrasive particles embedded in a soft bearing shell will wear the shaft, the effect is less than the scoring that might result from loose particles. A relatively thick oil film that will keep the parts separated is effective in preventing wear. Under conditions of boundary lubrication, the separation of the parts does not change appreciably with continued operation, and such wear will continue even after the initial surface roughness has been eliminated.

High-manganese or Hadfield steel possesses exceptional ability to resist abrasion. Its impact properties are also very good. The steel contains 10–14% manganese and 1.0–1.4% carbon. This material work hardens so rapidly, and is so tough, that machining is difficult and must frequently be done by grinding.

(c) *Adhesion or Galling.* This is a form of frictional failure that manifests itself as a welding of one surface to another. Portions of the surfaces stick together and when they are torn apart material is removed from one of them. A film of any type on the surfaces affords protection against galling. For galling to occur, the rubbing conditions must be such as to break down the film, at least locally, for the short time necessary to bring clean metal into

---

[7]See Dayton, R. W., "Mechanisms of Wear: Their Relation to Laboratory Testing and Service," in *Proceedings of Special Summer Conference on Friction and Surface Finish,* Cambridge, Mass.: Massachusetts Institute of Technology, June 1940, p. 127. See also Kragelski, I. V., *Friction and Wear,* Washington, D.C.: Butterworth Publishers, 1965; and Rabinowicz, E., *Friction and Wear of Materials,* New York: John Wiley & Sons, Inc., 1965.

contact with clean metal. Finely divided grit in the oil tends to clean the surfaces and promote galling. Like metals, when rubbed together, are more apt to gall than metals of different metallurgical structure. An exception to this rule is cast iron, which runs well with cast iron.

Any material which tends to form a tightly adhering film over the rubbing surfaces will decrease the tendency for galling. For example, extreme pressure, or EP, lubricants act in this way to prevent galling in hypoid gears. Coatings such as phosphate, oxide, tin, and cadmium are useful in preventing galling during the running-in periods.

(d) *Corrosive Wear.* Chemical attack will affect the wear rate—either making it less or making it greater. If the corrosion product is a tightly adhering film, the wear rate is reduced, as mentioned above for EP lubricants. When the resulting film is nonadhering, the wear rate is usually increased. Sometimes the film adheres but gradually builds up to such a thickness that it peels off, causing a sudden acceleration in the wear rate. Corrosion may be caused by the presence of water or certain small but active agents in the lubricant or atmosphere.

21. Short-Term Effects of High Temperatures

A rise in temperature generally causes a change in the physical properties of a metal. Tests on medium-carbon rolled and cast steels indicate that the tensile stress increases until the temperature rises to the blue-heat point at about 500–550°F. As shown by Fig. 14-14, this increase of strength is accompanied by loss of ductility; it is termed "blue brittleness." A further increase in temperature causes a rapid decrease of strength.[8] Other tests have shown that the impact strength at about 800–900°F is considerably less, but at 1300°F is somewhat greater, than at room temperature. A further rise in temperature, however, is accompanied by a rapid decrease in impact strength. The endurance limit for a change of temperature generally follows the same trend as the tensile strength. The modulus of elasticity drops for a rise in temperature, as is shown by Fig. 14-15.

Cast iron is but slightly affected by temperatures up to 800°F, but above 900°F a sharp decrease in strength occurs. Malleable iron, above 800°F, shows a decrease in strength and an increase in ductility with increasing temperature.

At subnormal temperatures, there will be an increase in the values for the tensile strength, yield strength, endurance limit, and hardness, and a

[8]See "Cooperative Short-Time High Temperature Tension Tests of Carbon Steel K6," *Proc. ASTM,* **33**(1), 1933, p. 213.

FIGURE 14-14   Tensile properties of cast, annealed or normalized, and of rolled, annealed or normalized, 0.28 percent carbon steel at elevated temperatures.

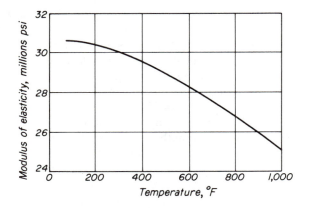

FIGURE 14-15   Reduction of modulus of elasticity with temperature for SAE 4130 annealed steel.

decrease in the ductility of both ferrous and nonferrous metals.[9] At low temperatures, however, many steels have a marked reduction in impact strength. Consequently, the designer must use special care in avoiding or reducing stress concentrations. A fine-grained steel is superior to a coarse-grained steel for such service. Three and one-half percent nickel steel, stainless steel of the 18 : 8 type, or nonferrous alloys are in widest use for impact loads at subnormal temperatures.

22. Creep of Steel at High Temperature

If a part is subjected to stress at high temperature for a long period of time, it will undergo a slow and permanent deformation called creep. When the part operates with close clearances, the designer must make a careful attempt to predict the amount of creep that will probably take place during the useful life of the machine. For applications where close clearances are not involved, the permissible deformations are limited only by the deformations that may cause embrittlement and failure.

The amount that a member deforms or creeps depends upon the stress, the temperature, and the kind of material. Materials which have the same properties at room temperature may differ greatly in creep characteristics at elevated temperature. Hence suitable values for working stresses at high temperatures can be determined only by experiments on each material. Reliable data can be secured only by conducting longtime creep tests which duplicate as nearly as possible the conditions under which the steel will be used. The testing equipment is expensive, and must embody great precision of temperature control and measurement of elongation.

The tensile test, in which the total elongation is plotted as a function of the time, is the most significant. A creep curve, as that shown by the typical examples of Fig. 14-16, consists of three parts. In the first stage, the material is becoming stronger because of the

FIGURE 14-16   Creep of mild steel.

[9]See "Impact Resistance and Tensile Properties of Metals at Subatmospheric Temperatures," ASME–ASTM Committee Report, Aug. 1941. See also "Behavior of Ferritic Steels at Low Temperatures," ASME–ASTM Committee Report, Dec. 1945.

strain hardening, and the creep rate is therefore decreasing. In the second stage, the strengthening due to strain hardening is counterbalanced by the weakening due to annealing, and the creep rate is approximately constant. In the third or last stage, the annealing effect predominates, and the deformations take place at an accelerated rate until failure results.

For sufficiently low stresses and temperatures, the third stage may not appear within the duration of the test. In fact, since it is usually impossible to conduct the test for as long a time as the service life of the part, designers usually extrapolate the straight portion of the curve beyond the duration of the test in order to predict the expected amount of creep at a future time. However, an element of danger lies in this practice, since the only way to be certain that the curve will remain straight for a service life of 10 or 20 years is to conduct the test for that length of time. Although experiments of such duration are manifestly impractical, it has been recommended that tests extending to 10% of the expected service life be performed whenever feasible. The reliability of short-term creep tests for predicting long-term deformations is decidedly doubtful at the present time.

The selection of a suitable steel for continuous service at high temperatures is a very difficult task. The following steels have been recommended[10] for use in oil refinery equipment: up to 900°F, 0.08–0.18 C killed (or degasified) steel; 1,000°F, 0.08–0.18 C plus 0.50 Mo; 1,100°F, C-Mo plus 1.25 Cr or 1.50 Si; 1,200°F, 2.0 to 5.0 Cr plus 0.50 Mo; 1,300°F, 18 Cr, 8 Ni; above 1,300°F, 16 Cr, 13 Ni, 3 Mo.

23. Cast Iron

Cast iron or gray iron is characterized by high percentages of carbon and silicon. The average composition of cast iron is shown by Table 14-15. Gray iron is the most extensively used cast construction material. It is cheap, easily cast, and readily machined. Although strong in compression, cast iron has the disadvantage of being weak in tension. When a casting must sustain high tensile loads, some other material, such as alloy cast iron, malleable iron, or cast steel, must be used. Such materials, however, are more expensive than

TABLE 14-15   Composition of Cast Iron

Carbon	2.00–4.00%
Silicon	0.50–3.00%
Manganese	0.20–1.00%
Phosphorus	0.05–0.80%
Sulfur	0.04–0.15%
Iron	Remainder

[10] See Fleishman, M., "Selection of Steel for High Temperatures," *Steel*, **102**, Jan. 17, 1938, p. 34.

gray iron. Cast iron is notable for its lack of ductility; the elongation for a standard tensile specimen is usually less than 1 percent. Cooling stresses can be relieved and machinability improved by annealing. For bodies of intricate shape, castings have inherent advantages over built-up or fabricated products.

It is general practice to classify cast irons with respect to tensile strength. Some properties[11] of cast irons are shown in Table 14-16. Suitable composi-

TABLE 14-16   Strength of Gray Cast Irons*

Class No.	Tensile Strength Min., psi	Average Transverse Load,† lb	Compressive Strength‡ psi	Average Shear Strength, psi	Modulus of Elasticity, psi	BHN	Usual Min. Wall Thickness, in.
20	20,000	1,800	80,000	32,500	11,600,000	110	1/8
25	25,000	2,000	100,000	34,000	14,200,000	140	1/8
30	30,000	2,200	110,000	41,000	14,500,000	170	1/4
35	35,000	2,400	125,000	49,000	16,000,000	200	3/8
40	40,000	2,600	135,000	52,000	18,100,000	230	1/2
50	50,000	3,000	160,000	64,000	22,600,000	250	1/2
60	60,000	3,400	150,000	60,000	19,900,000	275	3/4

*For SI units, divide stress values by 145, being the conversion factor for psi to MPa.
†Specimen 1.2-in. diameter, 18-in. supports, load at center.
‡Subject to variations up to ±10 percent.

tions to meet the strength requirements are chosen by the producing foundry unless the customer specifies otherwise. The cost of the iron increases with increase of strength.

Cast iron contains about 0.70% carbon combined with iron in a pearlitic structure. The remaining carbon appears as more or less finely divided graphite flakes. The flake graphite is responsible for the relatively low hardness and strength of cast iron: the larger and coarser the flakes, the lower the strength. Dispersal of the carbon is aided by the presence of silicon, but is retarded by sulfur. However, sufficient manganese is usually present to form MnS, which alleviates the effect of the sulfur. Phosphorus forms the hard constituent, steatite, which has no effect on the graphite dispersal. A finer dispersal and a higher tensile strength are secured by an increase in the cooling rate. Since the cooling rate is affected by the section size, the interior of thick sections is weaker than the material near the surface or thin sections of the same composition.

The microstructure for different forms of cast iron is shown[12] in Fig. 14-17.

[11]See "Symposium on Cast Iron," Proc. ASTM, 33(2), 1933, p. 115. See also ASTM Spec. A 48–64.
[12]See Moffatt, W. G., G. W. Pearsall, and J. Wulff, The Structure and Properties of Materials, Vol. 1, Structures, New York: John Wiley & Sons, Inc., 1964, p. 195.

FIGURE 14-17 Cast iron microstructures and summary of the phases coexisting at various temperatures.

When higher strength and hardness are desired, alloying elements such as nickel, chromium, molybdenum, and copper are added to cast iron. In general, the alloying elements serve to improve the characteristics of the graphite. Compositions are available which will produce castings suitable for resisting impact, abrasion, corrosion, and high and low temperatures. Modern research has developed alloys of cast iron which are suitable for service as gears, crankshafts, and forging and drawing dies. Heat treatments of quenching and tempering are usually practiced with alloy cast iron to obtain full benefit of the expensive alloy additions. The designer should consult with an experienced foundryman on the composition and heat treatment for specialized applications. The engineering staffs of firms which market the various alloys also render expert metallurgical advice.

Cast iron is difficult to weld. Its tendency to crack can be reduced by careful preheating before welding, followed by slow cooling. Bronze welding rods are extensively used on cast iron. Cast iron has excellent vibration-damping characteristics as well as good resistance to wear. The resistance to corrosion is only fair. It is difficult to secure satisfactory results with metal inserts cast in place.

White cast iron results if the cooling rate after pouring is very rapid. The carbon is retained in solution, and the metal is very hard and resistant to wear. White cast iron cannot be machined; if cast iron must be machined, care must be taken that thin sections of a casting are not chilled so rapidly that white iron forms. White cast iron has important applications. For example, the tread or outer surface of the rim of a freight-car wheel can be cooled so rapidly that white iron results while the rest of the casting remains soft and machinable. To do this, a portion of the mold, called a chill, is made of iron. The chill removes heat very rapidly from the molten metal. Other applications of white cast iron include rolling-mill rolls, railroad brake shoes, and plow shares.

Although the basic idea of forming castings is very simple, great skill and long experience are necessary to master the various techniques of producing quality castings of intricate form.[13]

24. Malleable Iron. Nodular Iron

As shown in Fig. 14-17, white cast iron, if completely free from graphite flakes can be converted into malleable iron by a prolonged annealing process. The long heating and slow cooling induces the carbon to form into clusters or rosettes called temper carbon. The resulting structure has better continuity

---

[13]See *Cast Metals Handbook*, 4th ed., Des Plaines, Ill.: American Foundrymen's Society, 1957. See also the author's *Design Engineering Projects*, Englewood Cliffs, N.J.: Prentice-Hall, Inc., 1968, Chapter 3.

and is much more ductile than gray iron. Mechanical properties are given in Table 14-17. The matrix surrounding the rosettes can be either pearlite or ferrite, depending on the rate of cooling. Grade 32510 is in widest use for malleable parts.

TABLE 14-17   Properties of Malleable and Ductile Cast Iron*

Material	ASTM Spec.	Grade or Type	Tensile Strength, psi	Yield Strength, psi	Elong in 2 in., %
Malleable	A47-61	32510	50,000	32,500	10
		35018	53,000	35,000	18
Nodular or ductile	A339-55	60-45-10	60,000	45,000	10
	A339-55	80-60-03	80,000	60,000	3
	A396-58	100-70-03	100,000	70,000	3
	A396-58	120-90-02	120,000	90,000	2

*For malleable iron, $E = 24,000,000$–$27,000,000$ psi; for ductile iron, $E = 25,000,000$ psi.

Ductile or nodular iron, as shown by Fig. 14-17, is obtained directly from the cupola and dispenses with the prolonged heating necessary to produce malleable iron. The addition of a small amount of magnesium causes the carbon to appear in spherical or nodular form. Like malleable iron it is much more ductile than gray iron, as indicated by Table 14-17. The use of this material is increasing rapidly as it can be used in many places, such as crankshafts, where a more expensive forging was formerly required. Nodular iron shrinks more than gray iron, but its melting temperature is lower than for cast steel.

The machinability of ductile iron is similar to that of gray iron of equivalent hardness. Ductile iron can be brazed, but, because of the high carbon content, should be given careful handling in much the same way as for gray iron. The damping capacity is only slightly better than that of steel. Annealed nodular iron has good shock resistance. It is suitable for subzero temperature service, and can also be used in applications at elevated temperatures. It has a low creep rate at high temperatures, and can resist corrosion as well as gray iron.

Meehanite iron is made by a patented process in which a fine dispersal of graphite is secured by the addition of a calcium-silicon alloy.

25. Cast Steel

Cast steels have approximately the same chemical composition as the wrought steels, except that the percentages of silicon and manganese are larger. Steel is difficult to cast. It shrinks about $\frac{1}{4}$ in. per foot and makes a rough looking casting. The casting first undergoes a reduction in size, known as solidification shrinkage, which is followed by a considerable thermal contraction as the casting cools to room temperature. The casting should be designed in such a manner that the solidification will occur progressively, starting in the portions farthest removed from the molten metal. Uniformity of sections or a gradual transition from thin to thick sections is desirable, and stress concentrations should be reduced as much as possible. Projections should be avoided in long castings, or collapsible molds and cores should be used, to prevent hot tears caused by the large amount of cooling contraction.

Steel castings are in wide use in railway equipment, heavy machinery, oil refineries, and rolling mills. The design of the part must be suitable to the casting characteristics and physical properties of the material. A satisfactory microstructure depends on the conditions present during solidification, and thus on correct foundry practice. Heat treatment, such as annealing, normalizing, or quenching and drawing, gives a more uniform internal structure and improves the physical properties. Castings are usually inferior to corresponding wrought metals in impact resistance. The shock or impact qualities are improved by heat treatment, and also by the use of alloys.

The physical properties of cast steel vary with the carbon content, as shown in Fig. 14-18. Smaller values for the strength and ductility will be found in the interior of castings of large section size, as is shown in Fig. 14-19.

Low-carbon steel castings, containing 0.15 to 0.30% carbon, have desirable elevated temperature and carburizing properties. Parts made of this material frequently compete with castings made of gray or malleable iron. Cast steel can be readily welded. The microstructure of deposited weld metal is essentially the same as that of cast steel. If the carbon content is high enough to cause hardening and loss of ductility from cooling in air, it may be necessary to preheat the casting before welding, and then follow this by an annealing treatment afterward.

Medium-carbon steel castings, in the range 0.30–0.40% carbon, have greater strength but somewhat less ductility than low-carbon castings. The increase in the amount of carbon effects an improvement in the casting qualities of the metal. Heat treatment gives improved physical properties over the as-cast condition. High-carbon steel castings contain 0.40% carbon and over. The greater hardness makes this material suitable for applications involving abrasion and impact. Machining must usually be preceded by an annealing heat treatment.

A wide variety of alloy steel castings is available for specific applications. Both low- and high-alloy castings are made. Heat treatment is usually practiced to secure the full benefit of the alloy additions. All the principal alloying elements used in wrought steel are employed in making steel castings. Certain alloys of high chromium and nickel content are suitable for high-temperature service or where corrosion conditions are present.

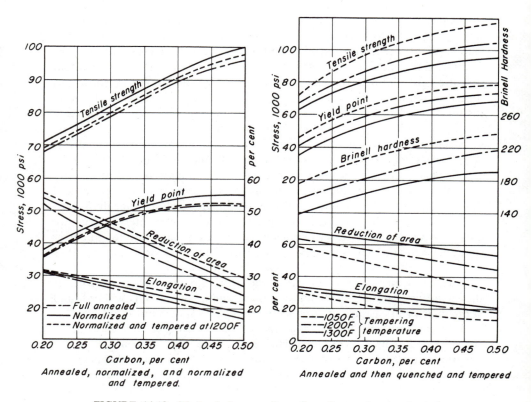

FIGURE 14-18 Mechanical properties of medium-carbon cast steel. (*Courtesy Steel Founders' Society of America.*)

FIGURE 14-19 Effect of mass on the mechanical properties of cast carbon steel specimens taken at center of test block.

Steel casting design is highly involved, and in general should be conducted with the cooperation of the foundryman.[14]

26. Tool Steel

Materials used for tools must be very hard and resistant to deformation and wear. High-carbon and alloy steels of high purity meet this requirement. Plain carbon steel of about 0.65% carbon is used for tools which must withstand considerable battering, such as hammers, heading dies, and hot forging dies. A carbon content of 1.10% is used for general purpose cutting and hand tools. Razors, drawing dies, files, and other tools requiring extreme hardness have percentages of carbon up to 1.40%.

The machining of such high-carbon steels must be preceded by normalizing, annealing, or spheroidizing heat treatments. The quenching and tempering of tool steels require expert handling. The tempering temperatures are low as compared to those of machine steels and are used only for relieving some of the quenching stresses.

The abrasion resistance and toughness of tool steel can be increased by the use of suitable alloys. Also, tool steel can be given a less rapid quenching, in oil for example, and residual stress and distortion will thereby be less. Low residual stress and low distortion are especially valuable if the body contains cross sections of both large and small diameters. Chromium, tungsten, and vanadium, either singly or in various combinations, are the principal alloys used. Compositions are available which hold their hardness and cutting edges even when raised to a red heat from the friction of the cutting.

High-speed steel of the 18-4-1 type contains 18% tungsten, 4% chromium, 1% vanadium, and 0.70% carbon, and has superior cutting qualities. Its manufacturing and heat-treating processes must be closely controlled. Stellite is a cobalt-chromium-tungsten-carbon alloy also used for cutting tools. Cemented carbide cutting tools are made by powdered metallurgy, and in general will outlast even high-speed steel.

27. Aluminum Alloys

Lightweight aluminum alloys have found wide applications in manufactured products. The high-strength alloys of this material have practically the same strength as mild steel. As is shown in Table 2-1, the weight per cubic inch is

[14]See *Steel Casting Handbook*, Cleveland, Ohio: Steel Foundrymen's Association, 1944. See also Briggs, C. W., R. A. Gezelius, and A. A. Donaldson, "Steel Casting Design for Engineers and Foundrymen," *Trans. Am. Foundrymen's Assn.*, **46**, 1938, p. 605.

only about one-third as great as for steel. The stiffness of a part made of aluminum is less and the deflection greater than that of a corresponding design in steel. Aluminum is a very versatile metal. It can be extruded, forged, or rolled into sheets. It is suitable for making sand, permanent mold, or die castings, and is also used for automatic screw machine products. Depending on the type of alloy, the strength can be increased by cold working or heat treatment.

Aluminum objects secure considerable protection from atmospheric corrosion by a coating of aluminum oxide, which forms naturally in air. Sometimes this oxide layer is built up artificially by anodic oxidation to give better protection. High-strength aluminum alloys are not so resistant to corrosion as the pure metal. In Alclad sheets, protection from corrosion is secured by rolling a coating of pure aluminum on both sides of a high-strength alloy core.

Soft aluminum is somewhat difficult to machine. The machinability of the alloys and the full heat-treated grades is superior to the machinability of the pure metal. Considerable experience and skill are required for the satisfactory welding of aluminum. Torch, metallic and carbon arc, and electric resistance methods are in use. Unlike steel, aluminum exhibits no change in color to warn the operator that the welding temperature is being approached. The presence of the oxide film adds to the difficulties of securing sound welds. The high strength secured by cold working or heat treatment is usually lost in the heated region. Cold working of welds to restore the strength is sometimes possible. Brazing techniques suitable for certain compositions have also been developed.

28. Alloy and Temper Designations

The numbering system for wrought aluminum alloys uses a four-digit number, the first of which designates the alloy type. The second digit identifies the specific alloy modification, and the last two digits identify the specific aluminum alloy or indicate the aluminum purity. The principal alloying elements and their digits are indicated by Table 14-18.

The temper of an aluminum alloy is one of the major factors governing its strength, hardness, and ductility. Some alloys are hardened and strengthened by cold work such as cold rolling, drawing, or stretching. Other alloys are heat treatable and their properties can be improved appreciably by appropriate thermal treatments. The temper designation is usually specified by a suffix to the alloy number, as indicated by Table 14-19.

The theory for the increased strength and hardness of heat-treated aluminum alloys differs in several respects from that for steel. After aluminum is

TABLE 14-18   Wrought Aluminum Alloy Designation System

Aluminum—99% min and greater	1xxx
Copper	2xxx
Manganese	3xxx
Silicon	4xxx
Magnesium	5xxx
Magnesium and silicon	6xxx
Zinc	7xxx
Other elements	8xxx
Unused series	9xxx

TABLE 14-19   Temper Designations for Aluminum

Temper	Process
-F	As-cast condition
-O	Annealed temper of wrought materials
-H14	Cold worked to tensile strength midway between tempers O and H18
-H18	Cold worked to full hard temper
-H34	Cold worked beyond temper range and stabilized
-H38	Same as H34 but with higher strength properties
-T3	Solution heat treated followed by strain hardening
-T4	Solution heat treated followed by natural aging at room temperature
-T6	Solution heat treated followed by artificial aging
-T8	Solution heat treated, strain hardened, and artificial aging

quenched, the material is a supersaturated solid solution of the alloying elements. At this time it is soft and ductile. If allowed to stand at room temperature the alloying elements are precipitated as hard particles which are effective in preventing movement within the crystal structure. Since considerable time may be required for the greatest strength to be reached, the process is called aging. For some alloys the process can be hastened by heating to a temperature that is known to give the optimum particle size for the maximum strength and hardness. After precipitation of the hardening constituents, the material is said to be in the T-condition.

29. Composition and Mechanical Properties

The composition and mechanical properties of several typical wrought aluminum alloys are given in Table 14-20. Alloys in the O-condition are suitable for forming, spinning, and deep drawing. Alloys in the work-hardened or fully heat-treated condition are not so workable. Suitable radii must be

TABLE 14-20   Composition and Mechanical Properties of
Wrought Aluminum Alloys*

Alloy	Nominal Composition Balance: Aluminum Plus Inpurities								Type
	Si	Fe	Cu	Mn	Mg	Cr	Zn	Ti	
Sheets and Plates									Non-heat Treatable
1100	1.0		0.20	0.05			0.10		Commercially pure aluminum
3003	0.6	0.7	0.20	1.0–1.5			0.10		Manganese-aluminum alloy
5005	0.4	0.7	0.20	0.20	0.5–1.1	0.10	0.25		Magnesium-aluminum alloy
Bars and Rods									Heat Treatable
2011	0.4	0.7	5.0–6.0				0.30		Copper-aluminum alloy
6061	0.4–0.8	0.7	0.15–0.4	0.15	0.8–1.2	0.15–0.35	0.25	0.15	Magnesium-silicon alloy
6063	0.2–0.6	0.35	0.10	0.10	0.45–0.9	0.10	0.10	0.10	Magnesium-silicon alloy
7075	0.5	0.7	1.2–2.0	0.30	2.1–2.9	0.18–0.40	5.1–6.1	0.20	Zinc-aluminum alloy

Alloy and Temper	Tensile Strength, psi	Yield Strength, psi	Elongation 2 in. %		Brinell	Shearing Strength, psi	Endurance Limit, psi
			1/16 in. sheet	1/2 in. round			
1100-O	13,000	5,000	35	40	23	9,000	5,000
1100-H14	18,000	17,000	9	20	32	11,000	7,000
1100-H18	24,000	22,000	5	15	44	13,000	9,000
3003-O	16,000	6,000	30	40	28	11,000	7,000
3003-H14	22,000	21,000	8	16	40	14,000	9,000
3003-H18	29,000	27,000	4	10	55	16,000	10,000
5005-O	18,000	6,000	30		30	11,000	
5005-H18	29,000	28,000	4		51	16,000	
5005-H38	29,000	27,000	5		51	16,000	
2011-T3	55,000	43,000		15	95	32,000	18,000
2011-T8	59,000	45,000		12	100	35,000	18,000
6061-O	18,000	8,000	25	30	30	12,000	9,000
6061-T4	35,000	21,000	22	25	65	24,000	13,000
6061-T6	45,000	40,000	12	17	95	30,000	14,000
6063-O	13,000	7,000			25	10,000	8,000
6063-T4	25,000	13,000	22			16,000	
6063-T6	35,000	31,000	12	18	74	22,000	10,000
6063-T835	48,000	43,000	8		105	30,000	
7075-O	33,000	15,000	17	16	60	22,000	17,000
7075-T6	83,000	73,000	11	11	150	48,000	22,000
7075-T73	73,000	63,000		13			

*Taken mainly from Alcoa Aluminum literature. Properties are typical; variations must be expected in practice.

provided on the forming tools, depending on the composition, the thickness of the stock, and the temper. Table 14-21 indicates some of the fabricating qualities and principal applications for these alloys.

The composition and mechanical properties of a number of widely used aluminum sand casting alloys are given in Table 14-22.

TABLE 14-21   Principal Characteristics and Use of Wrought
Aluminum Alloys

Alloy	Characteristics and Uses
1100	Spinnings, deep-drawn shapes, hand-hammered giftware
3003	Excellent formability with moderate strength; food utensils
5005	Good formability; fine grain size; panels, ductwork, decorative trim
2011	Highest machinability; most widely used screw machine stock
6061	Best heat-treated alloy for brazing and welding; boats, furniture
6063	Good formability; furniture tube, moldings, doors, windows, lights
7075	Forming requires careful attention; aircraft structures and skins, lock washers, fuel tanks, skis

TABLE 14-22   Composition and Mechanical Properties of
Aluminium Sand Casting Alloys*

Alloy	Composition† Cu	Si	Mg	Applications
43		12.0		General purpose casting alloy; cooking utensils, architectural and marine applications, pipe fittings
214			3.8	Food, chemical, marine and architectural applications
195	4.5	0.8		Machinery and aircraft structural members
220			10.0	High-strength and shock-resistant castings; aircraft structural members
356		7.0	0.3	Most popular casting alloy; automotive and railroad fittings, hardware, valves

Alloy and Temper	Tensile Strength, psi	Yield Strength, psi	Elong. in. 2 in., %	Compressive Yield Str., psi	Brinell	Shearing Strength, psi	End. Limit
43-F	19,000	8,000	8.0	9,000	40	14,000	8,000
214-F	25,000	12,000	9.0	12,000	50	20,000	7,000
195-T6	36,000	24,000	5.0	25,000	75	30,000	7,500
220-T4	48,000	26,000	16.0	27,000	75	34,000	8,000
356-T6	33,000	24,000	3.5	25,000	70	26,000	8,500

*Taken from *Alcoa Aluminum Handbook*. Values shown are typical; variations must be expected in practice.

†Percentage of alloying elements. Aluminum and normal impurities constitute remainder.

30. Magnesium Alloys

Magnesium, in its various cast and wrought forms, provides another medium for achieving lightweight designs. As shown by Table 2-1 the weight per cubic inch and the modulus are lower than for aluminum. Magnesium is less resistant to corrosion than aluminum alloys, and should be protected against a salt atmosphere and against constant contact with water. Proper provision for drainage or ventilation is necessary when moisture is present.

Magnesium has relatively low hardness, and is very easy to machine. No special composition is required for use in automatic screw machines. Magnesium can be hot forged with hammers or mechanical and hydraulic presses. It can be cold formed or drawn, but only with difficulty. The rapid work hardening can be relieved by annealing. Care should be exercised to avoid stress concentration such as caused by notches or sharp fillets. A sharp 60° V-groove will reduce the endurance limit to one-fourth or one-third the value for a standard specimen. When abrasion conditions are present, the soft magnesium must be protected by inserts such as plates, liners, sleeves, shoes, or bushings. Magnesium exhibits a high damping capacity which is effective in preventing vibrations from building up into large amplitudes.

When magnesium is welded with a torch, butt welds are preferable because the removal of the flux is usually easier than for lap welds, and the possibility of corrosion is reduced. All types of joints may be made with the helium arc process, which requires no flux. Resistance welding is satisfactory for magnesium, and may be applied to wrought products or thin-walled castings. Magnesium cannot be used in the form of rivets. Rivets of aluminum alloys 6053 and 6061 have been found to give satisfactory service. Brass and iron rivets should be avoided because of the possibility of galvanic action.

The composition and physical properties for a number of typical magnesium alloys are given in Table 14-23.

31. Copper Alloys

Copper is a very ductile metal, and may be spun, stamped, rolled into sheets, and drawn into wire and tubing. Mechanical working increases the hardness and strength, but decreases the ductility. The effects of cold work can be removed by annealing. Copper can be fabricated by soldering, welding, or brazing. Copper has excellent heat-conducting qualities, and is used in refrigerators, radiators, water heaters, and air-conditioning equipment. It is widely used as an electrical conductor. It cannot be hardened by quenching unless alloyed with beryllium. Copper resists certain types of corrosion.

The most important copper-base alloys are brass and bronze. Brass is a copper-zinc alloy, and bronze is composed principally of copper and tin. Zinc is cheaper than tin, so the cost of brass is less than the cost of bronze. Bronze, however, is usually considered the superior metal. It is commercial practice to designate certain reddish-colored copper-zinc combinations as "bronze" even though no tin is present.

TABLE 14-23 Composition and Mechanical Properties of Magnesium Alloys*

Alloy	Percentage Composition—Balance Magnesium and Minor Impurities					
	Al	Mn, min	Zn	Si, max	Cu, max	Ni, max
AZ 31B	2.5–3.5	0.20	0.7–1.3	0.30	0.05	0.005
AZ 31C	2.5–3.5	0.20	0.6–1.4	0.30	0.10	0.03
AZ 63A	5.3–6.7	0.15	2.5–3.5	0.30	0.10	0.01
AZ 81A	7.0–8.1	0.13	0.4–1.0	0.30	0.10	0.01
AZ 92A	8.3–9.7	0.10	1.6–2.4	0.30	0.10	0.01
HK 31A			Zirconium 0.50–1.00		Thorium 2.5–4.0	

Alloy and Temper	ASTM	Tensile Strength, psi	Yield Strength, psi	Comp. Yield Strength, psi	Shear Strength, psi	Elong. in 2 in. %	Brinell
Sand castings†							
AZ 63A-F	B80–63	29,000	14,000	14,000	18,000	6	50
AZ 63A-T4		40,000	14,000	14,000	18,000	12	55
AZ 63A-T6		40,000	19,000	19,000	21,000	5	73
AZ 81A-T4		40,000	14,000	14,000	18,000	12	55
AZ 92A-F		24,000	14,000	14,000	19,000	2	65
AZ 92A-T4		40,000	14,000	14,000	19,000	10	63
AZ 92A-T6		40,000	21,000	21,000	22,000	2	84
Sheets and plates†‡							
AZ 31B-H24	B90–65	42,000	32,000	26,000	23,000	15	73
AZ 31B-O		37,000	22,000	16,000	21,000	21	56
HK 31A-H24		37,000	29,000	25,000	21,000	8	57
Rods, bars, shapes†							
AZ 31B-F ⎱ AZ 31C-F ⎰	B107–65	38,000	29,000	14,000	19,000	15	49

*Data courtesy Dow Chemical Co.
†These values are typical. Minimum values are lower.
‡Strength values are for minimum sections. Values are somewhat less for larger sizes.

Brass and bronze are used in both the cast and wrought forms. The strength of brass increases with the zinc content. Most of the copper alloys can be work hardened to higher strengths than pure copper. In general, the machinability of the alloys is satisfactory. The free-cutting qualities are improved by the addition of small amounts of lead. Brass is about equal to copper in corrosion resistance, but bronze is superior to both. There are a great many copper alloys with a wide variety of commercial designations. A selected list of the wrought alloys, giving composition and mechanical properties, are shown in Tables 14-24 and 14-25. Several of the most commonly used casting alloys are listed in Table 14-26.

TABLE 14-24   Composition of Brass, Bronze, and Miscellaneous Wrought Copper Alloys*

No.	ASTM Spec.	Alloy	Cu, %	Sn, %	Zn, %	Other
1	B133	Copper				
2	B36 Alloy 3	Plates, sheets	85.0		15.0	
3	B36 Alloy 6	Plates, sheets	70.0		30.0	
4	B36 Alloy 8	Plates, sheets	65.0		35.0	
5	B134 Alloy 2	Brass wire	90.0		10.0	
6	B134 Alloy 7	Brass wire	65.0		35.0	
7	B16	FC brass†	60.0–63.0		Remainder	Pb 2.5–3.7
8	B21 Alloy A & B	Naval brass	59.0–62.0	0.5–1.0	Remainder‡	
9	B159 Alloy A	Phos. bronze wire	Remainder	4.2–5.8	0.30	P 0.03–0.35
10	B139 Alloy B2	Phos. bronze rod, shapes	Remainder	3.5–4.5	1.5–4.5	P 0.01–0.50§
11	B138 Alloy A	Manganese bronze	58.5	1.0	39.0	Fe 1.4, Mn 0.1
12	B171	Muntz metal	58.0–61.0	0.25 max.	Remainder	Pb 0.4–0.9
13	B150 Alloy 1	Aluminum bronze	80.0–93.0	0.60 max.	1.0	Al 6.5–11.0**
14	B196	Beryllium copper	Remainder			Be 1.8–2.05
15	B151 Alloy B	18% Ni silver	53.5–56.5	Ni 16.5–19.5	Remainder	Mn 0.5 max.
16	B127 (Monel)	Nickel-copper alloy	Remainder	Ni 63.0–70.0	Fe 2.5	Mn 1.25

*Some alloys have minor amounts of impurities. See specifications for exact compositions.

†FC, free cutting.

‡Pb, Alloy A, 0.20 max.; Alloy B, 0.40–1.0.

§Pb 3.5–4.5.

**Fe 4.0 max, Ni 1.0 max., Si 2.2, Mn 1.5

TABLE 14-25   Brass, Bronze, and Miscellaneous Wrought Copper Alloys

No.	Alloy	Condition	Tensile Strength,* psi	Rockwell, min.	Elong., %
1	Copper	Bars, all sizes	A 37,000		25
		Bars 0.188–0.375 in.	H 42,000		12
		Bars 0.50–2.0 in.	H 33,000		15
2	Brass, alloy 3	Plates, sheets, strips	1/2H 51–61,000,	56B	
			H 63–72,000	72B	
			Sp 78–86,000	82B	
3	Brass, alloy 6	Plates, sheets, strips	1/2H 57–67,000,	60B	
			H 71–81,000	79B	
			Sp 91–100,000	89B	
4	Brass, alloy 8	Plates, sheets, strips	1/2H 55–65,000,	57B	
			H 68–78,000	76B	
			Sp 86–95,000	87B	

TABLE 14-25 (cont'd)

No.	Alloy	Condition	Tensile Strength,* psi	Rockwell, min.	Elong., %
5	Brass wire, alloy 2	Round, hexagonal, octagonal, square }	1/2H 56–67,000, H 70–79,000 Sp 84,000		
6	Brass wire, alloy 7	Do	1/2H 79–94,000, H 92–117,000 Sp 120,000		
7	Free-cutting brass	Bars, 1 in. and under	A 48,000 YS 20,000		15
		Bars, 1–2 in.	A 44,000 YS 18,000		20
		Bars, 0.5 in. and under	1/2H 57,000 YS 25,000		7
		Bars, 0.5 and 1 in.	1/2H 55,000 YS 25,000		10
8	Naval brass, A, B	Bars, 1 in. and under	A 54,000 YS 20,000		30, 25
		Bars, 1–2 in.	A 52,000 YS 20,000		30, 25
		Bars, 1 in. and under	1/2H 60,000 YS 27,000		25, 18
		Bars, 1–2 in.	1/2H 58,000 YS 26,000		25, 20
9	Phos. bronze wire, A	Wire, 0.025 in. and under	Sp 145,000		
		Wire, 0.025–0.0625 in.	Sp 135,000		
		Wire, 0.125–0.250 in.	Sp 125,000		
10	Phos. bronze rod, shapes, B2 }	Rounds, 0.25–0.5 in.	H 60,000		10
		Rounds, 0.5–1.0 in.	H 55,000		12
11	Manganese bronze, A }	Bars, 1.0–2.5 in.	1/2H 70,000 YS 35,000		15
		Bar, 1.0–1.5 in.	H 76,000 YS 52,000		10
12	Muntz metal	Tube plates under 2 in.	50,000 YS 20,000		35
13	Aluminum bronze, 1 }	Bars, 0.5 in. and under	80,000 YS 40,000		9
		Bars, over 1 in.	72,000 YS 35,000		12
14	Beryllium copper	Rods, bars	A 60–85,000	46B–85B	
		Rods, up to 0.375 in.	H 95–130,000	92B–103B	
15	18 % Ni silver, alloy B }	Rods, 0.02–0.25 in.	H 90–110,000		
		Rods, 0.50–1.0 in.	H 75–95,000		
16	Nickel-copper alloy }	Plate, sheet, strip	A 70–85,000 YS 28,000	73B	35

*A, annealed; 1/2H, half hard; H, hard; Sp., spring; YS, yield strength.

TABLE 14-26   Typical Brass and Bronze Casting Alloys

| SAE | ASTM | Nominal Composition | | | | Commercial Designation |
		Cu	Sn	Pb	Zn	
40	B 62–63	85	5	5	5	Composition bronze or ounce metal
41	B 146–52 6B	67	1	3	29	Leaded yellow brass for general purposes; No. 1
43	B 147–63 8A	58	Fe 1.25	Al 1.25	39.25	Mn 0.25; high-strength yellow brass; managanese "bronze"
62	B 143–61 1A	88	10		2	Tin bronze; 88–10–2 or "G" bronze
64	B 144–52 3A	80	10	10		High leaded tin bronze; 80–10–10
68	B 148–65 9B	89				Al 10, Fe 1.0 Aluminum bronze; Grade B

ASTM	Tensile Strength, psi	Yield Strength, psi	Elong. in 2 in., %	General Properties and Applications
B 62–63	30,000	14,000	20	Good casting and finishing properties; free cutting
B 146–52 6B	30,000	11,000	20	Used where cheapness is main consideration
B 147–63 8A	65,000	25,000	20	Makes tough, strong castings
B 143–61 1A	40,000	18,000	20	Withstands heavy gear and bearing pressures
B 144–52 3A	25,000	12,000	8	Has low friction; withstands heavy loads
B 148–65 9B	65,000	25,000	20	BHN 110; for heavy service such as worms, gears, and similar applications

32. Alloys for Die Castings

Die castings are formed by forcing a molten alloy into metal molds or dies under high pressure. Only alloys of relatively low melting points are suitable for this process. Die castings are made principally from zinc, aluminum, and magnesium, and to a lesser extent from brass.

Die casting is strictly a machine operation and lends itself to very rapid production of parts. The chief disadvantage is the high cost of dies, which usually limits the use of the method to parts that are to be made in large quantities. Close tolerances on the finished product can be maintained. Consequently, very little or no machining is required. The die casting process is suitable for parts containing very thin sections or intricate shapes. Inserts such as studs or bushings can be included as a part of the casting. Die castings usually have a very smooth surface suitable for direct application of platings or organic finishes. Dimensions and weights, and the composition and mechanical properties of some of the better-known die casting alloys are listed in Tables 14-27 and 14-28.

TABLE 14-27 **Approximate Dimensional and Weight Limits for Die Castings**

	Zinc	*Aluminum*	*Magnesium*
Maximum weight of casting (lb)	30	25	10
Minimum wall thickness, large castings (in.)	0.050	0.080	0.080
Minimum wall thickness, small castings (in.)	0.020	0.050	0.050
Minimum variation from drawing dimensions per inch of diameter or length (in.)	0.001*	0.002*	0.002*
Cast threads (max no. per in. external)	24	12	12
Cast threads (max no. per in. internal)	24†	none	none
Minimum draft on cores (in. per in. of length or diameter)	0.003	0.010	0.010
Minimum draft on side walls (in. per in. of length)	0.005	0.010	0.008
Cored holes, min diameter (in.)	0.030	0.100	0.100

*Depends on conditions.
†Where cheaper than tapping.

Zinc die castings have gained extensive use because of their high strength, long die life, and moderate casting temperatures. The light weight of aluminum die castings is an important advantage in many applications. The casting temperature is higher than zinc, and die life is shorter. Magnesium castings also possess the advantage of low weight. Casting temperature is the same as for aluminum. The surface of magnesium castings is usually treated to inhibit corrosion and to serve as a base for the finish. Brass is cast at the highest temperature used for die casting, and die life is accordingly the shortest. High strength and hardness are advantages possessed by brass die castings.

The designer must give care and attention to all the details of a part to be made by the die-casting process. The thickness of the sections should be as uniform as possible throughout the casting. Solidification and cooling will thus take place uniformly, and the casting will be free from "hot spots" resulting from the slow cooling of large volumes of metal. Residual stresses are frequently found in castings of unequal sections. The parting line for the dies should be located at a point that permits easy removal of the flash resulting from the imperfect matching of the dies. If a bead can be located on the parting line, the flash can be very easily sheared off. Undercuts that require the use of slides in the die should be avoided whenever possible. The additional die expense for complex castings with undercuts, if made in sufficiently large quantities, is sometimes justified. Fillets should be used at reentrant

TABLE 14-28 Nominal Composition and Mechanical Properties of Die Casting Alloys (Percent by Weight)*

Alloy	Al	Cu	Mg	Mn	Si	Zn	Pb	Sn
Zinc: ASTM B86 AG40A (XXIII), or SAE 903, or Zamak† 3 AC41A (XXV), or SAE 925, or	3.5–4.3	0.25 max	0.02–0.05			Remainder	0.005	0.003
Zamak 5	3.5–4.3	0.75–1.25	0.03–0.08			Remainder	0.007	0.005
Aluminum: ASTM **B85–60** S12B, or Alcoa‡ 13	Remainder	0.6	0.10	0.35	11.0–13.0	0.50		0.15
SG100B, or Alcoa 360	Remainder	0.6	0.4–0.6	0.35	9.0–10.0	0.50		0.15
SC84B, or Alcoa 380	Remainder	3.0–4.0	0.10	0.50	7.5–9.5	3.0		0.35
Magnesium: ASTM **B94–58** AZ91A, or SAE 501, or Dow-metal§ R	8.3–9.7	0.10		0.13	0.50	0.35–1.0		
Brass: ASTM **B176–62** Z30A	0.25	57.0 min		0.25	0.25 max	30.0 min	1.50	1.50
Z331A	0.15	63.0–67.0		0.15	0.75–1.25	Remainder	0.25	0.25

Alloy	Tensile Strength, psi	Yield Strength, psi	Elong. in 2 in., %	Comp. Strength, psi	Shear Strength, psi	Endurance Limit, psi	Brinell Hardness	Melting Point, °F
Zinc AG40A (XXIII)	41,000		10	60,000	31,000		82	728
AC41A (XXV)	47,000		7	87,000	38,000		91	727
Aluminum	43,000	21,000	2.5		25,000	19,000		1065–1080
	44,000	25,000	2.5		28,000	20,000		1035–1105
	46,000	23,000	2.5		28,000	20,000		1000–1100
Magnesium: AZ91A	33,000	22,000	3	22,000**	20,000		60	1,105
Brass Z30A	45,000	25,000	10				120–130	1,650
Z331A	58,000	30,000	15				120–130	1,550

*Values of normal impurities not shown. Mechanical properties are typical; variations must be expected in practice.

†New Jersey Zinc Co.

‡Aluminum Co. of America.

§Dow Chemical Co.

**Compressive yield strength.

angles to increase the strength and prevent localization of cooling stresses. Threaded holes can sometimes be cored, and a saving in machining costs thereby effected. The designer should consult with a die casting firm on the design of the part in order to receive the guidance of specialists in this branch of engineering. Sometimes minor revisions can be made in the design which will result in a lowering of the cost of the dies and the production parts.[15]

Permanent mold castings are made by filling metal molds under gravity head. Both iron and aluminum castings are made by this process. Iron castings are usually annealed to improve the machining properties. Sand cores are used in the so-called "semipermanent mold method."

REFERENCES

1. BULLENS, D. K., *Steel and Its Heat Treatment*, Vol. I, New York: John Wiley & Sons, Inc., 1948.

2. BULLENS, D. K., *Steel and Its Heat Treatment*, Vol. III, New York: John Wiley & Sons, Inc., 1949.

3. DIETER, G. E., *Mechanical Metallurgy*, 2d ed. New York: McGraw-Hill Book Company, 1976.

4. EVANS, U. R., *Corrosion and Oxidation of Metals*, London: Edward Arnold (Publishers) Ltd., 1960.

5. FONTANA, M. G., and N. D. GREENE, *Corrosion Engineering*, 2d ed. New York: McGraw-Hill Book Company, 1978.

6. HEYER, R. H., *Engineering Physical Metallurgy*, New York: Van Nostrand Reinhold Company, 1939.

7. HOLLOMAN, J. H., and L. D. JAFFE, *Ferrous Metallurgical Design*, New York: John Wiley & Sons, Inc., 1947.

8. "Metals," Reference Issue, *Machine Design*.

9. *Metals Handbook*, Vol. 1, *Properties and Selection of Metals*, Metals Park, Ohio: American Society for Metals, 1961.

10. *Metals Handbook*, Vol. 2, *Heat Treating, Cleaning, and Finishing*, Metals Park, Ohio: American Society for Metals, 1964.

11. *Metals Handbook*, Vol. 3, *Machining*, Metals Park, Ohio: American Society for Metals, 1967.

12. *Metals Handbook*, Vol. 4, *Forming*, Metals Park, Ohio: American Society for Metals, 1969.

[15]See the author's *Design Engineering Projects*, Englewood Cliffs, N.J.: Prentice-Hall, Inc., 1968, Chapter 4. See also Barton, H. K., "Die Casting," *Eng. Digest*, **21**, Dec. 1960, p. 77; and Chase, H., *Handbook on Designing for Quantity Production*, New York: McGraw-Hill Book Company, 1944, Chapter 1.

13. SACHS, G., and K. R. VAN HORN, *Practical Metallurgy*, Metals Park, Ohio: American Society for Metals, 1940.

14. "Selection of Steel for Automobile Parts," *J. Soc. Auto. Eng.*, **57**, Aug. 1949, to **58**, Jan. 1950 (six articles).

15. WILLIAMS, G. T., *What Steel Shall I Use?*, Metals Park, Ohio: American Society for Metals, 1941.

16. WULFF, J., H. F. TAYLOR, and A. J. SHALER, *Metallurgy for Engineers*, New York: John Wiley & Sons, Inc., 1952.

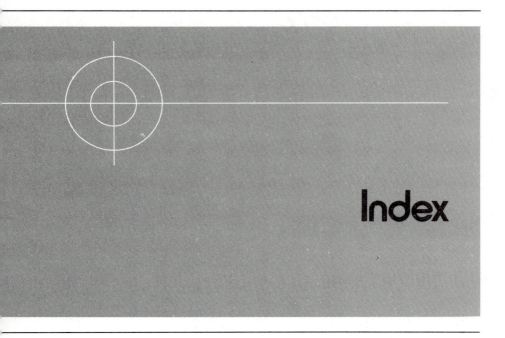

Index

SI PREFIXES

$1,000,000 = 10^6$ mega, M   lb or lbf, pound force

$1,000 = 10^3$ kilo, k   lbm, pound mass

$100 = 10^2$ hecto, h   psi, lb/in.^{2}

$10 = 10$ deka, da   J, joule ; newton-meter, Nm

$0.1 = 10^{-1}$ deci, d   W, watt ; joule per sec, J/sec ; Nm/sec

$0.01 = 10^{-2}$ centi, c   Pa, pascal ; N/m^2

$0.001 = 10^{-3}$ milli, m   MPa, mega pascal ; N/mm^2

$0.000\ 001 = 10^{-6}$ micro, μ

An asterisk follows each number below that represents an exact definition. Numbers not followed by an asterisk are approximate representations of definitions or the results of physical measurements.

LENGTH

1 in. = 25.4* mm = 0.0254* m   1 mm = 0.039 370 079 in.

1 ft = 304.8* mm = 0.3048* m   1 m = 39.370 079 in. = 3.280 840 ft

1 yd = 914.4* mm = 0.9144* m   1 m = 1.0936 yd

1 mi = 5,280 ft = 1,760 yd = 1.609 344* km   1 km = 0.621 37 mi

1 furlong = 220 yd = $\frac{1}{8}$ mi = 201.168* mi   1 micron = 0.000 001 m = 0.001 mm

1 rod = 16.5 ft   1 Angstrom = 0.000 000 1 mm

1 fathom = 6 ft

AREA

1 in.^{2} = 645.16* mm^2 1 mm^2 = 0.001 550 003 in.^{2}

1 ft^2 = 0.092 903 04* m^2 1 m^2 = 1,550.003 in.^{2}

1 yd^2 = 0.836 127 36* m^2 = 10.763 91 ft^2 = 1.195 99 yd^2

1 mi^2 = 2.589 988 11* (km)^2 1 km^2 = 0.386 10 mi^2

1 acre = $\dfrac{5,280^2}{640}$ = 43,560 ft^2 1 hectare = (100 m)^2

= (208.713 ft)^2 = 2.471 054 acres

= 4,046.856 m^2

= 0.404 865 6 hectares

VOLUME

1 in.^{3} = 16,387.064* mm^3 1 mm^3 = 0.000 061 024 in.^{3}

1 ft^3 = 0.028 316 846 592* m^3 1 m^3 = 1,000 liters = 61,023.74 in.^{3}

1 yd^3 = 0.764 554 857 984* m^3 = 35.314 67 ft^3

1 gallon (US) = 231 in.^{3} = 1.307 95 yd^3

= 3.785 411 784* liter   1 liter = 1.000 cc^3 = 61.023 74 in.^{3}

1 qt (US) = $\frac{1}{4}$ gallon   = 0.264 172 gallons = 1.056 688 qt

= 0.946 352 946* liter   1 cc = (10 mm)^3 = 1,000 mm^3

1 gallon (UK) = 277.42 in.^{3} 1 liter = 1,000 cc

= 1.200 95 US gallon = 4.546 09 liter   = 1,000,000 mm^3

1 bushel (US) = 2,150.42 in.^{3} 1 m^3 = (1,000 mm)^3

= 1.244 46 ft^3 = 0.035 239 07 m^3 = 1,000,000,000 mm^3

1 barrel (42 US gallons) = 158.987 liters   = 1,000 liters